Animal Behavior

ELSEVIER *science & technology books*

Companion Web Site:

http://www.elsevierdirect.com/companions/9780123725813

Animal Behavior, by Michael D. Breed and Janice Moore

Resources for Professors:

- **Answers to the end-of-chapter questions**
- **Figures and appendix material from the textbook**
- **Lab manual with fully developed and tested laboratory exercises available for courses that have labs (http://www.elsevierdirect.com/product.jsp?isbn=9780123725820).**

Instructor Resources site can be found at http://textbooks.elsevier.com/web/Login.aspx, and it is password protected. Please contact your sales representative at textbooks@elsevier.com for access to the Instructor Resources site.

TOOLS FOR **ALL** YOUR **TEACHING NEEDS**
textbooks.elsevier.com

ACADEMIC PRESS

Animal Behavior

Michael D. Breed
Ecology and Evolutionary Biology
The University of Colorado
Boulder, Colorado, U.S.A.

Janice Moore
Biology Department
Colorado State University
Fort Collins, Colorado, U.S.A.

AMSTERDAM • BOSTON • HEIDELBERG • LONDON
NEW YORK • OXFORD • PARIS • SAN DIEGO
SAN FRANCISCO • SINGAPORE • SYDNEY • TOKYO
Academic Press is an imprint of Elsevier

Academic Press is an imprint of Elsevier
30 Corporate Drive, Suite 400, Burlington, MA 01803, USA
525 B Street, Suite 1800, San Diego, California 92101-4495, USA
84 Theobald's Road, London WC1X 8RR, UK

Library of Congress Cataloging-in-Publication Data
Breed, Michael D.
Animal behavior / Michael D. Breed, Janice Moore.
  p. cm.
ISBN 978-0-12-372581-3
1. Animal behavior. I. Moore, Janice. II. Title.
QL751.B6836 2012
591.5—dc22   2010042369

British Library Cataloguing-in-Publication Data
A catalogue record for this book is available from the British Library.

For information on all Academic Press publications
visit our Web site at www.elsevierdirect.com

Printed in the United States of America

12 13   5 4 3 2

$79.95

3|19|14

To my wife Cheryl, with my thanks for all of her support, and my daughters, Ann and Elise, whose enthusiasm for learning has been a source of inspiration.

—Michael Breed

To my parents, Doyle and Tillie, with gratitude for decades of love, support, and common sense.

—Janice Moore

CONTENTS

vii

Contents

ix

Contents

In this text, we offer students an accessible approach to the major principles, mechanisms and controversies in the study of animal behavior. Throughout the book, we use Tinbergen's four questions—causation, survival value, ontogeny, and evolution—to frame animal behavior and to lend coherence to a diverse and highly integrative field of scientific inquiry. We see that same inquiry at the heart of the discipline, so we emphasize how to test hypotheses about animal behavior, and we encourage students to think critically about experimental evidence.

We take stands on controversial issues in this book and have clearly expressed our point of view in scientific interpretations. We do not expect faculty members who teach animal behavior courses to agree with all of our interpretations. (Indeed, the two of us have had some lively discussions about some of these topics as the work has progressed!) Instead, we see areas of disagreement as pedagogical tools to help students understand that not all scientific issues are settled. In fact, we have highlighted those unsettled areas, because a textbook is not a compendium of absolute knowledge; it is a snapshot of current scientific understanding. We encourage faculty and students who use this textbook to approach our statements critically, and to ask how further hypothesis testing will improve knowledge and understanding.

Every field has a history, but the history of animal behavior is particularly informative because it teaches us so much about how scientists sort through controversy and learn, collectively, how to think critically. It also tells us why we are only now asking questions that might have been off-limits to earlier workers. We therefore begin this book with a brief overview of that history. Then, because evolution is, famously, that thing without which nothing in biology makes sense, we follow history with a refresher on evolution. In our experience as teachers, we have realized that such a refresher is often desirable.

xi

After this introductory chapter, our approach flows from the physiological and genetic underpinnings of behavior (Chapters 2 and 3) through behavioral concerns of individual animals to the complexities of social behavior (Chapters 13 and 14). Chapters 2 and 3 provide ample foundation for mechanisms-oriented animal behavior courses and offer background for students in courses that do not emphasize mechanisms. Building on mechanisms, we consider behavioral homeostasis (Chapter 4). Learning (Chapter 5) and cognition (Chapter 6) are aspects of animal behavior that link neural processes with the behavior of the animal itself. An understanding of communication (Chapter 7) involves not only the underlying mechanisms, but the behavior of at least two participants. Midway through the book a study of orientation and migration (Chapter 8) builds a bridge from mechanisms to behavioral ecology. Behavioral ecologists will find contemporary coverage of the key elements of behavioral ecology in chapters on foraging (Chapter 9), self-defense (Chapter 10), mating systems (Chapter 11), parenting (Chapter 12), and social behavior (Chapters 13 and 14). Chapter 15 introduces the exciting—and essential—field of conservation behavior.

In our coverage, we recognize two emerging topics in animal behavior, cognition (Chapter 6) and conservation (Chapter 15), with full chapters. Cognition presents engaging and difficult hypotheses that will challenge our students, perhaps more than any other topic in animal behavior. Given societal debates over our relationships with animals and the ethics of maintaining animals in farms, zoos, and human households, our chapter on cognition provides a timely overview of the scientific evidence in that field. Cognition is one of the Next Big Topics in animal behavior and reveals some dimly lit areas of the discipline; we invite students to step onto the ground floor of an area of study that in the future will likely revolutionize our concept of and relationship with animals.

As for conservation, there is no denying that many species are in peril. Effective animal conservation programs can exist only with a thorough understanding of the behavior of the species that the programs seek to save. In our experience, community, ecosystems, and landscape-level ecologists often give short shrift to the importance of animal behavior in making conservation choices, and in so doing, they run the risk of failure. Effective conservation programs require knowledge of foraging behavior, mating systems, dispersal and migration. Successful release of captive-reared animals—an increasingly important strategy in saving endangered species—requires substantial knowledge of animal learning, ranging from imprinting to learned aspects of foraging and anti-predator behavior. We offer the last chapter of this text as a platform for integrating behavior and conservation, and hope to inspire a generation of students to use behavior as a conservation tool.

The majority of animal behavior students will not go on to careers in animal behavior, but most of them will enjoy the companionship of animals. We have highlighted the behavior of companion animals in special features called "Bringing Animal Behavior Home." This dimension of animal behavior is often important to students but may be overlooked in textbooks intended for animal behavior courses. Teachers of animal behavior courses may opt to cover companion animals, but even if a lecturer does not mention dogs or cats, including them in the text is one more reason for a student to read this book. This material also appeals to pre-veterinary students, who often enroll in animal behavior courses as preparation for their careers.

Because of our emphasis on evolution, we have not strictly excluded humans from our textbook, but neither is human behavior emphasized. Instead, when we do refer to human behavior, it is often in the context of questioning the traditional distinction between non-human animals and humans—a distinction that may be artificial in places, lacking sufficient scientific scrutiny. We explicitly reject an overly simplistic "my-genes-made-me-do-it" approach to behavior in general and human behavior in particular. In asking students to consider the continuum of all living things, we aim to promote critical thinking and a new consideration of traits that we may share with other species as a result of descent from common ancestors.

The laboratory manual that accompanies this text (*Field and Laboratory Exercises in Animal Behavior*) contains proven exercises. We encourage instructors who have not incorporated a laboratory in their course to consider doing so. A successful laboratory experience in animal behavior need not be expensive to present. The laboratory manual puts a strong focus on inquiry-based student participation, developing student strengths in hypothesis testing, and on giving the students hands-on experiences with topics covered in animal behavior courses. The exercises in the manual include a mix of field studies and laboratory studies. The targeted species of the field studies are widespread, and if the species mentioned in the manual does not occur locally, substitutions should be easy. The laboratory studies rely on easily-obtained study animals. The focus is slightly biased to experiments with invertebrate animals; this reduces the burden of obtaining IACUC approval for some of the laboratories. However, the most popular laboratory for our students is a multi-week exploration of *Betta* (Siamese fighting fish) behavior.

This book is an outgrowth of our decades of teaching animal behavior—what we have learned, what we have wished for, where we have found great joy. As we have worked through this immense undertaking, we have been impressed with the wide-ranging curiosity of our colleagues, friends, and students, and with their willingness to help in so many ways, surpassing our ability to thank them. We are particularly grateful to Ben Pless, Scott Altenbach, Jeff Mitton and Randy Moore, who generously contributed their outstanding photography to this effort, and whose images have proven that the appeal of animal behavior knows no academic boundaries. Thank you to all of our colleagues who took time to read all the chapters and provide us with expert feedback. Your insights and suggestions helped us create a better textbook. We also thank Rebecca von Gillern, whose editorial expertise greatly improved this book.

Michael D. Breed
Janice Moore

December 2010

CHAPTER 1

Of Cockroaches and Wolves: Framing Animal Behavior

1

LEARNING OBJECTIVES

Studying this chapter should provide you with the knowledge to:

- Understand that behavior, broadly defined, includes movement, social interaction, cognition, and learning.

- See that adaptive mechanisms provided by behavior give animals tools for adjusting to their environments and for manipulating the world around them.

- Learn that four central questions drive the study of behavior. They are causation, survival value, ontogeny (development), and evolution. This chapter should enable you to use these questions to form testable hypotheses about behavior.

- Be able to integrate the basic principles of evolution with an understanding of animal behavior.

- Appreciate that the roots of contemporary studies of animal behavior are in ethology, comparative psychology, sociobiology, and behavioral ecology.

1.1 INTRODUCTION: ANIMAL BEHAVIOR

Science is the outcome of human curiosity. We want to know the "why" and "how" of almost everything. In fact, all of biology can be addressed with two types of questions: proximate and ultimate. *Proximate* means "coming very soon," or "next"; *ultimate*, in contrast, means "coming at the end" of a process. In biological terms, proximate questions ask about mechanisms—how has something developed; how does it work?—and ultimate questions ask about how something has evolved—what is its selective advantage?

> **KEY TERM** Proximate questions ask about relatively imminent causes or mechanisms responsible for a trait.

> **KEY TERM** Ultimate questions ask about the evolution of a trait.

This curiosity and these types of questions have framed the study of animal behavior. All of us have watched animals, laughing at the antics of pets and marveling at the aerial acrobatics of birds. Shortly, we will discuss two animals in particular—the wolf and the cockroach—that invite us into the world of behavioral biology. That world differs from one in which we simply watch animals. Like the rest of science, it is grounded in questions—four of them, to be precise. Two are proximate and two are ultimate. We present them here, before we discuss wolves or cockroaches. These questions will be our constant guides, and if we allow them to, they will transform our "watching" into scientific exploration.

1. *What is the cause of a behavior?* **Causation** is the internal mechanism—nervous, hormonal, physiological—that results in a specific behavior. When we ask this type of question, we test hypotheses about how nerves, muscles, hormones, and physiology in general interact to produce behavior.
2. *How does a behavior develop?* Development, or ontogeny, involves both a genetic component and an environmental, or learned, component. When we ask this type of question, we ask where a behavior "comes from." We test hypotheses about the relative contributions of genetics and environment—"innate" tendencies and experience—to behavior.
3. *What is the survival value of a behavior?* Survival value addresses the usefulness of a behavior in terms of fitness, or its adaptive significance. When we ask this type of question, we test hypotheses about how behavior contributes to the survival and reproduction of an animal.
4. *How did the behavior evolve?* Evolution, or phylogeny, tells us about the origin of a behavior in time. When we ask this type of question, we test hypotheses about the beginnings of a behavior in ancestral organisms.

> **KEY TERM** Causation is the direct mechanism responsible for a behavior.

These questions were developed by one of the founders of the modern study of animal behavior, Niko Tinbergen. Later in this chapter, we will expand on these questions and how they might be applied. For now, we turn back to natural human curiosity about animal behavior.

Our Fascination with Animal Behavior

Long before humans envisioned science, animal behavior intrigued our distant ancestors. Understanding how and why animals behaved meant the difference between enjoying dinner and becoming dinner, between eating and being eaten (see Figure 1.1). Within a social group, it meant the difference between understanding and even manipulating that group, and being marginalized (see Figure 1.2). For these reasons and more, animal behavior is a course unlike any other. If scholarly interests have any genetic basis, we should all be intrigued by animal behavior!

FIGURE 1.1

Images from the Lascaux Caves in southwestern France, a home to Paleolithic humans some 16,000 years ago. The caves contain hundreds of images of animals, mostly horses and deer, and the paintings are so accurate that in many cases the animals they depict can be identified. The purpose of the paintings is lost to prehistory, but many people think that the humans who made these paintings might have used them in ceremonies that were meant to enhance hunting success. From left to right: an auroch (extinct in the Middle Ages), running horses and bulls, and a narrative scene involving a human. *Images: http://www.sacred-destinations.com/france/lascaux-caves, public domain.*

FIGURE 1.2

Observation and analysis of behavior are deeply engrained in human evolutionary history. This Japanese temple carving suggests investigation of behavior, along with analysis, and interpretation. *Photo: Michael Breed.*

The idea of innate human fascination when it comes to animal behavior is not as far-fetched as it sounds—although, as will become apparent, questions about intellect, predisposition, and genetics are difficult to answer in a definitive manner, especially with human subjects. Nonetheless, evolution, the foundation of biological thought, makes the concept of an innate interest in behavior a reasonable supposition.

DISCUSSION POINT: DEFINING BEHAVIOR

Behavior is broadly defined as movement, social interaction, cognition, and learning. In the scientific literature, the term *behavior* is surprisingly tricky to define. We think the definition works better as a discussion point than a factoid to be captured in a multiple-choice test. Some scientists' definitions of behavior focus on movement: behavior is the result of the activity of cilia, flagellae, or muscles. Interactions with other animals—in pairs or groups—are key to many complex behaviors. Learning and memory are important aspects of animal behavior, and it is possible to learn without moving a muscle. Thought and understanding—cognition—are also very much a part of animal behavior.

What value does behavior have for an animal? Behavior yields immediate flexibility. Using behavior, animals can change their locations, maximize their chances of survival by taking advantage of knowledge from previous experiences, and in some cases, gain the advantages of living in a group. Unlike a typical plant, rooted in the ground and limited to whatever conditions the environment creates for it, behavior gives animals tools for ecological choice. Even animals that are sessile (nonmoving) as adults have motile offspring that often exhibit preferences about where to settle in preparation for adult life. In sum, the richness of animal diversity yields an equally rich and enchanting variety of animal behavior.

The central precept of evolution is that organisms with traits that lead to increased survival and reproduction leave more offspring than do other such organisms without those characteristics.

FIGURE 1.3
Little Red Riding Hood has come to symbolize human fear of predators or, in this case, blissful ignorance.

If the advantageous characteristics have a genetic basis, those genes and the traits that they support will also increase in the population. In short, the ever-so-early human ancestor who was fascinated by animal behavior was much more likely to leave descendants than the one who was not terribly interested in where wolves were likely to hide or which watering hole was particularly appealing to deer.

Fast forward to the twenty-first century, and here you are, reading this book. Wolves are still with us, living at the margins of human habitat and in the wilderness beyond. It is no accident that wolves play a central role in legends, from the foundation of Rome by Romulus and Remus, human children reared by wolves, to the tale of Little Red Riding Hood, a girl with a red cape and remarkably bad judgment (see Figure 1.3). Like us, wolves are social animals. Like us, they seem clever at times; they may even solve problems and use past events to predict future ones (this is part of what we call *cognition*; see Chapter 6). In fact, domesticated wolves that we call dogs have done quite well for themselves. They have fared a lot better than their wild cousins, who inspire so much fear and resentment that they have been forced to near-extinction.

DISCUSSION POINT: GENES AND ENVIRONMENT IN BEHAVIOR

Consider how we might test the hypothesis that some of our behavioral preferences may be largely inherited, whereas others are strongly influenced by the environment. Chapter 3 will offer some good hints about how to do this for a given preference, but testing this hypothesis is not a trivial exercise. Also, note that we have called the idea that humans may have a genetic predisposition to be interested in animal behavior a "supposition" and "not … far-fetched." In other words, we are speculating. Is our speculation warranted? Why or why not?

1.2 WOLVES: LESSONS IN SOCIAL BEHAVIOR

Human fear of wolves notwithstanding, wolves reveal much about social behavior and how it functions in ecological associations, such as those between predators and prey.

Evolutionarily, predators are under strong natural selection to eat, but they are faced with prey that are under strong selection to escape or defend themselves (see Chapter 10). This leads to a kind of predator–prey arms race.

For prey species, the evolution of large body size can be an extremely effective defense. Large size, while not in itself a behavioral trait, has major behavioral implications. We should not be surprised to discover that the largest herbivores in most ecosystems (elephants, baleen whales, and bison come to mind) are many times the size of the largest carnivores. A large, strong animal can be difficult or impossible for a predator, no matter how well armed, to take down.

What is the evolutionary response of predator species? Chapters 13 and 14 reveal that social behavior has many costs and many benefits. One selective advantage for predators that cooperate as a group is that such cooperation allows them to overcome prey that would be impossible for an individual to handle. Add social cunning—the ability of predators to work together against prey—and the result is truly formidable carnivores like killer whales, African lions, and of course, wolves.

How does cooperation help predators when they hunt animals three to five times their size? One highly effective strategy employed by wolves is "relay running." Tired prey, gasping for oxygen and their muscles weakened, are easier to kill than rested prey. If one wolf were to chase an elk for miles, the two animals would tire more or less equally. But what if the wolf can trade the lead in the chase with another wolf? The elk gets no respite, forced to run at full speed, while each wolf can slow its pace when another takes its turn. Over time, the elk can no longer outrun the wolves, which then group for the kill (see Figure 1.4).

Wolves also are clever at ambushing prey; watching the direction that a herd of prey is moving, they can hide near that possible pathway (see Figure 1.5). This means that when the wolves spring out of hiding, they will be near to striking distance of the prey and can catch the prey before prey defenses are organized. Indeed, it is this inclination that is thought to be the basis for what has become, through artificial selection, the highly skilled herding ability seen in herding dogs. They work with the shepherd, circling around the "prey," giving them the "eye" (a stare that immobilizes sheep), and generally cooperating with the shepherd to move and control the sheep. It is the final sequence—the kill—that is thought not to be a desirable attribute of a herding dog. This general sequence can be discerned in many wild cooperative canid hunters.

5

FIGURE 1.4
Wolves are cooperative hunters and work in packs; these photographs offer just a glimpse into the complex communication that allows successful cooperation. *Photo: Frank Wendland, W. O. L. F. Sanctuary, www.wolfsanctuary.net, www.facebook.com/wolf.sanctuary.*

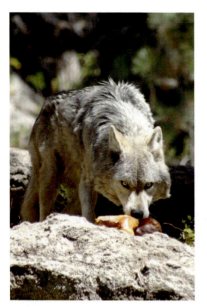

FIGURE 1.5
This photograph of a wolf dining on "prey" is from a wolf sanctuary, but nonetheless allows the viewer to understand how "eye" might be an effective tool to use against potential prey. It is never a good idea to run from canid.
Photo: Frank Wendland, W. O. L. F. Sanctuary, www.wolfsanctuary.net, www.facebook.com/wolf.sanctuary.

It is worth mentioning that prey defenses reach well beyond running to escape (see Chapter 10 on self-defense). In the co-evolutionary race between wolves and their prey, social behavior is as powerful a weapon for elk and bison as it is for the wolves. In fact, being the target of social hunters like wolves may be one of the driving evolutionary forces leading ungulates (hooved mammals, like deer) to live in groups. Given the opportunity, herds of some ungulates will circle like a wagon train in a Western movie, with the more vulnerable young members of the herd in the center of the herd where they can be protected. (Later in this book, you will discover how some herds act in the best interests of the young members, whereas other herds behave more selfishly.)

Wolves can take smaller prey, such as rabbits and hares, as well, but this is more of an individual proposition than social foraging is. The ability of wolves to switch prey preferences, depending on the availability of each prey species and competition with other predators, gives them the flexibility to respond to changing environmental conditions. Wolves therefore serve as one of the classic examples of how predators can control prey populations, particularly when a given prey species population increases. To survive, the average Minnesota wolf needs to eat the nutritional equivalent of 15–20 deer a year. Deer are most vulnerable in the winter when the snow is deep, as they may be weakened from lack of food and the snow makes it more difficult for them to escape. This means that in a winter with little snow the wolves may have a hard time, and their population levels can be affected. Despite the claims of nature programs on television, old and sick deer are not sacrificed for the "good of the species." Instead, these are the easiest deer for wolves to take. For that reason, deer populations where wolves are present are likely to contain mostly healthy animals; in contrast, when wolves are absent, sick or malnourished deer may survive.

Wolf behavior came to the forefront in the conflict over reintroduction of wolves into the Greater Yellowstone Ecosystem of Wyoming and Montana. Like most animals, wolves do not recognize human boundaries, whether those are the limits of protected areas, state lines, or international borders. Radio collars reveal that wolves can move hundreds, or even thousands, of kilometers in startlingly short periods of time. These long-distance movements have resulted in wolf populations far from their protected reserves and have created strong conflict among conservationists, ranchers, and developers.

Wolves capture the imagination, and because of the devotion of scientists who study them, we understand something of their behavior. Unfortunately, they do not lend themselves to the close-in scrutiny required to answer some behavioral questions, and most of us do not have a lot of personal experience with wolves. We may be more familiar with, say, crickets or cockroaches. In fact, why it is almost impossible to step on a cockroach? No matter how fast we think we are stomping, the cockroach will probably be faster. While this animal behavior question is not quite as gripping as how to avoid being a wolf's dinner, for most people in the modern world it may be more familiar and, in some respects, more puzzling. After all, humans are "smarter" than cockroaches (an assumption we may have to challenge later in this book)—why this lack of talent when it comes to smashing them? Surprisingly, exploring a bit about cockroach behavior will reveal some fundamental concepts that apply to all of animal behavior.

OF SPECIAL INTEREST: JUST-SO STORIES AND THE INTERPRETATION OF BEHAVIOR

A favorite childhood game is telling "just-so" stories. All that is needed is an observation and a vivid imagination. This process was captured eloquently by Rudyard Kipling. In "How the Camel Got His Hump," a genie turned the "hmmph" of a resentful camel with nothing to do into a hump (see Figure 1.6). Kipling's engaging story is an excellent example of how we can look at a behavior and imagine how it came to be, without any proof at all. The story concludes with a moral about the effects of laziness on humans:

The Camel's hump is an ugly lump
Which well you may see at the Zoo;
But uglier yet is the hump we get
From having too little to do.

Of course, we know that the camel's hump did not evolve as a result of sloth, but thinking critically about Kipling's construction of these stories is a good exercise in scientific reasoning. Scientifically, we rely on comparative or experimental tests of hypotheses, and because of that, we should always ask if the controls in the experiment were appropriate, if all variables were accounted for, and if additional hypothesis testing is needed to strengthen our conclusions. In studying animal behavior it is extremely important not to imagine an interpretation for a behavior. It is difficult to avoid telling just-so stories about behavior, but learning to recognize hypotheses that have actually been tested is key in practicing the science of animal behavior. In sum, behavioral hypotheses need to be accurately stated and tested with sound scientific methods.

FIGURE 1.6
"The Camel's hump is an ugly lump. . ." —Rudyard Kipling.

DISCUSSION POINT: THINKING BACK TO THE FOUR QUESTIONS

Review Tinbergen's four questions at the beginning of this chapter. How would you investigate them using wolves? What would your hypotheses be and how would you test them? What difficulties might arise?

1.3 COCKROACHES: MODELS FOR ANIMAL BEHAVIOR

We can argue that natural selection has favored cockroaches that do not wait around to be obliterated when a foot is descending on them—that such cockroaches leave more offspring with that valuable trait than more sluggish conspecifics do—and we would be correct. That still does not tell us exactly how the cockroach manages to be so perceptive and fast; that is, it does not address causation. For that answer, we turn to the world of neuroethology—the study of the interaction of the nervous system and behavior.

Cockroaches are wonderful research animals. *Periplaneta americana*, the American Cockroach (a misnomer—the insect was introduced to the rest of the world from Africa), can be raised easily and in large numbers. Indeed, there are thousands of species of cockroaches (most do not live in houses!), allowing for fascinating interspecific comparisons. They are hardy and large enough to allow significant experimental tinkering; while determined entomologists can and have performed surgery on mosquitoes, cockroaches are more convenient candidates for the operating table.

Given the luxury of simply staring at a cockroach before it skitters away, we would see that there are two small appendages on the posterior end of the animal. These are called *cerci* (the singular is *cercus*, from a Greek word meaning "tail") and they are covered with little "hairs" (see Figure 1.7). The hairs move when there is a sudden shift in the air currents around them. This does not have to be a breeze; it can be as subtle as a change in air pressure in front of a descending foot or predator's flicking tongue.

The direction of movement of the hairs tells the cockroach nervous system about the direction of air current, and thus the source of the threat, in the following manner: An individual hair can move most easily in one direction only, so given an air current from a specific direction, only certain hairs will move. Each hair is connected to a nerve; the direction of the air current will determine which hairs move and, in turn, which nerves fire. The nerves are connected to a ganglion at the posterior end of the cockroach; the ganglion contains large nerve cells called *giant interneurons* (GIs). These GIs extend from the posterior of the cockroach, through the thorax, up to the head of the insect. In the thorax, they are connected to neurons that control leg motion and running. Action potentials (neuronal messages; see Chapter 2) can race up these giant interneurons and communicate with ganglia in the thorax, which then cause legs to move (see Figures 1.8 and 1.9).

FIGURE 1.7

American cockroaches, *Periplaneta americana.* Note the pair of small posterior appendages on the cockroach in the lower righthand corner of the photograph. These are the cerci, important predatory-avoidance sensors. *Photo: Jeff Mitton.*

8

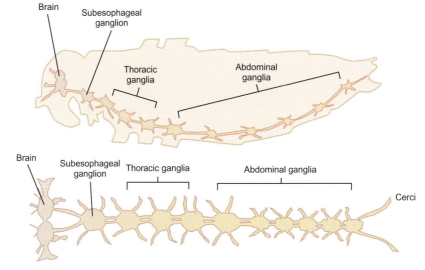

FIGURE 1.8

The nervous system of an insect. The nerve cord is ventral (true of arthropods and their relatives), and the spacing of the ganglia reflect the segmentation of the organism.

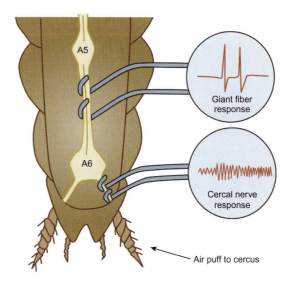

FIGURE 1.9

The cerci with neural connections. Changes in air pressure move the "hairs" on the cerci, initiating action potentials (nervous impulses) that travel up the giant interneurons. These impulses override all other nervous activity and cause the animal to jump, orienting away from the air pressure source, and run away.

There are two behavioral consequences of all the wiring just described: (1) A few milliseconds after the hairs move, the cockroach turns away from the perceived threat, and (2) after turning, it runs. The turn-away direction is informed by which hairs are moving, that is, the direction of the air current; the impulse to run then overrides everything else that the cockroach might have been doing. After all, that is the nature of escape—it is a life-or-death matter.

Because of this behavior, cockroaches have a remarkable story to tell us. Not only do we understand the mechanism that causes escape, but we understand how natural selection favors this behavior. We might then ask if very young cockroaches are as skilled as older ones at escape—in other words, how does the escape response develop? As expected, given the importance of the escape response, the hatchling cockroach escapes just as effectively as an adult does. Even though the newly hatched cockroach has far fewer hairs on its cerci, the ones that are there do their job perfectly. Moreover, molting does not seem to alter the escape response very much. Recall that cockroaches are hemimetabolous insects; that is, an immature cockroach, called a *nymph*, looks very much like an adult, with the exception of wings and sexual morphology. In the process of becoming an adult cockroach, the nymph grows and molts its exoskeleton several times. It is common for the process of molting to affect at least some other processes in insects, but the cockroach escape response is not affected in any significant way by this profound event in the life cycle. (There are some changes in the response, but they are on the order of a few milliseconds, and may not be ecologically significant.)

The fact that the escape response remains the same throughout cockroach development probably reflects the fact that effective escape is advantageous for cockroaches of any age. This contrasts with a cockroach's ability to produce sex attractant, which is an adult feature. Although immature cockroaches look very similar to adults, they are not sexually mature; in cockroaches, as in vertebrates, morphological and behavioral development is influenced by hormones. The ability to produce sex attractant, therefore, does not develop until adulthood and, in at least some species, is timed to coincide with egg production.

Finally, if we look at relatives of cockroaches, we can put the escape response in a context framed by evolutionary history, or phylogeny. Cockroaches are not the only animals with giant interneurons; many arthropods have them, as do other organisms that are considered relatives of arthropods by some scientists. In most of the animals that have been investigated, one of the functions of the GIs is escape; for instance, the flick of a crayfish tail and the scamper of a cockroach are both mediated by the GIs and they both accomplish the same thing: predator avoidance.

Although the phylogeny of arthropods and other invertebrates is being reconsidered—a common occurrence in science as we increase our understanding—we might reasonably hypothesize that a ventral nerve cord, with GIs that assist in escape, is a shared trait among arthropods and their kin, inherited from a common ancestor. In light of this evolutionary context, the absence of such structures in, say, spiny lobsters becomes much more intriguing than if it were simply a nugget of information dropped out of someone's isolated laboratory. How do spiny lobsters cope without GIs? What led to their loss? We leave this question about evolution for contemplation, but it may well have everything to do with the interaction of behavior, the nervous system, and evolution.

1.4 THE FOUR QUESTIONS REVISITED

At the beginning of this chapter, we introduced four attributes—causation, survival value, development, and evolution— that frame animal behavior. They are actually decades-old questions put forth by a behavioral biologist, Niko Tinbergen, in 1963 (see Figure 1.10).

9

If we can answer Tinbergen's questions about a given behavior, we will understand that behavior at every level is addressed by science. Because of the core importance of these questions, first introduced at the beginning of this chapter, we will now consider them in greater detail.

The wolf and the cockroach offer good examples of how different organisms lend themselves to different questions. The wolf invites us to think about behavior from the perspective of ultimate questions—how social behavior evolved—which in turn causes us to ask about the survival value of social behavior. Cockroaches, because of their size, their compatibility with laboratory settings, and their overall versatility, allow a more detailed approach to development and causation. Both species offer windows into how Tinbergen's four questions can be applied in analyses of animal behavior.

FIGURE 1.10
Niko Tinbergen, a key figure in the establishment of ethology. Tinbergen's four questions: causation, survival value, development, and evolution, still form the foundation for studies of animal behavior.

10

OF SPECIAL INTEREST: PROXIMATE AND ULTIMATE QUESTIONS IN ANIMAL BEHAVIOR

Classical ethology research was usually conducted under natural conditions and posed questions that we consider "ultimate"—that is, questions aimed at understanding evolutionary considerations. Behavioral research conducted on wild animals under field conditions tells us a lot about how animals behave in nature but also has some disadvantages: knowledge of the animal's previous experience—perhaps even its sex or age—is unavailable. Moreover, the observer has no control over weather, predators, and other influences that can interfere with comparisons. Comparative psychology solved these problems; the subjects—including their diets, rearing conditions, age, and history—were well known and the conditions of observation could be controlled and replicated. However, the laboratory conditions may bear small resemblance to the field conditions under which the behavior might actually be expressed. Comparative psychologists tended to focus on "proximate" questions—questions about mechanism and other more immediate influences on behavior.

Each type of research has strengths and weaknesses. Neither is "right" or "wrong"; the choice of methods depends on the organism and the question. A combination of field and laboratory work often provides the most convincing evidence with which to test a hypothesis.

Causation: What Is the Cause of a Behavior?

To understand causation, we focus on the animal's behavioral responses to stimuli from the external environment. Does the animal handle stimuli the same way every time they are presented? Or does the nature of the response depend on the animal's internal state? Consider the difference in behavior between hungry and satiated (well-fed) animals. The sight of a potential prey animal, such as a deer, may elicit stalking, chasing, and pouncing by a hungry mountain lion, but a well-fed lion may have little interest in a passing deer. Studying causation allows us to delve into how behavior is regulated and how animals make choices about their activities. In the case of the wolf, this might mean understanding the relationship of hunger signals (e.g., nutrients in the blood), inclination to hunt, and effort expended in hunting. In the case of the cockroach, the exquisite connection between the cerci, the GIs, and the escape behavior, activated by a small change in air pressure, is a classic example of causation.

Development: How Does a Behavior Develop?

How does a behavior change during the lifetime of the animal? To survive and reproduce, an animal must change its behavior throughout its life. Animals that undergo metamorphosis are extreme examples of this: larval stages are devoted to foraging and growth, while adults disperse and reproduce. Many birds and mammals acquire skills while still under parental protection; young animals learn through experiences that shape their adult behavior. Sexual maturation plays an important role in development, as well. Physiological and morphological changes, coordinated by hormones, prepare animals for the behavioral challenges of adulthood. Hypotheses about ontogeny (development) postulate learning and physiological development as driving forces in shaping the behavior of adult animals.

Survival Value: What is the Survival Value of a Behavior?

In other words, how does the behavior help an animal to survive—what is the adaptive significance of the behavior? Much of behavior is clearly related to the day-to-day survival of animals. Finding food, water, and shelter are continuing behavioral challenges for animals. Because of this, it is all the more puzzling to observe a behavior and find, as we often do, that its survival value is not immediately clear. For instance, why does a dog circle three or four times before lying down? The immediate survival value of this behavior may not be obvious, but we would hypothesize that the behavior is rooted in an ancestral behavior that did improve survival. (See the next question: evolution!) Testing hypotheses about survival value plays an important role in developing an accurate understanding of behavior.

Evolution: How Did the Behavior Evolve?

Behaviors do not arise spontaneously when animals need them. As with body structures, behaviors evolve through modification of previously existing actions. Sometimes an action that is used in one context is "co-opted" for use in a different behavioral context; for example, the predatory behavior of a venomous snake may be co-opted and modified to serve as defensive threat. Displays, such as those used in courtship, are often assembled over the course of evolution from a seemingly unrelated set of movements that the animal performs in other contexts. Testing hypotheses concerning behavioral evolution often requires constructing a "family tree," or cladogram, of related species and tracing how a behavior has changed in either its form (the way in which it is produced) or its function over evolutionary time.

11

1.5 EVOLUTION: A REVIEW

Behavior, like any other biological trait, evolves. Evolution is the foundation of biological thought in all fields, and understanding evolution paves the way to understanding animal behavior. Because of the explanatory power of the theory of evolution by natural selection, this book is grounded in evolutionary principles; they are the backbone of our understanding of animal behavior and of biology in general. Indeed, without those principles, biology shrinks to a collection of facts about the living world, much like a collection of cute stuffed animals—nice enough to look at, but not particularly coherent, and certainly no source of predictions about other collectible items.

Given the importance of evolution, then, it would be a mistake not to review that theory. In fact, we need to begin by visiting the word *theory*. To many nonscientists, the word *theory* can be used interchangeably with the word *hunch*. Scientists, on the other hand, have a clear definition for *theory* that goes well beyond the notion of a hunch. For scientists, a theory is an overarching concept that explains a large number of facts and observations about the natural world and that can be used to make predictions about future observations.

> **KEY TERM** A theory is an overarching concept that explains a body of facts about the natural world and that generates testable predictions.

A theory has such weight and scope and explains so many facts that it is unlikely to be refuted, although it may be refined as one tests the predictions it generates.

Darwin's idea—the theory of evolution by natural selection—is one such theory, supported by evidence from geology, paleontology, agriculture, embryology, biogeography, anatomy, and molecular biology. It is based on fairly simple observations about populations of organisms: (1) Organisms can often produce far more offspring than those required to replace the parents—more than available resources can support.

> **KEY TERM** Natural selection is a process that results in increased survival and reproduction compared to that of competing organisms.

(2) In some species, this excess results in competition for those limited resources, and the outcome of that competition determines which individuals survive and reproduce. (3) These competing individuals are not alike; they have a variety of different traits. (4) The traits that belong to the successful competitors will be overrepresented in the next generation compared to traits that conferred fewer benefits. This process resulting in differential survival and reproduction is called *natural selection*. (5) Some of this variation among traits can be inherited; it has a genetic basis.

When the representation of the traits changes in the population—usually because of the action of natural selection—gene frequencies change. This change in gene frequency is called *evolution*. (Darwin did not know about genes, so he called evolution "descent with modification from a common ancestor." We now know that gene frequency is the thing that is modified.)

DISCUSSION POINT: NATURAL SELECTION

Note that there are two components of the process of natural selection: survival and reproduction. While natural selection favors traits that improve survival and reproduction relative to those that do so to a lesser extent, sometimes traits that improve survival do not improve reproduction, or vice versa. How might increased reproduction, for instance, affect survival? Could increased survival have negative consequences for reproduction? What happens then? What kinds of experiments or observations might tease these apart?

Genetic Variation: A Necessary Ingredient of Genetic Change

In thinking about these observations, we see that genetic variation is needed for evolution, or genetic change. After all, if animals are genetically identical, there are no genetic advantages or disadvantages to be had, no distinctions to be made. What are some causes of genetic variation?

Three of the most important causes of genetic variation are mutation, gene flow, and genetic mixing:

1. *Mutation* is a change in an organism's DNA. If it occurs in the DNA of eggs and sperm, it can be inherited. The effects of mutations are random. Some will be beneficial, some detrimental, some will be of no consequence.
2. *Gene flow* occurs when organisms move into a population, bringing their genes with them. In that way, new genes can be introduced, and genetic variation can increase.
3. *Genetic mixing*, or recombination, is a third cause of genetic variation. This occurs when chromosomes from parents line up during gamete production. Chunks of information from one parent may switch places with homologous (similar) chunks from the other parent, thus creating new combinations of genetic traits. (See Chapter 3 for a more detailed discussion of these genetic mechanisms.)

Changes in Gene Frequencies

Given these causes of genetic variation, how do gene frequencies change; that is, how does evolution itself happen? Scientists are still exploring this question, and new discoveries are not in short supply, but four mechanisms are particularly important: mutation, genetic drift, migration, and natural selection.

Mutation not only serves as a source of genetic variation, but when it occurs, it also alters gene frequencies. Because of this, it is one cause of evolution, albeit a limited one. Mutations occur at very low rates, and because they are random, we do not expect them to produce directional change.

Genetic drift is another accidental shift in gene frequencies, based on the fact that some individuals fail to survive simply because they are unlucky. They build their nest in the path of a tornado or in low-lying areas just in time for a particularly bad hurricane season. They are eliminated not because they are poorly adapted to the environment, but because of unpredictable misfortune. Nonetheless, the genes that these unfortunate organisms carry are removed from the population and gene frequencies change. As might be expected, this is especially influential in small populations, where a random event can eliminate some genotypes completely. Thus, genetic drift can actually reduce genetic variation. Once genetic variation is reduced, it may be restored by processes such as mutation and migration, but these processes work very slowly.

Migration is another source of genetic variation that can also change gene frequencies and result in evolution. When organisms join a population and interbreed with residents, the subsequent generation will exhibit gene frequencies that differ from those in the population prior to the arrival of the migrants.

Mutation, genetic drift, and migration can all result in evolution, but such changes in gene frequencies do not necessarily proceed in any particular direction. Mutation and genetic drift have random consequences, and migrants are not necessarily better suited for the environment than the host population is. In short, there is no reason to expect increased survival and reproduction associated with traits that change as a result of this kind of evolution.

Natural selection, the fourth mechanism of evolution, is different. It is the only mechanism that causes an increase in the frequency of adaptations—the inherited traits that promote survival and reproduction. For natural selection to work, there must be genetic variation. Because of that variation, some individuals are better at surviving and reproducing than others. They leave more offspring, and those offspring carry the genes that are associated with the beneficial traits. In the process, gene frequencies change.

At this point, please note that although genetic variation occurs in an unpredictable fashion, and although mutation, genetic drift, and migration can lack any recognizable direction, the products of natural selection are anything but random. This is why evolution may be based on random events (e.g., mutation, genetic drift), but it does not often generate random results. Natural selection, a powerful engine of evolution, increases the frequency of adaptations; it does not result in a random mix of traits, even though the material on which it acts may be randomly generated.

This, then, is *fitness*—the ability to contribute genes to the next generation. The popular media would have us believe that fitness means winning reality shows, getting rich, winning races, and doing other spectacular things. In contrast, biologists say that fitness has two main components: survival and reproduction.

> **KEY TERM** Fitness is the relative ability of an organism to contribute genes to the next generation.

13

Adaptation and Behavior

Although natural selection results in adaptation, not every trait is an adaptation. Some folks are eager to explain almost everything they see in terms of adaptation, but a bit of skepticism rarely hurts a scientist. Recall that not every change in gene frequency (i.e., evolution) results from natural selection; changes that result from genetic drift or mutation are probably not going to be adaptive. In addition, some traits might have been adaptive at an earlier time but are no longer adaptive; the pelvic bones of whales are examples of such vestigial traits. These bones served the terrestrial ancestors of whales, but are of little use to whales now. Yet other traits might not be adaptive in a perfect world, but may exist as a result of constraints imposed by different traits that are adaptive. Thus, although it might seem to be invariably adaptive to produce a large number of supremely healthy offspring, resource limitation or other environmental challenges may thwart this simple expectation. Instead, they may favor organisms that limit the number of offspring they produce, investing heavily in each one, or organisms that produce many offspring, each of which receives little investment.

Is every trait that increases survival and reproduction an adaptation? No, because some beneficial traits are not the products of natural selection. The fact that we do not wander aimlessly across busy highways certainly increases our probability of survival and reproduction, but that trait is learned, not inherited; in addition to the fact that an adaptation increases survival and reproduction, an adaptation is inherited.

Finally, an adaptation may be a beneficial trait for which the ancestral function has been modified. An original trait is often co-opted to result in something more beneficial. Some interesting terminology has grown up around this concept, but the core idea is that original traits are often co-opted. For instance, in animal behavior studies, signals used in communication are often co-opted from noises, movements, or odors that the animal already produces (see Chapter 7). A trait or structure that is evolutionarily malleable and can, over the course of natural selection, be modified for a new purpose is often termed a **preadaptation**.

> **KEY TERM** A co-opted trait, in evolutionary terms, is one that served a different function in an ancestor than the one it does today. These are also called *exaptations*.

> **KEY TERM** A preadaptation is a trait that is subject to modification for a new function by natural selection.

This reinforces an important evolutionary principle: natural selection acts on available variation, not the best variation imaginable. Thus, organisms do not get the adaptations they "need" in some evolutionary version of online shopping; their adaptations are a result of natural selection acting on existing traits.

Optimality and Behavior

A consideration of the limitations of adaptation and natural selection leads to the idea of optimality. An optimum, from a Latin word meaning "best," is a fairly unequivocal notion: it is not almost-best or sort of good; it is unsurpassed—the best. This idea will be covered in more detail in Chapter 9, but for now, given the discussion so far, it is probably apparent that there are many explanations for why adaptations are not always optimal. For a variety of reasons, ranging from developmental constraints to ecological trade-offs, the genetic variation available to natural selection may not include the ideal trait; in that case, the product of natural selection will not be optimal. In addition, the world is not a homogenous place, either in time or space. What might be optimal today, or in this location, may be suboptimal tomorrow, in another location. In fact, it is precisely the heterogeneous nature of the world at large that favors a particular and powerful adaption: sexual reproduction.

The thought of sexual reproduction as an adaptation might sound strange at first, but consider the fact that many organisms do not depend completely on sexual reproduction. Instead, they often clone themselves when reproducing. Plants frequently do this;

14

strawberries, aspen trees, and daylilies all generate other plants from underground roots. In the world of animals, pieces of some flatworms can grow into entire individuals, some fish eggs can develop and hatch without being fertilized, coral reefs grow when individuals bud and create new individuals. Initially, it seems that such organisms achieve high fitness; after all, 100% of their genes are passed on to each offspring compared to the 50% that sexually reproducing parents donate, often called "the cost of meiosis." Given those differences, how can sexual reproduction be an adaptation?

Sexual reproduction is a major source of genetic variation. In addition to the genetic mixing that can occur between chromosomes when gametes are produced, fertilization also generates variation among offspring of animals such that each offspring carries a unique combination of genetic material bequeathed by its parents. Thus, while sexually reproducing animals must produce twice the number of offspring found in asexually reproducing animals to be able to convey the equivalent amount of genetic information to the next generation, the variety among sexually produced offspring means that at least some of them may be able to survive and reproduce in a changing world. Of course, if conditions are stable, then asexually reproducing animals have an advantage: Their own combination of traits succeeded, and as long as nothing changes, that combination should work well in subsequent generations.

In a world threatened by plagues, blistering droughts, and unrelenting floods, it is difficult to argue that the earth is a particularly stable environment. Indeed, the advantage of sexual reproduction is so significant that even asexually reproducing animals frequently participate in sexual reproduction as well. Chapter 12 reveals how all sexual reproduction is not equal, and how mates are evaluated and selected. As will become apparent, the value of producing variable offspring has fueled much animal behavior, from the flash of peacock tails to the chorusing of frogs.

Speciation and Behavior

Finally, the formation of species (called *speciation*) provides a powerful tool for asking questions about the history of traits, including behavioral traits. So far, this review of evolution has addressed changes in gene frequency and the manner in which those changes happen. How can such genetic shifts within a population of animals result in the biological diversity we see today? Something else has to happen: the flow of genes within that population must be interrupted. This can occur when a population is subdivided and parts of that population are isolated from each other. Eventually, if environments of the subdivided groups differ, natural selection will favor different traits in the two new populations. As time passes, differences will accumulate, and the two populations will no longer be able to interbreed were they to have that opportunity. By the way, complete isolation is not necessary for species formation; a small amount of gene flow will not counteract the accumulation of differences.

Such reproductive isolation can result from a variety of causes:

1. *Geographic barriers.* If a population is subdivided by the emergence of a mountain range, river, or other inhospitable habitat, animals on one side of the barrier will be unable to breed with animals on the other side. The same effect occurs if part of the population moves away.
2. *Resource shifts.* For animals that live and reproduce on a resource, the ability to colonize new resources decreases the likelihood that they will encounter or mate with individuals in the parent population.
3. *Mate choice.* If females diverge in their preferences for male characteristics, for instance, and if that divergence has a genetic basis, then eventually there will be two distinct gene pools, each sporting one or the other preferred male trait.
4. *Genetic change.* Mutations that prevent proper meiosis can produce individuals that cannot mate with other members of the population. This is thought to be the origin of about 4% of plant species.

15

Phylogeny and Behavior

Such isolating events produce splits in lineages; if these divergences persist, two species will result from one. Tracing this history reveals a branching pattern—a "tree of life." (Actually, it is more like a disorganized bush.) The relationships among the branches—which branch came from what stock—are hypotheses to be tested, and the resulting bush-like representation is called a *phylogeny*, a combination of two Greek words meaning "phylum" or "race" and "origin." (The phylogeny shown in Figure 1.11 is not particularly bushy, but it does show the relationships clearly.) Because these relationships are hypotheses, phylogenies will be refined as we understand more about them. Keep in mind that every phylogeny is a working hypothesis.

KEY TERM A phylogeny is a hypothesis of evolutionary relationship among types of animals (often species).

16

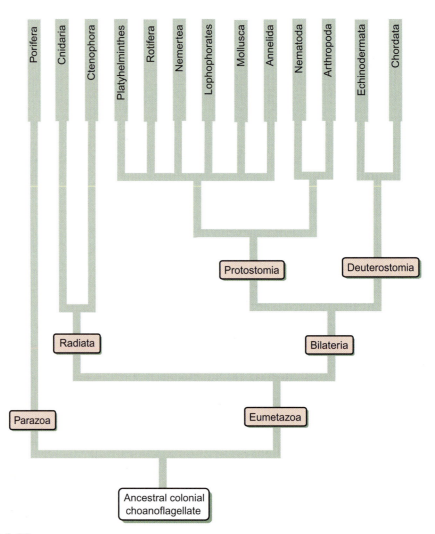

FIGURE 1.11

A phylogeny of Animalia, showing the profound effect of behavior (movement) on the evolution of all animals. The Radiata have poorly developed musculature and nervous systems; many are sessile, and the ones that do move are often at the mercy of currents and tides. Because of this, the radiate animals must be equipped to confront the world as it approaches them from any direction. Bilateral symmetry (found in the Bilateria), along with a suite of other traits, is favored when locomotion is a more important part of the behavioral repertoire, allowing more controlled movement through the environment. Radially symmetrical members of the Bilateria (e.g., sea stars) are either sessile or thought to be descended from sessile ancestors.

Phylogenies yield a variety of important information. Three nuggets to value in this course are the following:

1. Phylogenies are radiations over evolutionary time, not ladders of evolutionary progress. There is no "high" or "low," and the fact that humans can do calculus, and a sea star cannot, does not make us any better at living. If anything, it probably means that humans are a good deal worse than a sea star when it comes to life underwater. (The notion of a ladder-like progression of life forms has been with us since Aristotle proposed the Great Chain of Being. During medieval times, angels were inserted between humans and God, so humans were almost at the pinnacle, but not quite.)

2. Because phylogenies are radiations, and not ladders, anything alive today represents a line that has been subject to natural selection for as long as the lineage of any other living organism. Sea stars are not "older" than humans, nor are they more "primitive," and they are in no way our ancestors. Instead, all living things share common ancestors to greater or lesser extents. A phylogeny reveals when that shared ancestry ceased, relatively speaking.

3. The places in a phylogeny where branching occurs are called *nodes*. These nodes represent the last common ancestor that the organisms in the branches shared. If organisms share a common ancestor, then they share characteristics (*characters*, in scientific terms) that they have inherited from that ancestor. The more characters that organisms share as a result of descent from a common ancestor, the more closely they are related. Based on the hypotheses of relationships portrayed in phylogenies, we can ask if traits, including behavioral traits, are influenced by evolutionary history and shared ancestry. If they are, we may be able to predict their occurrence in other animals.

Because of this, it is not surprising that the GIs of arthropods work in similar ways and that they often have similar functions. The ragged bush of evolutionary relationships will make sense of discoveries about animal behavior. It will also generate more questions. This is how science thrives.

> **Note**
> From ancient times through the Enlightenment, scientists were called "philosophers"! This ancient tradition continues today with the advanced degree "Doctor of Philosophy" conferred upon scientists as well as scholars of other disciplines.

1.6 THE STUDY OF ANIMAL BEHAVIOR: WHERE DID IT COME FROM?

In a world that is alive with the flashes of color, the buzzing and the splashing that come from animal behavior, why do we bother with history? When compared to other aspects of biology, animal behavior comes from unusual, and sometimes seemingly incompatible, places. Those origins influence what we study today, and how we study it. They offer important considerations: are field studies better than ones in the laboratory? How do we ask questions about development, causation, evolution, or survival value in each setting? Who are the people who wrested our knowledge away from the "just-so" story tellers and placed it into the realm of science? History answers these questions, and many more, and helps us understand our work today and our aims for the future.

FIGURE 1.12
Aristotle contributed much to understanding the natural world and to forming a scientific basis for studying animal behavior. His works laid the foundation for accurately recorded natural history and for scientific testing of hypotheses about behavior.

17

Ancient Greece

Given the hypothesized hard-wired fascination with behavior, it will not be surprising that scholars and philosophers have written about behavior since antiquity. Aristotle (384–322 BCE) (see Figure 1.12) probably thought most deeply about animal emotions and intelligence, writing, "Many animals have memory, and are capable of instruction; but no other creature except man can recall the past at will" (*The History of Animals*, Book I). These words, centuries old, capture a current debate in animal behavior. Is Aristotle correct, or are at least some animal species capable of "mental time travel"; that is, can they call on past experiences to shape future plans?

Darwin and the Victorians

Another highly notable figure in the science of animal behavior, Charles Darwin (1809–1882; see Figure 1.13), lived centuries later. Darwin's thinking about evolution has had a tremendous effect on studies of animal behavior. His 1862 book on orchid pollination showed great insight into insect behavior and the co-evolution of pollinator behavior with orchids. Later, he turned his full attention to behavior in his 1872 book *The Expression of Emotions in Man and Animals*. Articulating the fascination with social behavior that is shared by many people, Darwin explored how humans and other animals convey their emotions and act impulsively based on emotion. He argued,

"That the chief expressive actions, exhibited by man and by the lower animals, are now innate or inherited,—that is, have not been learnt by the individual,—is admitted by every one. So little has learning or imitation to do with several of them that they are from the earliest days and throughout life quite beyond our control; for instance, the relaxation of the arteries of the skin in blushing, and the increased action of the heart in anger."

FIGURE 1.13
Charles Darwin had a keen interest in animal behavior and wrote extensively about his observations and hypotheses concerning behavior.

This gives great weight to the role of evolutionary history in shaping behavior, and it also poses some profound questions, for public debate over the relative contributions of genes and the environment to behavior has raged from Darwin's time to the present. The issue of nature versus nurture and how these shape behavior will be examined in more detail in Chapter 3.

The late 1800s and early 1900s were the heyday of natural history exploration of the world. Earlier explorations, such as Lewis and Clark's of the American West, Darwin's voyage on the Beagle, and Alfred Russel Wallace's discoveries in the Amazon and the islands of the southwestern Pacific Ocean laid the groundwork for public fascination with natural history in Victorian England, the United States, and Europe. Published accounts of explorations found large and eager audiences of readers; the writings of natural historians fed the public imagination and had much the same effect on Victorian society that modern-day videos of animal behavior have on current public perceptions of wildlife and animal behavior. This era also saw the public display of exotic animals in zoos develop as entertainment and education; in zoos the public could directly see some aspects of the behavior of animals that they would never encounter in nature (see Figure 1.14).

FIGURE 1.14
Zoos gained popularity as public entertainment in Victorian times. Live animal exhibits allowed the public to observe the behavior of exotic animals.

The Transition to Modern Science

Georges Romanes (born in 1848 in Canada, studied and worked in England, died 1894) was the academic successor to Darwin, bridging Darwin's work and contemporary evolutionary biology. Romanes is credited with founding comparative psychology as a field of study, and contributed to our early understanding of cognition (thought processes) in pigeons. Romanes was a creature of his time; he used anecdotal information and attributed attitudes such as "unselfishness" and "courage" to his study animals. The attribution of human traits to animal behavior began long before Romanes, of course, but Romanes was explicitly anthropomorphic in his work, claiming that he could infer the mental state of an animal from its behavior, based on what his own mental state would be if his behavior were similar. This was, as one writer claims, "poetic," but it did not serve the scientific study of animal behavior well.

OF SPECIAL INTEREST: CLEVER HANS

Clever Hans was indeed a clever horse (see Figure 1.15). According to the popular press of the early twentieth century, he was able to spell, to read, and to do arithmetic. He answered questions posed by strangers by tapping his hoof, and was always correct, even when his trainer was absent. How could this be?

That was the question asked by the director of the Psychological Institute of the University of Berlin, who in 1907 sent a student, Oskar Pfungst, to find out what Clever Hans could and could not do. Clever Oskar devised a mixture of trials; in some, the questioner knew the answer, and in others, the questioner did not. Pfungst showed that Clever Hans was able to answer correctly only when the questioner knew the answer. It seems Hans had learned that questioners would make very small, involuntary movements as his taps neared the correct answer, and he used that to gauge his response. Hans was clever indeed, but he did not read and do arithmetic. This revelation did considerable damage to the idea that animals shared human mental abilities.

FIGURE 1.15
This drawing represents an old photograph of Clever Hans, the horse that was clever enough to learn a lot about human behavior. He learned, for instance, when to stop tapping his hoof—that is, when human handlers unconsciously exhibited signs as his foot tapping neared the right answer to an arithmetic question. Hans never learned arithmetic, but he knew a lot about human behavior.

C. L. Morgan (1852–1936), Romanes' successor, began the process of setting things aright. He is recognized as the first biologist to question the anthropomorphic approach that was so popular in the nineteenth century. In 1894, he wrote what Galef has called "possibly the most important single sentence in the history of the study of animal behavior" in a book called *An Introduction to Comparative Psychology*:

> In no case may we interpret an action as the outcome of the exercise of a higher psychical faculty, if it can be interpreted as the outcome of one which stands lower in the psychological scale.

This statement is called *Morgan's Canon*. A canon is a rule or a law, and this rule has had both beneficial and regrettable effects on the study of animal behavior. It was, of course, essential to the development of animal behavior as a science rather than a storytelling device. On the other hand, it was sadly misinterpreted in some cases, leading to profound underestimation of the cognitive abilities of many animals. Such was the backlash against attributing human qualities to animals on the basis of behavior that even the study of cognition was frowned upon for decades. There is a difference between arguing for the most parsimonious explanation (as Morgan did) and denying the possibility of any other explanation.

FIGURE 1.16
A. Karl von Frisch, a pioneer in the study of sensory physiology and behavior. He solved the puzzle of the dance language of honeybees. B. Konrad Lorenz, followed by some of the geese that helped secure his position as one of the founders of animal behavior. Lorenz was deeply interested in instinctive behavior; among other things, he studied imprinting in goslings.

FIGURE 1.17
A Skinner box, named after B. F. Skinner, who designed a similar apparatus. The classic box contains rewards (the food), signals (lights and speaker), and negative stimuli (the electric grid, in this case). Using rewards and punishments, he studied animal learning.

Parsimony is difficult to implement in a world of anecdotal data. We have Edward Thorndike to thank for directing animal behavior away from that world. Thorndike studied learning in cats under laboratory conditions at Columbia University; he was also a major contributor to early work in human learning. He insisted that the study of animal behavior be quantitative and subject to rigorous scientific standards. He standardized his testing conditions and took care to expand sample sizes and do systematic comparisons.

European and American Traditions

In Europe, public fascination with natural history evolved into scientific inquiry about the mechanisms underlying animal behavior. Karl von Frisch, an Austrian biologist born in 1886 (died 1982; see Figure 1.16A), studied the sensory biology of honey bees. Among his discoveries were the existence of ultraviolet perception, the ability of bees to perceive the polarization of light, and the use of color vision by bees. His observations of bees ultimately led to his decoding the dance language of honey bees.

Konrad Lorenz (born in Austria in 1903, died 1989; see Figure 1.16B) and Niko Tinbergen (born in Holland in 1907, died 1988; see Figure 1.10) joined von Frisch in being leaders in the study of animal behavior in natural environments, and this developing scientific field became known as ethology. Von Frisch's, Lorenz's, and Tinbergen's contributions come up repeatedly in this book, and the three were awarded the 1973 Nobel Prize in Physiology or Medicine for their roles in developing important theories and conceptual approaches for the study of behavior. The focus of ethology was on the study of animals in natural, or minimally modified, environments. A critical common thread in the work of ethologists was a focus on innate, or instinctual, factors that shape behavior. This is an important point that we will revisit.

In America the development of animal behavior studies as a scientific endeavor took a very different path. J. B. Watson (1878–1958) founded "behaviorism" as a discipline of study; behaviorism focused on external manifestations of behavior, rather than the internal emotional or innate lives of humans and animals, and viewed modification of behavior by reinforcement as the predominant mechanism underlying behavior. B. F. Skinner (1904–1990) was a Harvard professor whose approach emphasized the ability of animals to learn and modify their behavior based on experience. Using animals such as rats and pigeons, which were easy to maintain in the laboratory, he developed paradigms for how animals could learn skills based on rewards and consequences. He is famous for developing the "Skinner box" (see Figure 1.17), an isolated environment in which the animal could be presented with stimuli, rewards, and punishments, and the resulting behavior could be studied. Skinner's work was in the field of comparative psychology. In contrast to the ethologists, Skinner emphasized learning and developmental flexibility in shaping behavior. Interestingly, the work of the Russian behaviorist Ivan Pavlov (born 1849, winner of the 1904 Nobel Prize in Physiology or Medicine for his research on digestion) on conditioning in dogs had more influence on Skinner than on the European ethologists.

Daniel Lehrman (1919–1972, a professor at Rutgers University in New Jersey) became the focal point, on the American side

Speaker
Signal lights
Lever
To food dispenser
Food pellet
Electric grid
To shock generator

of the Atlantic, of the split between the approaches of the ethologists and comparative psychologists. The nature versus nurture argument (see Chapter 3) became highly personal, as well as academic. The two sides of the debate finally came together to form a consensus at a meeting in Palo Alto, California, in 1957. However, the nature–nurture debate resurfaces from time to time, as it did in the bitter conflict between E. O. Wilson, a Harvard biologist who championed sociobiology as integrating neurobiology, physiology, behavior, ecology, and evolution, and Richard Lewontin (also at Harvard) and others who attacked sociobiology as social determinism, or a disguised version of social Darwinism.

Modern Science

Contemporary studies of animal behavior have outgrown much of this debate, as nearly all scientists recognize the intertwining of genes and environment in shaping an organism, including its behavior. The scientific field of animal behavior occupies much of the ground envisioned by Wilson for sociobiology, encompassing studies that range from the genetic and physiological mechanisms underlying behavior to those that place behavior in its ecological and evolutionary contexts. In the past three decades, *behavioral ecology* has succeeded *ethology* as the term used to characterize studies of behavior in natural environments. *Comparative psychology* has fallen into disuse, as animal research in many psychology departments has turned to the neurobiological bases of behavior.

DISCUSSION POINT: WHAT'S IN A NAME?

What do we—the folks who study animal behavior—call ourselves? It turns out that the answer depends on whom you are talking to. In this book, for convenience, we often use the term behaviorist. Is a person who studies wolves in the field a behaviorist? B. F. Skinner might disagree. What about *behavioral biologist*? Psychologists who study animal behavior might disagree. Much like the definition of animal behavior, the name of a person who studies that topic can be hard to pin down! The difficulty in defining the animal behavior and the difficulty in naming those who study it may well be a reflection of the spotted history of the field, gathered in from different traditions in field and lab. You will see a variety of identifiers applied to people who study animal behavior. The important point is what they do and the scientific rigor with which they do it.

In 1976, Donald Griffin courageously established the field of cognitive ethology, with a book titled *The Question of Animal Awareness: Evolutionary Continuity of Mental Experience*. This was no small task; scientists had embraced Morgan's Canon with fierce enthusiasm and were not inclined to deal with things they could not measure directly. Basing his argument in solid evolutionary theory, Griffin questioned the prevailing assumptions about the cognitive superiority/uniqueness of humans and opened the door for further scientific investigation. Griffin was not new to the field of animal behavior. Griffin made other significant contributions to the study of behavior; almost 40 years before his first book about cognition, he and a fellow student at Harvard discovered that bats use sonar when navigating in the dark.

The world of animal behavior continues to change, reflecting novel scientific discoveries as well as emerging social concerns about animals. For instance, conservation behavior is a new and rapidly developing field; as the name implies, *conservation behaviorists* bring their expertise to important issues such as preservation of endangered species, design of natural reserves, and management of human–animal interactions in urban and suburban settings. From its early beginnings with clever horses and dancing bees, the study of animal behavior is helping to redefine the way we look at how behavior works, how it benefits animals, how it develops, and how it has evolved.

1.7 *UMWELT*: THE WORLD IN WHICH ANIMALS BEHAVE

This unlikely history, from Clever Hans to the contrasting worlds of the ethologists and the comparative psychologists, leads us to the central and continuing challenge in the study of animal behavior. Surprising as it may sound, this challenge is not about observing secretive organisms, not about discovering predator–prey linkages or the real meaning of some chirp or bellow. Those are indeed challenges, but our foremost obstacle is understanding the sensory-perceptual world of the study organism—that is, its *Umwelt*—in a way that allows us to ask meaningful questions. This German word meaning "environment" was used by Jakob von Uexkull (1864–1944), sometimes called "the father of ethology." He explained this word by referring to how questing ticks locate hosts. They needed three cues—butyric acid (secreted from mammalian sebaceous glands), warmth, and then hair—in order to succeed in locating their next blood meal. For a tick, then, those three stimuli are its entire world (see Figure 1.18).

FIGURE 1.18

The idea of Umwelt articulates one of the central ideas in animal behavior: each animal has a sensory-perceptual world. The most important task of the behavioral biologist is also the most difficult—to penetrate that world. A study of ticks inspired Jakob von Uexkull to think about Umwelt. Many animals have contributed to our knowledge of animal behavior, but few have contributed more than the tick! *Photo: Charles Schurch Lewallen.*

Never mind the green leaf it may be sitting on, never mind the color of the sky—butyric acid, body temperature, and hair inform the behavior of the questing tick. Thus, if we try to determine if a questing tick prefers a long, thin leaf or a lobed leaf, we will probably be wasting our time. To ask sensible questions in animal behavior and to sensibly interpret the results, we must understand the *umwelt* of the animal we study.

We stress that this is not as easy as it seems. Our main impediment is the fact that we are trapped in our own *umwelt*. If there is any doubt, try counting how many times words like *see* or *look* are used in conversation—"I'm not sure I see the difference between the political philosophies of Smith and Jones," or "How does that vacation schedule look to you?" Is the philosophical difference really visible, or is the speaker using *see* to mean something like *comprehend*? Is the question really about the appearance of the schedule, or is it about the appeal of the proposed schedule? We, as humans, use *look* and *see* because vision is a large part of our sensory-perceptual world. It would be just as reasonable to sniff the difference between two philosophies, but no one ever says that; sniffing is not a huge part of the human *umwelt*.

> **KEY TERM** *Umwelt* is the overall context in which an animal behaves, including its sensory environment and its behavioral capabilities.

This is perhaps the fuel for the excitement that has surrounded animal behavior since before it was named—the challenge, never quite met, of getting out of our own skins and perceiving the world in a completely different way. The animals around us are fellow-travelers; that much has been clear for ages. Why do they do what they do? How do they "see" the world?

SUMMARY

The science of animal behavior is young, and Tinbergen's questions remain unanswered for many interesting behaviors. That is expected in the world of science. What is important is learning a bit more about how to ask the questions, and what questions are appropriate. The goal is to place observations of behavior in a framework of concepts. Using the four central questions of animal behavior—causation, development, survival value, and evolution—we can form testable hypotheses and come to a much deeper understanding of behavior.

Behavior includes movement, social interaction, cognition, and learning. Behavior is nonetheless difficult to define, and there may be definitions better than the one you encounter here. At its core, behavior provides animals with adaptive mechanisms for adjusting to changes in their environment and for manipulating the world around them.

We have provided you with two intriguing examples of animals that have contributed greatly to our understanding of animal behavior, and we have offered a quick refresher tour of evolution. This should help you to apply the basic principles of evolution to your understanding of animal behavior. You will probably want to refer to this section as you read the rest of this book.

This chapter concludes with a brief overview of the history of the study of animal behavior. Contemporary studies of animal behavior are rooted in ethology, comparative psychology, sociobiology, and behavioral ecology. As historians agree, the value of history is not in memorizing names and dates, but in understanding our collective path to the present day. This is as true of science as any other realm of history, and grasping how thought about animal behavior has developed over the past two and a half millennia will illuminate our current and future perspectives on how animals behave.

STUDY QUESTIONS

1. Make your own attempt to define behavior. What are the important elements? Do you see why movement alone does not capture all of behavior?
2. Without looking back in the chapter, list the four central questions of animal behavior (Tinbergen's questions).
3. Explain the importance of each of the four questions.
4. Wolves exemplify complex social behavior. Can you see how hunting in groups makes wolves more efficient predators and increases the range of prey they can exploit?
5. Cockroaches help us understand causation and adaptive value. Can you explain how cockroaches can so easily escape when you try to squash them?
6. Read some scientific papers about animal behavior that were written in the early and middle parts of the 20th century. Would you view them as ethological or psychological? Why? Do you recognize influences from Lorenz, Skinner, and others?

23

Further Reading

Burghardt, G.M., 2010. Comparative Animal Behavior – 1920–1973, In: Michael D. Breed and Janice Moore (Eds.), Encyclopedia of Animal Behavior, Academic Press, Oxford, Vol. 1, pp. 340–344 (http://www.sciencedirect.com/science/referenceworks/9780080453378).

Darwin, C., 1872. The Expression of Emotions in Man and Animals. John Murray, London.

Drickamer, L.C., 2010. Animal Behavior: Antiquity to the Sixteenth Century. In: Michael D. Breed and Janice Moore (Eds.), Encyclopedia of Animal Behavior, Academic Press, Oxford, pp. 63–67 (http://www.sciencedirect.com/science/article/B6MV5-50G4S80-6B/2/8dc30829a6175ac6490be06ddf6aa7a8).

Drickamer, L.C., 2010. Animal Behavior: the Seventeenth to the Twentieth Centuries, In: Michael D. Breed and Janice Moore (Eds.), Encyclopedia of Animal Behavior, Academic Press, Oxford, pp. 68–72 (http://www.sciencedirect.com/science/article/B6MV5-50G4S80-66/2/15d6769a4160781c02e8015f8da6bd1a).

Espinoza, S.Y., et al., 2006. Loss of escape-related neurons in a spiny lobster, Panulirus argus. Biol. Bull. 211, 223–231.

Galef Jr., B.G., 1996. Historical Origins, In: Houck and Drickamer Foundations of Animal Behavior. University of Chicago Press, Chicago, pp. 5–12.

Houck, L.D., Drickamer, L.C. (Eds.), 1996. Foundations of Animal Behavior: Classic Papers with Commentaries. University of Chicago Press, Chicago.

Kays, R, Curtis, A, Kirchman, J.J., 2010. Rapid adaptive evolution of northeastern coyotes via hybridization with wolves. Biol. Lett. 6, 89–93.

Kellert, S.R., Wilson, E.O. (Eds.), 1993. The Biophilia Hypothesis. Island Press, Washington D.C, 484 pp.

Kojola, I., Aspib, J., Hakalaa, A., Heikkinena, S., Ilmonic, C., Ronkainend, S., 2006. Dispersal in an expanding wolf population in Finland. J. Mammal. 87 (2), 281–286.

Kruuk, H., 2003. Niko's Nature: A Life of Niko Tinbergen and His Science of Animal Behaviour. Oxford University Press, Oxford, 391pp.

Libersat, F., Leung, V., Mizrahi, A., Mathenia, N., Comer, C., 2005. Maturation of escape circuit function during the early adulthood of cockroaches Periplaneta americana. J. Neurobiol. 62, 62–71.

Mech, L.D., 1981. The Wolf: The Ecology and Behavior of an Endangered Species. University of Minnesota Press, 384 pp.

Mech, L.D., 1993. The Way of the Wolf. Voyageur Press, 120 pp.

Mech, L.D., Boitani, L. (Eds.), 2007. Wolves: Behavior, Ecology, and Conservation. University Of Chicago Press, Chicago, 472 pp.

Moore, R., Moore, J., 2006. Evolution 101. Greenwood Press, Westport, CT.

Ritzmann, R., 2010. Visuomotor Control: Not So Simple Insect Locomotion. Curr. Biol. 20 (1), R18–R19.

Sorabji, R., 1993. Animal Minds and Human Morals: The Origins of the Western Debate. Cornell University Press, Ithaca, NY, 267 pp.

Taborsky, M., 2010. Ethology in Europe, In: Michael D. Breed and Janice Moore (Eds.), Encyclopedia of Animal Behavior, Academic Press, Oxford, Vol. 1, pp. 649–651, (http://www.sciencedirect.com/science/referenceworks/9780080453378).

Tinbergen, N., 1963. On aims and methods of Ethology. Z. Tierpsychol. 20, 410–433.

Verhulst, S., Bolhuis, J., 2009. Tinbergen's Legacy: Function and Mechanism in Behavioral Biology. Cambridge University Press, 262 pp.

Wilson, E.O., 1984. Biophilia. Harvard University Press, Cambridge, MA 157 pp.

CHAPTER 2

Neurobiology and Endocrinology for Animal Behaviorists

LEARNING OBJECTIVES

Studying this chapter should provide you with the knowledge to:

- Learn how axons carry electrical signals from point to point within the nervous system and the role of neurotransmitters in carrying information among neurons.

- Place the relative roles of the central and peripheral parts of the nervous system in coordinating behavior and see how reflex loops provide a basic model for coordination of behavioral responses.

- Learn that hormones are produced by glands and transported to target organs or tissues, where they have immediate or long-term effects.

- Gain a comprehension of hormones and behavior in vertebrates, with a focus on hormonal effects on development and behavior, and gain a similar comprehension of hormones and behavior in invertebrates, with a focus on understanding the roles of juvenile hormone and ecdysone in invertebrate behavior.
- Learn that perception of the external world is essential for all forms of life, including bacteria, and that information gained from perception is critical in behavioral decisions.
- Appreciate that each species' sensory world may be unique and understand that assumptions about perception based on human senses may be dramatically misleading in studies of animal behavior.
- Know that animals may perceive a constellation of stimuli that includes light, polarization of light, temperature, sound, molecular energy (smell, taste), magnetic fields, electric fields, and inertia (touch) and understand how each stimulus is sensed.

2.1 NEUROBIOLOGY, ENDOCRINOLOGY, AND SENSORY SYSTEMS: AN OVERVIEW

This chapter is devoted, in large part, to Tinbergen's first question: causation (mechanisms underlying behavior). This chapter will cover basic background about neural, endocrine, and sensory systems that is necessary for understanding animal behavior. An animal's neural system is composed of neurons, which collect information about the animal's internal and external environments and organize responses to the environment. An endocrine system relies on transmitters, hormones, that carry signals from glands via the body fluids or blood to target organs. Sensory systems, which are a subset of the neural system, are the interface between an animal and its external environment; sensory organs gather information about the world outside an animal. Animals are also capable of sensing their internal world, through perceptions of states such as pain and hunger.

These systems provide the foundation on which animal behavior is built; a basic understanding of these systems is essential for an animal behaviorist. For instance, consider sensory systems: What is their ecological role? True, they are extensions of the nervous system, but they are also gateways through which animals collect essential information about their surroundings. In turn, the abiotic nature of those surroundings can have profound effects on the way sensory systems function. A failure to comprehend the sensory world of animals sadly limits any understanding of their behavior. This is the message of *umwelt*, the concept of understanding an animal's behavior within its environment introduced in Chapter 1. Tinbergen was right when he included causation in his list of essential questions for those who want to understand animal behavior.

Likewise, even when one is studying behavioral ecology or conservation, a basic knowledge of how the nervous and endocrine systems come together to shape animal behavior is imperative. The conservation of migratory birds, for example, is intricately linked to the hormones that shape their migratory behavior and how climate change or environmental pollutants might affect the endocrine triggers for migration. Some knowledge of how neurotransmitters shape personality and interact with social roles is fundamental to understanding social behavior. Fortunately, as a general rule, neural, endocrine, and sensory mechanisms use the same underlying framework across all animals, so mechanisms found in mammals are also likely to exist in some form in insects.

Much of the material in this chapter reviews what is normally covered in a general biology course, so referring to a general biology textbook, or to a beginning physiology textbook, for additional information may be helpful. What follows is a brief refresher, a look at that "old" material with a behavioral twist.

Behavior is *modulated*, or adjusted to fit conditions, on two very different time scales: short-term, moment-to-moment change in activity and longer-term change, perhaps over seasons. Hormones are important agents of modulation that act at both time scales. A hormone is a molecule produced by a gland that affects a target organ within an animal.

> **KEY TERM** Modulation is the ability to vary a behavior along a range of possible responses. For example, aggressiveness might be modulated depending on the strength of an opponent.

Shifts in behavior, such as the transition from sleeping to searching for food, or feeding to mating, are usually mediated by changes in neurotransmitter levels. Neurotransmitters are small molecules used to convey signals within the nervous system. Neurotransmitters can be viewed as hormones that tend to act quickly. Age-related or seasonal changes in behavioral state, on the other hand, are associated with changes in the underlying hormonal state of the animal. Thus, when a bird or mammal enters into its daily inactive period (in other words, when it sleeps), the level of gamma-aminobutyric acid (GABA), a neurotransmitter, goes up in its brain. Drugs used by humans to treat insomnia, such as Ambien or Lunesta, increase the levels of this neurotransmitter.

A good example of longer-term behavioral changes induced by hormones is the female bird or mammal preparing a nest for her young; she likely has elevated levels of prolactin, a hormone that is generally associated with parental behaviors. The following sections begin with a discussion of minute-to-minute regulation of behavior and continue with long-term seasonal and lifelong developmental changes.

2.2 WHAT DOES AN ANIMAL BEHAVIORIST NEED TO KNOW ABOUT NEUROBIOLOGY?

The nervous system takes information from cells and tissues, organizes it, and prioritizes the animal's behavioral responses. It is the physiological key to animal behavior; all behavior is organized by the nervous system. Sensory cells, discussed in Section 2.3, gather information from the environment and send it to the central nervous system via axons. The central nervous system (CNS), which in most animals includes the brain and a central nerve cord (the spinal cord in vertebrates), organizes incoming information. The CNS sets priorities for the animal's use of its time and energy and sends signals to effector cells, such as muscles, which translate the decisions made in the central nervous system into behavioral and physiological actions. Learning and genetics meet in the nervous system; it is the storehouse for learned information, while genetic information in nervous system cells plays key roles in organizing the development and expression of behavior. Interconnecting cells in the nervous system house memory, decision-making, and other integrative functions.

The neuron is the basic unit of the nervous system. A neuron is a type of nerve cell, consisting of an axon, a cell body, and dendrites. Neurons carry electrical impulses between locations via axons. Neurons form networks of varying complexity, with the brain, a dense cluster of neurons, being the most complex. This section begins with insights into neurons and how they work, moves on to how neurons communicate with each other and with tissues and organs within an animal, and concludes with how the central nervous system functions as a coordination center for behavior.

CASE STUDY

A Simple Nervous System

Take a look at an organism that is, at once, seemingly simple and complex. The roundworm, *Caenorhabditis elegans*, has 302 neurons (compare this with the billions of neurons possessed by humans) and 81 muscle cells (see Figure 2.1). Because of the anatomical simplicity of *C. elegans* and extensive knowledge of its genetics and development, it has become a model for understanding the basics of behavior. The role of each neuron can be investigated by looking at the behavior of mutants in which the activity of a single neuron is changed, by staining worms for specific neurotransmitters and observing which neurons contain those transmitters, and by measuring the metabolic activity in neurons. Genetic techniques can also be used to modify the production or reception of neurotransmitters, to disable sensory input, or to eliminate specific behavioral responses. Neurons can be removed or disabled using lasers. The combination of this vast array of techniques, the ability of investigators to rear many generations of worms in short periods of time, and the large number of laboratories focusing on *C. elegans* mean that this worm joins fruit flies and mice in the pantheon of well-known research organisms.

The nervous systems of roundworms are more complex than those of cnidarians and echinoderms in many ways, but not as complex as those of arthropods (crustacea, spiders, insects, and their relatives) and vertebrates. Roundworm behavior includes mating, egg-laying, foraging, and movement; in other words, they do everything an animal would be expected to do, and it is all organized by those 302 neurons. *Caenorhabditis elegans* senses the environment in at least two ways: chemosensation (smell, taste) and mechanosensation (touch or feel). "The worm" (as its students like to call it) also responds to temperature and light, even though there are no known sensory structures in this animal for these two inputs. This suggests that much is yet to be learned about the sensory world of *C. elegans*. It may not be as simple as it looks!

Mating behavior in *C. elegans* is complicated by an arrangement of sexes that is unusual from a human perspective. The worms are either male or hermaphroditic. The hermaphrodites pass through a male phase during their larval development and later function as females, but they have the possibility of self-fertilization. Males are attracted to females by chemical secretions and exhibit a stereotyped sequence of mating behavior, culminating in ejaculation of sperm into the female's vulva and deposition of a copulatory plug (which may prevent subsequent matings by other males). About 300 progeny are produced during the worm's roughly two-week lifespan. Like many other nematodes, *C. elegans* exhibits an interesting twist when unfavorable environmental conditions persist; larvae can enter the "Dauer" ("continuing" in German) stage, in which they cannot eat, do not age, and are protected from environmental stresses. This can prolong the worm's life for weeks or months.

One surprise to emerge from studies of *C. elegans* is that much the same suite of biogenic amines—octopamine, tyramine, dopamine and serotonin—serve as neurotransmitters in *C. elegans* as in insects and vertebrates. This suggests that the basics of communication among neurons and between neurons and other types of cells appeared very early in animal evolution and that the fundamental framework of communication in the nervous system has been maintained through the development and diversification of neural systems. Dopamine is involved in mechanosensation (touch or feeling the environment), locomotion, and learning. Octopamine seems to be involved in regulating egg-laying behavior. Another neurotransmitter, gamma-aminobutyric acid (GABA), provides a good example of how the intimate details of nervous system function can be teased from such a simple system. Of the 302 neurons found within any *C. elegans* worm, 26 have GABA. GABA inhibits locomotion, movements of the head used in foraging, and stimulates defecation.

If so much can be accomplished with so few neurons, why do animals like octopi and honeybees have millions of times as many neurons? As new sensory modes, more muscles, and more possibilities for learning and cognition have been added over the course of evolution, neurons for these functions have been built on top of the basic scaffold similar to that found in *C. elegans*. This principle, that evolution builds on existing structures, means that the neural systems can exhibit a fair amount of complex layering. While no existing animal is an ancestor of any other existing animal, features—in this case, the nervous system—may have been subjected to different selection pressures in different evolutionary lineages. Thus, the nervous system of *C. elegans* may resemble that of an ancestor more closely than the nervous system of a grasshopper does.

FIGURE 2.1

This is *Caenorhabditis elegans*, the "lab rat" of the nematode world. It is a typical nematode in many regards—typical shape, typical features, typical anatomy. Note the tubular gonads in this female, tubular intestine, and tubular body wall. The tube-within-a-tube body plan of nematodes, together with an incredibly resistant cuticular covering, means that they are well-adapted for a wide range of environments, famously ranging from the beer-soaked mats in European pubs to polar permafrost. Beginning with the work of Butschli in the late 1800s, nematodes have been in the forefront of developmental studies. It is therefore no surprise that "the worm" continues to answer our questions about function—and stimulate more!

Axons Carry Information To and From the Central Nervous System

NEURONS AND THE TRANSMISSION OF INFORMATION

Neurons are the basic cellular units of the nervous system. Neurons come in many sizes and shapes, but they all have the same essential structure. A neuron consists of a cell body, or soma, with a nucleus. Branching structures called *dendrites* extend from the neuron and interact with adjacent neurons. The axon is a signal-carrying extension that targets other neurons, muscles, or organs. The cell body is the portion of the neuron containing the nucleus. The dendrites are branchlike structures of neurons that interface with other nerve cells or with organs and tissues, and an axon is the elongated portion of the neuron that carries the nervous signal.

Axons transmit electrical or chemical signals from one location in an animal's body to another. When the nerve cell is at rest—that is, when it is not transmitting information—there is an electrical difference between the inside of the cell and the surrounding fluid; this is called the **resting potential**. The nervous impulse, or **action potential**, starts when the cell membrane is stimulated by a neurotransmitter or by a sensory input. This stimulation causes the cell membrane to depolarize; the electrical difference between the outside and the inside of the cell reverses because of a temporary change in the permeability of the membrane. The depolarization can be visualized as a wave of electrical charge moving along the axon as shown in Figure 2.2. As one point on the membrane depolarizes, the depolarization spreads to adjacent parts of the membrane. In this way, the action potential travels the length of the axon and then is transmitted to another nerve, or to a muscle. The action potential can only travel in one direction, from the soma to the end of the axon. The connections with other nerves or with muscles are called *synapses*.

> **KEY TERMS** An action potential is the spike of electrical activity when a nervous signal is propagated. It results from depolarization of the axon membranes. The resting potential is the electrical state of a neuron when no signal is being carried.

This discussion has centered on neurons, which are the information-carrying and -storing cells in the nervous system. They are not the only nervous system cells, however. Given the focused role of neurons in the nervous system, it is no surprise that they need support to be able to survive. This support is provided by glial cells. Glial cells are the helper cells of the nervous system. They ensure that the metabolic needs of the neurons are met and provide structural support for the neuronal tissues. Two types of glial cells merit special mention: Astrocytes are important in the metabolic support of neurons in the central nervous system,

29

FIGURE 2.2
A myelinated neuron, with its basic elements labeled. In this diagrammatic representation of a nerve cell, the myelin sheath is blue. The myelin sheath greatly enhances speed of transmission; this allows for faster reaction times. Some invertebrates lack myelin on their axons, although they do have giant neurons that can transmit impulses quickly. The action potential, a wave of depolarization, travels from the cell body to the dendrites at the end of the axon.

Dendrites

Soma

Axon

Myelin sheath

To synapse

and in vertebrates, Schwann cells provide metabolic and mechanical support to the axons that extend from the central nervous system to the muscles.

What determines the speed of a nervous impulse moving down an axon? Why does speed matter? If the action potential moves along the axon, as shown in Figure 2.2, then its speed is determined by the diameter of the axon. This is similar to an electrical wire; a skinny wire has more resistance to electricity than a fat wire. Speed matters because a millisecond can make a life-or-death difference if a predator threatens. Thus, axons are expected to be large in more speed-critical parts of the nervous system—where axons need to carry information over long distances to organize escape from predators. In invertebrates, this is exactly the case; axons that have critical functions in fast escape responses have large diameters and are called *giant axons*.

> **KEY TERM** Giant axons are found in invertebrates, particularly in escape systems. Their large size speeds the transmission of action potentials.

Chapter 1 introduced cockroach giant interneurons (GIs) and the key role they play in a cockroach's ability to avoid being stomped on. The strategy of using GIs reaches a pinnacle in squid; those neurons are up to 1.0 mm in diameter and carry signals at about 20 meters per second. Large diameter gives GIs less electrical resistance so that action potentials move faster in GIs than in other invertebrate nerves. Giant axons are used in nervous circuits in which speed of response is critical, such as those involved in organizing escape behavior (refer to Figure 2.2).

Another solution to the problem of speed is to "insulate" the neuron. Some animals—all vertebrates and a variety of invertebrates—do this with myelin sheaths. The myelin sheath consists largely of fatty materials and has been best studied in vertebrates, where it is maintained by Schwann cells. The myelin sheath is segmented by the nodes of Ranvier (see Figure 2.2; action potentials jump from node to node, speeding transmission. This is called saltatory conduction.). Because the myelin sheath depends on fat metabolism in the glial cells, damage to the sheath, as happens in the human disease multiple sclerosis, impairs transmission of the signals along the axon. A myelinated vertebrate axon 20 microns in diameter can carry signals up to 100 meters per second.

NEUROTRANSMITTERS SHUTTLE INFORMATION FROM NEURON TO NEURON

Neurotransmitters are the messengers of the nervous system. They are relatively small molecules that carry information across synapses from a nerve cell to its neighboring cells and are a critical part of the internal machinery controlling animal behavior. Generally speaking, the neurotransmitter is held in membrane-bound vesicles near the synapse. The nerve cell with these vesicles is the presynaptic cell. When stimulated, the vesicles merge with the cell membrane of the presynaptic cell, and the neurotransmitter is released into the synapse, or "synaptic space." The neurotransmitter molecules cross the synaptic space and match with receptor molecules in the membrane of the postsynaptic cell, causing depolarization in that membrane and continuing the transmission of the impulse. These receptors are critically important; for each neurotransmitter there are several receptor molecule types, guaranteeing a transmitter-specific message. Each neurotransmitter has many different functions in the nervous system, and the receptor type involved in regulating a behavior often tells us more about the behavior than the identity of the neurotransmitter might tell us. The most common neurotransmitter is *acetylcholine*, which often is the messenger between axons

> **KEY TERM** Neurotransmitters are small molecules that carry messages among axons and between the nervous system and other tissues and organs.

> **KEY TERM** Acetylcholine is a neurotransmitter that acts, in many animals, at synapses between nerves and muscles.

A — Serotonin, 5-HT

B — Dopamine

C — Acetylcholine

FIGURE 2.3
The chemical structures of three common neurotransmitters: (A) serotonin, sometimes called 5-hydroxy tryptamine, (B) dopamine, and (C) acetylcholine. These molecules share small size, the presence of a nitrogen molecule, and polarity, that is, having a chemical charge difference across the compound. Collectively such compounds are sometimes referred to as biogenic amines. Their small size allows them to be easily transported across cell membranes, but their polarity reduces unintended diffusion across non-polar cellular membranes. Insecticides like malathion, which is commonly used in mosquito control, are acetylcholine esterase inhibitors. This means the insecticide prevents the enzyme that breaks down acetylcholine in the synapses from acting. The resulting accumulation of acetylcholine results in uncoordinated firing of nerves.

Note
Insecticides like malathion, which is commonly used in mosquito control, are acetylcholine esterase inhibitors. This means the insecticide prevents the enzyme that breaks down acetylcholine in the synapses from acting. The resulting accumulation of acetylcholine results in uncoordinated firing of nerves and leads to death of the insect.

and muscles as well. Other common neurotransmitters are octopamine, serotonin, and dopamine; they usually function in the central nervous system. All of these neurotransmitters are found in both vertebrates and invertebrates.

For this system to work, the neurotransmitter must be removed from the synapse after the signal is no longer needed. This happens by either cleaving the neurotransmitter to inactivate it or by re-uptake of the neurotransmitter into the presynaptic cell. For instance, a specialized enzyme called *acetylcholine esterase* breaks down acetylcholine in the synapse. The components can then be recycled. In contrast, serotonin is taken up directly by the presynaptic cell (see Figure 2.3).

31

THE CENTRAL NERVOUS SYSTEM, WHEN PRESENT, ORGANIZES BEHAVIOR

The complexity and size of the central nervous system varies from simple *nerve nets* in organisms like jellyfish (Cnidaria) to highly centralized systems (see Figure 2.4). Cephalization is the concentration of elements of the nervous system, especially those involved in sensory activities and coordination of responses, into an anterior head. Cephalization is strongly linked to behavior, because it is usually found in mobile animals, and characterizes the part of the animal (anterior) that encounters the environment first. Sessile (nonmoving) animals are less cephalized; they encounter the environment from all directions, which favors a distributed nervous system.

FIGURE 2.4
Mussels, a common intertidal mollusc. Their sessile, filter-feeding lifestyle is associated with a nervous system that lacks any cephalization. *Photo: Michael Breed.*

Cephalized animals have brains of varying complexity. Complex brains are typically compartmentalized, with separate but interconnected structures for functions such as olfaction, vision, and integration. The structures of highly sophisticated brains that evolved independently, such as those found in octopi, insects, and mammals, share this compartmentalization.

Does brain size matter? This is an intriguing but surprisingly complex question. In humans, people over a broad range of brain sizes seem to have roughly equal intelligence and cognitive abilities, a point that was vigorously argued by Stephen Jay Gould, one of the deepest evolutionary thinkers of the past

KEY TERM A nerve net is a simple, decentralized, multipolar multipolar nervous system found in animals like jellyfish.

century. Gould, in his brilliant book, *The Mismeasure of Man*, argued that the temptation to assess people's intelligence by looking at brain size was based in racism and deeply held assumptions about the superiorities of some human cultures over others.[1] Gould's argument is reasonable, yet within a larger (multispecies) taxonomic group such as birds or mammals, it does appear that cognitive abilities correlate with brain size across a wide range of species; in mammals, cognition and cerebral cortex volume have a particularly strong relationship. (Note that such broad interspecific comparisons differ from intraspecific comparisons.)

The highly intelligent representatives of the molluscan phylum, squid and octopi, have very large brains relative to their more sedentary cousins, such as clams, oysters, and scallops. But the "smartest" insects, at least as measured by navigational and communication abilities, like the honeybees, have tiny brains compared to any squid, octopus, bird, or mammal. Adult human brains weigh 1300–1400 grams and have over 100 billion neurons. Octopus brains have about 300 million neurons and weigh up to a gram. Honeybees have slightly less than a million neurons packed into less than a cubic millimeter (about 1 milligram in weight). Some insects are so tiny as to be barely visible to the naked eye, yet they carry on all the basics of insect life—feeding, flight, mating—with a brain that is 1000-fold miniaturized when compared to a honeybee's brain. How basic functions are maintained in miniscule brains is a fascinating unanswered question.

DISCUSSION POINT: INTELLIGENCE

The idea of honeybees being the "smartest" insects, based on navigation and communication abilities, raises a question that has challenged biologists and psychologists for decades: How do we measure intelligence? We will return to this question in later chapters, but now would be a good time to begin thinking over this puzzle. What is intelligence? Is it the same for all animals—is dog intelligence the same as human intelligence? What aspects of intelligence are favored by natural selection in different habitats or lifestyles? After pondering this question (for which there is probably no single correct answer), think about how intelligence can be measured and whether it can be defined and measured in a way that would allow comparison among species. How does *umwelt* affect "intelligence"?

The brain receives information about the animal's physiological state and serves as a command center for translating physiological needs into behavioral responses. Cnidarians and echinoderms have neurons but no centralized brain. They still manage to organize fairly complex behaviors, as do protists. The echinoderms' lack of cephalization is taken to reflect sessile ancestry—will mobile echinoderms be cephalized millions of years from now? Cnidarians are anatomically different from most other animals, and their nervous system is no exception. First of all, while some cnidarians do move (e.g., jellyfish), that movement is largely at the mercy of currents; they are not powerful swimmers. In addition, many other cnidarians are mostly sessile (e.g., corals, anemones). In cnidarians a net of neurons lies between the inner and outer body walls. Unlike nerves in most other animals, these nerves can transmit impulses in any direction. Sensory nerves project to the surfaces of the animal, and nerves also attach to contractile cells, which function similarly (if weakly) to muscles in other animals. This nerve net allows transmission of information from one part of the animal to another and coordination of simple movements. The stinging organs of cnidarians, nematocysts, discharge without nervous control. Well-fed cnidarians are less likely to discharge their nematocysts. Nervous connections to the nematocysts may affect the likelihood of discharge, depending on the need for food (see Figure 2.5).

Comparing Brains

Molluscs are a particularly interesting case of brain evolution and how natural selection has acted on locomotion and the nervous system to influence that evolution. They are one of

FIGURE 2.5
The anemones in the two upper photos are underwater, with their tentacles and mouths exposed. Below, the mass of anemones sits above the tideline, exposed to the sun and air. They have retracted their tentacles, thus minimizing their surface area and reducing risk of dessication and other threats. *Photos: Michael Breed and Tricia Soares (top left).*

33

three major groups of animals to have evolved a highly sophisticated brain, the other two being arthropods and vertebrates. Molluscs (slugs, snails, clams, squid, and octopi) have ganglia—collections of neurons—associated with the mouth, foot, and gut. Cephalopods (squid and octopi) are active predators; they have complex brains and are capable of sophisticated movements and color changes, along with learning. Mussels and other bivalves such as clams have limited mobility, and consequently natural selection has not favored a complex nervous system.

The nervous system of the insect reflects its segmented body plan. Each of the insect's segments has its own nerve center, called a *ganglion*, which is connected by a pair of nerves to the ganglia of the adjacent segments. The insect's head consists of a set of fused segments; scrutiny of the head capsule of a grasshopper reveals some of the sutures between the segments. Inside the head, the segmental ganglia have fused to form a brain.

> **KEY TERM** A ganglion (pl. *ganglia*) is a concentration of axons that integrate sensory input and locomotory output, and which also report the status of a portion of the animal to the brain.

OF SPECIAL INTEREST: INSECT BRAINS

The insect brain performs nearly all of the same functions as the brain of a bird or mammal but is much more compact. It is also organized much differently. The major parts of an insect's brain (shown in Figure 2.6) are as follows:

- Optic lobes, which provide a link between the receptor surfaces of the eyes and the brain. The optic lobes perform the first steps in interpreting visual information for the insect.

- Ocelli (singular = ocellus), simple, accessory eyes that are important in perceiving the intensity of illumination. Many insects have three, although some have only two or even none.
- Antennal lobes. The antennal lobes receive input from millions of olfactory receptors and reduce it to manageable taste and smell information.
- Mushroom bodies. If an insect "thinks," this is where it happens. These are the "higher" centers in the insect brain, where learning, memory, and integration occur.
- Neurosecretory cells. Peptide hormones produced by the neurosecretory cells regulate endocrine and homeostatic functions in insects. These cells are analogous to the hypothalamus of the vertebrate brain.
- Corpora cardiaca and corpora allata. Analogous to the pituitary in vertebrates, these organs secrete hormones into the circulation. The corpora allata produce juvenile hormone, which plays a critical role in the regulation of development and behavior in many insects.
- Suboesophageal ganglion. The input for this structure comes from the insect's mouthparts; this ganglion coordinates the action of the mouthparts when the insect feeds.

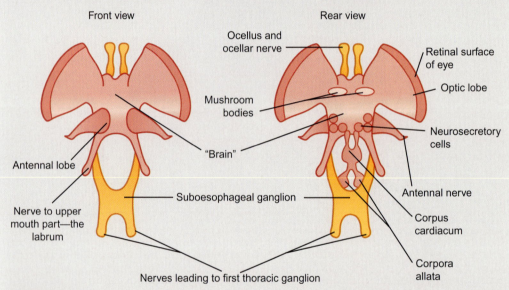

FIGURE 2.6

The insect brain is a fusion of anterior ganglia, with distinct areas devoted to vision (e.g., ocellar nerves, optic lobes), antennae, and even memory (mushroom bodies). Note the subesophageal ganglion—the first separate ganglion leading to the ganglionated ventral nerve cord typical of arthropods. It forms the ventral part of the nerve ring that circles the esophagus—hence subesophageal—and attaches the dorsal brain to the nerve cord. The corpora allata (sing. corpus allatum) synthesize and release juvenile hormone. The corpus cardiacum (which is actually a fusion of two corpora cardiaca) synthesizes some other hormones and stores other neurohormones that are synthesized in the brain.

Vertebrate brains appear to have been built in an evolutionary sequence starting with fish, which have olfactory and optic lobes, whose basic functions are for smell and vision. The cerebral hemisphere, which becomes the center of tasks such as learning and cognition in birds and mammals, is less well developed in fish. The cerebellum coordinates physical movements. The hypothalamus coordinates physiological functions using neurohormones, while the medulla oblongata has axonal connections to control the cardiac and respiratory systems (see Figure 2.7).

Comparison of reptilian, avian, and mammalian brains (see Figure 2.8) reveals that the neocortex of the cerebrum becomes progressively larger and more dominant in function.

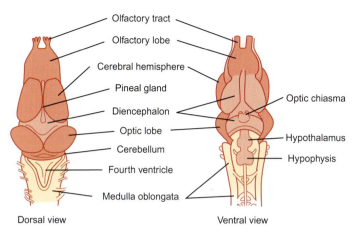

FIGURE 2.7
Many of the basic elements of the fish brain can be recognized in the brains of other vertebrates.

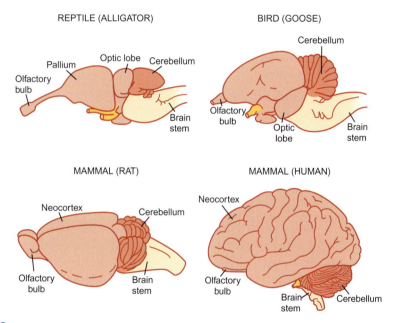

FIGURE 2.8
These diagrams provide a comparative look at vertebrate brains, which in turn can reveal much about differences in behavior. For instance, the rat's olfactory bulb is significantly larger than that of humans; rats have a much keener sense of smell than humans do.

DISCUSSION POINT: BRAINY AND BRAINLESS ANIMALS

How do the differences between the nervous systems of the "brainy" predatory octopus and the brainless filter-feeding clam reflect the differences in their behavior? Is it surprising to find such different nervous systems in animals that are both molluscs?

How Behavior Is Generated

Quick responses are necessary in critical situations. *Reflex loops* are a simple integrating mechanism for organizing quick responses. Reflex loops require no "thought" or decision making on the part of the animal. Sensory input is transmitted to the central

> **KEY TERM** A reflex loop is a simple connection from a sensory cell to the spinal cord and back to a muscle. An example is the human patellar reflex (see Figure 2.9).

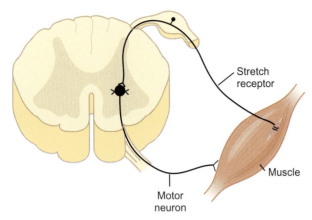

FIGURE 2.9

A diagram of a simple reflex loop in a vertebrate. Stretching the muscle sends a signal via an axon to the spinal cord. This sensory input then connects with a motor neuron that signals the muscle to move. When a physician taps a knee tendon with a mallet the tendon stretches, and just such a signal is sent to the patient's spinal cord. The motorneuron then fires, causing the leg to jerk. In general, the function of this reflex is to reduce the stress on the tendon; in the physician's office, it serves as a basic test of the nervous system.

nervous system, it is linked to the appropriate output, and a signal to act is then sent to a muscle (see Figure 2.9). Sometimes animal behavior, particularly the behavior of insects, is characterized as being entirely instinctive, or *reflexive*. Another term for this view of behavior is *hard-wired*. When behavior is characterized as being hard-wired, the animal is not credited with possessing decision-making processes to shape that behavior. The relationship between hard-wired behavior and thought, or cognition, is interesting and complex; we will discuss this issue in detail in Chapter 6.

Connecting the nervous system to the muscular system is critical to making all this work. Motorneurons have axons that carry signals from more central locations in the nervous system to the muscles that will execute the intended actions. In most animals, acetylcholine released from the end of the axon stimulates the muscular contraction (see Figure 2.10).

Responsiveness and Decision-Making Processes within the Nervous System

Neurotransmitters modulate general behavioral state at the level of the central nervous system. We have already seen how acetylcholine functions in the transmission of nerve impulses in the peripheral nervous system. Acetylcholine also has a major function in the central nervous system, where its levels are critical in learning and memory. Drugs that inhibit the activity of acetylcholine esterase can enhance memory functions in humans and animals. Drugs for Alzheimer's disease, such as Aricept, are acetylcholinesterase inhibitors; they interfere with the breakdown of acetylcholine in the brain, causing levels to rise so that normal learning and memory functions can operate.

Neurotransmitter levels are also associated with what is called *personality* and *emotion* in humans. Is there potential for nonhuman animals to display states that look very much like personality and emotion? Behavioral expression is affected by neurotransmitter levels, the quantity of neurotransmitter receptors, and the presence of enzymes that activate or deactivate the transmitters. The "personality triangle" depicts how three major neurotransmitters interact in determining mood (see Figure 2.11). This concept is applied

FIGURE 2.10

The electrical impulse (action potential) travels through the axon to the neuromuscular junction. Acetylcholine is released from vesicles in the nerve and initiates depolarization of the muscle membrane, which in turn causes the release of calcium, triggering muscle contraction.

most frequently to humans, but pharmacological studies and veterinary practice indicate that some of the same principles apply to many, perhaps most, mammals, although the exact behavioral effects can vary dramatically. These same neurotransmitters are present in the brains of other animals, such as birds, but their role in regulating behavior is less well understood in nonmammals.

Serotonin is the most commonly manipulated neurotransmitter in pharmacological interventions in animals. Serotonin levels in synapses can be regulated by drugs called *selective serotonin reuptake inhibitors* (SSRIs). Drugs such as Prozac®, Zoloft®, and Lexapro® are SSRIs; they alleviate depression in humans by increasing the levels of serotonin in the brain. The action of these drugs in regulating behavior is not, however, limited to humans; many mammals respond to SSRIs in ways that are reminiscent of human responses.

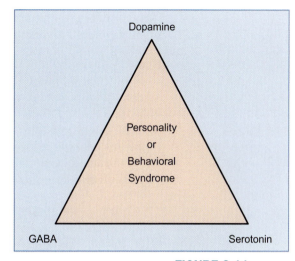

FIGURE 2.11
Considerable contemporary research on animal behavior has focused on personality—behavioral variation among individual animals such as being "shy" or "bold". Personality is sometimes called a behavioral syndrome to make the concept less anthropomorphic. The brain concentrations of GABA, dopamine, and serotonin can interact in determining the expression of personality. This was developed to describe human states, but the same interactions occur in other mammals and in birds.

37

BRINGING ANIMAL BEHAVIOR HOME: DOGS AND ANTIDEPRESSANTS

SSRIs are effective in moderating aggression in dogs. Clomipramine, an older drug that also affects serotonin-based behavioral responses, is commonly used in dogs to treat anxiety. In humans, clomipramine has been used to treat depression and obsessive compulsive disorder, although it (and other antidepressant drugs in this general class, which are called *tricyclics*) has largely been replaced by SSRIs in the treatment of human depression. If canine aggression is sometimes induced by anxiety—and that is a reasonable hypothesis—this would explain the efficacy of serotonin-related treatments for aggression in dogs. In mice, the expression of "anxiety" is higher in strains that have been genetically selected to be nonaggressive. While this anxiety–aggression relationship is the opposite of that suggested for dogs, the existence of a correlation between these two factors does imply a similarity in how these behaviors are controlled in the two species.

GABA, discussed previously, is most strongly associated with wakefulness/sleepiness. Sleep aids, sometimes called *hypnotics* by physicians, such as Ambien®, activate GABA receptors, imitating the effect of increased GABA concentrations. Low GABA may also be associated with depression. In animals, low GABA levels are generally associated with aggression, and treatments that increase GABA lower aggression. Binding GABA receptors mimics the effect of increasing GABA levels.

Another neurotransmitter, octopamine, merits mention because it is often implicated in the regulation of behavior in invertebrates. Examples of behaviors associated with octopamine levels include foraging activity in honeybees, olfactory responsiveness in fruit flies, and firefly flashing. Octopamine appears to be associated with general arousal in insects; injecting or feeding octopamine to insects results in high levels of activity.

Dopamine is most commonly thought of in the context of Parkinson's disease in humans. Parkinson's disease, a progressive loss of muscular control due to death of the cells in the brain which produce dopamine, can be counteracted to a certain extent by treatment with compounds that supplement brain dopamine, such as L-dopa. Dopamine is known to modulate mood and behavior, especially pleasure. Endorphins, a class of neuropeptides that occur in the brains of many animals, regulate pleasurable sensations by stimulating dopamine release. Endorphins are found in many animal species, including insects, but their function outside of vertebrates is not well known.

Opioids, such as heroin, mimic the internal effects of endorphins and cause massive dopamine releases, resulting in similar, but stronger, pleasurable sensations. In dogs, selegiline hydrochloride, a drug developed as a Parkinson's treatment for humans, may enhance training. Generally, behavioral rewards stimulate dopamine receptor systems, so learning that is in itself rewarding, or is rewarded by a trainer, involves dopamine. Because of the link between reduced dopamine and neuromuscular control, veterinarians may be wary of lowering dopamine concentrations. Dopamine also plays a very significant role in regulating obsessiveness and impulsive behavior in dogs.

2.3 WHAT DOES AN ANIMAL BEHAVIORIST NEED TO KNOW ABOUT ENDOCRINOLOGY?
Hormones and Regulation: The Basics

The study of hormones and behavior offers an opportunity to examine all of Tinbergen's questions from a single scientific viewpoint: endocrinology. By way of review, hormones are the messengers that help integrate physiological functions and behavior, communicating from glands to organs and tissues; they are chemical signals between tissues/organs and other tissues/organs. Typically, hormones regulate longer-term changes in behavior, such as seasonal differences or development from juvenile to adult. There are exceptions, such as the fast-acting short-term effects of adrenaline on fight-or-flight responses.

Hormones mold development and in doing so illuminate behavioral ontogeny. Many hormones are conserved evolutionarily so that the same molecules appear across a wide range of taxa. For example, insulin, a regulator of sugar metabolism in humans, is present in many insects and has similar functions! This conservation of molecules across vast evolutionary distances is yet one more statement of how evolutionary history and phylogenic relationships shape the survival mechanisms of animals.

Before beginning this review, consider a general principle, true across all of science—perhaps all of life—but particularly relevant to the study of hormones and behavior: correlation should never be confused with causation. Simply said, the fact that events are correlated does not mean that one has caused the other.

A straightforward approach to studying behavioral neurobiology or behavioral endocrinology is to sample hormone levels and associate those levels with behavior. For example, blood samples can be collected from male birds during the mating season and compared with samples collected from the same birds in the fall. Does the testosterone level differ between the two samples? Likely, the answer is yes. Does this mean that testosterone caused the difference? Probably not. The existing information does not indicate if the behavioral change caused the hormone level to change, if the hormone caused the behavior to change, or if some other variable(s) caused either or both to change. This experiment demonstrates a correlation between testosterone levels and behavior, but offers no information about causation.

What are some ways to test hypotheses about causation? For hormones, usually the approaches involve some combination of the following manipulations: (1) removal of the organ or tissue that produces the hormone, (2) supplementation/addition of the hormone in experimental animals' bloodstreams, (3) blockage of the receptors for the hormone or neurotransmitter using drugs, (4) use of RNA interference techniques to block the synthesis

38

OF SPECIAL INTEREST: THE RING DOVE

Consider the ring dove (see Figure 2.12). It is a domesticated bird with wild relatives—a wonderful choice for a study that addresses causation with an eye to ultimate questions. The courtship and reproduction cycle proceeds as follows: If a male and female are together, the male will display and court the female. After a day or two, the couple begins to build a nest. They copulate, continue to work on the nest, and eggs are laid. Both parents incubate the eggs. Upon hatching, the young are fed with crop milk, a special substance that both parents produce in the linings of their crops (a pouch in the upper part of the digestive tract in birds used for food storage and production of food for nestlings). This wanes as the young reach fledging age, and the parents return to courtship.

FIGURE 2.12

Ring doves (*Streptopelia risoria*) have played a central role in our understanding of the relationships among hormones, behavior, and environment.

All this may seem unremarkable, and in one sense, that is correct. Mating and producing offspring are not particularly exceptional occurrences. But how do the two doves decide to build a nest at the same time? How do they coordinate incubation and crop milk production? Hormones coordinate these events; the production of the hormones is influenced by the environment, including the behavior of the pair of doves, along with endocrine feedback. What is exceptional about this behavioral sequence is not the fact that it happens, but the endocrine-behavior-environment synthesis that allows it to happen—in synchrony—in the first place.

Let's revisit that behavioral sequence and mark where these influences are important. The preceding text is repeated here in bold, with the hormonal influences inserted in italics.

> **If a male and female are together, the male will display.** *Androgens are necessary for male display.* **After a day or two, the couple begins to build a nest.** *Male courtship display causes the release of FSH from the pituitary. FSH—follicle-stimulating hormone—initiates a cascade of development in the female reproductive tract, including estrogen release. Ovaries and uterus grow and develop.* **They copulate, continue to work on the nest, and eggs are laid.** *The nest's presence stimulates progesterone secretion; progesterone in both sexes promotes incubation behavior. (Other hormones are involved in egg production as well.)*
>
> **Both parents incubate the eggs.** *Incubation stimulates the release of prolactin from the pituitary. Prolactin inhibits release of previous hormones, so that sexual behavior wanes. Prolactin also stimulates crop milk production.* **Upon hatching, the young are fed with crop milk, a special substance that both parents produce in the linings of their crops.** *Prolactin levels fall during the course of nestling rearing. Hormones that promote courtship and reproduction once again increase, no longer inhibited by prolactin.*
>
> **This wanes as the young reach fledging age, and the parents return to courtship.**

Hormones are not the only influences on this cycle, however. For instance, males that are deaf or males that are with an ovariectomized female do not produce as much testosterone as males that can hear or males that are with a female in reproductive condition. Indeed, it seems that both males and females need to hear vocalizations during courtship for necessary endocrine changes to ensue and result in egg development. Visual cues are also important. If males and females are separated after nest construction and held behind a glass partition, males that can see the female incubate eggs will proceed to produce crop milk; males that do not see mates incubate will not produce crop milk.

Ring doves have been carefully studied for decades, and this summary is a fraction of what is known. The purpose here is not to provide an exhaustive summary, but rather present a hint of the riches that are just being discovered in the biology surrounding the interaction of behavior, hormones, and the environment.

of the neurotransmitter or hormone. These types of direct experimental manipulation provide stronger evidence for causative effects than the observation of a correlation.

Hormones, Physiology, and Behavior

Returning to the review of endocrinology, organs and tissues cannot operate effectively without information from other parts of the body. For instance, in birds and mammals, reproduction is usually seasonal, so the ovaries need to be told the appropriate time of year for egg production. Males of the same species may compete intensely for mates when females are ovulating; male muscles and nervous systems need to be primed for this competition. Longer-term behavioral changes, of the sort that happen as an animal matures or between seasons of the year, are usually coordinated by the endocrine system. The endocrine system produces hormones, messenger molecules that communicate among organs and tissues. Glands are frequently the source of hormones; the adrenal glands of vertebrates are a good example. Hormones have effects over short time periods (these are called *releasers* because they have an immediate effect in releasing a response) and over long ones (these are called *primers*, because they prepare the body for a response, even if that response is not immediate). An immediately obvious advantage of hormonal control over neural control is that changes can be easily maintained over long periods of time by continued release of a hormone.

> **KEY TERM** A receptor is a molecule, typically a protein, that receives the hormone and passes the information carried by the hormone on to metabolic or genetic mechanisms within a cell. Hormone receptors can be embedded in the cell membrane or may be found in the nucleus.

Hormones are usually relatively small molecules; this allows them to be carried as signals from tissue to tissue in body fluids. Every hormone has unique *receptor* molecules. These receptor molecules transduce (change) the hormonal signal into an intracellular (within cell) signal that triggers a physiological or genetic response to the hormone.

Some hormones act on receptors on the surfaces of cells; others can pass through cell membranes and act directly on receptors in the cell nucleus. This ability to penetrate cell membranes can be critical, as the "blood-brain barrier" separates the nervous system from some types of circulating hormones, but not others. Steroids, a class of hormones that includes testosterone and estrogen, generally enter cells and act directly to affect gene expression. Peptide hormones, such as oxytocin and vasopressin, rely on receptors on the surfaces of cells, which then cause a cell to produce secondary messengers (such as cyclic adenosine monophosphate, cAMP) that act within the cell to affect its function.

Hormones and Behavior in Vertebrates

In this overview of some of the major ways in which hormones affect behavior in vertebrate animals, the focus is on development, a topic that exemplifies how hormones and behavior interact. Two additional major topics will appear in later chapters. Hormonal priming for sexual behavior in vertebrates and the relationships among steroid hormones, aggression, and territoriality are covered in Chapter 11 on mating. Chapter 12, on nesting and parental behavior, includes the hormonal bases of parenting in vertebrates.

Among the vertebrate hormones, two *steroids* are most familiar: testosterone, best known for its role in male reproduction and behavior, and estrogen, which plays similar roles in females. Steroids are structurally related to cholesterol; because they dissolve in fats and oils, steroids easily pass through cell membranes and can work in the nucleus of cells by turning gene expression off and on. These molecules are not easily degraded, so steroid and steroid-like compounds persist in

> **KEY TERM** Steroids are a class of hormones that include testosterone, estrogen, and ecdysone. Chemically, they are related to cholesterol.

the environment and build up in the fatty tissues of animals; this is a problem from a conservation standpoint and is covered in more detail later in this chapter.

In addition to steroids, there are other kinds of hormones, including peptides, relatively short chains of amino acids. These peptides act by triggering receptors on cell membranes. They are more fragile than steroids; the most familiar one is insulin, which helps to regulate the energy metabolism of animals. Other major vertebrate hormones, along with their sources and their action in regulating behavior, are summarized in Table 2.1 and Figure 2.13.

The hormones involved in sexual development and behavior in vertebrates are summarized in Figure 2.13. The environment and an animal's internal condition, including its age, come together to determine when the brain (specifically, the hypothalamus) should produce gonadotropin-releasing hormones (GnRH). These hormones act on the pituitary, which in turn produces two hormones, follicle-stimulating hormone (FSH) and luteinizing hormone (LH), which act on the gonads. The primary hormone of the female gonads (ovaries) is estrogen, which is a steroid. The primary hormone of the male gonads (testes) is testosterone, also a steroid. These steroid hormones then regulate, among other things, physical and behavioral readiness to mate (see Figure 2.14).

TABLE 2.1 Vertebrate Hormones and Behavior: An Overview

	Produced By	Target Organs/ Tissues	Steroid, Biogenic Amine, or Peptide?	Behavioral Effects
Luteinizing hormone (LH)	Hypothalamus	Gonads	Peptide (glycoprotein)	Indirect, via stimulation of estrogen or testosterone production by gonads
Follicle-stimulating hormone (FSH)	Hypothalamus	Gonads	Peptide (glycoprotein)	Indirect, via stimulation of estrogen or testosterone production by gonads
Estrogen	Gonads, primarily the ovaries	Brain, tissues, and organs involved in reproduction	Steroid	Female sexual receptivity, maturation, female secondary sexual characteristics
Testosterone	Gonads, primarily the testes	Brain, tissues, and organs involved in reproduction and dominance/ aggression	Steroid	Dominance, aggression, male sexual receptivity, maturation, male secondary sexual characteristics
Progesterone	Gonads (the ovaries) and the placenta	Brain, mammary glands	Steroid	Reproduction (particularly postfertilization), parental behavior
Oxytocin	Hypothalamus, released from the pituitary	Uterus, mammary glands, brain	Peptide	Birth, lactation, bonding
Vasopressin	Hypothalamus, released from the pituitary	Brain, kidneys (for regulation of hydration)	Peptide	Bonding
Prolactin	Pituitary	Mammary glands, brain	Peptide	Parental behavior
Insulin	Pancreas	Metabolic tissues, possibly the brain	Peptide	Possibly hunger
Epinephrine (adrenalin)	Adrenal	Brain, vascular, and respiratory tissues	Biogenic amine	Fight or flight—vascular and respiratory control, sense of alertness/awareness

41

FIGURE 2.13
Distribution of vertebrate endocrine glands in hypothetical (hermaphroditic) rat; the hermaphroditic rat, which does not exist in nature, is drawn to show both male and female endocrine organs. The hypothalamus is the main neuroendocrine interface in vertebrates. It coordinates the production of many hormones and can have a central role in the control of behavior. Neuropeptides from the hypothalamus induce the pituitary gland, which is suspended from the hypothalamus to produce a number of hormones.

During development, the endocrine system plays a critical role in the expression of age-appropriate behavioral patterns. For vertebrates, estrogen and testosterone have critical roles, while juvenile hormone is important in insects. These hormones affect both physical and behavioral development; because they affect both of these developmental areas, physical and behavioral development are appropriately coordinated.

FIGURE 2.14
The brain, pituitary, and gonads form a unified system that regulates sexual maturation and reproduction. Steroid hormones (estrogen and testosterone) provide negative feedback, indicated by the arrows, to the brain, signaling that the gonads have been activated.

42

Behavioral changes during growth and development are key to an animal's survival. Very obviously, immature animals do not behave like adults do. Steroid hormones, such as estrogen and testosterone, help to regulate behavioral development in vertebrates. Hormonal changes drive striking seasonal patterns in sexual receptivity, conflict and aggression, and parental behavior. In vertebrates, behavioral changes of this type are strongly associated with the levels of steroid hormones. Seasonal changes are equally marked in some invertebrates, but their physiological bases are less well understood.

After an animal is hatched or born, its nervous system often is incompletely formed. Nevertheless, it coordinates the activities needed to sustain the young animal's life, such as seeking food or its mother, hiding from potential predators, and finding the appropriate environmental conditions. Vertebrate species are termed either *altricial*—needing intensive parental care at hatching or birth—or *precocial*—being relatively independent at birth (see Figure 2.15). Mouse pups are altricial; they have incompletely developed eyes and lack the ability to feed themselves; they must be carefully nurtured by their mother. Altricial young cannot move about or thermoregulate, and their visual systems may not be fully developed. They require intensive parental care, usually centered around a nest or den where the young are protected from the environment. Marsupials, songbirds, bats, rodents, carnivores, and humans are other examples of animals with altricial young.

FIGURE 2.15
Canada goose young are able to swim and feed themselves soon after hatching. While they rely on their mother for protection and for information about feeding, their mother does not pass food to them. Contrast these precocial goslings with the intensive parental care needed by altricial young. *Photo: Michael Breed.*

FIGURE 2.16
Major areas of the human brain.

Species in which the female lays eggs and then leaves or dies have precocial development. For example, baby sea turtles are precocial. Upon hatching on a beach, they are programmed to seek the water, where they then can maintain themselves without parental assistance.

What selective forces lead the young of some species to be semi-independent, whereas the young of others are so poorly developed that they need intensive parental care? One hypothesis is that having semi-independent offspring allows the mother to have a larger clutch. Alternatively, some habitats or feeding preferences may favor offspring independence. In mammals the young of herbivores are more fully developed at birth than the young of carnivores. This difference may reflect a number of selective forces, such as the fact that many herbivores must continually be on the move to find suitable forage. Herbivore young may also have fewer challenges in terms of learning how to select and process food than do carnivores, which must learn complicated strategies of prey capture.

Vertebrates begin gender differentiation early in development. Sex steroids (estrogen, testosterone, and progesterone) play critical roles in the embryonic development of males and females (see Figure 2.16). In the earlier stages of development, all embryos are female-like in their morphology. In male embryos, testosterone production from the testes induces male external genitalia. Testosterone may also induce the development of male-specific details of brain structures and sets the stage for male behavior later in life. Embryonic cells in the hypothalamus and telencephalon have working estrogen receptors; the function of these receptors is not well understood, but exposure of these cells to estrogen probably influences later behavioral development.

One of the most interesting aspects of endocrine influences on behavior is the effect that developing embryos have on each other in mammals. Field studies of mammals often yield observations of behavioral variability among family members; one of the factors that may cause these differences in personality (see Chapter 6) is uterine position. Most mammalian species produce litters of two or more young; these young shared their mother's uterus during development. In the uterus, the embryos are lined up so that except on the ends of the rows, each embryo has two neighbors (see Figure 2.17). Because steroid hormones easily move through biological membranes, hormone produced by one embryo may pass in the uterine fluid to adjacent embryos. Testosterone from male embryos is particularly potent; in addition to controlling the development of the male genitalia/reproductive system, it also

FIGURE 2.17
A uterus dissected open, revealing the order of the embryos.

apparently induces the development of testosterone receptors in the developing male brain so that later in life high numbers of receptors make the male brain more susceptible to the behavioral influences of testosterone. Female fetuses produce estradiol, but less is known about the effects of fetal estradiol on neighboring fetuses.

The effects of uterine position are best known in rodents, but these principles seem to apply broadly across the mammals. An embryo can have two male neighbors (2 M), two female neighbors (0 M), or one of each (1 M). The 2 M young, whether genetically male or female, are more masculinized, and 0 M young are more feminized. The 2 M embryos (male or female) are much more sensitive as adults to the effects of testosterone; for example, they hold larger territories, and when injected with testosterone, they are more aggressive. 2 M mice urine mark more frequently, and male mice urine mark more often when exposed to a 2 M female than to a 0 M female (presumably because the 2 M female smells more masculine and induces a greater territorial response from the male mice). When treated with testosterone, female mice that are 2 M are more likely than 0M females to exhibit male sexual behavior (mounting). On the other hand, 0 M females are more likely to exhibit female sexual behavior (lordosis) and to copulate successfully. In addition to these hormonal influences, embryos in some positions in the uterus have better access to nutrients, and are born larger and stronger than other embryos. These size and strength differences carry over into juvenile and adult development, yielding animals that are more likely to be dominant.

BRINGING ANIMAL BEHAVIOR HOME: TESTOSTERONE, CASTRATION, AND MALE BEHAVIOR

Male animals can be strong and behaviorally difficult to manage; male strength and behavior are largely under the control of testosterone, which is produced in the testicles. Removal of the testicles from male birds (caponization) and mammals (castration) is a technique that was established early in human history, many years before scientists demonstrated the existence of hormones, defined their chemical structures, or discovered their physiological effects.

Castration results in lowered levels of aggression, decreased muscle mass, and an overall fattier body composition. Castrated animals have more trainable, malleable personalities. In agriculture, the most commonly castrated animals are horses (the castrated stallion is a gelding), cattle (a castrated bull is a steer), sheep, and chickens (the rooster becomes a capon).

Companion animals, particularly dogs and cats, are often castrated, with the goals of making male animals more manageable and reducing the potential for both population growth and health issues, including hormone-sensitive cancer. Testosterone primes the expression of male behavior by affecting development of the brain; males castrated after male behavior has developed often retain male behavioral patterns.

A simple example of this is leg-lifting (territorial marking) during urination by male dogs. Lifting the leg is a common sign of maleness; young male dogs develop this behavior as they mature. Males castrated before the behavior develops rarely express it, whereas males castrated after the behavior develops retain the behavior even though the hormonal signal has been removed. Contemporary studies of the effects of testosterone on male behavior often employ chemical castration or supplemental testosterone delivered by implants; typically, territoriality and aggression can be manipulated with these treatments.

As indicated earlier, establishing linkages between hormones and behavior often involves, as a first step, measuring hormone levels and looking for correlations between shifts in hormones in the blood and behavior. This approach has been powerful in suggesting a variety of behavioral roles for steroid hormones in vertebrates, and in fact some of the relationships, such as that between testosterone and aggression in males, seem to hold up under further scrutiny. Other relationships, such as the hypothesized role for hormones in regulating paternal behavior, are not so well supported. This points out the danger of leaping from a demonstrated correlation to an assumption of causation.

OF SPECIAL INTEREST: FIGHT-OR-FLIGHT RESPONSES

When confronted with a life-threatening event, an animal needs to mobilize many systems at the same time to save itself. The hormone epinephrine (adrenaline) does exactly this. Epinephrine is produced by the adrenal glands, which are associated with the kidneys, in response to startling stimuli or fearful reactions. Notable effects of epinephrine include increased heart and respiration rates; vasocontraction, particularly in the peripheral circulation, and heightened sensory and cognitive awareness (see Figure 2.18).

FIGURE 2.18
The external appearance of an animal clearly indicates whether its mode is "fight" or "flight." Inside the skin, endocrine changes are just as dramatic.

In animals with seasonal mating/migratory cycles, the testes typically enlarge as mating season approaches, corresponding to increasing intermale competition, courtship displays, and willingness of males to copulate. Migration seems, at first glance, to be a strong candidate for hormonal control. Spring migrations precede mating seasons, and winter migrations culminate the reproductive cycle for many animals. Yet if there is a single hormone that regulates migration, it has so far eluded discovery; the effects of treatments such as gonadal ablations and hormone supplements do not support a consistent model for hormonal regulation of migration itself, although events surrounding it are under hormonal control. For instance, animals must store substantial fat in preparation for migration, and hormonal regulation of food intake and fat deposition plays a role in readiness for the migratory journey.

Territoriality is more clearcut. Testosterone injections or implants result in males that hold larger territories or have higher dominance status. However, hyperaggressive behavior does not always translate into more mating success because males may shift their time budget to expend too much time and energy on territoriality and dominance, and consequently have too little time and energy to mate or feed young successfully. Testosterone levels probably represent the result of an evolutionary balance that maximizes mating potential, and not hyperaggression. This topic receives more scrutiny in Chapter 11.

The manipulation of human hormones has a long, if somewhat undignified, history. Some of the more extravagant efforts were fashionable for several decades during the late nineteenth and early twentieth centuries (see Figure 2.19). These efforts included radiation of *in situ* human gonads, injection of various concoctions (e.g., seminal fluid, blood, minced testes), and transplantation of testes from a variety of species. (Some practitioners were fastidious in their choice of donors, restricting them, for instance, to one variety of goat; others were more catholic in their approach and included, say, a broad range of ungulates or great apes.) Patients/customers were the rich and the famous of the day, including one of the greatest poets in the English language, W. B. Yeats, and Sigmund Freud, the founder of modern psychoanalysis. The best of these attempts, carefully experimental by the standards of the time, paved the way for the modern understanding of reproductive endocrinology, a field that, as you might imagine, was nonexistent in the first part of the twentieth century. Possibly the worst was typified by the "goat gland doctor," "Dr." John Brinkley. Brinkley was a radio commentator, evangelist, and would-be governor of Kansas. (After 16,000 goat transplants into humans, he went bankrupt and fled to Mexico, where he continued his practice.)

FIGURE 2.19
The Gland Stealers satirizes the work of Serge Voronoff, a Russian physician working in France who reported success with testicular transplants. In this book, wealthy men of a certain age head into Africa in search of 100 gorillas—or more precisely, their testicles. While this may not qualify as the Great American Novel, it is evidence that testicular transplants were a significant part of the culture in the early 1900s.

Hormones and Behavior in Invertebrates

Among invertebrates, the regulation of behavior by hormones is best understood in insects. The endocrine systems of other arthropods, such as spiders, millipedes, centipedes, and crustacea, are similar to that of insects, and many of the principles presented here apply to arthropods in general. Beyond arthropods, invertebrates are a heterogeneous lot, and the neuroendocrine functions of bag cells in molluscs stand out as being well-enough understood to merit some discussion.

JUVENILE HORMONE AND ECDYSONE

KEY TERM Juvenile hormone (JH) is a hormone that regulates development, reproduction, and behavior in arthropods. Chemically, it is a terpenoid. It is produced by the corpora allata, glands behind the brain that are connected to the brain by neurosecretory axons.

For invertebrates, the best-studied hormone with behavioral effects is *juvenile hormone* (JH), a nonsteroidal terpenoid and a product of the corpora allata (CA) in insects. Juvenile hormone affects mating behavior, pheromone secretion, and parental behavior in adult insects, as well as worker behavior in eusocial insects, such as honeybees. Although juvenile hormone is the best known of insect hormones, there are other important hormones as well.

KEY TERM Ecdysone is a hormone that regulates molting and reproduction in arthropods. It is a steroid and is produced by several glands, including the prothoracic glands and the ovaries.

For instance, the other primary hormone of arthropod growth and development is *ecdysone*, a steroid. Ecdysone is primarily known for its role in regulating the cycle of exoskeleton molting (ecdysis) in growing arthropods, but it also regulates egg production by stimulating yolk synthesis in adult females. Ecdysis is a phase in molting; this is the time when arthropods shed their exosekeleton in order to molt. A neurohormone, *prothoracicotropic hormone* (PTTH), released from the brain stimulates the prothoracic glands to secrete ecdysone. Interestingly, PTTH is a peptide that is evolutionarily related to insulin. The insect neuroendocrine system and the hormones of behavioral interest are shown in Figure 2.20.

PTTH is secreted at a certain time of day, in conjunction with the circadian rhythm of the insect (see Chapter 4). Transplanting the brain of an insect reared on one light-dark cycle into the body cavity of an insect on a different cycle will cause the recipient to follow the molting rhythm of the donor.

In addition to its many behavioral functions, juvenile hormone regulates two important developmental and physiological processes in insects: maturation and reproduction. In insects that undergo complete metamorphosis, JH determines whether larval, pupal, or adult characteristics develop during the molt cycle. As ecdysis approaches in a larva, if JH levels are high, the larva will molt to the next larval stage. If JH levels are diminishing, the pupal transformation will ensue. Absence of JH results in the development of the adult insect. In fact, experimental removal of the corpora allata results in premature development of adult characteristics, and application of JH to final-stage larva will inhibit pupation and result in "supernumerary" instars, that is, unusually large larvae that have undergone more molts than the normal number for that species.

> **KEY TERM** Prothoracicotropic hormone is a peptide hormone produced by neurosecretory cells in the arthropod brain. It stimulates the production of ecdysone by the prothoracic gland.

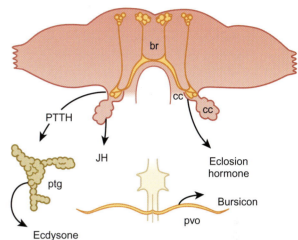

FIGURE 2.20

A dorsal view of the insect brain and endocrine glands. Neurosecretory cells in the brain have axonal connections with the corpora cardiaca (CC; sing., corpus cardiacum), small paired glands behind the brain. The CC, in turn, have axonal connections with the corpora allata (CA; sing., corpus allatum). Juvenile hormone (JH) is secreted from the CA. A releasing hormone, prothoracicotropic hormone, stimulates ecdysone production by the prothoracic glands. Ecdysone is also produced by the ovaries.

47

OF SPECIAL INTEREST: JUVENILE HORMONE, LABOR, AND THE HONEYBEE

In an intensively studied species, the honeybee, many scientists think JH regulates the behavioral activities of workers throughout their lives. Aggressiveness of guard bees is correlated with their blood JH levels, for example. Even though guards have high JH levels, their ovaries are relatively undeveloped. Guarding behavior prevents robbing of honey from the nest by bees from other colonies. It is not an easy trick for the guards to tell which incoming bees to attack and which to tolerate. They do this by smelling odors on the surface of the incoming bees; the bees that do not smell like nestmates are attacked. Nestmate recognition is similar to kin recognition, and is expressed by most eusocial insects—termites, ants, bees, and wasps. A guard bee may attack a bee from another colony, preventing her from entering the guard's nest. JH titers in worker honeybees progressively increase through the first 15 or so days of the worker's life. During this period, workers perform tasks inside the hive, such as nursing larvae, constructing comb, and cleaning cells. JH titers peak around day 15; workers this age guard, remove dead bees from the colony, and fan at the colony entrance to cool the nest. Older workers forage for pollen and nectar.

Juvenile hormone has numerous behavioral roles, including the induction of courtship behavior, sex pheromone production, diapause, and migration, in many insects, along with influencing caste differences in eusocial insects. Juvenile hormone is particularly interesting, in contrast with vertebrate hormones, because it regulates so many systems (see Figure 2.21).

Juvenile hormone also regulates reproductive cycles in adults. Because mating behavior is often synchronized with the ovarian cycle, it makes sense for mating behavior and pheromone production to be linked with JH (see Figure 2.22). In some insects, including

FIGURE 2.21

Juvenile hormone is produced by paired glands, the corpora allata, that are attached by neurosecretory axons to the insect brain (refer to Figure 2.20). This figure illustrates the interaction of juvenile hormone and a steroid hormone, ecdysone, in the development and molting of insects. Ecdysone peaks in correspondence with the molt, and juvenile hormone peaks correspond to developmental changes from juvenile to adult. Behavioral changes also occur during development and may also be linked to these hormones. Removal of the corpora allata from an adult disrupts mating behavior, while often treatment of an insect with juvenile hormone stimulates mating behavior and sexual receptivity. *Adapted from: Nijhout, H. F., 1994. Insect Hormones. Princeton University Press, Princeton.*

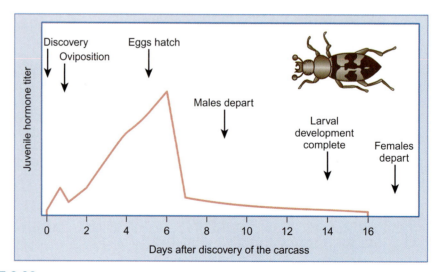

FIGURE 2.22

Changes in juvenile hormone titers through a reproductive cycle of a burying beetle. These beetles find a small carcass, bury it underground, and then lay their eggs on the carcass. Both parents remain with the carcass and their offspring until after the eggs hatch, and the females remain until their larvae finish developing. Juvenile hormone increases in the period between egg-laying and egg-hatching, and experiments suggest that JH plays a role in regulating parental behavior in this species. *Adapted from Scott and Panaitof, 2004, Hormones and Behavior 45: 159–167.*

some species of cockroach, this is exactly the case. In other species, the role of JH has evolved one step further so that the linkage with ovarian activity is lost.

MOLLUSCAN HORMONES

Bag cells (see Figure 2.23) provide an interesting noninsect endocrine example from among the invertebrates. The bag cells are specialized neurons found in the abdominal ganglion of molluscs such as the sea slug, *Aplysia*. These cells produce an egg-laying hormone that coordinates reproductive behavior in these molluscs.

> **KEY TERM** Bag cells are hormone-producing secretory cells in molluscs.

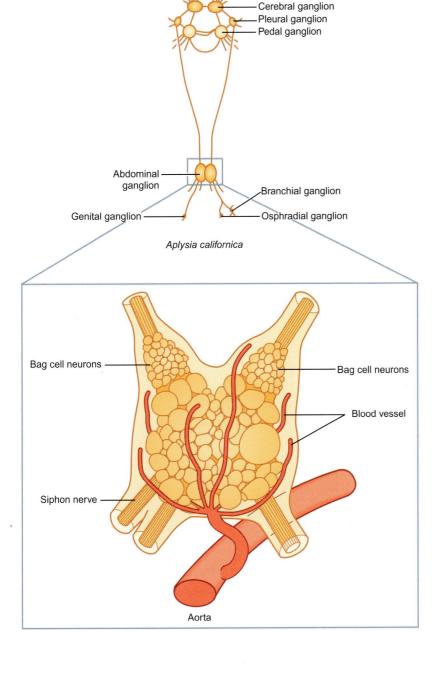

Aplysia californica

49

FIGURE 2.23
The nervous system of the sea slug *Aplysia californica* (top) and a detailed drawing of the abdominal ganglion (bottom). Signals from the cerebral ganglion cause release of egg laying hormone from the bag cells in the abdominal ganglion. This hormone coordinates egg-laying behavior, causing the sea slug to stop feeding and moving, to initiate contractions for egg-laying, and to deposit the eggs on the substrate.

FIGURE 2.24

Estrogen contaminated sewage affects the reproductive behavior of many aquatic vertebrates. In this example, the ability of fathead minnows (*Pimephales promelas*) to hold nests and reproduce was measured. The pink bars represent males that were not exposed to environmental estrogens; the light pink areas on those bars show how many not only held nests but had eggs in the nest. Males exposed to the environmental estrogens (blue bars) rarely nested and never succeed in fostering eggs. *After Martinovic et al., 2007, Environmental Toxicology and Chemistry 26: 271–278. Drawing: Duane Raver, USFWS.*

Conservation: The Behavioral Effects of Environmental Endocrines

Before leaving this review of endocrinology, consider that the widespread dissemination of estrogen-like compounds into ecosystems (particularly aquatic environments) has had important behavioral effects. The hormones in birth control pills, for example, are excreted in urine and move, largely unaltered, into wastewater systems and ultimately into streams and rivers. These estrogens affect morphological development, so intersex (partially female, partially male) fish and amphibia have become common in some locations. Exposure to estrogenic compounds during development also affects mating behavior and has the potential to alter parental behaviors. This is an important emerging issue in applied animal behavior and conservation (see Figure 2.24).

2.4 WHAT DOES AN ANIMAL BEHAVIORIST NEED TO KNOW ABOUT SENSORY SYSTEMS?

Sensing the Environment

Perception of the external world is essential for all forms of life, including bacteria. Animals gather information about many aspects of the environment. To do this, they use organs which act as transducers, converting various kinds of energy from the environment into action potentials. Because there are indeed various kinds of energy, depending on the type of information being received, there are different kinds of transducers, each specialized to convert a specific type of energy into action potentials. For instance, the rods and cones in eyes transduce photons into nervous impulses. This information then moves along the nerves in the form of action potentials until it reaches the central nervous system, where it is processed and used in determining the animal's next behaviors. In contrast, the sensory mechanisms used with sound differ depending on the sound's wavelength. The following sections contribute to an overall discussion of the sensory systems used by animals to take advantage of these different forms of environmental information.

Sensory Biology: Dangerous Assumptions

Humans, scientists included, have had difficulty accepting that animals can live in sensory worlds that differ remarkably from that of humans. Only in recent decades have people generally accepted the fact that animals can use sensory information impossible for humans to perceive. Some of the first breakthroughs came from studies of honeybees and established that bees can see ultraviolet wavelengths of light and perceive the plane of polarization of light.

OF SPECIAL INTEREST: ULTRASOUND AND BATS

Sound is a particularly interesting communication medium because it can be used in such varied ways. Infrasound—very low-pitched sound—can travel great distances. On the other end of the sound spectrum, bats use ultrasound—very high-pitched calls—for communication and echolocation. Because bats prey on insects, many insect species are attuned to bat echolocation calls and take evasive measures if they hear a bat call. Males produce a calling song to attract females in greater wax moths (*Galleria mellonia*) and lesser wax moths (*Achroia grisella*); they stop calling if a calling bat approaches; presumably the bat can orient to the mating call of the moth (see Figure 2.25).

FIGURE 2.25
Bats not only capture flying insects using echolocation, but they also can sense ground-dwelling prey, like this large centipede. In fact, they can even use echolocation in fishing. (More about that in Chapter 7.) *Photo: Scott Altenbach.*

As may be apparent by now, it takes quite a leap of the imagination to comprehend that bees see this vastly different world and that, in fact, the colors of flowers may include colors invisible to humans. These colors evolved under natural selection for visibility to pollinators; whether humans could see them or not had no evolutionary consequence. Sensory biologists and animal behaviorists now recognize that the colors, odors, and sounds of an animal's world can extend far beyond what humans perceive, and entire sensory realms, such as electrical fields, are apparently closed to humans. Or are they? What are humans perceiving unconsciously, and are there ways to investigate that?

OF SPECIAL INTEREST: *UMWELT* AND SENSORY BIOLOGY

We introduced the central concept of *umwelt* in Chapter 1. *Umwelt* is a core concept in animal behavior because the price of ignoring it is extremely high; the price that we pay for ignoring it is accuracy and understanding. In studying animal behavior, we must never assume that an animal can only perceive what we think we can perceive. An animal may have no perception of a sound, odor, or wavelength of light that seems obvious to the student or scientist but may have keen perception of information in its environment that is outside the human sensory realm. Seemingly good experiments have been confounded and misinterpreted because an investigator failed to realize that the study animals could sense something in the environment that was unknown and imperceptible to the scientist.

In the words of the poet Mary Oliver, "A dog can never tell you what she knows from the smells of the world, but you know, watching her, that you know almost nothing."[2]

51

Transduction

Animals perceive energy, in its many forms, by transducing energy from the environment into signals within the nervous system. Transduction is an extremely important process in biology as well as in engineering. Transduction is the conversion of energy from one mode to another, such as light to electricity, or sound to vibration. When a laser "reads" the information on a CD or DVD, the information recorded on the disk is transduced from minute bumps on the surface of the disk to electrical signals that are then amplified into sound or pictures. The same process takes place at the sensory interface between an animal's external and internal worlds. Information comes to the animal in the form of energy—light, sound, magnetic, and so on. The external energy causes changes in specially designed cells, which send the energy along in the form of action potentials in the nervous system. Thus, the receptive cells serve as transducers, converting one form of energy into another.

Generally, the first physiological response to perception is in the realm of specialized molecular receptor molecules on cells. Examples of types of energy that animals perceive include light, polarization of light, temperature, sound, molecular energy (chemoreception), magnetic fields, electric fields, and inertia (touch).

CHEMORECEPTION

Assessing chemical information in the environment was probably one of the earliest sensory elements to evolve. Animals that lack complex sensory organs are nonetheless able to respond to chemicals in their environment. It is thought that what chemosensory structures lack in anatomical complexity—all that is really needed is a fluid-covered membrane—they may make up in receptor variety. In essence, in the course of chemosensation (the perception of chemicals in the environment), a chemical dissolves in the fluid that covers a membrane, and eventually that chemical or an intermediate messenger system encounters an appropriate receptor. Transduction occurs when the chemical and the receptor produce an action potential. In many invertebrates, chemosensory tissue can be found in specialized depressions on the animal's surface.

The importance of chemosensation to even the earliest life forms becomes more clear upon considering how all living things "leak" chemicals into the environment. Imagine a halo of chemicals surrounding each organism, the unavoidable result of ongoing metabolic activity. To other organisms—everything from potential mates to potential predators—this halo is an advertisement to those with chemoreceptors to perceive it (see Figure 2.26).

Taste is a form of chemoreception dedicated to chemicals associated with food. The microvilli at the distal end of the taste cells are exposed to saliva through a pore in the taste bud. There are five types of taste recognized for humans: meat (umami), salt, sweet, bitter, and sour. Meat and sweet guide us to nutrition, bitter and sour steer us away from toxins, and salt helps us maintain osmotic *homeostasis*. These types of taste are then modulated considerably by the sense of smell. Likely other mammals have similar arrays of taste sensors, although the number and responsiveness of each type of sensor evolves to fit the dietary needs of any given species.

In invertebrates, chemosensors can occur anywhere on the surface of the animal; natural selection favors their development where they are most useful. Thus, many insects have chemosensillae on their front legs; they can "taste" with their feet (see Figure 2.27). Vertebrate chemosensory structures are

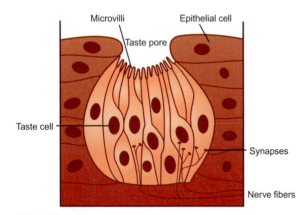

FIGURE 2.26
A chemoreceptive sensor—in this case, a taste receptor.

Microvilli · Epithelial cell · Taste pore · Taste cell · Synapses · Nerve fibers

KEY TERM Homeostasis is the maintenance of constant internal conditions in an animal. This is usually done by using feedback loops, in which the internal state is monitored and physiological or behavioral adjustments are made.

52

generally concentrated in the upper part of the respiratory system and are divided among the tongue (taste buds), the vomeronasal organ (on the roof of the mouth in vertebrates) and the olfactory epithelium, which is somewhat anterior in the respiratory system. Sometimes the vomeronasal organ is called *Jacobson's organ*. The vomeronasal organ may be specifically sensitive to pheromones, and in snakes, tongue flicking brings molecules sampled from the air into the vomeronasal organ. Until recently, birds were thought to have limited chemosensory abilities, but at least some birds, such as homing pigeons, are now known to have well-developed chemosensory abilities.

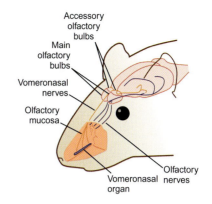

DISCUSSION POINT

If flies taste with their feet, why don't we taste with our fingers?

LIGHT AND VISUAL PERCEPTION

Vision begins with receptive cells that transduce light (generally considered to range from longwave infrared radiation to shortwave ultraviolet radiation) into action potentials. Structurally, visual receptors may appear complex, but the underlying principles are simple. When light is perceived, photons are absorbed, and this causes a pigment called *retinal* to change its energetic state. The energized pigment combines with a protein, *opsin*, the primary light-sensitive pigment in the eyes, which leads to depolarization of the membrane of a nerve cell. One photon/retinal/opsin chain is not enough to stimulate an action potential to the brain, but the combined action of many pigment molecules in the cell can generate an action potential.

Light receptors are widespread in metazoan animals. Even cnidarians and flatworms have light-sensitive pigment cells and collections of cells that allow them to discriminate light and dark. In animals with vision that allows some image formation, light-receptive cells are usually concentrated in a light-receptive surface, the retina, and light is focused on the retinal surface by a lens or lenses (see Figure 2.28 and Figures 2.29A and 2.29B). The processing of light information is complex and involves filtering and enhancement of the image in several stages, beginning at the level of individual receptor cells and proceeding next to interactions among adjacent receptors. Image formation and pattern recognition are functions of the central nervous system.

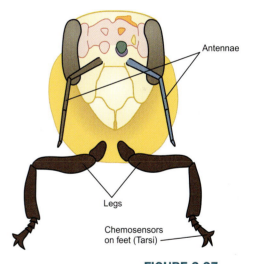

FIGURE 2.27
Locations of chemosensors in a mammal (top), a lizard, and a honeybee (bottom).

53

MONOCHROMATIC VERSUS COLOR VISION

One of the basic ways in which most animals differ from humans is in how their visual receptors respond to "color." What is color? "Color" is the name for the internal representation of the spectrum of wavelengths reflected by objects. Light itself seems colorless to humans, who perceive color when some of the wavelengths or spectra that comprise light are reflected from a surface. Thus, a surface that appears to be green is in reality reflecting light that is mostly in the green part of the spectrum (see Figure 2.30).

There is a wealth of information available in ultraviolet and infrared spectra, and although humans cannot see these wavelengths, many animals can. Bees see ultraviolet, but not

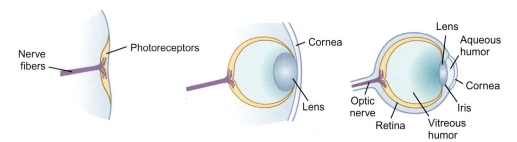

FIGURE 2.28

A variety of eyes, or photoreceptors. Some invertebrates (e.g., flatworms) have photosensitive pigment cups that do not form images. Others have eyes that form images (e.g., octopus) or that are sensitive to the smallest movement (e.g., many insects). The eye on the right is typical of a vertebrate. Animals that live in dark places such as caves or intestines usually have no eyes at all, leading to the inference that the production and maintenance of an eye might be energetically expensive.

A

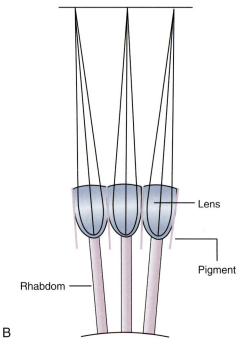

B

FIGURE 2.29A

The general organization of the vertebrate retina. The photoreceptive cells form a layer behind the nerve (ganglion) cells. In vertebrates with color vision, cone cells provide wavelength-specific responses. (The cone cells illustrated here follow the human pattern of red-, blue- and green-sensitivity.) The rod cells are sensitive across a broad range of the visual spectrum, providing monochromatic (grayscale) vision. Visual pigments are stacked in layers in the rods and cones.

FIGURE 2.29B

In the insect eye, hundreds or thousands of photoreceptive structures called ommatidia are packed together making a compound eye. Three are pictured here. Each ommatidium corresponds, more or less, to a rod or a cone in a vertebrate eye. While the array of ommatidia in a compound eye can be impressively beautiful, the number of photoreceptors is much smaller than what is found in a typical vertebrate retina, resulting in a lower resolution image. Because each ommatidium essentially captures a point of light, representations of insect vision as an octagonal grid overlaid on a picture are incorrect.

red. Not surprisingly, flowers pollinated by bees are rarely red, but many have ultraviolet patterning that is invisible to the unaided human eye. Most birds can see near ultraviolet wavelengths, as well as the entire human visual spectrum.

Some primates, including humans, gain color vision from having three types of cone receptors. These receptors respond to different wavelengths and have broadly overlapping

FIGURE 2.30
(Left) An image shown in grayscale, as if an organism perceiving this image has only one type of visual receptor that is sensitive across a broad range of wavelengths. (Right) An image shown in color, representing the visual world for an organism with several types of light receptors, each most sensitive at a different wavelength.

response curves, so some primate color perception results from the combined effects of two receptors responding to an intermediate wavelength. "Purple" is the name assigned to the perception of the combination of red and blue receptors. No one wavelength produces purple, and in that sense, such a combination can be called *nonspectral*.

OF SPECIAL INTEREST: THE DISCOVERY OF COLOR VISION

Thomas Young (1773–1829) was the first person to hypothesize that humans could see colors with only three types of receptors (see Figure 2.31). When he was not busy figuring out the nature of light, discovering the cause of astigmatism, learning how the lens of the eye worked, or practicing medicine in London, he was deciphering the Rosetta Stone, a task that was no doubt enhanced by his knowledge of at least six languages.

FIGURE 2.31
Thomas Young, a London physician who also deciphered the Rosetta Stone, was the first person to develop a fruitful hypothesis about color vision in humans.

Color perception by non-primate animals can differ from primates in several important ways: different numbers of color receptors, more broadly or narrowly tuned receptors, use of different parts of the wavelength spectrum, or absence of perception in parts of the primate visual range.

For instance, most nonprimate mammals have only two color receptor types: one in the blue–violet (or even in the ultraviolet) range and the other in the green. (So much, by the way, for red being a meaningful warning coloration for the majority of the mammals.) Most birds and some fish (e.g., goldfish) have four color receptor types. Some insects have five or even more receptor types. Having multiple color receptors creates more possibilities for nonspectral colors, as each receptor type can pair with each other receptor type to yield a non-spectral color. Having four receptor types translates into a mind-boggling minimum of four nonspectral colors, illustrating the difficulty of imagining the visual world of other species.

DISCUSSION POINT: COSTS AND BENEFITS OF COLOR VISION

Can you come up with evolutionary hypotheses about the costs versus the benefits of having a certain number of color receptor types? What might impose an upper limit on the number of receptor types? (*Hint:* Animals that live in very dim light usually have only one receptor type, giving them grayscale vision. This allows them to take advantage of minimal light energy. Each time a different receptor type is added, the light energy is split among the receptors.)

Broadly overlapping receptor ranges, such as those found in humans, result in the perception of intermediate colors, and intermediacy reduces contrast. More narrowly defined receptor ranges, such as the honeybee's, increase the perceived contrast between colors. Contrast may be critical for picking small items, such as flowers, out of monotonous green-colored backgrounds. The answer to the question "What is color?" is difficult. The most important point is that no one should assume that any animal species perceives the world in the same way that humans do (see Figure 2.32).

FIGURE 2.32
The spectral sensitivity of the European Starling, showing this bird's four color receptors and its visual responsiveness to light.

Most crepuscular (active at dawn or dusk) or nocturnal animals have monochromatic vision. This means they have only one type of visual receptor in their retina, which responds to the presence or absence of light energy. Under low levels of light, human vision is monochromatic, using receptors called *rods*. Because objects reflect different wavelengths of light and different amounts of light, objects that reflect a variety of wavelengths—that is, that are different colors— will produce "pictures" that the animal can discriminate, even if an animal has only one type of visual receptor. The example of a seahorse photographed against a colored background (refer to Figure 2.30) shows that in a monochromatic rendition, contrast between the light and dark areas preserves an interpretable image.

SHAPES AND IMAGES

Despite the difficulty of imagining how nonhuman animals perceive and remember images, there are some clear general principles. First, most animals respond to shapes within their visual fields and to the spatial relationships among those shapes. This observation suggests that "snapshot" images of the environment may be formed and stored. Second, interactions between adjacent visual receptors (ommatidia in the compound eyes of arthropods, or rods and cones in vertebrate retinas) emphasize edges at the margins of shapes. (This will become ecologically relevant in the study of defenses against predators.) The perimeter of a square, for example, is more strongly perceived than the center part. The acuity of vision (the detail of the image) depends on the number of receptors in a given area of receptive surface. An insect eye has a few thousand ommatidia in a few square millimeters; the center of the human eye has 160,000 cones per square millimeter; and the hawk's eye has over 300,000 cones per square millimeter. The hawk has by far the highest visual acuity.

Flicker fusion is about how rapidly an animal can process motion as separate images. For humans, flicker fusion occurs at something less than 60 images per second; images flashed at a rate faster than the flicker fusion rate blur together, while slower rates result in the perception of flicker. In the United States, alternating current electricity is 60 cycles per second, and generally no flicker is perceptible. A strobe light flashing at a lower rate yields a flickering image. Film or video images are shown at slower rates—16 to 25 images per second—but the similarity of image from frame to frame eliminates the perception of flicker. Animals that rely on seeing small movements by prey, such as cats and hawks, have much higher flicker fusion rates, over 300 cycles per second. This allows them to pick out minute prey movements and perhaps accounts for the fascination of cats with motion.

SOUND AND SOUND PERCEPTION

Humans usually experience sound as the result of vibrations in air or water. While sound that humans can sense is usually carried through these media, vibrations can also travel through soil, including rocks. Thus, sound can travel through a variety of substances with different densities, and the physical characteristics of the medium through which the sound travels have a major influence on how the sound can be used. For instance, it requires more energy to make water vibrate than to vibrate air, and it requires a great deal of energy to make soil vibrate. Thus, the use of vibrations in communication depends on the ability of the sender to make a substance vibrate. Because of this, large animals such as elephants are more likely than small animals to use vibrations in the soil for communication. In addition, the speed at which sound travels depends on the density of the medium through which it is traveling. This, in turn, varies with physical conditions such as temperature, chemical composition, and the like.

OF SPECIAL INTEREST: COMMUNICATION IN THE OCEAN

The speed of sound in the surface waters of oceans is approximately 1520 m/sec, almost five times faster than its speed in air (334 m/sec at 20°C). In some ocean depths, cold temperatures and moderate pressures yield ideal conditions for transmission; these ideal conditions are called the *SOund Fixing And Ranging Channel* (SOFAR), and sounds under these conditions can travel thousands of kilometers. Marine mammals can communicate across oceans via the SOFAR channel.

Vibrations in soil move even faster—up to 13,000 m/sec. The frighteningly rapid travel of tidal waves (tsunamis) and the even faster travel of earthquake tremors through the earth's crust reflect the speed at which any vibration if strong enough will travel in these media.

Analytically, sound is usually treated as waves, analogous to ocean waves or ripples in water caused by dropping a stone into a pond or puddle. The reason is that sound, in essence, is vibration; vibrations that are sensed with ear-like structures are called *sound*.

CHARACTERISTICS OF SOUND

Sound has a number of characteristics and properties that affect its perception. Wavelength is the period between waves of sound, also known as *pitch* or *frequency* (see Figure 2.33). Expressed in *Hertz* (Hz), cycles per second, the human ear

KEY TERM A Hertz is a unit of frequency that is approximately one cycle per second. It is named after Heinrich Hertz, who discovered electromagnetic radiation in 1888. Electromagnetic radiation is characterized by very long wavelengths.

57

areas, sometimes affiliating with other animals. Female members of a family tend to stay together, and of course, their juveniles travel with them. These female-centered groups may merge with other such groups periodically. Adult males are less likely to join groups (see Figure 2.35).

FIGURE 2.34
African elephants move in social groups that band and disband over time. Contact among members of social groups is maintained, in part, by infrasonic communication. *Photo: Pen-Yuan Hsing.*

FIGURE 2.35
The large ear of the African elephant serves several functions, including infrasonic hearing and thermoregulation. *Photo: Pen-Yuan Hsing.*

THERMAL PERCEPTION

Heat and cold receptors play critical roles in habitat choice, avoidance of dangerous conditions (such as overheating), and behavioral thermoregulation. Cold receptors respond to cool temperatures by release of calcium and sodium ions into a signaling pathway when the receptor cell is cooled. These are familiar to us as the receptors that respond to menthol as if it were cold.

Heat perception can also rely on TRP channel receptors (TRP stands for transient receptor potential). TRP channel receptors are the primary transducers for heat and cold. Some heat-sensitive TRP receptors respond to capsaicin, the chemical that gives hot chili peppers their heat. This is not the case for all heat-sensitive TRP receptors, however, and birds, for example, seem not to suffer from eating capsaicin-laced seeds.

In the most dramatic example of thermal detection, pit viper snakes, heat is perceived as infrared radiation. The pit of a pit viper is lined with infrared sensitive receptors that are protected by a membrane. Lore among biologists and snake hobbyists falsely states that detailed images, as in the vertebrae eye, are produced by the receptors in the pits. The images are much less detailed, and more subject to interference, than this belief asserts. Pit vipers need a cool background to provide contrast for perception of their prey; this accounts for the fact that most pit vipers hunt at night, when surfaces are cooler, and choose hunting locations with fairly uniform surfaces.

MECHANORECEPTION

Mechanoreception, also termed *tactile sense*, or *touch*, is an important source of information for animals. Touch provides information when light is unavailable, when noise interference obscures sound, or when vocalizations might attract predators. In mammals, touch is an important aspect of affiliative behaviors in social groups. Touch also provides important information about proximity of food, predators, and other environmental features. The tactile sense in vertebrates is supported by pressure-responsive nerve cells near the surface of the skin. These nerve cells act directly as transducers.

Tactile receptors are not evenly distributed over the animal's surface. They are in higher concentration, and therefore closer together, on surfaces that are used for exploration, manipulation, expression, or that are critical to protect. A standard neurological test in humans involves touching the skin with two needles. On the hands or face, the two stimuli are felt as separate points when only a few millimeters apart. On the upper arm or back, the needles will be felt as a single point when close together and will be felt as separate stimuli only when they have been moved one or two centimeters apart. The upper arm and back are less critical areas and are not involved so much in exploration, manipulation, or expression; consequently there are not as many tactile receptors in these areas. In insects, hair-like sensillae play the same role. Again, there are many touch receptors in critical areas, such as the antennae, and many fewer on the thorax and abdomen.

BRINGING ANIMAL BEHAVIOR HOME: EXTRAORDINARY SENSORY ABILITIES

Dogs and cats are startlingly able to sense some things about their environment that humans sense poorly or not at all. Dogs perform much better than humans in olfaction and detecting ultrasound. Cats also hear ultrasound and are much more sensitive to motion than humans (or dogs), and in their play respond very strongly to moving items.

Selective breeding in particular has heightened olfactory abilities in "scent hounds"—dog breeds that include beagles, bloodhounds, and coonhounds. Scent hounds can reliably track the odor of a specific animal or human, and can detect odors left when an animal passed hours or even days earlier. The olfactory epithelium of scent hounds is 6 to 50 times as sensitive as that of humans. In addition to being employed in hunting, scent hounds work in criminal investigations, including tracking suspects and in searches for drugs or other contraband. Scent hounds are often used at disaster scenes as "cadaver dogs," in searches for the remains of fatally injured humans.

Ultrasonic sensitivity in dogs is most commonly exploited in training whistles. Sound from a high-pitched, or ultrasonic, whistle carries long distances and is easy for the dog to pick out from background noise. A dog's hearing may extend to 60,000 Hz, well above the human range; the ability to hear at high frequencies may have assisted the ancestors of domestic dogs in locating prey or in communicating.

Motion sensitivity in cats is likely an adaptation for seeing the movements of small prey, such as mice, when the cat is hidden some distance from the potential prey. The optical processing regions of cat's brains are highly sensitive to movement across the visual field. Motion also seems to feed into a cat's sense of alertness or perhaps pleasure; cats seek play with objects that move rapidly (such as a string that is made to "dance" by a human).

ELECTRORECEPTION

Electroreception is common in fish and amphibia, which live in water, a good electrical conductor. It is absent (or at least poorly understood) in terrestrial vertebrates and invertebrates. Animals cause local distortions of electrical fields and produce electrical currents whenever a nerve cell depolarizes (sends a signal) or a muscle contracts. Usually, there are enough ions in the surrounding water to make the medium a good electrical conductor, and electrical currents and fields are excellent sources of information. Additionally, the interaction of water currents and the earth's magnetic field produces electrical fields that can be used in navigation.

Electroreceptors of fish include ampullary receptors, which are canals opening from the surface of the fish into cavities lined with nerve cells, and tuberous receptors, found on weakly electric fish (mormyriforms and gymnotoforms). The tuberous receptors have a similar structure to the ampullary receptors but are designed to reduce the loss of the incoming signal.

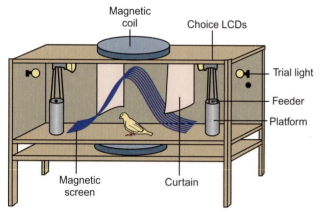

62

FIGURE 2.36
In this clever experiment, pigeons were trained to respond to a light in one way when the magnetic field was turned off and in another when it was turned on. They were then tested with and without the magnetic field and responded correctly, showing that pigeons can perceive the magnetic field. This experimental apparatus is derived from the Skinner Box, which is discussed in more detail in Chapters 1 and 5.

MAGNETORECEPTION

Many bacteria are able to perceive magnetic fields and follow lines of strength of those fields when they move. Magneto-orientation may provide bacteria a way of achieving straight-line movements, which otherwise would be difficult or impossible. Bacteria also may combine magnetic field information with other environmental factors, such as oxygen concentration, to find favorable habitats. Magneto-receptors, which are iron-rich, membrane-bound structures in bacterial cells, apparently evolved more than once in bacteria.

In multicellular animals, the first suggestion of magnetic field perception came from work on homing pigeons. Pigeons seemed to retain their navigation abilities under adverse conditions for compass orientation, such as cloudy skies. William Keeton, an influential figure in the study of pigeon homing, found that, at least in some experiments, equipping pigeons with magnets on their heads disrupted their orientation.[3] He took this to implicate perception of the earth's magnetic field in pigeon orientation. Magnetite crystals were later found at the base of pigeon skulls, suggesting a magnetic perception organ (see Figure 2.36). The realization that fluctuations of the earth's magnetic field are correlated with systematic errors in honeybee dances (*Missweisung*) led investigators to explore whether honeybees, as well, might have the ability to perceive the earth's magnetic field. This prompted the discovery of magnetite crystals in the abdomen of honeybees.

The distribution and importance of magnetoreception organs among animals has yet to be appreciated. The typical experimental demonstration of such organs includes placing magnets in or on an animal and monitoring the animal for changes in orientation. Recent experiments with salmon show that they have magnetite crystals in their olfactory organs, but implanting magnets in salmon did not affect their orientation. Similarly, sea turtle migration was not disrupted by magnets placed on the animals. In field orientation experiments, animals have access to a wide range of information and may discard magnetic influences in favor of other stimuli when the magnetic field is disrupted.

KEY TERM *Missweisung* are systematic errors in bee dances due to fluctuations in the earth's magnetic field.

SUMMARY

The nervous system receives sensory inputs, coordinates movements, controls physiology, and stores information. The basic unit of the nervous system, the neuron, carries electrical signals from point to point within the nervous system and works in a similar manner across all species. Neurotransmitters carry information among neurons. There are a variety of neurotransmitter molecules, including acetylcholine, dopamine, serotonin, and gamma-aminobutyric acid. Each neurotransmitter has several distinct functions; acetylcholine, for example, carries signals from nerves to muscles but is also involved in learning and memory. Neurotransmitters can set an animal's behavioral responsiveness and prime an animal to behave appropriately in a given context. They are responsible for states that resemble mood in animals.

In some animals, the nervous system amounts to a net of interconnections that allows communication among locations in the body and basic coordination of movements. Most animals, though, have a brain, an anterior mass of neurons. Concentration of neural function in a brain allows for more control over complex functions, and brain complexity is correlated with behavioral complexity. The central and peripheral parts of the nervous system coordinate behavior; reflex loops provide a basic model for coordination of behavioral responses.

The interaction between hormones and behavior is intricate. Hormones are produced by glands and transported to target organs or tissues. Hormones can either prime the tissue or organ for activity or stimulate it into action. Many hormones are steroids, whereas others are peptides (short amino acid chains). For vertebrates, numerous hormones, including estrogen, testosterone, progesterone, prolactin, oxytocin, and vasopressin, affect behaviors as well as physiology. Juvenile hormone and ecdysone are especially important in the behavior of some invertebrates. Juvenile hormone promotes many behaviors, including mating and labor in social insects. Human use and disposal of endocrines and endocrine-like compounds in the environment have important behavioral effects. Steroid hormones and chemicals that mimic steroid hormones persist in the environment and affect development, reproduction, and reproductive behavior in vertebrates. Aquatic organisms, such as fish and amphibia, have been particularly hard hit by these effects.

All living things must perceive the external world. Even bacteria have chemoreceptors and use that information. Information from the surrounding world is critical to behavioral decisions as well as survival, and animals have evolved sensory mechanisms that transduce a wide range of energy from the environment. Animals may perceive the environment in a variety of modalities. As a result, each species' sensory world may be unique. Assumptions about perception based on human senses may be dramatically misleading in studies of animal behavior. Animals transduce energy from their environment into signals within their nervous systems. Sensory systems are derived, evolutionarily, from systems that enable unicellular organisms to perceive molecules in their environments.

How do neurobiology, endocrinology, and sensory biology come together in shaping behavior? These three topics form the core of studying causation in animal behavior. Neurobiology gives us insight into the inner workings of an animal's brain and into the interaction of the brain with the rest of the body. Endocrinology shows us how the development of organ systems is coordinated and how seasonal behavioral changes are orchestrated. Sensory biology tells us about an animal's windows into the external world; sensory systems provide a bridge between the external world and the internal neural and endocrine mechanisms.

Generally speaking, evolution is a relatively slow, multigeneration process that affects regulation of behavior over very long time scales. Hormones stand at an intermediate point in the time scale, operating within the animal's lifespan. Neurotransmitters support immediate change, as when a nerve signals a muscle to twitch, or when the brain needs

63

FIGURE 2.37
A time scale of behavioral regulation from evolutionary responses (between generations) to rapid responses of animals to immediate needs or stimuli. This figure captures the general trend from genetic modifications during evolution to neurotransmitters effecting immediate change. Some behaviorally active messenger molecules, such as epinephrine and norepinephrine are considered to be hormones but coordinate rapid responses.

TIME SCALE FOR MODIFICATION OF BEHAVIOR

Between generations	Within generation long-term developmental change or response	Within generation, rapid change or response
Genes	**Hormones**	**Neurotransmitters**
1. Respond to natural selection. 2. Provide blueprint for nervous, muscular, and skeletal systems. 3. Turning activity of specific genes on and off during development is major regulatory tool.	1. Serve as messengers between tissues and organs. Are often able to pass freely across membranes (particularly steroid hormones), and are broken down slowly 2. Interact with genes (can regulate gene activity, or can be regulated by gene activity. 3. Are the primary means by which genetically programmed developmental changes are implemented. Example: Genes code for releasing hormones from pituitary that stimulate testes and ovaries to produce testosterone and estrogen. These hormones then feed back to affect how much releasing hormone is produced	1. Are small molecules that are easily produced and broken down. 2. Cannot freely pass through cell membranes, so are limited in their scope of action. 3. Serve as messengers between nerve cells, or between nerves and muscles.

to organize a response to a specific stimulus. This line is blurry, though, because some hormones, such as adrenalin and insulin, have almost instantaneous effects, while the overall levels of neurotransmitters and their receptors in an animal's brain determine much about the animal's behavioral potential (see Figure 2.37).

Moving forward with study of animal behavior, you will probably need to occasionally refer to this chapter or to explore some of the books listed in the Further Reading section. Even the most field-based, ecologically oriented animal behaviorist needs to have a basic mastery of these topics as they provide the keys to full understanding of animal behavior.

STUDY QUESTIONS

1. What are the parts of a neuron, and how is an action potential transmitted by a neuron?
2. What are the roles of neurotransmitters in behavior?
3. Describe the mood triangle. Does this approach to understanding behavioral state apply only to humans or is it more broadly applicable?
4. What glands produce estrogen and testosterone? What are the behavioral effects of the removal of the male glands in a developing bird or mammal?
5. What kind of molecule is juvenile hormone and what gland produces it? What behavioral effects does juvenile hormone have?
6. How does the pit of a pit viper work? How is this different from heat and cold perception in most animals?
7. Can you explain how scientists can test for the presence of magnetoreception as a sensory mode?

Further Reading

Heiligenberg, W., 1990. Electrosensory systems in fish. Synapse 6 (2), 196–206.

Nijhout, H.F., 1998. Insect Hormones. Princeton University Press, Princeton, New Jersey, 280 pp.

Norris, D.O., 2006. Vertebrate Endocrinology, fourth ed. Academic Press, San Diego, 560 pp.

Rudy, J.W., 2008. The Neurobiology of Learning and Memory. Sinauer, Sunderland, Massachusetts, 380 pp.

Smagghe, G. (Ed.), 2009. Ecdysone: Structures and Functions. Springer, Berlin/Heidelberg, 590 pp.

Verhulst, S., Bolhuis, J., 2009. Tinbergen's Legacy: Function and Mechanism in Behavioral Biology. Cambridge University Press, Cambridge, 262 pp.

Zupanc, G.K.H., 2010. Behavioural Neurobiology: An Integrative Approach. Oxford University Press, Oxford, 400 pp.

Notes

1. Gould, S. J. 1996. The mismeasure of man. W. W. Norton & Co., New York, 448 pp.
2. Oliver, M., 2004. "Her Grave," new poems (1991–1992). In: New and Selected Poems, Vol. 1, Beacon Press.
3. Keeton, W. T., Larkin, T. S., Windsor, D. M. 1974. Normal fluctuations in earths magnetic-field influence pigeon orientation. Journal Of Comparative Physiology 95:95–103.

Behavioral Genetics

LEARNING OBJECTIVES

Studying this chapter should provide you with the knowledge to:

- Realize that knowledge of the genetic underpinnings of behavior is essential to understanding how behavior evolves.

- Comprehend the difference between proximate and ultimate causes in behavioral genetics and be able to apply these concepts.

- Understand how behavioral genetics helps to explain the physiological and neurobiological systems that control behavior.

- Know that most behavior is shaped by a combination of genetic and environmental factors.

- Be aware that single-gene effects on behavior should not be confused with an erroneous idea that complex behavior is "controlled" by those genes.

- Realize quantitative genetics provides better explanations than single-gene models do for most animal behavior traits.

- Understand that molecular approaches to behavior genetics provide important techniques for exploring the regulation of behavior.

- Appreciate the impressive array of behavioral genetic tools available for testing hypotheses about the evolutionary and genetic foundations of behavior.

3.1 INTRODUCTION: PRINCIPLES OF BEHAVIORAL GENETICS AND THE EVOLUTION OF BEHAVIOR

Before Mendel saw his first pea plant, people knew that traits could be inherited. They also knew that the environment could affect inherited traits. Thinkers at least as far back as Shakespeare understood that behavior is usually the outcome of intrinsic influences (*nature*) and environmental ones (*nurture*). Based on this understanding, in *The Tempest* (IV.i.188–189), Prospero bemoans his attempts to reform the intrinsically savage Caliban: "A devil, a born devil, on whose nature/Nurture can never stick...." (In other words, Prospero cannot modify Caliban's nature.) Unlike Shakespeare, scientists have the tools to study the relative contributions of genetics and environment to behavior. What follows in this chapter will not help reform the Calibans of this world (it is to be hoped that there are few!), but it will introduce the genetic basis of behavior, which, in turn, is the key to understanding how behavior evolves.

> **KEY TERM** In the nature–nurture debate, nature is genetics. Nature is the extent to which genetics influence behavior.

> **KEY TERM** In the nature–nurture debate, nurture is environment. Nurture is the extent to which environment influences behavior.

The goal of this chapter is to provide a toolbox of genetic approaches to behavior.[1,2] Genetic studies of behavior take a much stronger problem-solving approach than other areas of behavioral science. An important part of solving a problem is choosing the right tool, and progress through this chapter can be measured by the facility with which the tools are understood and used. Knowledge of genetics gives students of behavior windows into both the evolution of behavior and the physiological regulation of behavior. Assimilating this material gives students powerful problem-solving abilities that can be applied in the study of almost any type of behavior.

CASE STUDY

The "Killer" Bee

The "killer" bee story is a classic example of plans gone awry. In the early 1950s, Brazilian geneticist Warwick Kerr came up with a well-intentioned scheme to introduce honeybees into the tropical Americas. Honeybees provide an outstanding subsistence source of food and wax for farmers, and these products can also be cash crops that supplement income from products such as bananas and papayas. Kerr planned to keep tropically adapted honeybees, imported from southern and eastern Africa, isolated and to use controlled crosses to develop hybrid bees that would be manageable and productive in the Brazilian tropics. Unfortunately, an unintended release of the African bees in 1957 resulted in the spread of very dangerous, unmanageable insects across the Western hemisphere.

One of the most extreme examples of ecotypic differentiation is the Western honeybee, *Apis mellifera*. Because honeybees provide many of the classic examples in behavioral genetics, we'll give some background here on their *ecotypes*. The geographic distribution of this species, from southern Africa to northern Europe, through the Middle East and into central Asia, encompasses both tropical and temperate environments, as well as arid and mesic (moist) environments. More than two dozen subspecies of *Apis mellifera*—ecotypes differentiated by color, body size, and behavior—have been recognized. This species has been introduced in the Americas and is the common honeybee seen in many North American habitats. One subspecies is also, famously, the "killer" bee, the highly defensive bees now found in much of Central and South America.

> **KEY TERM** An ecotype is a genetically differentiated population within a species that is adapted for a particular habitat.

Honeybees were already present in the Americas, but these bees originated in Europe and are not well adapted to tropical environments. Honeybees in most of North America are the descendants of bees brought by settlers from Europe beginning in the mid-1600s. They are mostly derived from *Apis mellifera*

(Continued)

CASE STUDY (CONTINUED)

ligustica, the ecotype that is characteristic of the Italian peninsula, but they also have genetic influences from a variety of other European ecotypes. The "Africanized" or "killer" bee is *Apis mellifera scutellata,* but it has hybridized with other ecotypes as it has spread through South and Central America and, more recently, into Texas, Arizona, and California.

The differences among honeybee ecotypes have a genetic basis; this was well established by beekeepers long before scientists approached the question, but **common garden** and controlled crosses of ecotypes by scientists support this conclusion.

> **KEY TERM** Common gardening is a technique in which animals (or plants) are maintained in the same environment while observed for differences. If the animals differ in behavior across their native habitats, then keeping them in a common garden helps to separate environmental from genetic influences by standardizing environmental influences. If they all behave the same when in the common garden, then the behavioral differences observed across their native habitats likely are due to environment. If the differences persist in a common garden, then a genetic hypothesis is supported.

FIGURE 3.1
Distribution map of the major ecotypes of the Western honeybee.
Adapted from http://www.sicamm.org/images/Map.jpg.

The most striking difference between "Africanized" bees and the honeybees to which most North Americans are accustomed is the heightened defensive response of the "Africanized" bees. With bees of European origins, under most conditions you can walk to within a few meters of a hive without risk of being stung, and if approaching the hive from behind, you can probably walk up to it and sit on it without risk. In contrast, "Africanized" bees often respond to an animal's movement 50 to 100 meters from their hive by flying and stinging. Each "Africanized" bee sting is no more potent that a European bee sting, but mass stinging can be lethal to large animals such as humans, dogs, horses, and cattle. A conservative estimate is that several hundred people have died in South and Central America from mass stinging events since the release of the bees in 1957. "Africanized" bees exhibit extreme responses in all phases of nest defense when compared with their European counterparts; the differences between the ecotypes are largely due to genetics (see Figure 3.1 and Table 3.1).[3]

These observations raise the question of why the root African stock from which the "Africanized" bees are derived exhibits such extreme responses. The best hypothesis is that honeybees

69

TABLE 3.1 Comparison of the behavioral attributes of the African ecotype, *Apis mellifera scutellata*, which probably gave rise to "African" bees in the Americas, with those of the Western European ecotype *Apis mellifera mellifera*

	Apis mellifera scutellata	Apis mellifera mellifera
Geographic origin	South and Eastern Africa	Northwestern Europe
Reproduction	Many swarms, absconds from nests to migrate when climate turns unfavorable	One or possibly two swarms a year, rarely absconds
Honey storage	Low	Very high
Defensiveness	Very high	Moderate
Size and color	Small, black	Medium, black

(Continued)

in Africa have faced a combination of predators that includes humans and their evolutionary predecessors. This amounts to strong selection favoring extreme defense. The *scutellata* ecotype has other adaptations for its tropical environment, such as propensity to invest in reproduction rather than storing large quantities of honey, which makes it less attractive to beekeepers; its one strongly positive attribute from a human point of view is its ability to do well in tropical climates, where honeybees of European origin usually fail.

A key lesson from this story is that containment and isolation of potentially dangerous or ecologically damaging introduced animals need to be extremely secure. Knowledge of the behavior of species in their native habitats does not always predict how well they will do in a new habitat, and species as diverse as the "killer" bee, cane toads, Burmese pythons, European starlings, and English sparrows have demonstrated unexpectedly high behavioral abilities to reproduce and disperse in new (to them) habitats.

Why Study the Genetics of Behavior?

Behavior, like all characteristics of animals, is shaped by a combination of genes and environment. This chapter presents techniques used by animal behaviorists to discover how genes and environment come together to determine behavior. This is a surprisingly controversial enterprise, and because of that, the nature–nurture debate and the political overtones possible with such discussions are included.

How Does Behavioral Genetics Inform Studies of Evolution?

Knowing how and why behavior has evolved is a central goal in the study of animal behavior, one that is greatly informed by an understanding of the genetic influences on behavior. Studying genes that seem to influence behavior gives insight into the physiological and neurobiological underpinnings of behavior. In addition, separating genetic from environmental effects allows both identification of the selective forces that shape behavior and determination of how behavior responds to those forces over evolutionary time. In effect, behavioral genetics allows observation of evolution in action.

Natural selection acts on genetic variation; if there is no genetic variation, then traits cannot change over time and evolution cannot occur. It follows from this that behavioral geneticists often focus on understanding how genetic variation affects a behavioral trait. Inevitably, studying the effects of genetic variation leads to considering the effects of environmental variation on the behavior, as well. The behaviors an animal expresses are partly a reflection of the animal's environment and partly a reflection of the animal's genes. Neither environment nor genes *determine* an animal's behavior, but both environment and genes contribute to the behavioral phenotype. A major goal of behavioral genetics is to understand the extent of these environmental and genetic correlations with the behavioral phenotype. Not surprisingly, some behavioral phenotypes, such as signals involved in courtship,[4] often show little variance among environments, whereas others, such as foraging behavior, are typically quite responsive to environmental differences.

Understanding Single-Gene Effects on Behavior

What is the path from genes and their products to the behavior of an animal? Causation is one of the key issues in animal behavior; new techniques allow linking genes, neural mechanisms, and behavior in exciting ways. Genes affect behavior at many levels. While it is tempting to think of genes as guiding behavior ("my genes made me do it"), examples of direct genetic instructions for behavior are rare. Instead, genes most often affect behavior by setting limits, or restraints, on the range of behavior expressed by animals. Thus, behavioral options are most frequently limited by genes, not directed by them.

> **OF SPECIAL INTEREST: PHENOTYPE AND GENOTYPE**
>
> Quick refresher: *Phenotype* is the outward manifestation of a trait, that is, the actual behavior, morphology, or physiology. *Genotype* is the characterization of the genes associated with the phenotype. Phenotype does not always reflect genotype in the same way because of environmental influences on the phenotype. The interplay between genes and environment is often measured using *heritability*, a key concept in genetics. Heritability is the extent to which variation in phenotype in a population of animals is correlated with genetic variation. Calculating heritability allows scientists to investigate the genotypic and environmental roots of phenotypic variability.

Genetically based human diseases rank high in human awareness. One such disease, Huntington's, results from the modification of a repeated nucleotide sequence in a single gene, *huntingtin*. The function of the protein for which *huntingtin* codes is not fully understood, but the modification of the gene has devastating degenerative neuromuscular effects for its human carriers. Many of the symptoms of Huntington's, such as unstable gait, are behavioral, and because of this, it is tempting to suggest that locomotory stability is under the control of a single gene. In fact, modification of the function of a single gene can indeed dramatically affect behavior, but this does not mean that the particular behavior is under control of that gene. Rather, the failure of the gene means that a needed step in the machinery underlying the behavior is absent. That absence causes multiple systems to fail. If any one of many genes other than huntingtin ceases to function, the results might be equally catastrophic.[5]

Fruit flies, *Drosophila melanogaster*, offer another example of one gene that ultimately affects behavior. Fruit flies can be rendered less receptive to mating by modifying the gene (the *Icebox*, or *ibx*, mutation) that is involved in normal formation of brain structures. Unable to respond to a potential mate, the modified flies cannot mate. Does that mean that the *ibx* gene "controls" mating? Not at all. It merely controls one small but crucial step in the mating sequence.[6]

In fact, virtually every behavior is shaped by genes acting in sequence or in a coordinated fashion to produce that behavior. Most investigators in behavioral and neural genetics now feel that genetic control of any one behavioral trait is dispersed over a large number of neural locations, rather than being coordinated by "executive neurons" that integrate the behavior. If many neural elements are involved in the production of a behavior, it logically follows that many genes must underlie the production and regulation of those neural elements. Because behavior is typically shaped by a large number of genes acting in concert, one of the main goals of contemporary behavioral genetics is to understand how multiple interacting genes can shape specific behavioral patterns.

3.2 THE NATURE VERSUS NURTURE DEBATE

What is the balance between genetics (instinct or nature) and learning (nurture) in shaping behavior?[7] Few topics have wasted more emotional energy and created more futile academic fury than the question of behavioral plasticity. An extreme view holds that all animal (including human) behavior is instinctive, with little room for learning and flexibility of response. At the other extreme, some scientists argue that behavior is plastic, modifiable over a great range of possibilities, and that individual experience determines behavior. As is typically the case in such debates, the truth lies between the extremes.

In the animal world, behavior can be envisioned on a continuum between learning and instinct. Depending on the behavior and in some cases, the individual, the explanation may (rarely) involve one of the ends of the continuum, or (more likely) any one of an infinite regulating or opposing biological forces so that human societies can function. Understanding the complex relationship of biology and culture is beyond current understanding of either. Intellectual surrender in the face of this complexity is wrong-headed—this interplay forms one of the great questions of the twenty-first century!—but caution is reasonable when making sweeping "scientific" statements about the biological bases of human social behavior.

3.3 EVOLUTION AND BEHAVIOR

While a fine-grained discussion of the genetics that underlie behavior may address Tinbergen's causation question, his questions about survival value and evolution are not far away. As discussed in Chapter 1, evolution is a change in frequencies of genes in populations. Such a change normally takes place over the course of generations; the environment, however, may change within the course of a single lifetime. In the face of short-term environmental change, behavior often provides the route for an individual animal to adapt to that change. Migration, for example, is usually a behavioral adaptation to seasonal fluctuations that occur during an animal's lifetime. Similarly, learning the location and types of available food resources allows animals to accommodate shifts in food organisms—again, during a single lifetime. Thus, within-generation phenotypic flexibility is very important to animal survival.

Equally important to the survival of lineages is response to longer-scale environmental change, such as a slowly shifting climate. This type of change invokes evolutionary responses by favoring the survivorship of some phenotypes over others, which in turn leads to shifts in gene frequencies that then produce physiological, morphological, and behavioral modifications. All of these modifications are phenotypic; that is, they are all outward manifestations of genetic traits.

If an animal's behavior allows flexibility, it can be a short-term tool with which an individual animal can respond to a variable environment. If it has a genetic basis, it can also be molded over generations by natural selection, in response to long-term environmental variation.

Evolution is often perceived as taking many generations to have a visible effect. This may be generally true, but there are many examples of rapid evolutionary change in response to strong selection. Strong artificial selection, such as that applied by animal breeders to domestic species (e.g., rabbits, chickens, dogs,[14,15] cats, and cattle), can have substantial effects over three to five generations. This suggests that populations of species in new environments (such as invasive species) or species that are experiencing rapidly changing environmental conditions could have the flexibility to exhibit rapid evolutionary responses if sufficient genetic variation is present.

FIGURE 3.2
Marbled murrelet
(*Brachyramphus marmoratus*).

It is also possible to use genetics to understand evolutionary history. The history, even the recent history, of sea birds is particularly difficult to infer because of their highly mobile lifestyle (see Figure 3.2). Genetic studies of these animals in the Pacific Northwest of North America have revealed that there are two genetically distinct populations: one on the Aleutian Islands and one from the eastern Alaskan Peninsula. Although they nest in different locations, the genetic distinctions can be traced back to a single ancestral population that expanded in the early Pleistocene period and then was separated by later Pleistocene glaciations.[16]

3.4 THE BEHAVIORAL GENETICS TOOLBOX

Behavioral geneticists have a toolbox, a set of largely experimental approaches that can be used to unravel how genes shape behavior (see Table 3.2). Some of the basic tools come from comparative evolutionary studies; for example, constructing a *cladogram*, or phylogeny, allows an investigator to follow changes in behavioral patterns in evolutionary

TABLE 3.2 The Behavioral Genetics Toolbox

Tests of Ultimate Causes: The Evolution of Behavioral Phenotype	Proximate Causes: Whole Organism Studies of the Genetic Bases of Behavior	Proximate Causes: The Genetic Dissection of Mechanisms Underlying Behavior
1. Geographic variation, subspecies, and ecotypes	1. Mutational studies	1. Microarrays and other assessments of gene expression
2. Comparative phylogenetic studies	2. Heritability	2. Candidate gene approaches
3. Cross-fostering	3. Quantitative trait locus (QTL) analysis	3. RNA knockout
4. Twin analyses		
5. Artificial selection and inbred lines		
6. Natural selection and behavior		

time and to identify key behavioral innovations.[17] Other tools, such as calculations of heritability, are borrowed from quantitative genetics. Powerful methods for identifying genes that may regulate behavior come from molecular methods of genetic mapping and identification of *quantitative trait loci* (QTLs).[18] *Expressed sequence tags* (ESTs) provide a direct window into the relationship between gene activity and behavior. This section explores the behavioral geneticists' toolbox and shows how these tools are applied to the central issues of evolution and causation in animal behavior.

As noted in Chapter 1, when biologists talk about *ultimate causes*, they are thinking about evolution and the long-term selective pressures that shape an organism's phenotype. Studies of ultimate causes often boil down to understanding where, in the course of evolution, a trait first appears, or where a structure first is used for a particular function. All mammals have digits—fingers and toes—but few have opposable thumbs and the high manipulative ability that comes with having a thumb-like digit. Knowing when opposable digits evolved, and how they were adaptive, helps biologists to understand the ultimate questions associated with these structures, that is, how they evolved. Because genes evolve and help to determine behavioral phenotypes, understanding ultimate causes plays a key role in behavioral genetics.

Often, studies of ultimate causes suggest approaches to finding *proximate causes*, in this case, the present-day physiological or behavioral expressions of gene activity. Biologists who focus on how genes regulate the functions of organisms study the proximate causes of traits. They discover the sequence of events that start with the transcription of DNA and end with the expression of a behavior. The most elegant behavioral genetic studies link the evolutionary roots of behavior to its proximate causes.

KEY TERM A cladogram represents a hypothesis of evolution within a group of species with a tree-like drawing.

KEY TERM A quantitative trait locus (QTL) is a gene that contributes, with other genes, to a phenotype. Because multiple genes contribute to the phenotype, no one gene "determines" the phenotype.

KEY TERM An expressed sequence tag (EST) is a genetic marker that is linked to the gene being studied. When the gene of interest is expressed, the EST "reports" that activation. This allows an investigator to see how gene activation correlates with physiological and behavioral activity.

Geographic Variation, Subspecies, and Ecotypes

Humans figured out long ago that behavioral, as well as physical, traits run in families.[19] Behavioral genetics sheds the light of modern science on this observation. One starting point is a search for associations between genetic differences and behavioral differences among animals. Comparing families, subspecies, and species and looking at differences at those levels help to indicate the extent to which genes influence behavior.

Strains, sometimes called *ecotypes* or *subspecies*, can be produced naturally or by artificial selection. When subspecies are formed naturally, they can be compared to see if the subspecies differ behaviorally; if so, there is a possibility of genetic influence on those behaviors. In the case of artificial selection, animals are bred to produce strains with differing characteristics, including behavior; comparisons among these strains are also important sources of behavioral genetic information. Because an animal's behavioral phenotype is determined by a combination of environmental and genetic factors, scientists can estimate genetic influences by eliminating environmental variation in the animal's life; if they are successful in doing so, then any between-strain differences can be attributed to genetics. For instance, if different subspecies are kept in shared or identical environments, but behavioral differences nonetheless persist, then one must seriously consider the genetic basis for those differences.

This procedure, called a common garden technique, is frequently used for testing the hypothesis that there is an underlying genetic basis for phenotypic differences (see the Case Study and key term on page 69 for more on this technique). There are several ways to eliminate environmental variation. Moving the animal to a new environment and asking if a behavior persists is one way. If behavior changes with the change in environment, then a strong genetic influence on behavior is doubtful. However, conclusions are more difficult to come by if the behavior persists. Instead of using translocated parents, another method of eliminating environmental variation is to focus on offspring (F1 individuals) reared in the shared environment. This approach may be informative if the offspring are not subject to parental influence, but if parent–offspring interactions are important in shaping the behavior of the F1 generation, then a more sophisticated approach, such as cross-fostering or planned matings between strains (see below), may be required to test the hypothesis that behavioral differences among ecotypes have a genetic basis.

The gray wolf (*Canis lupus*) is an excellent example of a species with population-level behavioral differences that may reflect selective effects of differing environments on those populations. Prior to the spread of humans from Europe into North America, wolves were widely distributed on the continent and could be easily separated into subspecies based on their habitats. The gray wolf ranged across the north-central part of the continent, while the Mexican wolf (*Canis lupus baileyi*) was found in the arid Southwest and was half the weight of the gray wolf (see Figure 3.3). Thus, gray wolves are widespread, and we find much more

FIGURE 3.3
A gray wolf (left), a Mexican wolf (center), and an Arctic fox (right). *Photos: (Left) Frank Wendland, W.O.L.F. Sanctuary, www.wolfsanctuary.net, www.facebook.com/wolf.sanctuary; (Center) U.S. Fish and Wildlife Service; (Right) Keith Morehouse, U.S. Fish and Wildlife Service.*

behavioral variation among such widely distributed populations than among populations with more limited distribution and narrower habitat requirements (e.g., the Arctic fox, *Alopex lagopus*, which is specifically adapted to cold northern climates).

Larger patterns of behavioral evolution can sometimes be visualized by superimposing behavior on a phylogeny or cladogram of a taxonomic group. A phylogeny is a "tree" that expresses supposed relationships among taxa (see Chapter 1). Phylogenies can be constructed in a variety of ways and may follow the intuition of the scientist studying the evolution of the taxa in question. A cladogram is also an evolutionary "tree," but it is a well-defined hypothesis about relationships that is constructed following explicit rules about the analysis of evolutionary relationships; oftentimes, computers are required to implement the "cladistic analysis" of large data sets composed of multiple traits belonging to many species. Cladograms are typically constructed using a set of morphological characteristics, molecular characteristics, or both. Behavioral characteristics can be included in a cladistic analysis, but investigators interested in the evolution of behavior usually avoid circular reasoning by constructing a cladogram using morphological or molecular (nonbehavioral) characteristics and then observing how behavioral patterns fit into the cladogram. Cladistic analysis is most important in helping to test whether a particular behavior or syndrome of behaviors evolved only once in evolutionary history or if it evolved multiple times (convergent evolution). It can also help piece together the evolutionary sequence of events that lead to a behavior; for instance, there may be a shift from a simple behavioral pattern in basal species in the tree to a complex behavioral pattern in more derived species, or the behavior may have become simplified over evolutionary time, with derived species exhibiting only the core attributes of a behavior.

The same principles of cladistic analysis apply to the study of social evolution in sweat bees (family Halictidae). Some species of sweat bees are solitary; females establish nests of their own and do all the work without assistance. Others are social; several females

OF SPECIAL INTEREST: BIRD'S NEST SOUP—THE EVOLUTION OF NESTING BEHAVIOR IN SWIFTLETS

Swiftlets are abundant in South and East Asia. The nests of some swiftlet species are used as the basis for bird's nest soup in Asian cooking. Many swiftlet species build nests that are glued to a tree or a rock face (see Figure 3.4A). The glue is a salivary secretion that is alleged to have marvelous nutritional and medicinal properties and is the key ingredient in the expensive soup. In a few species, the main nesting material is small twigs or feathers. In other species, the glue gains more prominence as a construction material, and in a few species, the glue is the predominant material. Lee et al.[20] chose four characteristics of swiftlet nests: whether glue or a rock ledge serves as the primary support for the nest, whether feathers are used in construction, whether twigs or other vegetation are used, and proximity of nests to each other (whether the birds nest in colonies).

FIGURE 3.4A
Variation in swiftlet nests. Some are made almost entirely with salivary glue and adhere to a rock, whereas others sit on rock and are made of twigs or other plant material.

They then constructed a cladogram for the bird species and examined how well the behavioral variables fit with the pattern of swiftlet evolution (see Figure 3.4B). If the use of salivary glue as the main support for the nest had evolved only once in swiftlets, the expected result would be for this behavior to appear on only one branch of the cladogram. In fact, the behavior appears on three different branches. This evidence shows how a behavior can evolve more than once within a single group of related species.

FIGURE 3.4B
A cladogram of swiftlets, with the behavioral traits shown on the right. *Source: Lee, PL, Clayton, DH, Griffiths, R & Page, RD 1996, Does behavior reflect phylogeny in swiftlets (Aves: Apodidae)? A test using cytochrome b mitochondrial DNA sequences. PNAS, Vol. 93, pp. 7091–7096. Photo: Dale Clayton.*

(usually a mother and her daughters or a group of sisters) occupy a nest, and labor in the colony is divided among the colony members. Did this shift from solitary nesting to complex social behavior happen only once in the course of sweat bee evolution, or has it happened repeatedly? Is it possible for the reverse to happen—for a solitary species to be derived from social ancestors? The cladogram (see Figure 3.5) helps to answer these questions; social behavior has evolved more than once in sweat bees, and there is at least one case of reversal, with a solitary species derived from social ancestors.[21]

Phylogenetic inertia is our final stop in thinking about the genetics that underlie the ultimate causes of behavior. Why, for example, does a dog turn around a few times before lying down (see Chapter 1, page 12)? The adaptive roots, or ultimate causes, of this behavior may lie in ancestral preparation for sleep by trampling vegetation or in checking for potential parasites; in a domestic dog, the behavior is charming but meaningless. It is always possible that an observed behavior or phenotype had a function in evolutionary history that is now lost; the behavior persists because it is engrained in the genetic instructions the animal receives from previous generations.

> **KEY TERM** Phylogenetic inertia is the carryover of traits that evolved in previous habitats, even if those traits have little relevance in the current habitat.

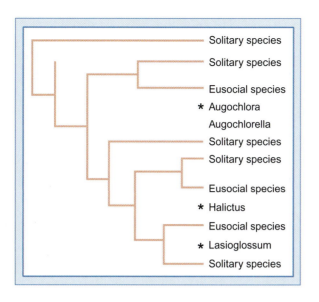

FIGURE 3.5

A cladogram of sweat bees, showing the points at which the worker caste has evolved. This cladogram shows that even very complex behavior can evolve multiple times within a larger taxonomic group. *Source: Brady, S.G., Sipes, S., Pearson, A., & Danforth, B.N., 2006, Recent and simultaneous origins of eusociality in halictid bees, Proceedings of the Royal Society B-Biological Sciences, 273 (1594), pp. 1643–1649.*

Proximate Causes and Correlations: Whole Organism Studies of the Genetic Bases of Behavior

CROSS-FOSTERING

Cross-fostering is a simple and highly effective technique for separating the effect of rearing environment from genetic (instinctive) influences on behavior. Consider two species of swiftlet: species A, which grows up in nests that resemble weaving, and species B, which grows up in well-constructed mud cups. If members of species A are reared in nests of species B, then how they build their own nests will depend on whether experience is important shaping nest-building behavior. If the behavior is entirely genetically programmed, then species A animals will weave nests, even if they are reared in a species B nest. If the behavior is shaped by learning and experience, then species A will learn and adapt to species B's mud construction.

The basic procedure of *cross-fostering* involves transferring some newly born or hatched young of species A from their parental nest to the nests of species B. Ideally, if there are multiple offspring from a given mating, the clutch or litter would be subdivided, with some going to B, and some staying home with A. The converse happens as well—that is, some young of species B are moved to homes with A, while others remain with species B. This results in four groups: species A raising A, species A raising B, species B raising A, and species B raising B. This sets the stage for asking what happens to the behavior under these four conditions. Does the behavior reflect their genetic background (the behavior of their parents) or their social background (the behavior of the family that reared them)? After a period of time, during which the behavior of the focal animals develops, the behavior of the transferred (fostered) animals and the non-fostered controls is documented, measured, and compared. If the

KEY TERM Cross-fostering is transfer of young between mothers. This technique is used to separate genetic from environmental influences on behavior.

79

transferred animals behave like the controls of their species, then the usual conclusion is that genetics dominates, and if they behave like their adopted "family" in the host nest, then environmental influences are predominant.

A recent study highlighted the utility of cross-fostering experiments in separating genetic from environmental (learned) components in bird song (see Figure 3.6).[22] Johannessen and his colleagues wondered if great tits and blue tits, both common in Europe, learned songs from tutors (parental birds) in their own nest. Because these species nest in the same habitat, it would be possible for young birds to pick up songs from birds in their own nest, from neighboring nests of the same species, or from neighboring nests of other species. The investigators located nests and transferred some eggs between nests of the two species. This resulted in a classical cross-fostering design, in which treatment individuals

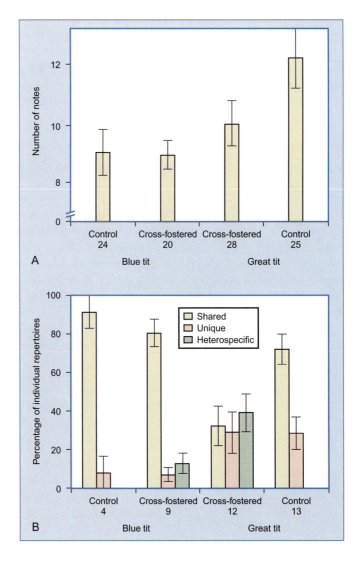

FIGURE 3.6

Results of cross-fostering experiments between blue and great tits in which song-learning of the nestlings was measured. Cross-fostered birds use songs that are intermediate between the two species. (A) Cross-fostered great tits have fewer notes and shorter songs than control birds of the same species. (B) Heterospecific songs appear as part of the repertoire of cross-fostered birds of both species. *Adapted from Johannessen et al., 2006, Animal Behaviour 72: 83–95.*

are reared by birds of other species and control animals are reared by their own species. Because the experiment involved two species, the test had the added dimension of allowing the scientists to ask if song learning in these birds is flexible enough to accommodate songs from a different species. Blue tit songs were very similar between control and cross-fostered birds, whereas the songs of cross-fostered great tits took on many of the characteristics of blue tit foster parents. Elements of the song of the biological (same species) parent remained, but because the experiment was conducted in the field, the results do not indicate if these song elements are genetically derived or if they are learned from birds in neighboring nests. Further experimentation using a cross-fostering design in a more controlled setting would sort this out. After leaving the nest, cross-fostered birds in this experiment appeared to prefer the company of their foster species to that of their own species, suggesting that song learning is part of a larger syndrome of species-specific imprinting (see Chapter 5).[23]

DISCUSSION POINT: CROSS-FOSTERING

What needs to be done to make sure a cross-fostering experiment is adequately controlled so that genetic and environmental components of a behavior can be separated?

Although intrauterine conditions are often not considered when studying environmental effects, they are some of the first environmental conditions than an animal encounters. Can the intrauterine environment affect the later behavior of animals? (See Chapter 2 for more discussion of this topic.) How can maternal influence and genetic effects be distinguished? Cross-fostering can play a role here as well. Embryos implanted into recipient uteri can serve as a control for prebirth effects of maternal physiology on the behavior of offspring. This technique has been used in mice, with the intriguing result that cross-fostered embryos behave like the recipient strain of mice, yet if cross-fostering is performed after birth, then the mice behave like their biological parents (see Figure 3.7). This result does not invalidate cross-fostering experiments done after birth or hatching, but it does raise the interesting caution that prenatal effects need to be considered when designing cross-fostering experiments. In humans, some investigators think that altered maternal well-being (e.g., stress) changes the uterine environment, resulting in lifelong effects on the behavior of the offspring.

In other mammals, placental position can have the same sorts of effects on adult behavior; in mammals with three or more embryos in the uterus, offspring behavior after birth is affected by relative position (front to back) in the uterus and by the gender of adjacent embryos in the uterus. The influence of maternal effects and interactions between embryos on gene expression in offspring is one example of "epigenetic" effects, that is, variable functions of genes that do not involve actual changes in DNA. This variation can be caused by a broad array of environmental influences that are only beginning to be understood but that are clearly critical to attempts to parse the phenotypic contributions of "nature" and "nurture" (see Section 2.3 for more on this topic).

FIGURE 3.7
B6 mice embryos were implanted in mice of two strains (B6 or BALB). The offspring, which were all genetically identical, were raised by either B6 or BALB mothers. The only offspring that exhibited BALB behavioral traits were those that had been both prenatally and postnatally fostered by BALB mothers. *Adapted from Crabbe and Phillips, 2003, Nature Neuroscience.*

81

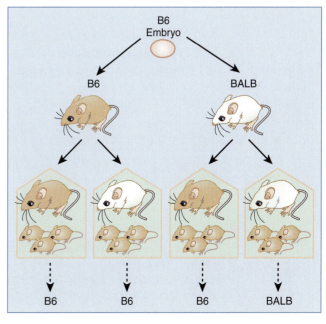

TWIN STUDIES

Twin studies are similar to cross-fostering tests of the relative importance of genes and environment in shaping behavior. These studies are most commonly performed in humans, for whom cross-fostering experiments are ethically questionable. They take advantage of the fact that sometimes human twins are monozygotic (MZ) and sometimes they are dizygotic (DZ). Monozygotic twins come from the same fertilized egg and are genetically identical (see Figure 3.8); DZ twins come from separate fertilized eggs and thus are no more related to each other than if they were siblings born at different times. Assuming that both twins in a pair experience the same rearing environment, then experimenters can compare similarities (concordances) and dissimilarities (discordances) in behavior between MZ and DZ twins; if MZ twins show greater degrees of behavioral concordance than DZ twins do, those differences are probably due to genetics because rearing environments are controlled. For instance, Deater-Deckard and colleagues recently found that behavioral attributes such as task persistence, anger/frustration, and conduct problems in children were much more similar in MZ twins than in DZ twins, although environmental influences also have substantial importance. Perfect concordance of behavioral traits is rare even in MZ twins, suggesting that genes are not the sole influence on behavior; environment and chance play large roles as well. Monozygotic twins occur in at least some nonhuman primates[24] and perhaps in other animals, so the possibility of using twin techniques for behavioral studies extends beyond humans, although detection of MZ offspring in nonhumans requires genetic testing.

> **KEY TERM** Twin studies can segregate environmental from genetic influences because twins are genetically identical, but may be reared in different habitats.

Fertilization   Two cell stage   Morula

Blastocyst   Blastocyst splits to form two embryos

FIGURE 3.8
Embryogenesis of monozygotic twins.

82

CLONED ANIMALS

Cloned animals offer a major potential for application of the principles of twin analyses in unique ways: a clone offers the opportunity to observe the genetic equivalent of MZ twins born at different times! If a favorite pet is cloned, will the clone's behavior resemble the original pet's behavior in the desired ways? Human monozygotic twins are often quite similar behaviorally, sometimes eerily so in behavioral quirks and mannerisms, but also can differ in significant ways; if cloning of pets gains widespread acceptance, it will be interesting to see how well pet owner expectations are met.

> **KEY TERM** Cloned animals are genetically identical. Use of cloned animals extends the idea of twin studies by allowing larger sample sizes and more controlled conditions.

ARTIFICIAL SELECTION

Artificial selection and *inbred lines* allow exploration of behavioral genetics by testing the responses of behavior to selection or to reduction of genetic variation. Recall from Chapter 1 that genetic variation is necessary for either natural or artificial selection to produce shifts in gene frequencies, and the

> **KEY TERM** An inbred line is a population in which closely related animals, such as siblings or parents and offspring, have been repeatedly mated so that nearly all genetic variation is lost. This is similar in effect to cloning.

only traits that can be selected are those found within the range of variation genetic variation present in the population. The potential for selection to modify a trait is assessed by measuring its heritability (see below), one estimate of genetic variation.

Artificial selection, in scientific laboratories and in animal husbandry, has dramatic effects on behavior. Perhaps the broadest range of artificially selected behavior is seen in domestic dogs, which display a wide variety of behavioral attributes. These behavioral patterns are the result of selection for dogs that assist humans in work (e.g., retrievers, shepherds) or as companion animals. Most domestic livestock (such as chickens, horses, cattle, sheep, goats, and swine) reflect the results of artificial selection for manageability in confinement, ease of training, and docility (see Figure 3.9).

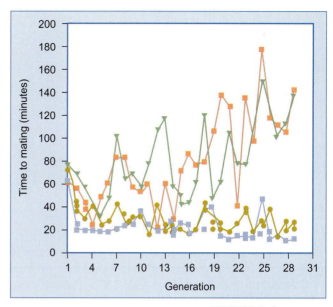

SINGLE-GENE MUTATIONS

A single-gene mutation is a genetic change in just one gene. Such a mutation, spreading rapidly through a broad geographic area, has dramatically changed the social behavior of the black imported fire ant, *Solenopsis richteri*, and the red imported fire ant, *Solenopis invicta* (both introduced via cargo ships to the United States in the twentieth century). The behavior of animals in their natural setting results from many generations of natural selection; the results of this natural selection are observed more often than is natural selection itself in action. Examples of rapid evolutionary change in behavior are relatively rare, but the introduction of exotic and animals (and plants) into ecosystems creates new opportunities for natural selection to act. Animals in new (to them) environments face dramatically changed regimes of selection, which can lead to strikingly rapid evolutionary change.

Colonies of *Solenopsis invicta* are normally monogynous, meaning each colony has a single queen and the workers in the colony are daughters of that queen. Growth of monogynous colonies is limited by the number of eggs this single queen can lay, and by how much food can be collected to feed to the developing larvae. For this ant in its native habitats in Argentina and southern Brazil, the reproductive capacity of a limited number of queens is adequate for colonies to be competitive. Once this species was introduced into the southeastern United States, it encountered a habitat devoid of its natural predators, parasites, and competitors. These conditions favor colonies with the capacity for explosive growth. This growth can be attained by having many queens per colony (polygyny), each laying large numbers of eggs.

The difference between monogynous and polygynous fire ants lies in the expression of a single gene, Gp-9, that codes for a pheromone receptor molecule. The mutation appears to rob the worker ants of the ability to discriminate among queens (see Figure 3.10); consequently, they tolerate a large number of queens, including queens that are genetically unrelated to each other or the workers. This is an excellent example of how a change in a single genetic component that underlies a complex social system can have major effects on the function of the entire social structure. Recent discoveries of the polygynous form of *Solenopsis invicta* in Taiwan, Australia, and China underscore the effectiveness of the mutation in facilitating ant invasions.[25]

Another invasive ant, the Argentine ant (*Linepithema humile*), has been highly successful, possibly because it exhibits reduced intraspecific aggression. A genetic study using microsatellite DNA markers showed that these ants experienced a population bottleneck

FIGURE 3.9
The response of fruit flies to artificial selection for mating speed. Each symbol represents a line of flies selected for high or low speed. Over generations the lines diverge; after roughly 7–10 generations the differences are apparent and by the 30th generation the differences are extreme. The end-of-chapter discussion of the application of microarrays to behavioral genetics will include this example. *Adapted from Mackay T.F., et al., 2005 Proc Natl Acad Sci U S A May 3 102 Suppl 1:6622–9*

FIGURE 3.10
A fire ant queen with workers and eggs. *Photo courtesy of Sanford Porter/USDA-ARS.*

during introduction to California that reduced genetic diversity. This is consistent with the hypothesis that reduced genetic variability may impair nestmate or colony recognition, allowing the invaders to behave like one large colony. This, in turn, allows them rise to higher densities than they do in their native habitat.[26]

The fruit fly, *Drosophila melanogaster*, is the workhorse of laboratory genetics.[27] It is raised easily in large numbers in the laboratory and has a fast generation time. This makes it a good "model system" (see Chapter 1). Many genetic mutations affect fruit fly mating behavior. Male behavior is inhibited by genes named *Nerd* and *fruitless* and is enhanced by *Voila*. Female receptiveness to mating is impaired by genes such as *dissatisfaction, spinster,* and *chaste*. The ways that these genes might affect courtship and mating become more apparent when the sequence of courtship behavior and related sensory modes are considered. Visual, auditory, and chemosensory systems may all be affected.

For instance, mutations that affect the visual system reduce fruit fly male mating success by making it difficult for males to find females (see Figure 3.11). Auditory mutations make females less receptive to mating because they cannot hear the male courtship song. Chemosensory mutations can affect the courtship performance of sexes. In fact, *Voila* seems to affect chemosensory cells in the male's front legs. Mutations that affect processes in the brain are even more interesting because they can give insight into the organization of the central nervous system and the generation of complex behaviors such as mating. Fruit flies offer insights into this level of behavioral control as well.

Orientation   Following   Single-wing vibration

Attempted copulation   Copulation

FIGURE 3.11
Drosophila courtship; these are examples of behaviors that mutations may affect.

HERITABILITY

Heritability provides a powerful tool for measuring the variation in a behavior and for allocating it to variation correlated with genes or variation correlated with environment.[28] Heritability is the proportion of phenotypic variation in a population due to genetic variation. *Variation* is the operative word here: It is important to understand and remember that heritability is all about variation such that studies of the heritability of a behavior are actually studies of the variability of that behavior within a population.

Understanding heritability and how it is used rests on the following key concepts:

1. Heritability is a population-level measure, not an individual measure; only populations have heritability.
2. Heritability is NOT a measure of the degree of genetic control of a behavioral trait.
3. Heritability can differ among environments; the same population may show a different heritability for a trait if the environment is changed.
4. The behavior in question must vary among individuals in the population for it to have a measurable heritability; invariable behavior has no heritability.
5. Strong selection (natural, sexual, or artificial) on a behavioral trait reduces the heritability of that trait (because it reduces the variation).

Keep these principles in mind through the following discussion of heritability.

A recurring theme in this chapter has been the fact that phenotype is determined by the collaboration of an animal's genes and the effects of its environment, past and present.

84

The study of heritability allows students of behavior to parse those influences: heritability indicates how much of the variation in phenotype can be ascribed to genetic variation. Consider a population of animals and its heritability, using the following notation:

V_p = variation of the phenotype
V_e = variation due to the environment
V_g = variation due to genetic effects

With these three notations, a simple equation can be generated that describes the relationship among phenotype, environment, and genes: $V_p = V_g + V_e$. This is a restatement of a central concept: phenotypic variation is the sum of environmental and genetic variation.

There are two types of *genetic variation* (V_g): additive and nonadditive. Recall that animals can differ genetically because they have differing alleles (forms of a gene) at loci (locations on the chromosome) that influence the trait in question. Usually, the effects of the different alleles working together is the sum of their activity, so genetic variation due to allelic differences is called *additive genetic variation* (V_a). Sometimes one allele is dominant to another in determining the effect of a locus, or loci may interact in ways that change the action of the genes (*epistasis*). Variation due to dominance, epistasis, and other types of interactions among alleles and loci is *nonadditive* (V_{na}).[29] Total genetic variation, V_g, is the sum of the additive and nonadditive components, $V_g = V_a + V_{na}$.

Taking another approach, recall that heritability is the proportion—or percentage—of the phenotypic variation that can be attributed to genetic influences. Like any other percentage, to get this proportion, divide the genetic variation by the phenotypic variation (V_g/V_p). This yields a measure called broad-sense heritability. Notice that *broad-sense heritability*, H^2, involves both additive and nonadditive genetic variation. When only additive genetic variation is used in the calculation, then the result is *narrow-sense heritability*, $h^2 = V_a/V_p$.

Given this explanation, how might heritability be used in studies of animal behavior? What have studies of heritability of behavior found? Heritability is useful in two ways:

> **KEY TERM** Genetic variation is a measure of the variation in phenotype that is due to all variation in genotype.

> **KEY TERM** Additive genetic variation is the proportion of genetic variation that is due to simple additive effects among genes.

> **KEY TERM** Nonadditive genetic variation results from interactions between genes and from gene dominance.

> **KEY TERM** Epistasis occurs when genes interact.

> **KEY TERM** Broad-sense heritability is the proportion of phenotypic variation that is explained by genetic variation.

> **KEY TERM** Narrow-sense heritability is the proportion of phenotypic variation that is explained by additive genetic variation.

85

- Broad-sense heritability is used as a measure of the magnitude of genetic influences on a trait.
- Narrow-sense heritability is particularly useful in predicting how animals will respond to artificial or natural selection. If a trait has a high heritability, selection or controlled breeding can change that trait because high heritability is based on a high level of genetic variation that can be subject to selection. (Remember, low genetic variation does not give selection much "wiggle room.")

BRINGING ANIMAL BEHAVIOR HOME: HERITABILITY AND THE BEHAVIORAL GENETICS OF DOGS

The fact that much of dog behavior has genetic underpinnings is patently obvious. Differences in temperament and ability among breeds are well known. These differences are generally associated with the purposes for which the breed was developed. This brief summary, which is typical of how dog breeds can be classified (see Figure 3.12), suggests some of the traits that have been subject to selection:

1. Sheepdogs (shepherds, collies, and the like) are selected for their keen ability to focus on sheep or cattle, their ability to manipulate the behavior of these animals, and their ability to learn and follow their handler's commands. Some think that much of the herding behavior is derived from predatory behavior, but the actual killing behavior has been suppressed.

2. Terriers are energetic hunters, very attracted to small animals. They are quite willing to follow their prey down burrows, hence their name—derived from *terra*, or earth—not terror, as some owners insist. They dig.

3. Scent hounds (including beagles, bassets, fox and coonhounds) are able to behaviorally exploit their keen sense of smell in tracking prey.

4. Retrievers (including labrador and golden), known also as gun dogs, are selected for retrieving ability. A specific behavior that has been selected is a "soft mouth," the ability to handle prey without damaging the item or attempting to consume it.

5. Companions and Toys (miniature and toy poodles, Pekingese, Chihuahua) display behavioral traits that make them attractive household pets.

6. Sighthounds (afghans, borzois, greyhounds) use their distance vision to track prey. They are also selected for high running speed and endurance.

FIGURE 3.12

The domestic dog, *Canis familiaris*, is strong testimony to the variation that can exist within a biological lineage.

In general, dog breeds that are recognized by groups such as the American Kennel Club (AKC) "breed true." This means that pairing any male and any female in the breed will result in pups with the breed-specific characteristics. If you think about this, you will recognize that the only way to accomplish "breeding true" is through reduction of genetic variation related to those breed-specific traits. Any one breed of dogs will have less genetic variation within the breed than you find if you look across all dogs. At some point, breeding of this sort will eliminate most or all of the additive genetic variation in the breed for some traits. At the point at which the additive genetic variation for a trait has been exhausted, no further "improvement" of the breed (depending on the trait) is possible even through carefully designed pairings.

Thinking about how heritability is calculated, it follows that within any dog breed you would expect to find low heritabilities, particularly for the traits that are thought to characterize the breed. Does this mean that the traits do not have a genetic underpinning? No—not at all. It simply means that further attempts to select for the trait will be futile. Following this line of reasoning, you would expect to observe higher heritabilities for traits (including, of course, behavioral traits) if you include a variety of dog breeds in a study, and lower heritabilities if you focus on only one dog breed.

Most measures of heritabilities of dog behavioral traits are from studies of single breeds. A wide range of behaviors have been measured, such as "willingness," fighting the leash, hare tracking, and "obedience." Nearly all of the studies were performed within breeds (the alternative would be to do controlled matings between breeds). Generally, heritabilities for behavioral traits range from 0 to 0.25. Heritabilities for personality traits of German Shepherds are 0.24, or less.[30]

Does this mean, then, that mixed-breed dogs have more hybrid vigor? A careful breeder of purebred dogs will have used genetic testing and careful pedigree scrutiny to minimize the occurrence of genetic diseases in her kennel's lineage. If the same scrutiny is not applied to the production of a mixed-breed litter, there is no reason to expect such a healthy outcome. A puppy-mill breeder of purebred dogs will often use inbreeding and back-crossing, with no concern for genetics, to produce unhealthy puppies. In short, it is the exclusion (or inclusion) of undesirable genes, and not the heritability of other traits, that contributes to healthy or unhealthy puppies.

87

How is heritability measured? Assessing heritability often boils down to how much parents resemble their offspring, or offspring resemble each other. Generally, heritability is calculated by looking at associations between the expression of the trait in families. Mechanically, this is easy to do; the average value of the two parents (not surprisingly, this is called the *midparent value*) is graphed on the x-axis, and the value for the offspring goes on the y-axis. If enough parent–offspring pairs are analyzed, the slope of the resulting line is the heritability. In some cases for traits that relate to sexual behavior, such as antlers in deer, the trait is expressed in only one gender; if single-parent values, rather than midparent values, are used, the heritability is twice the slope. Similar techniques can be used if data are available for groups of siblings. The result is a number between 0.0 and 1.0, with the extremes being no phenotypic variation due to additive genetic variation (zero or 0.0) and no phenotypic variation associated with environmental variation (1.0). The most common way of calculating heritability of a trait is by using linear regression analysis, a statistical technique that yields the equation for a line describing the effect of one variable on another. When the phenotypic values for offspring are regressed on the parental values, the slope of the resulting line is the heritability.

Another way of looking at heritability is to explore how a behavioral trait responds to either natural or artificial selection. If selection on the trait results in change from generation to generation, this suggests that the trait is heritable because selection can act only if there is genetic variation for the trait. On the other hand, if selection has no effect on a trait, then its heritability is probably low. Another way of saying this is that a heritable trait is a selectable trait.

OF SPECIAL INTEREST: CALCULATING HERITABILITY

Figure 3.13 shows the heights of 50 male and 50 female students, in centimeters. We also asked these students to give us their parents' heights. We averaged the father's and mother's height; this gives us the *midparent* value for each student.

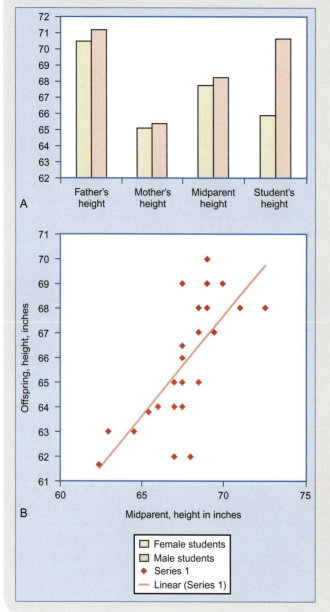

FIGURE 3.13

(A) Bar graph of male student heights and female student heights compared to the height of their parents. Midparent is the average of the two parents' heights. (B) Regression analyses of the male and female student heights against their midparent values. The slope of the regression line and the heritability is 0.82.

Heritability studies have many applications in animal behavior (see Figure 3.14). Indeed, using artificial selection (10 generations) and house mice, Swallow and colleagues[31] showed that voluntary wheel-running could be increased by 75%. Interestingly, these increases were expressed as increased revolutions per minute, not increased minutes per day, meaning that the mice did not run longer, but ran faster.

Following these thoughts about heritability and selection, it is reasonable to predict that strong selection reduces heritability to zero. Why? In theory, strong natural selection should eliminate all additive genetic variation because the alleles favored by selection will be the only ones remaining in the population. Without additive genetic variation, the narrow-sense

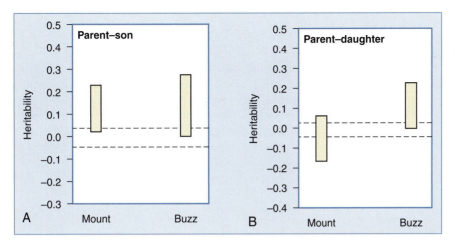

FIGURE 3.14
Heritability analysis of courtship in houseflies. For both behaviors, buzz and mount, the heritability is significant for the parent-son analysis, but only buzz is significant for the parent-daughter treatment. *Adapted from Meffert et al., American Naturalist 160: S198–S213 Suppl. S DEC 2002.*

heritability is zero, and no further evolutionary change in the trait would be expected. This theoretical prediction is supported by data in some cases, whereas in others there is substantial heritability of traits that nonetheless seem to have been under strong selection. In animal behavior, one particular concern is what this prediction means for traits related to behavior such as dominance and mate choice.

> **KEY TERM** Balancing selection occurs when extreme phenotypes are at an advantage, so selection favors the maintenance of both phenotypes. For example, very large and very small animals may be at an advantage in mating, so selection would favor these phenotypes.

Why might additive genetic variation persist over evolutionary time in the face of seemingly strong selection? Four mechanisms commonly explain this (refer to the evolution primer in Chapter 1):

- *Balancing selection.* This type of selection occurs when more than one phenotype is favored. It could occur among animals in competition (as in calling males and satellite males in a mating chorus) or between generations, as when environmental conditions shift, so that more than one phenotype is favored over time.
- *Epistasis.* These are interactions among genes. While one phenotype may be favored, the alleles' effects on other genes may be deleterious, preventing fixation of the allele. Fixation is elimination of all but one allele; when a character or trait is fixed, no genetic variation remains.
- *Correlated characteristics in males and females.* If a trait that is important in one sex, either for competition for mates (as in horns or antlers in male ungulates) or in mate choice, is produced by both sexes, then the deleterious effect on the other sex may outweigh selection for maximizing the trait. For example, in barn swallows tail length varies in both males and females but is a sexually selected trait only in males (see Figure 3.15).[32]
- *The handicap principle.* If producing a phenotype is costly enough, selection against extreme individuals may counterbalance selection favoring that phenotype. This is a form of balancing selection, driven by the cost of the phenotype. The basis of this idea has been most strongly promoted by Amotz Zahavi, but it has gained general acceptance in this somewhat modified form.

FIGURE 3.15
Male and female barn swallow tails.

What does high or low heritability mean, in terms of history of selection? Low heritability results from strong selection on a trait, so key features for an animal's survival are expected to have low heritability. High heritability, correspondingly, reflects a more benign selective environment; genetic variation is tolerated if selection is not high. The calculation of heritability of traits in Japanese quail is an outstanding example of the use of heritability in a behavioral study.

> **KEY TERM** The handicap principle states that production of a phenotype may be costly, making that phenotype an honest signal of an animal's condition. An honest signal is a phenotype that conveys information about the signaler's genotype or current condition. Dishonest signals may deceive the recipient about the genotype or condition of the signaler.

89

OF SPECIAL INTEREST: DOMINANCE AND HERITABILITY IN JAPANESE QUAIL

Both male and female Japanese quail (*Coturnix japonica*) start pecking other birds soon after hatching. In adults, males and females establish dominance hierarchies (see Chapter 9). In an interesting study, heritability for a large number of behavioral traits was determined in Japanese quail (see Figure 3.16). Dominant males have more mating success, supporting the hypothesis that strong selection has reduced additive genetic variation for dominance in males. Dominance rank has a higher heritability for males than for females. Heritability may be maintained in females because the trait has no fitness consequences or because there are balancing factors.[33]

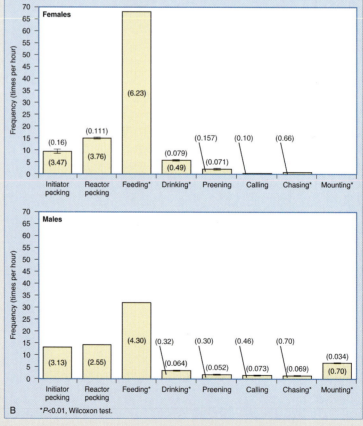

FIGURE 3.16

Heritabilities of behavioral traits of Japanese quail (*Coturnix japonica*). In young birds pecking is heritable in both sexes (0.74 overall). In adults, however, heritability for pecking is high in females (1.33) but 0 in males. Other traits have moderate or low heritabilities. *Adapted from Nol et al., 1966, Animal Behaviour 52: 813–820.*

QUANTITATIVE TRAIT ANALYSIS

What is a quantitative trait? A quantitative trait varies continuously, in contrast to qualitative traits, which have discrete forms. Another aspect of a quantitative trait is that it is determined by a number of genes acting together. A good example of a quantitative trait is height in humans; "normal" adult humans range in height over a span of more than half a meter (more than 2 feet). Human height is greatly influenced by genes, but no one gene is solely responsible for height. Usually, quantitative traits are normally distributed; a graph of the trait results in a bell-shaped curve.

In these cases, a normally distributed trait, whether it is physical, such as height, or behavioral, such as pollen collection, is usually the result of the contribution of a number of genes. Recent advances in genetics now allow scientists to map the genes having the greatest influence on the trait. These genes, as described previously, are called quantitative trait loci, or QTLs. QTL analysis is now a standard format for studying genetic influences on behavior.[34,35]

It follows that a QTL is a location on a chromosome that is thought to regulate an organism's phenotype for quantitative trait. Other experimental approaches will give the same result, but all QTL analyses rely on a linkage map and good behavioral measures. Investigators find numerous markers such as generally single-nucleotide polymorphisms (SNPs), amplified fragment length polymorphisms (AFLs) or random amplification of polymorphic DNA (RAPDs) and determine their linkage group map locations by cross-breeding.

OF SPECIAL INTEREST: HOW IS A QTL ANALYSIS DONE?

To perform a QTL analysis, a behavioral biologist needs three critical sets of information: a linkage map, a behavioral assessment, and a breeding experiment.

First there must be a linkage map of the genome of the study species. This is a significant limiting factor for QTL analyses. In construction of a map, the first step is to find physical (such as the bands on *Drosophila* chromosomes) or molecular markers scattered throughout the animal's genome. This map needs to be well saturated; that is, the markers must be distributed evenly and frequently enough to assort during recombination.

Many types of molecular markers are available; the linkage map for honeybees, which will be our main example in this section, was constructed using RAPDs; these are small DNA segments that can be identified using polymerase chain reaction (PCR) techniques.

To visualize how the data are analyzed, think of a chromosome, remembering that in most animals chromosomes come in pairs. The scientist has genetic markers (short sequences of DNA) scattered along the chromosome; at a given location, the DNA sequences of these markers vary between the two copies of the chromosome. During meiosis, the two members of the chromosome pair cross over and segments of the chromosomes are exchanged, resulting in genetic recombination. Think of two different markers along the chromosome.

Most importantly, the chances of recombination occurring are high if the markers are far apart, and low if the markers are close together. Now what do you think happens to genes that regulate behavior? If they are close to a marker, they likely stay linked with that marker during recombination. If they are far from a marker, then they are much less likely to stay linked. Hundreds, perhaps thousands, of markers are needed to produce a well-saturated map of an animal's genome. In honeybees, mapping has been facilitated by the fact that recombination rates are much higher than in most organisms.

Second there must be a good method available to measure how the behavior varies among animals.

Third all that remains is a breeding experiment. Typically, animals whose behavior has been measured and that have been genotyped are mated. These crossed animals with different behavioral phenotypes allow us to follow how the behavioral phenotype is correlated with the markers. The behavior and genotypes of their offspring are then determined, as well.

The following questions are usually addressed in a QTL analysis:

- How many genes influence the expression of a quantitative trait?
- What is the level of influence of each gene on the trait?
- Where are the genes located on the chromosomes?
- What is the function of each gene?

Answers to these questions will be discussed in the following section. Note that the last question, about function, is the most difficult to answer.

Using a rather elaborate computer program, scientists can ask whether the variation in behavior is correlated with each marker. They do this by focusing on one "family" (a male, female, and their offspring) and asking if the behavior is always high with a certain sequence of a marker and always low with another sequence. In that case, the gene for the behavior is likely to be close to the marker on the chromosome. If the variation in the behavior is more or less random in relation to the marker, then there is little or no linkage between the behavior and that chromosomal location. Most quantitative traits, including behavioral traits, have high correlations with from two to four locations in an organism's genome; that is, they are close to a marker. There are, of course, exceptions, but this is a good rule of thumb. The higher the correlation, the more important a gene may be in regulating the behavior. These chromosomal locations probably do not have genes that code directly for a particular behavior, but rather, ones that code for factors that shape the behavior. For example, a behavioral trait may be influenced by three genes, one of which affects activity (or arousal), another that affects sensory perception, and a third that influences latency to respond to a stimulus.

Many behaviors are quantitative traits. Aggressiveness, for example, varies among individuals in a wide range of animals, such as honeybees, rodents, horses, dogs,[36] and various primates. Another good example is expression of play behavior in mammals. Intelligence, to the extent that it can be measured, is also a quantitative trait. Activity levels are particularly good examples of quantitative traits.

Figure 3.17 shows a typical set of activity measurements—in this case, a QTL analysis of aggressive behavior (stinging) in honeybees. The purpose of this analysis was to identify the locations in the genome of loci responsible for the large difference in stinging behavior between the "African" honeybees in Mexico and the gentler bees found farther north. In an attempt to further focus in on the regulation of aggressiveness in honeybees, a later study sequenced the genome around the sting-2 QTL; this is the *candidate gene* approach, which is discussed in more detail in the next section of this chapter. [37,38]

> **KEY TERM** A candidate gene is a gene that has been identified as having a strong possibility of playing a role in regulating a phenotype.[39]

What does knowing the locations of QTLs accomplish? In working with domestic animals and wanting to select for certain behavioral traits, the knowledge of the QTLs would allow the design of an efficient breeding program to achieve the desired result. This has not actually been done in animals, but the principle is commonly applied in crop plants. By knowing which chromosomal locations are important, scientists can, conceivably, track how their genetic manipulations are affecting the regulation of the trait.

BRINGING ANIMAL BEHAVIOR HOME

Canine compulsive disorder (CCD), in which behaviors are executed repeatedly and with some distress, shows a high resemblance to human obsessive-compulsive disorder (OCD). Dogs may lick, tail chase, flank suck, engage in pica (consumption of indigestible substances), and pace or circle. Using SNPs from Dobermans diagnosed with this malady, scientists have recently determined that a gene on chromosome 7 may be responsible for some susceptibility to CCD.[40] Sixty percent of dogs with multiple compulsive behaviors have the allele thought to confer risk (*CDH2*), compared to 22% of control Dobermans. Such studies open the door for early intervention into compulsive disorders and perhaps even treatment.

FIGURE 3.17
(A) Chromosomes and QTLs for stinging behavior in honeybees. (B) The crossing scheme used to map these behaviors, along with LOD analyses. The Roman numerals indicate chromosome number; each QTL for the behavior corresponds to a peak in the LOD analysis. *Adapted from Hunt et al. 1998. Genetics, Vol. 148, 1203–1213.*

93

Proximate Causes and Correlations: Genetic Dissections of Mechanisms Underlying Behavior

The basic goal of searching at the genetic level for correlations with behavior is to uncover the genes that "cause" the behavior. In other words, scientists can use a reductionistic approach to move from the organismic to the molecular and reveal the pathways by which a behavior is regulated. Generally, the first step is to identify candidate genes that are hypothesized to regulate the behavior. QTLs are revisited here because they can be used to develop candidate gene hypotheses. Other candidate gene approaches include microarrays (see page 99) of expressed sequence tags. Once candidate genes have been identified, experiments can determine if, indeed, that gene has a regulatory function for the behavior. Experimental studies establish the role of these genes in actually regulating phenotype.

BRUTE-FORCE APPROACHES TO FINDING CANDIDATE GENES

In some animals, such as fruit flies (*Drosophila melanogaster*), the roundworm (*Caenorhabditis elegans*), and house mice (*Mus musculus*), so much is known about the species' genetics that powerful mutagens, such as certain chemicals or radiation, can be applied to laboratory populations of the animal, and then animals of the next generation can be tested for "interesting" behavioral abnormalities. This technique has been used, for example, to find fruit flies whose biological clocks did not work properly. Once these mutants have been identified, backcrosses with "normal" laboratory strains can be used to map the location of the gene that causes the abnormality. This type of work takes many hours of concentrated laboratory effort because thousands of animals must be mated and tested, but this remains a powerful tool for identifying genes that code for proteins or regulatory elements that affect specific behaviors.

DISCUSSION POINT: CAUSE AND CORRELATION IN BEHAVIORAL GENETICS

We pointed out at the beginning of this chapter that single-gene mutants can give misleading information about genetic regulation of a behavior. A standard technique for producing mutants involves treating animals with either radiation or a chemical mutagen and then screening their offspring for behavioral abnormalities. Genetic investigation can then pinpoint the mutant gene that correlates with the abnormality. Is this proof that the gene "controls" the behavior? Why or why not?

FROM QTLS TO ASSESSING GENE EXPRESSION AND BEHAVIOR

For organisms with known linkage maps (e.g., nematodes, fruit flies, honeybees, mice,[41] humans), QTL analyses have provided intriguing windows into how variation in a small set of genes can explain much of the phenotypic variation in behavioral traits.[42] QTL analyses then open the door to identification of specific genes that are involved with a trait, and move the discussion from whole-organism approaches to looking at proximate causes at the level of the gene. Molecular techniques can pinpoint gene candidates for control of behavior. The exact location on the chromosome of a QTL location is not precisely known; the data show only that it is between two markers. If the distance between the two markers is not too large, molecular geneticists can overcome this problem by sequencing the chromosome from the two markers ("upstream" from one, "downstream" from the other). By comparing the DNA sequences with sequences of known genes, the scientist may determine the identity of the genes between the markers. The scientist can then form a hypothesis about which of the genes in the QTL region has the behavioral effect; this is a candidate gene hypothesis.

FIGURE 3.18
Junglefowl, the wild ancestor of domestic chickens.

A recent study of chickens[43] illustrates how interesting candidate genes are identified. Chickens are the domesticated descendants of junglefowl, a Southeast Asian bird species. Junglefowl (see Figure 3.18) are wilier and less manageable than their domestic relatives, in part because they are more aggressive. In this study, the investigators hypothesized a relationship between genes for growth, identified through QTL analyses, and domestication.

The investigators measured dominance and inspection of strangers (birds new to the social group). Differences among birds in these behaviors suggest that the arginine vasopressin receptor 1a (AVPR1a) gene may be involved in these social behaviors. Thus, AVPR1a is a candidate gene for social behavior in chickens. This is a fascinating discovery because vasopressin is the same hormone discussed in Chapter 2, Part II as being important in pair-bonding and lifetime mating in voles. To verify the function of vasopressin in chicken social behavior, the investigators will now have to perform experiments directly addressing the effects of vasopressin on the behavior, but this finding is certainly highly suggestive. The candidate gene approach is only now coming into its own, in terms of being useful in behavioral investigations, but it holds extremely high promise in drawing links between genes and behavior and in allowing scientists to see similarities in behavioral regulation among species.

MICROARRAYS AND ASSESSMENTS OF GENE EXPRESSION

Are there are different patterns of gene expression between cells of animals with different behaviors? This question can be answered by looking at expressed sequence tags (ESTs). An EST is a tag based on a known functional gene sequence from an animal. An EST can be amplified—reproduced many times—and then be placed as part of a microarray (see Figure 3.19). A microarray holds each amplified EST in a small well, like an indentation on a microscope slide. A sample of tissue from an animal is then stimulated to produce proteins, which are labeled fluorescently. If a protein matches an EST, it binds, and the well becomes a fluorescent dot in the microarray, as shown in Figure 3.19. ESTs can then be compared between tissues, between animals, or between species. In the simplest analyses, the number of ESTs fluorescing is compared—for example, between sleeping and awake animals. But if details about the ESTs and which genes they derive from are known, this analysis can also provide clues about specific genes, leading to candidate gene hypotheses.

RNA KNOCKOUTS

RNA knockouts are a relatively new tool in the exploration of genes and behavior. This technique involves synthesizing oligonucleotides that are complementary to the RNA products of genes thought to influence a behavior. When the synthetic oligonucleotide is introduced into the organism, it binds with the RNA from the target gene, effectively inactivating (or knocking out) that gene. Key to this approach is having a candidate gene with a hypothetical role in regulating a behavior.

95

FIGURE 3.19
An EST microarray, showing strong expression of some ESTs and little or no expression of others.

SUMMARY

Behavioral genetics forms a critical component of nearly all behavioral studies. Understanding genetics establishes a thread that runs from the ultimate evolutionary causes of behavior to the behavior's proximate underpinnings, and knowing the genetic underpinnings of behavior is essential to understanding how behavior evolves. Behavioral genetics also helps in understanding the physiological and neurobiological systems that control behavior. While the approach of this chapter is genetics, most behavior is shaped by a combination of genetic and environmental factors. This chapter and the following two chapters—Chapter 4 on behavioral homeostasis and Chapter 5 on learning—work together to shape a picture of how behavior is controlled.

To understand the evolutionary roots of behavior (ultimate causes), behavioral geneticists often employ phylogeny. Genetics brings a broad range of tools to behavioral investigations. These tools are used to establish the phylogeny of animal groups, which can then be used to understand patterns of evolution for specific behaviors, such as nest construction by birds or colony defense by bees.

Peeling away the layers of genetic and physiological regulation of a behavior using behavioral genetics—starting with differences among species and moving down to the regulation of

26. Suarez, A.V., Tsutsui, N.D., Holway, D.A., Case, T.J., 1999. Behavioral and genetic differentiation between native and introduced populations of the Argentine ant. Biol. Invasions 1, 43–53.

27. Yamamoto, D., Nakano, Y., 1999. Sexual behavior mutants revisited: Molecular and cellular basis of *Drosophila* mating. Cell. Mol. Life Sci. 56, 634–646.

28. Ruefenacht, S., Gebhardt-Henrich, S., Miyake, T., Gaillard, C., 2002. A behaviour test on German Shepherd dogs: Heritability of seven different traits. Appl. Anim. Behav. Sci. 79 (2), 113–132.

29. Meffert, L.M., Hicks, S.K., Regan, J.L., 2002. Nonadditive genetic effects in animal behavior. Am. Nat. 160 (Suppl. S DEC), S198–S213.

30. Ruefenacht, S., Gebhardt-Henrich, S., Miyake, T., Gaillard, C., 2002. A behaviour test on German Shepherd dogs: Heritability of seven different traits. Appl. Anim. Behav. Sci. 79 (2), 113–132.

31. Swallow, J.G., Carter, P.A., Garland Jr., T., 2004. Artificial selection for increased wheel-running behavior in house mice. Behav. Genet. 28, 227–237.

32. Cuervo, J., de Lope, F., Moller, A., 1996. The function of long tails in female barn swallows (*Hirundo rustica*): An experimental study. Behav. Ecol. 7, 132–136.

33. Nol, E., Cheng, K., Nichols, C., 1996. Heritability and phenotypic correlations of behaviour and dominance rank of Japanese quail. Anim. Behav. 52, 813–820.

34. Gleason, J.M., Nuzhdin, S.V., Ritchie, M.G., 2002. Quantitative trait loci affecting a courtship signal in *Drosophila melanogaster*. Heredity 89, 1–6.

35. Ruppell, O., Pankiw, T., Page, R.E., 2004. Pleiotropy, epistasis and new QTL: The genetic architecture of honey bee foraging behavior. J. Hered. 95 (6), 481–491.

36. Netto, W.J., Planta, D.J.U., 1997. Behavioural testing for aggression in the domestic dog. Appl. Anim. Behav. Sci. 52 (3–4), 243–263.

37. Guzman-Novoa, E., Hunt, G.J., Uribe, J.L., Smith, C., Arechavaleta-Velasco, M.E., 2002. Confirmation of QTL effects and evidence of genetic dominance of honeybee defensive behavior: Results of colony and individual behavioral assays. Behav. Genet. 32 (2), 95–102.

38. Hunt, G.J., Collins, A.M., Rivera, R., Page, R.E., Guzman-Novoa, E., 1999. Quantitative trait loci influencing honeybee alarm pheromone levels. J. Hered. 90 (5), 585–589.

39. Hunt, G.J., Guzmán-Novoa, E., Fondrk, M.K., Page Jr., R.E., 1998. Quantitative trait loci for honey bee stinging behavior and body size. Genetics 148, 1203–1213.

40. Dodman, N.H., Karlson, E.K., Moon-Fanelli, A., Galdzicka, M., Perloski, M., Shuster, L., Lindblad-Toh, K., Ginns, E.I., 2010. A canine chromosome 7 locus confers compulsive disorder susceptibility. Mol. Psychiatry 15, 8–10.

41. Gershenfeld, H.K., Neumann, P.E., Mathis, C., Crawley, J.N., Li, X.H., Paul, S.M., 1997. Mapping quantitative trait loci for open-field behavior in mice. Behav. Genet. 27 (3), 201–210.

42. Jordan, K.W., Morgan, T.J., Mackay, T.F.C., 2006. Quantitative trait loci for locomotor behavior in *Drosophila melanogaster*. Genetics 174 (1), 271–284.

43. Wiren, A., Gunnarsson, U., Andersson, L., Jensen, P., 2009. Domestication-related genetic effects on social behavior in chickens—Effects of genotype at a major growth quantitative trait locus. Poult. Sci. 88, 1162–1166.

Homeostasis and Time Budgets

99

LEARNING OBJECTIVES

Studying this chapter should provide you with the knowledge to:

- Appreciate how homeostasis describes the behavioral and physiological processes that animals use to maintain appropriate internal and external environments.

- Use contemporary versions of motivation theory to understand how animals make behavioral decisions to help maintain their homeostasis.

- Employ measures of displacement behavior and redirected behavior, in the form of self-directed behaviors, as powerful tools for assessing animal welfare.

- Understand how biological clocks, pain, sleep, appetites for food or mating, and fear influence animals' choices among possible activities.

- Use time budget analyses to quantify and interpret how and why animals allocate their time in maintaining homeostasis.

4.1 INTRODUCTION

This chapter addresses a seemingly simple set of questions: How and why does an animal behave in a given way at a given time? What causes an animal to stop one activity and start another? What, for example, moves a sleeping dog to get up, stretch, and solicit play or food? What prompts bouts of sexual activity? Why do animals behave one way in the spring and another in the winter?

Beneath these simple questions—all facets of Tinbergen's core issue of causation—lies a wealth of complexity and unsolved puzzles in animal behavior. This chapter introduces concepts that are at the root of much of the contemporary study of animal behavior. Some of the answers to the questions reflect deep evolution—the biorhythms shared by every living thing, for instance.

> **KEY TERM** Drive theory postulates that animals have a store of drive energy that is expended when a task is performed. This explains how animals allocate their efforts among tasks.

Coverage begins with a concept from the early days of ethology: *drive theory*. Drive is an animal's internal motivation to initiate or complete a task. Drive theory is useful in developing ways of measuring *animal welfare* and for assessing techniques for ameliorating the stresses of captivity for animals.

CASE STUDY

Have no Fear

What happens when fear is unnecessary? Fear is useful but restricting. It keeps animals out of trouble but costs them opportunities. A fearful animal may survive but may not forage or mate as often as a fearless competitor. In this chapter we discuss how fear affects animal activity.

Islands are places where fear seems to disappear. Islands tend not to have large predators, so herbivores and small predators live out their lives in peace. Fearlessness in island animals may result from generations of relaxed selection for avoiding predators. Visitors to the Galapagos Islands, including Charles Darwin, have noticed the remarkable "tameness" of the island's animal residents. Antipredator fear can also be a matter of learning; geese and deer famously adjust their behavior during hunting season; late in the hunting season experienced animals assiduously avoid hunting blinds, decoys, and areas that they associate with gunshots.

What happens when much-needed fear is absent? This is a major conservation concern as humans introduce predators to islands. Marine iguanas on the Galapagos are remarkably placid, allowing visitors to walk nearby and even to touch them (see Figure 4.1). The introduction of dogs and cats to the Galapagos creates a threat that the iguana populations have not experienced for millions of years. Galapagos marine iguanas show some flexibility in response, even though adaptation has largely taken away their antipredator behavior. When threatened by predators, they have higher levels of stress hormones than unthreatened animals and are more likely to attempt to flee, albeit inefficiently and largely unsuccessfully.[1,2]

FIGURE 4.1
Galapagos Marine Iguana. These organisms have lived in a predator-free environment during evolutionary history since colonizing these distant islands. As a result, they are fearless around potential predators (such as humans) that cause animals in most other environments to flee. Lack of fear occurs in many insular organisms, and as humans and their associates (cats, rats, dogs, etc.) have colonized the earth, fearlessness has been the cause of extinction in more than one instance. *Photos: Randy Moore.*

FIGURE 4.2A
A lizard basking in a sunfleck on the floor of a tropical rainforest. Basking behavior is a common mechanism for increasing body temperature and improving muscle performance and digestion. Basking can also generate "behavioral fever," elevated body temperature that helps the animal fight infections. (Photo: *Michael Breed*)

FIGURE 4.2B
Cold-blooded animals, like the lizard in Figure 4.2A, are typically assumed to benefit from behavioral thermoregulation, but basking is also common in warm-blooded birds and mammals, such as these cormorants. (Photo: *Michael Breed*)

A discussion of drive theory is followed by an overview of contemporary models for *behavioral homeostasis*. *Time budgets* are introduced as an analytic tool for understanding trade-offs between homeostatic demands. The concluding section of this chapter focuses on personality and differences among animals in behavioral propensities.[3,4,5]

Behavioral regulation is part of the general concept of homeostasis, or maintenance of appropriate internal states in animals. In many ways, behavior is the gateway between the external, or ecological, world of an animal, and its internal, or physiological, world. Homeostatic regulation generally refers to maintenance of constant internal state, but to achieve this, animals must continually modify their behavioral state. Behavior is the necessary homeostatic link that gives animals the means to obtain food, regulate their internal temperature, and locate themselves in an appropriate habitat. Behavior is also the major dividing line between the homeostatic strategies of animals, which often depend on mobility, and the homeostatic strategies of most relatively immobile plants. This can be as simple as behavioral thermoregulation, causing an animal to bask in the sun when it is cold or seek shade when it is hot, or as complex as migrating thousands of miles to find a seasonally appropriate habitat.

Neurobiology and endocrinology provide the regulatory foundations on which homeostasis is built. Homeostatic regulation requires information relayed in *feedback loops*; the basking lizard in Figure 4.2 moves into the sun to warm itself and out of the sun when it is too warm. Thus, simple feedback (skin and blood temperature) regulates basking behavior. Physiology texts often use the analogy of a home furnace regulated by a thermostat. If the house temperature falls below a fixed point,

> **KEY TERM** Animal welfare is the well-being and health of animals. Welfare includes behavioral welfare, so that management of anxiety, fear, and pain are major issues in animal welfare. Humans take responsibility for the welfare of companion, captive, and farm animals.

> **KEY TERM** Behavioral homeostasis is the maintenance of appropriate living conditions by using behavior to achieve those conditions by minimizing their fluctuations. It comes from two Greek words meaning "similar" and "stoppage."

> **KEY TERM** A time budget is a description and quantification of how an animal divides its available time among its activities.

> **KEY TERM** A feedback loop occurs when an animal continually assesses its condition and modifies its behavior or physiology to achieve a targeted condition.

101

the furnace turns on. Once the desired temperature is reached, negative feedback from the thermostat turns off the furnace. Generally, negative feedback loops provide the most effective behavioral regulation. When an animal's stomach is full, negative feedback inhibits further feeding behavior. However, positive feedback, such as a food reward or praise during learning, plays a key role in shaping behavior and also fits the general model for feedback loops. (See Chapter 5 on learning for more details on this topic.)

4.2 BEHAVIOR AND HOMEOSTASIS

Drive, or motivation, is among the most interesting concepts in animal behavior.[6,7,8,9] Ethologists developed the concept of drive to answer questions about behavioral choice and switching. The study of drive has gone out of style in animal behavior, in part because it addresses "deep motivation," those urges that even highly articulate human beings may not be able to describe or explain. Clearly, these impulses do not lend themselves to crisp, testable predictions, and scientists rightly shy away from ideas that are untestable. Drive theory enjoys prominence here because the concepts it has yielded, such as *displacement behavior* and *redirected behavior*, are important in interpreting the welfare of animals. When taken to the extreme, these behaviors become pathological and manifest as *self-directed behavior* and *repetitive*, or *stereotypical*, *behavior*.[10,11] Using tools developed from drive theory, applied animal behaviorists can assess the condition of captive animals or companion animals and improve the welfare of those animals.

> **KEY TERM** Displacement behavior is the release of drive energy in an irrelevant task when the desired behavior cannot be completed. Animals often groom as displacement behavior.

> **KEY TERM** Redirected behavior is the direction of behavior to a third party or an inanimate object. If you kick the wall when you're angry at your roommate, you have redirected your aggression.

> **KEY TERM** Self-directed behavior is behavior such as grooming that an animal performs on itself. Self-directed behaviors can become pathological in anxious animals.

> **KEY TERM** Repetitive behaviors (sometimes called *stereotyped behaviors*) such as pacing are often pathologies resulting from the displacement or redirection of energy when an animal can't perform a desired task.

In 1972, McFarland and Sibly suggested analyzing drives as vectors, much as vectors of force might be analyzed by physicists.[12] This suggestion has not gained much attention, but it remains the most recent approach to reconciling drive theory with a modern understanding of homeostatic regulation. Drive models are at once crude and clever. The graphical representation of two hydraulic models of drive (see Figure 4.3) shows the crudeness of drive models from physiological and neurobiological standpoints. In contrast, contemporary biology generally views homeostatic regulation as a complex interaction among control systems. The cleverness of drive models derives from the attempt by McFarland and Sibly to explain, with little knowledge of animals' inner workings, how competing needs are assigned behavioral priorities.

Drive theory asks us to imagine the source of an animal's internal energy for behavior (refer to Figure 4.3). In simple models, this energy flows into a reservoir, which then empties as the animal expends "motivational energy" in its activities. In the model, activities are ranked by positions of valves on the tank, so less-important activities receive energy only if higher-priority behaviors have already been accomplished. But what happens if excess energy accumulates or if an animal is conflicted between expressing two behaviors? Approach–avoidance conflict is particularly common: if a food item is attractive, but collecting it is risky, an animal may sneak forward and then jump back nervously as it attempts to snatch

102

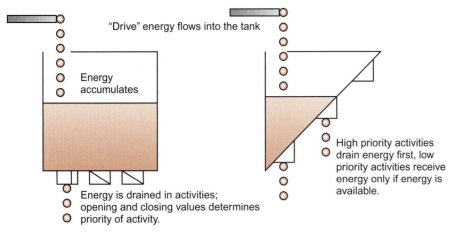

FIGURE 4.3
Hydraulic models of drive. Early ethologists thought of animals as having a supply of energy that could be allocated to tasks. Over time, this energy accumulates and ultimately is released through performance of a behavioral act. Energy models of drive gave ethologists tools for explaining how animals prioritized their behavior.

the item. Similarly, the relative risks and benefits of fighting a territorial intruder versus fleeing from it result in a fight-or-flight dilemma. In both instances, the animal is caught on the cusp of indecision between approach and avoidance, or fight and flight. This fits well with the model of drive, as energy for action builds up but is unexpended.

How might this energy be dissipated? This is the point at which displacement behavior and redirected behavior come into play. If a conflict occurs between allocating energy to one of two or more behaviors (such as occurs in approach–avoidance conflict), energy might be dissipated in a third irrelevant act, such as grooming.

Given this explanation, it becomes clear that animals engaged in activities such as displacement resulting from approach–avoidance conflict may be under a good deal of stress. Because of this, drive theory offers value for the assessment of animal welfare. It is the foundation for important tools—introduced in the following sections—that explain and evaluate some behaviors in both natural and confined habitats. These tools can provide a window into the internal state of animals in captivity. Given advances in contemporary science, many behavioral states (hunger, motivation to court and mate, migration, for example) are best understood on physiological and neurobiological levels, as functions of learning, hormones, and neurotransmitters. Although drive theory was developed decades ago, its utility in welfare assessments is strong justification for contemporary students to understand the theory.

In the sections that follow, drive theory will be incorporated into contemporary explanations for everyday behavior of animals as they resolve motivational conflicts. It is also the basis for understanding the expression of animal behavior at the extremes of these conflicts, along with resulting pathology.

Displacement and Redirection

Understanding how animals deal with internal conflicts contributes to animal welfare assessment. Studies in this area provide a larger picture of what occurs in abnormal circumstances, such as captivity. These responses stem from mechanisms animals exhibit in response to conflicting motivations. Such situations can become welfare issues at the extremes, but under more moderate circumstances, they are part of everyday life for every animal. Drive theory attempts to explain how such conflicts are resolved. A young animal may be conflicted, torn between the desire to play and the desire to be near its mother's protection. The baby cannot resolve the conflict, so it vacillates between the two activities.

103

Observations often reveal that conflicted animals groom themselves; this self-directed behavior (SDB) may be a reflection of the anxiety caused by this conflict.[13]

Displacement Behavior

Displacement behavior occurs when an animal performs an act that is irrelevant to the behavioral context. Animals caught in approach–avoidance conflict often groom; this is displacement behavior. Displacement behavior is obvious in humans under socially stressful situations. When attracted to a person at, say, a party, shy humans have been known to groom their hair with their fingers or touch their faces as a form of displacement behavior. Much of this behavior is hypothesized to be energy dissipation associated with conflicting drives.

Displacement behavior includes self-directed behaviors such as self-grooming, touching, or scratching, displayed when an animal has a conflict between two motivations, such as the desire to approach an object while at the same time being fearful of that object. Many, perhaps most, birds and mammals groom in similar ways when faced with a conflict between approaching and avoiding another animal (see Figure 4.4). In social hierarchies, lower-ranking animals groom more frequently than do higher-ranked animals, possibly reflecting the conflict between attraction to the social group and avoidance of the higher-ranking animals in the group.

FIGURE 4.4
Displacement behavior occurs when an animal experiences conflicting drives, which in some cases is called approach-avoidance conflict. The animal may do something irrelevant, such as scratch or preen, in these situations.

104

Redirected Behavior

Redirected behavior is usually aggressive; because it cannot be directed at the target object, it is directed elsewhere. A person who kicks his dog or flings her pillow after an argument with a roommate has redirected the aggression that developed as a result of an event or behavior that occurred previously. Classically, most discussion of redirected behavior centers on aggression; redirection can also emerge in behaviors such as nest building or cleaning.[14]

An animal that has been attacked by the dominant individual in its social group may not retaliate (due to the possibility of stimulating a more severe attack by the dominant) but may instead redirect its aggression to other less-dominant members of the social group. Humans who have older siblings often experience their brother's or sister's redirected aggression when their sibling has been punished by a parent. Musth elephants redirect aggression to other species or to inanimate objects.

In animal welfare assessments, redirected behaviors may not be aggressive. They frequently involve behaviors such as *cribbing* (licking or chewing cage or stall bars) and repetitive pacing. Sometimes these behaviors, particularly repetitive pacing, are termed *stereotypical* behaviors, but this is potentially confusing because *stereotypy* has a distinctly different meaning in the study of the evolution of communication (see Chapter 7).

OF SPECIAL INTEREST: MUSTH

Bull elephants annually cycle between a state of heightened aggressiveness, called *musth*, and nonmusth. Excess energy is dissipated in displacement behavior or redirected behavior by the elephants. In elephants, musth presents an extreme example of redirected aggression; often an elephant in this condition will attack any object or animal that happens to be in the way.

A musth elephant is primed to mate, and fights other bull elephants, attacks other animals, and may destroy inanimate objects in its way. Musth bulls produce a distinctive low-frequency vocalization, the musth rumble; have thick secretions from their temporal glands (the duct from the temporal gland opens between the eye and the ear); and continuously dribble urine. Testosterone levels peak in musth males and probably regulate this extreme form of reproductive behavior. Legendary "rogue" elephants were probably musth bulls, redirecting their aggression at random objects, including villages and their inhabitants.

Young bulls do not go into musth and may be inhibited from doing so by older males. Asian elephant males start their musth cycles by the time they are 20, but African elephants do not reach this stage of maturity until they are 25 or so years old. Musth has implications for maintenance of captive populations of elephants in conservation programs. Musth males are unmanageable and extremely hazardous to elephant handlers (see Figure 4.5).

Musth is an interesting reproductive strategy. Presumably, males cannot maintain themselves in this physiological state for extended periods of time. In a sense, it might be a costly advertisement of mate quality, as only well-fed, healthy males may be able to physiologically support the energetic demands of being in musth. When more than one bull is in musth, fights result in risk of serious injury. Bulls may adopt a strategy of avoiding fights by coming into musth out of synchrony with stronger or more powerful bulls. Because female ability to conceive is highly seasonal, this strategy can be costly; an elephant whose musth is poorly timed may sire few offspring. Bulls that exhibit off-season musth are displaying a type of satellite strategy in which their mating efforts are peripheral to the main mating competition. This and other mating strategies will be covered in more detail in Chapter 11 on mating.

FIGURE 4.5
During mating season, mature male elephants redirect their reproductive energy and enter a condition of musth, an extremely aggressive state in which they are likely to attack whatever is in their path, including inanimate objects.

105

Self-Directed Behaviors and Repetitive Behaviors

Self-directed behavior can be another result of dissipation of behavioral energy. Self-directed behaviors are behaviors that an animal performs on itself, without a social partner. Strictly speaking, self-directed behavior ranges from benign activities to potentially destructive behavior such as feather plucking, masturbation, and obsessive licking. Self-grooming, or auto-grooming, is a normal part of the behavioral repertoire of a wide range of animals and usually serves to clean the surface of the animal of parasites. In behavioral evaluations of domestic or caged animals, self-directed behaviors often are grouped with redirected behaviors, sharing the umbrella term of self-directed behavior (SDB).

> **KEY TERM** A focal animal is an animal within a group that an investigator targets for observation.

Social stress in natural habitats can also give rise to SDBs. SDBs increased in wild olive baboons when the animal nearest the *focal animal* was dominant.[15] The proximity of dominant animals was associated with a 40% increase in self-directed behaviors, indicating a higher level of social anxiety in animals near dominant individuals.

Assessment of Animal Welfare

As drive theory fell out of favor, studies of displacement behavior suffered collateral damage. Animal behaviorists paid little attention to displacement behavior until a study by University of Chicago primatologist Dario Maestripieri pointed out that it might be a good measure of anxiety levels.[16] Since that time, a growing body of literature has argued that self-directed and repetitive behaviors accurately indicate anxiety. Particular attention has been paid to primates, including humans. Measures of displacement behavior, for example, have been applied in psychiatric studies of human anxiety.

Measuring self-directed behavior gives animal behaviorists powerful tools for assessing the welfare of animals in captivity. For assessment purposes, both SDBs such as grooming, and stereotypies, such as repetitive pacing are evaluated. Quantification of SDBs allows an animal behaviorist to measure both the level of stress caused by a captive environment and the effectiveness of interventions intended to ameliorate that stress. If SDBs are reduced after the intervention, then it is judged to be effective.

Pathological SDBs most often are grooming to the point of self-damage, such as feather plucking in birds or paw-licking in dogs. Domesticated ungulates sometimes display redirected behavior by cribbing, in which they lick or chew on the bars of their enclosure. Pacing, excessive grooming, and cribbing are all common in zoo animals and can be interpreted as symptoms of suboptimal conditions for the animal. Pacing and excessive grooming occur most often in carnivores and primates but can be seen in a wide range of other birds and mammals.

Measuring SDBs in captive animals offers a window into the animals' "emotional" state. Animals with high rates of SDBs require changes in their conditions. The most common response in zoos to high rates of SDBs is habitat enrichment. Depending on the species, there may be opportunities to search, "hunt," or play with toys. Additional space (larger cage size) can help, but the key factor seems to be giving the animal something interesting to do (see Figure 4.6). In general, habitat modifications that increase the complexity of the animals' environments result in reduced expression of SDBs; this reduction can be interpreted as a result of reduction of the stresses inflicted by the captive environment. Zookeepers may present primates with food that is hidden or disguised, requiring the captive primates to solve simple problems to be able to find their food. Similarly, the presence of objects to play with and manipulate reduces the stress of captivity for primates.[17] This modified understanding of the needs of captive animals has also influenced the companion

(A) (B) (C)

FIGURE 4.6
Modern zoos work to enrich habitats for captive animals by (A) offering animals their natural foods—in this case, nuts that the hooded capuchin monkey must crack, (B) giving carnivores like tigers meals in "puzzle" containers that require some effort prior to dinner, and (C) providing "toys," such as this elephant toy. *Photos: (A, C) Michael Breed.*

animal world and is reflected in the increased number of dog toys that challenge dogs to find and dislodge a treat; most owners find this preferable to dislodging, say, pillow stuffing.

Chimpanzees are an excellent example of animals that display higher levels of SDBs in anxiety-inducing situations (see Figure 4.7). In the case of responses to vocalizations from animals in neighboring cages, captive chimps displayed SDBs when housed in groups but did not do so when socially isolated. If the vocalizations suggest to the chimps that an attack is imminent, isolated animals may realize that no other chimps are in their cage and therefore may feel safe even when hearing the vocalizations.[13]

Problem-solving tasks can also induce SDBs in captive chimps, depending on the difficulty of the problems and way they were introduced. If the chimps started with an easy problem and then progressed to more difficult problems, they displayed more SDBs when confronted with the difficult problems. Chimps who received only difficult problems did not display more SDBs and positive auditory reinforcement during the problem-solving reduced SDBs.[17]

In addition to mental challenges, social environment is important. Animals reared in captivity require a social environment that mimics the richness and complexity of the species' normal social environment. Because of this, social animals reared in social isolation or in socially deprived environments typically express pathological SDBs. Harry Harlow's famous experiments on rhesus macaque monkey infants, mostly conducted in the 1950s and 1960s, showed that social isolation resulted in physiological and behavioral pathologies related to stress (see Figure 4.8).[18,19] Today, scientists have an increased understanding of how non-humans experience the world (in part, thanks to Harlow's work!), and Harlow's experiments would not be done. Nonetheless, they have given us important windows into how social deprivation during development and the stress associated with isolation affect behavior later in life. Moreover, Harlow's research itself is part of what informs current judgment about the effects of social deprivation on young macaques. (Note that rearing in social isolation differs from the condition of socially isolated chimpanzees in the previous study; the animals in the first study were not necessarily reared in isolation.) Take heart from the fact that Harlow also did rehabilitation experiments; in the right social conditions, deprived youngsters could regain normal behavior in relatively short order.

In canids, such as wolves, or felids, such as lions or tigers, larger enclosures and complexity in food discovery are key to reducing social stress (optimally, live prey that must be killed, although few zoos are willing to do this). The discussion point invites consideration of the importance and role of zoos and other captive environments for nondomesticated animals.

FIGURE 4.7
Grooming serves many functions, from parasite removal to emotional expression. In the case of social animals such as these monkeys, it can also reinforce social bonds. *Photo: Michael Breed.*

107

FIGURE 4.8
Harry Harlow asked how social deprivation and aspects of mothering such as warmth and food influenced the development of young rhesus monkeys. The effects were wide-ranging, and tended to reflect the severity of the treatment, providing important insights into the psychological and social requirements of young social mammals. Fortunately, the experimental subjects suffered no lasting effects after social rehabilitation.

DISCUSSION POINT: ZOO ANIMALS, ANIMAL WELFARE, AND THE EDUCATIONAL FUNCTION OF ZOOS

The continued existence of zoos is one of the most passionately debated issues in animal behavior and conservation. Here, we present both sides of the argument; a class discussion of this topic will probably reveal a wide range of opinion on this issue. Thus, drive theory, far from being out of date, provides critical tools in evaluating captive animal welfare.

Proponents of zoos argue that maintaining animals in captivity serves educational and conservation missions. Zoos remain the primary point of public exposure to exotic animals, and people who become engaged with what they see in the zoo are likely to favor conservation of natural habitats. Children, in particular, love zoos and may be drawn into a lifelong commitment to conservation by zoo visits. For species that are in serious trouble in their native habitats—many of which are being destroyed or diminished—zoos provide captive breeding programs, which may be the last defense against extinction and may provide animals for reintroduction into natural habitats.

Opponents of zoos argue that no positive benefit of zoos outweighs the harm done to animals by their conditions in captivity. Based on observations of captivation-induced behavioral pathologies, opponents argue that no animal can be maintained humanely in a zoo. Conservation goals can be better met by preserving natural habitats and perhaps by establishing populations on large reserves outside the native habitat where visitation is limited. Modern media allow observation of a wide range of animals from around the globe in movies and television, making zoos unnecessary for public education. In the extreme, this argument extends to the idea that it would more ethical to allow a species to go extinct than to maintain its only population in captivity.

Do the educational and conservation benefits outweigh the animal welfare issues? Is the ability to make behavioral assessments of welfare and to use these measures to evaluate the success of enrichments adequate to ensure the welfare of animals in captivity?

The presentation of SDBs and animal welfare so far has mostly focused on wild animals kept in captive conditions. How do these principles extend to domesticated animals, and what is the neural and endocrine basis of these behaviors? When observing an animal expressing SDBs, most people would characterize the animal's state with words such as *anxious*, *nervous*, *tense*, or *distressed*. While there are serious pitfalls associated with using terms that have a human emotional context to describe the internal state of a nonhuman animal, the evolutionary relationships among humans and other mammals establish a scientifically sound argument for homologies between the internal states (emotions) of humans and nonhuman mammals.

Neuroscience and endocrinology tell us that the same suite of neurotransmitters and hormones underpins behavior across a much broader range of animals than mammals. Given this line of thinking, it is reasonable to hypothesize that birds and mammals with SDB-type behavioral pathologies are "anxious." In addition, it is not difficult to see that repetitive pacing and similar behaviors resemble obsessive compulsive disorder in humans.

Is there a neuroscience-based argument for this homology between human and nonhuman anxiety? How can a hypothesis about the similarities between apparent animal anxiety and human anxiety be tested? Anxious mammals, such as domestic dogs or cats, may lick their paws obsessively, claw furniture, chew destructively, or drool excessively. These behaviors fit in the general classifications of SDBs and redirected behavior. Humans do not claw furniture, and drooling is frowned upon, but humans do indeed feel anxious; if a human reports an overwhelming sense of anxiety to her or his physician, she or he would likely be prescribed a drug such as fluoxetine (Prozac). These drugs act as selective serotonin reuptake inhibitors (SSRIs), increasing the amount of serotonin in the space between neurons; behaviorally, taking such a drug has the effect of ameliorating symptoms of depression. The relationships between neurotransmitters, such as serotonin, and behavioral state were discussed in detail in Chapter 2. If this type of treatment does not work, that person might be prescribed a medication like Xanax, which enhances production of the neurotransmitter gamma-aminobutyric acid (GABA).

All of this should make it no surprise that anxious animals respond to therapy with drugs that increase the action of serotonin, a neurotransmitter implicated in depression and obsessive-compulsive disorder (OCD) in humans. This is consistent with the hypothesis that animal anxiety is similar to human anxiety. Fluoxetine and similar drugs are commonly used to treat dogs and cats for separation anxiety and similar behavioral issues. Interestingly, these drugs are effective in reducing certain kinds of aggression in dogs and cats, as well, supporting another hypothesis discussed later in this chapter: that much aggression in animals is fear based. Typically, medication of a dog or cat for anxiety-related behavioral problems is accompanied by behavioral training, and for long-term treatment success, it is usually as important, or more important, to teach the animal's owner to modify its behavior as it is to medicate the pet.

Pathological behaviors arise in other domesticated animals, such as swine, horses, and cattle. Generally, these behaviors are regarded as symptomatic of welfare problems, such as keeping horses confined in stalls for long periods of time. The same solutions used with zoo animals—increasing habitat complexity and making the animals' lives more interesting—apply to treatment of anxiety-related problems in these species. The best-known non-mammalian behavioral pathology associated with captivity is feather-plucking in birds. Again, reducing behavioral stressors in the environment is important, but birds also respond to drugs developed for treatment of obsessive-compulsive disorder in humans. Knowing that the neurobiological analogy with human anxiety extends to birds puts an exclamation point on the argument that many behavioral pathologies have the same neurobiological roots across a broad range of vertebrates.

FIGURE 4.9
An American goldfinch captured in a mist net during a study of migratory birds. The behavioral effects of such captures on wild animals are largely unknown, but the corticosteroid response to events such as mist-netting or close encounters with predators is probably costly to the animal, especially during an energetically demanding season. *Photo: Michael Breed.*

Before leaving this topic, let's note an important point regarding stress and the endocrine system. Stress is a physiological response to difficult conditions, including hunger, fear, or pain. It also may involve the social stress of being low in a dominance hierarchy or the individual stress of being caged. Stress is usually accompanied by high levels of corticosteroids, hormones produced by the adrenal glands in response to stress. In humans, corticosteroid levels are associated with internal symptoms of anxiety, such as high blood pressure. Chronically high corticosteroid levels can result in a variety of physiological pathologies. For nonhuman animals, handling can dramatically raise corticoid steroid levels as part of a stress or fear response. In agricultural settings, repeated handling can reduce growth rates of poultry and other animals; this is probably due to the effects of corticosteroids in raising metabolic rates. Less is known about handling effects on wild animals (see Figure 4.9), but it is a safe assumption that handling, via the action of elevated corticosteroids, causes the animal to use valuable energy resources and may disrupt future behavior of the animal. Corticosteroids can be measured in fecal samples from animals, providing a tool for assessing stress on a physiological level.

109

4.3 BIOLOGICAL CLOCKS AND CIRCADIAN RHYTHMS

How does a pet seem to know what time it is? This behavior becomes especially apparent if, say, she gets a treat on a regular schedule. If someone has forgotten the treat, the pet assuredly has not, and there will almost always be a reminder in the form of a wet nose or an insistent purr. As strange as it may sound, animals—and even protists and fungi—can tell time, in a manner of speaking. In fact, the study of biological rhythms—activities that are linked to regular environmental cues and occur cyclically—may be the only topic in animal behavior that began with the study of plant behavior!

In the early 1700s, a French astronomer, Jean-Jacques d'Ortous de Mairan (see Figure 4.10), realized that the leaves of the heliotrope plant opened and closed with a daily rhythm. By placing the plant in continuous darkness, he showed that the leaf action continued; it was not a response to sunlight. Instead, it was something now called *free-running behavior*, the maintenance of daily rhythm in the absence of any cues. Almost all organisms have *biological clocks*, that is, endogenous timing mechanisms (pacemakers) that allow an organism's activity to occur in a rhythmic manner, linked to external events. The external expression of a biological clock is called a *biological rhythm*.

Note
Even Darwin was intrigued by de Mairan's work and studied daily rhythms in plants. In *The Power of Movement in Plants*, he hypothesized that "sleeping" plants conserved energy, a hypothesis that has yet to be rejected as knowledge about rhythms in plants increases.

KEY TERM A biological rhythm is a behavioral or physiological attribute that changes over time on a predictable cycle.

Because the earth rotates on its axis every 24 hours, many environmental cues and significant environmental changes occur with regularity (see Figure 4.11). The ability to predict these changes and to prepare for them, be they daily or annual, is clearly adaptive. Daily rhythms are called *circadian*, from a pair of Latin words, *circa* (about) and *dies* (day). A circadian rhythm runs on a 24-hour cycle and is entrained to the day–night cycle of the natural world. Behaviors that happen at certain times of day—for instance, insect molting or rodent peak activity—are circadian. *Circannual rhythms* are also important and include seasonal behaviors such as hibernation or migration. Intertidal animals often exhibit *lunar rhythms* (28-day cycles); their activities coordinate with the presence or absence of water in their immediate environments.

No one has seen a biological clock; current hypotheses hold biological clocks to be molecular in nature. The evidence of their existence, however, is more apparent. Free-running rhythms are one type of evidence. An animal placed in constant darkness, without any environmental cues, will nonetheless exhibit an activity rhythm that approximates (but is not quite equal to) 24 hours.

FIGURE 4.10
Astronomer Jean-Jacques d'Ortous de Mairan was among the first scientists to document biological clocks. Source: *http://www.hhmi.org/biointeractive/museum/exhibit00/index.html*.

Genetics provides additional evidence. In 1971, Ron Konopka and Seymour Benzer discovered a gene, one they called *period*, or *per*, that occurred in fruit flies (*Drosophila melanogaster*).[20] Variants of this gene were associated with variations in circadian activity in the flies. Animals that have mutations in the genes associated with biological clocks will breed true for those mutations and pass the mutant rhythms on to their offspring. Thus, a mutant might free-run at a rhythm of 22 hours, instead of 24. Genomics has brought a much deeper understanding to biological clocks.[21]

Even jet lag is a form of evidence for a biological clock. When an animal that feeds at a certain time of day every day is moved to a different time zone, it will continue to feed based on the original rhythm, at the time of day that corresponds to the feeding time in the previous time zone.

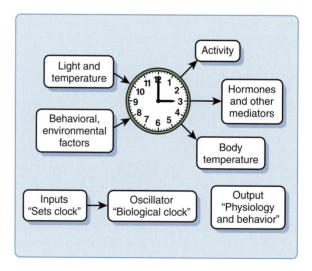

FIGURE 4.11
This diagram summarizes how a circadian clock regulates the activity of an animal. The *period* is the time it takes to complete an entire cycle—in this case, 24 hours. When light and dark alternate in a 24-hour cycle (left half of figure), activity peaks during the light part of the cycle (called the *photophase*) and decreases to a minimum during the dark phase (*scotophase*). When the photophase is removed and the animal is kept in constant darkness (right side of figure), cyclic activity continues but becomes free-running. Thus, the cyclical presence of light does not create the activity cycle, but serves as a *zeitgeber* to keep activity on a 24-hour pattern. The activity cycle shown in this figure typifies a *diurnal* (day-active) animal; it would be reversed in a nocturnal animal.

Of course, animals do recover from jet lag—that is, they become synchronized with their new time zone—and circadian rhythms in the real world track 24-hour cycles much more accurately than the free-running rhythm does. This happens because the endogenous pacemaker can be entrained by external cues. The external cue is called a *zeitgeber*, or time-giver; the most common *zeitgeber* in the terrestrial world is the day–night cycle. The light part of that cycle is called the *photophase* or *photoperiod*. Many environmental events have been shown to act as *zeitgebers*—even the regular sound of a school bell can serve that function—but the predominance of photoperiodic *zeitgebers* is thought to reflect the dependability of that cue.

The circadian rhythm is the building block for the other rhythms that overlie it. Thus, the circannual rhythm regulates migration (see Chapter 8), but it can do so only if the organism perceives seasonal photoperiodic changes and, indeed, uses its clock mechanism in conjunction with other cues to discern compass direction. Sleep/wake cycles, feeding bouts, and foraging compasses all depend on daily rhythms. Seasonal changes in reproductive status, migration, and hibernation are also tied to biological clocks (see Figure 4.12).

The genetic basis for clock function varies across animals, but the underlying mechanism appears to be universal. To function, a biological clock depends on the presence of a molecular cycle that can be reset to stay in synchrony with the 24-hour solar cycle. The basic principle underlying biological clocks is the daily buildup of a protein, which is phosphorylated as the day progresses. This buildup is reinforced by positive feedback at the molecular level. When an upper threshold of phosphorylation is reached, negative feedback then causes the protein to degrade until a lower threshold is reached and the cycle restarts. In birds, mammals, and fruit flies, genes like *Period, Clock, Timeless,* and *Cryptochrome* play key roles in the clock (see Figure 4.13).

In mammals, the clock is found in the suprachiasmatic nuclei in the hypothalamus region of the brain. The daily oscillation of the molecular clock is linked with the endocrine system of the animal, which directly regulates behavioral and physiological changes over the 24-hour cycle.

111

FIGURE 4.12

How a circadian clock regulates the activity of an animal. When light and dark alternate in a 24-hour cycle (left side of figure) activity peaks in the middle of the light part of the cycle and drops off to a minimum in the earlier part of the dark phase. When the light portion of the cycle is removed (right side of figure), activity continues to cycle, but it becomes free-running. Thus, the presence of the light-dark cycle does not create the cycle but serves as a zeitgeber to keep activity on a 24-hour pattern. This figure represents a diurnal (active during the day) animal. The activity cycle for a nocturnal animal would be reversed. *Adapted from Bell-Pederson et al.,* Nature Reviews Drug Discovery, *doi: 10.1038/nrd1633.*

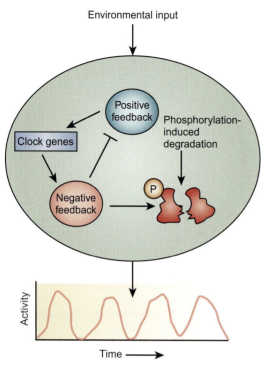

Environmental input

Positive feedback

Clock genes

Negative feedback

Phosphorylation-induced degradation

P

Activity

Time

FIGURE 4.13

This schematic of the molecular workings of a biological clock shows the build-up of a phosphorylated protein during the day until a threshold is reached that elicits negative feedback and protein degradation. The degradation process continues until it reaches a lower limit of phophorylated protein, whereupon phosphorylation begins again. Environmental input of some type is required to synchronize this process with environmental events (e.g., solar cycle, lunar cycle, etc.).

While biological clocks have a general role in regulating animal activity cycles, they also play key roles in daily and seasonal movements (see Chapter 8).[22] For example, the sandhopper, a small crustacean found on European beaches, uses the moon to set its direction when it swims from the open water to the wet part of the beach at night.[23] Moon compasses are common in nocturnal animals. On a moonlit light, the position of the moon allows the sandhopper to find the beach. But what about times of the month when the moon is not visible in the night sky? The sandhoppers use their biological clocks to calculate the expected position of the moon and are still able to find the beach. Similarly, honeybees can use their biological clocks to compensate for the apparent movement of the sun in the sky as they fly to and from food, or when they dance to demonstrate the location of food to other bees. The ability to use biological rhythm to compensate for the "movement" of the sun is called *clock-shifting*, and will be revisited in Chapter 8.

4.4 MODERN CONCEPTS OF HOMEOSTATIC REGULATION

Some areas of homeostatic behavior, such as sleep, feeding and appetite, and pain, lend themselves more readily to scientific exploration. As scientific understanding of the hormonal and neurobiological regulation of behavior has increased, animal behaviorists have shifted from discussions of motivation and drive to examinations of those physiological regulators. After all, behavior is the result of physiological needs (e.g., thirst, hunger) or biological imperatives (e.g., reproduction) regulated via neurochemical and hormonal mechanisms. Consequently, many behaviors directly improve an animal's physiological condition and have direct effects on internal homeostasis.

Sleep

Sleep plays a key role in time budgets and behavioral decision making. A sleeping animal is immobilized (reversibly, not paralyzed), is not easily disturbed, and generally expresses a preference for a sheltered or protected location before entering sleep. Sleep, or sleep-like states, occur in all vertebrates, including fish, and have also been observed in some insects.[24] Sleep probably has three adaptive functions: (1) A sleeping animal conserves energy; this is particularly true of endotherms (birds and mammals), which lower their metabolic rates while sleeping. (2) Sleep deactivates an animal during phases of the daily cycle when activity might put it at risk. (3) Sleep allows a time when brain repair and memory consolidation can take place.

DISCUSSION POINT: SLEEP

We have put forward three adaptive hypotheses for sleep: energy conservation, risk avoidance, and brain repair/memory consolidation. Are these hypotheses mutually exclusive? Are they testable, and if so, how would you go about testing them? Can you think of other possible adaptive advantages of sleep?

KEY TERM Sleep is a period of inactivity during which brain activity is depressed or modified.

While the details of how sleep is regulated are not fully worked out, the basic framework for sleep starts when an animal's circadian cycle primes it to sleep at appropriate times. This priming sets the stage

112

for regulation of overall arousal/drowsiness. Deep sleep is also physiologically regulated. Melatonin, a hormone of the vertebrate pineal gland, plays a key role in regulating daily activity cycles; high levels of melatonin are associated with drowsiness.[25] A peptide hormone, called *orexin* or *hypocretin*, also has a major regulatory role in the sleep cycle. Orexin causes wakefulness and drives appetite; in its physiological absence (or in genetic mutants lacking the ability to produce orexin), an animal has a low state of arousal, or is *narcoleptic*. Gamma-aminobutyric acid (GABA) plays in important role in the transition between light sleep and deep *rapid eye movement (REM) sleep*. Human sleep aids, such as zolpidem (Ambien), promote the effect of GABA in suppressing brain activity and inducing sleep. The importance of sleep in the biology of most animals is emphasized by the pathological, impaired mental states of animals that have been sleep deprived. Sleep deprivation is a powerful torture device among humans; this fact demonstrates the importance of sleep in homeostasis.

> **KEY TERM** A narcoleptic is a hormone, neurotransmitter, or drug that induces sleep. It comes from narcolepsy, a nervous disease that is characterized by excessive sleep. (Gr. narco = "numbness"; L. epilepsia = to take hold of)

> **KEY TERM** Rapid Eye Movement (REM) sleep is a deep sleep state that is key to the neural repairs and memory consolidation that occur during sleep.

Sleep is, then, a key behavioral trade-off. The sleep state is beneficial to the organism, but the increased arousal threshold during sleep puts the animal at high predation risk. In addition to finding preferred locations for sleep, animals use a variety of behavioral strategies to reduce risk while sleeping. Birds such as wrens and warblers, which migrate to tropical climates during the nonbreeding season, tend to sleep far out on the ends of branches. This makes it harder for foraging snakes to find them, and the flexibility of the branch makes it likely that the snake will vibrate the branch, arousing the bird. Some birds are known to "sleep with one eye open," so that one hemisphere of the brain is in a sleep state while the other side of the brain is aroused enough to perceive threats. Many animals become more physically tolerant of close contact with family, flock, or herd members during sleep. This close contact allows shared vigilance and ensures that if one member of the group is aroused by a disturbance, the others will be as well.

Feeding and the Regulation of Appetite

Some phases of an animal's life pose particular challenges for energy uptake.[26] These phases include growth periods, preparation for overwintering or migration, competition for mates, and in the case of mammals, pregnancy and lactation for females. Control of food intake, and consequently of foraging behavior, is extremely complex and involves three interrelated control mechanisms. First is the sense of satiety, or fullness. Second is the responsiveness of the body, and of behavioral reactions, to the insulin control system. Third are seasonal and developmental changes that affect fat storage (and consequently feeding behavior) for use in egg production, overwintering, or migration. These three control mechanisms are detailed next.

SATIETY

Satiety is a sense of fullness after eating. The key neurotransmitters controlling appetite, at least in vertebrates, are serotonin (5-HT) and catecholamine. These neurotransmitters act to reduce feeding behavior and consequently food consumption. Carbohydrate intake can result in both increased levels of serotonin and catecholamine and also changes in the number of their receptors. One effect of increased carbohydrate intake is to elevate raw materials available in the brain for neurotransmitter synthesis; this feeds back in the form of suppressed appetite. However, the relationship is complex and carbohydrate intake does not always result in satiation.

The picture concerning satiation in insects is much simpler. Stretch receptors detect the degree of stomach expansion that accompanies feeding. At least in the most-studied insects for satiation (mosquitos, houseflies, and honeybees), feeding stops when the stomach is fully stretched. Making a small incision through the body wall and piercing the stomach so that ingested fluids drain out results in continuous unregulated feeding, as does severing the nerve that serves the stretch receptors.

INSULIN CONTROL SYSTEM

Sugars are the basic energy substrate of living things, so it is not surprising that sugar circulation and storage are critical to animal life. Perhaps more surprising is that virtually all animals seem to share a common metabolic pathway for regulating circulating sugars—the insulin signaling system, or *insulin control system*. (Insulin is a small peptide that is produced by the vertebrate pancreas.)

> **KEY TERM** The insulin control system regulates carbohydrate metabolism in nearly all animals. Insulin is a peptide hormone, secreted by the pancreas in vertebrates, that is released in response to rising blood sugars; it stimulates sugar storage or conversion to fats.

Some systems are so fundamental to organismic function that they show little, if any, change over the course of evolutionary time. These systems regulate basic components of life. The functions of the families of genes that code for these highly conserved processes are intricately linked to animal behavior. Here, the insulin signaling system (ISS) illustrates the complexity of a highly conserved signaling system.

As implied previously, the genetic code for insulin, in some form, is found in all animals. Correspondingly, all animals also have the genetic codes for insulin receptors—proteins that respond to the presence of insulin. While it may not be surprising that insulin from pigs or cattle has strong effects on human physiology, it is startling to learn that insulin from cattle has physiological effects in very distantly related organisms, such as locusts, flies, and honeybees. The reason is that the genetic code for both insulin and the insulin receptor molecule are so important to survival—so closely linked to the function of cells and cell membranes—that they have remained relatively unchanged over millions of years of evolution.

In addition, the function of insulin has been conserved across a wide range of taxa. Recent studies have implicated the insulin signaling pathway (or insulin-like signaling) in regulation of feeding behavior in a range of animals, including roundworms, insects, and, of course, vertebrates. The sugars themselves may vary across taxa. For instance, in vertebrates, blood sugar (glucose) is regulated by the insulin signaling pathway. The insulin signaling pathway is also important in metabolic regulation in insects, even though the blood sugar in most insects is trehalose, which is composed of two glucoses.

How does blood sugar relate to behavior? Blood sugar regulation leads to a cascade of behavioral events, beginning with the internal perception of hunger or satiation, and ultimately leading to food searches. The primary physiological role of insulin is to regulate blood sugar.

Insulin is the hormone of glucose storage; it is secreted in response to elevated blood sugar levels and triggers cellular storage of glucose in the form of complex carbohydrates or fats. Lowered blood sugar then triggers further feeding behavior. In vertebrates, another hormone, glucagon, mobilizes stored energy substrates. The insulin signaling pathway appears to affect feeding and related behaviors through the action of neuropeptide Y (NPY), a neurotransmitter produced as a result of activation of the insulin signaling pathway (see Figure 4.14). Like many molecules with deep evolutionary roots, insulin has also evolved functions that differ from that of the original molecule. Insulin-like growth factors play key roles in the development and maintenance of the nervous system in vertebrates. Age-related loss of insulin-like growth factors negatively affects neural functions such as learning and memory. Mutations in the insulin signaling pathway interfere with heat-seeking behavior in roundworms, a behavior important in foraging for some of these animals. In vertebrates, the insulin signaling pathway is tied into learning and memory, playing a key role in behavioral integration.

FAT STORAGE

Fat storage is essential for egg production, migration, or prevention of starvation during lean times. It also carries potential costs, particularly in predation risk. (Fat animals may move more slowly and can be easier and more attractive targets for predators.) Regulation of appetite through satiety control and homeostatic control of blood sugar results in fat storage if the animal overeats relative to its current energy demands. Excess energy intake triggers the liver (in vertebrates) or the fat body (in insects) to metabolize sugars into fats, which can then be stored. Behaviorally, this means that feeding must be triggered by factors that do not completely depend on blood sugar regulation. The reason is that an animal must take in more food than is necessary to maintain blood sugar.

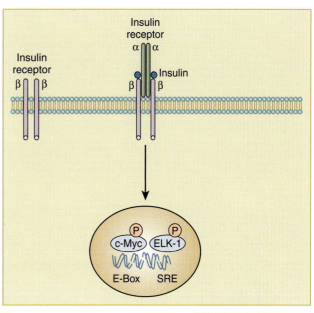

FIGURE 4.14
Insulin is produced when blood sugar is elevated. It activates receptors on cell membranes that in turn produce messenger compounds within the cell that trigger the storage of glucose. Blood sugar then drops, motivating more feeding behavior.

115

Pain

Pain is a homeostatic mechanism that allows animals to recognize and avoid potential injuries. A painful experience may cause an animal to withdraw from a risky location and may also cause it to avoid that location in the future. Pain is hard to study and quantify because pain is a subjective human characterization of a response to bodily damage. Some scientists and philosophers argue that pain is restricted to humans because the experience of pain suggests self-knowledge or self-awareness of a type that might be restricted to humans. Yet even casual observations of injured mammals and birds suggest a commonality of experience, convincing most animal behaviorists that at least some vertebrates feel pain in the same way as humans do. Going beyond vertebrates, it is also reasonable to suspect some avoidance response to harmful stimuli on the part of most animals. Pain is emphasized here because of its link to the behavioral welfare of animals and the use of behavioral techniques for assessing pain in animals.

Most pain results from stimulation of *nociceptors*, specialized pain receptors found throughout the body. There are three types of nociceptors: thermal (sensitive to potentially damaging temperature differences), chemical, and mechanical. (Because the thermal receptors respond to temperature differences, extreme hot and extreme cold yield the same sensation.) When nociceptors are stimulated, the signals travel to the primary somatosensory cortex of the brain. Substance P, a peptide neurotransmitter, is important for the transmission of signals in nociceptive nerve fibers. The pain is represented in the brain, which assigns the pain to the appropriate location so the animal can, if possible, remove the part of its body experiencing pain from the cause of the pain (see Figure 4.15).

Pain can be subdivided into two categories. The first is sharp or immediate pain, which is the instant result of an injury such as a bite, sting, stepping on sharp object, and so on. These stimuli directly activate the nociceptive system and can cause a "reflex" avoidance action that is processed in the spinal cord, well before the signal reaches the brain. The second type of pain is due to inflammation of tissues. Inflammation is a complex immunological response including the dilation of small blood vessels that increases circulation to damaged tissue, but it has the additional adaptation of stimulating the perception of pain.

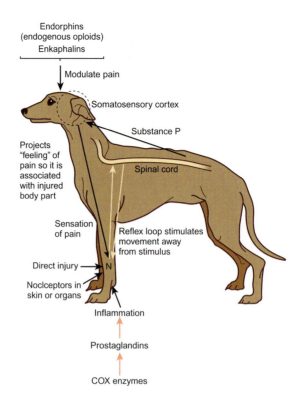

Endorphins
(endogenous oploids)
Enkaphalins

Modulate pain

Somatosensory cortex

Substance P

Projects
"feeling" of
pain so it is
associated
with injured
body part

Spinal cord

Sensation
of pain

Reflex loop stimulates
movement away
from stimulus

Direct injury

Noclceptors in
skin or organs

Inflammation

Prostaglandins

COX enzymes

FIGURE 4.15

In this schematic diagram of pain perception pathways, an injury to the dog's ankle (lower left in figure) results in stimulation of nocioceptors, which relay the signal to the spinal cord. A reflex loop (labeled in yellow) causes quick movement of the leg away from the painful stimulus, and at the same time the impulse travels up the spinal cord to the somatosensory cortex of the brain, which projects the "feeling" of pain to the appropriate part of the body. An immune response to infection or continued irritation of the wounded area causes inflammation (pathway labeled in red).

Key elements in the perception of pain from inflammation are the *cyclooxygenase enzymes* (COX enzymes) that catalyze a critical step in the synthesis of prostaglandins. Prostaglandins have numerous functions; they are important in mediating inflammatory pain and sensitizing nerve cells to pain. The link from inflammation to pain via the action of COX enzymes and prostaglandins is present in at least birds and mammals, and perhaps other vertebrates.

> **KEY TERM** Cyclooxygenase enzymes (COX enzymes) play a key role in perception of pain by helping to synthesize prostaglandins.

Inflammatory pain in animals can be modulated pharmacologically. Steroidal anti-inflammatories, including cortisone, reduce inflammation and, consequently, decrease pain. Over time steroids have potentially serious side effects; they may best be used for short-term reactions, such as insect stings. Nonsteroidal anti-inflammatory drugs, such as aspirin, ibuprofen, and naproxen inhibit prostaglandin synthesis and release. COX-2 inhibitors (brand names include Rimadyl, Celebrex, and Vioxx) specifically inhibit the COX enzyme that catalyzes production of the type of prostaglandin most directly involved in inflammation and pain. Controversy over potential cardiovascular side effects has greatly reduced prescription of COX-2 inhibitors for humans.

In the vertebrate brain, opioid receptors modulate overall feelings of pain and euphoria. Endorphins, enkephalins, and related compounds are internal messengers in the opioid system. Opioid receptors in the brain respond to opiates (opium-like compounds) and are responsible for feelings of pleasure. Endorphins are the body's internal opiates; production of endorphins results in pleasurable sensations. Drugs, such as morphine, serve as analgesics by binding with opioid receptors. Opioid-like receptors have also been discovered in invertebrates, but because opioid receptors and somatostatin (a growth hormone) receptors are similar, it is difficult to assign a function to these receptors.

BRINGING ANIMAL BEHAVIOR HOME: ARTHRITIS IN DOGS

A commonly observed type of inflammation that causes pain in animals is arthritis. Anyone who has cared for an arthritic dog will have little doubt that the dog is in pain. Dogs with mild arthritis are often prescribed aspirin or naproxen under a veterinarian's supervision. A COX-2 inhibitor is commonly prescribed for dogs with more serious arthritis or with hip dysplasia (a painful congenital misalignment of the hip joint). Some reports suggest that COX-2 inhibitors may have unfortunate side effects for some dogs, but they remain in fairly widespread veterinary use.

OF SPECIAL INTEREST: DO ANIMALS FEEL PAIN?

Here are the lines of evidence that scientists generally consider when testing hypotheses for the perception of pain in nonhuman animals:

1. Possession of Neurotransmitters, Biochemical Pathways, or Neural Systems Homologous to Nociceptive Pathways in Mammals

Arguments for pain perception in animals can be built on genome searches for genes that code for products such as Substance P or prokineticins, or for receptors that may be involved in pain modulation, such as opioid receptors. A flaw in this argument is that DNA sequences may be deeply rooted in evolution, but the function of the products from similar DNA can differ substantially across organisms. In other words, the presence of matching DNA sequences does not guarantee matching function. Similarly, the presence of matching neurotransmitters does not prove identical signaling function. Prokinecitins and opioid receptors, in particular, occur in many diverse animals, and their presence has been used to argue for perception of pain as a general phenomenon among animals.

The anatomical version of this type of approach is to investigate the functional anatomy of the nervous system. In salmon, receptors on the skin map to the brain in much the same way that nociceptors do in mammals. Low-intensity electric shocks result in smaller brain responses than high-intensity shocks, suggesting that fish can perceive painful stimuli, based on structures that are similar in function to the mammalian somatosensory system.

2. Sensitivity/Response to the Painkilling Effects of Known Analgesics

One standard test for response to pain is the application of acetic acid, which is relatively mild, to an animal's skin. The animal typically responds by wiping the spot with a leg; water, applied as a control, does not induce wiping. In frogs treated with analgesics, tolerance of acid was increased. This result shows that amphibians respond behaviorally as if the acid causes pain. The fact that analgesics reduce the pain response in both frogs and mammals supports the hypothesis that the biochemical basis for pain response is similar in the two groups.

3. Behavioral Avoidance/Response to Situations Thought by the Investigator to be Painful

Recent studies show that crustacea avoid circumstances under which they have previously experienced conditions that can be defined as painful. Application of mildly noxious chemical stimuli to the surface of crustacea cause grooming movements that suggest nociception.

In sum, birds and mice (and by extension, presumably all mammals) meet all three of these criteria for pain perception; the most commonly used nonhuman model system for the study of pain is mice. Other vertebrates, including fish and amphibia, also meet the criteria. Among the invertebrates, roundworms meet the first criterion, and crustacea meet the third.

The evidence for pain as a physiological response to allow an animal to avoid injury is widespread and convincing. Does it "feel" the same to nonhuman animals as it does to humans? We will probably never know, just as, in truth, we do not know if "sad" feels exactly the same to every human being. Nonetheless, attempting to answer this question, and to develop objective methods of assessing pain in animals, has great significance to the societal debate about ethics and the treatment of companion and farm animals. If well-intentioned avoidance of anthropomorphism has overstepped reasonable bounds in science, this is the area in which it has done so.

117

Fear

Fear, like pain, is a homeostatic safeguard that can keep an animal from lethal risk. But, also like pain, fear is a subjective characterization drawn from human emotions. What is the evidence for fear in animals? Fearful behavior in domestic animals is easily recognized; fearful dogs have a typical tail down, ears back, mouth and face neutral posture when moderately afraid. This can progress to a more defensive posture (teeth bared) or to a submissive posture (rolled on their back, stomach exposed) depending on the type of threat and the dog's likelihood of winning a direct encounter with the threat. Many other mammals, such as cats and horses, give signals that can be interpreted as reflecting fear.

> **KEY TERM** Fear is more easily recognized by behavior—cowering, defensive threats, flight—than it can be defined. We define fear as a set of adaptive behaviors and their corresponding internal state (emotion) that help an animal to avoid danger.

Fear has a very real adaptive value in an animal's life. Fear is what keeps animals from harm's way. Like much of what has been discussed in this chapter, it is easier to spot fearful-seeming behaviors in birds and mammals than in animals that share fewer communicatory channels with humans. In order to remove subjectivity from assessments of fearfulness, animal behaviorists sometimes refer to a certain kind of fear called *neophobia*—the avoidance of the new and unfamiliar in the environment. Neophobia is discussed in more detail in Chapter 15 (on conservation and behavior); as members of species such as coyotes, raccoons, and bears have overcome neophobia, they tolerate closer contact with humans. This close human–wildlife interaction in suburban environments presents difficult conservation challenges.

Fear can, of course, be a response to abuse (either at the hands of a human or by a dominant animal in a social group), and veterinarians and animal welfare agencies often use measures of fearfulness when assessing prior treatment of animals. Fearful animals can sometimes be dangerous to work with because the boundary between fearfulness and defensive aggression can be difficult for a human observer to predict. Some animal workers believe that fear is the cause of most aggression problems in domestic animals. For many pets, patience and a positive social environment can overcome fearfulness that results from an abusive history.

4.5 TIME BUDGETS AND TRADE-OFFS: BALANCING DEMANDS IN HOW ANIMALS BUDGET THEIR TIME

An animal's time budget is the nexus among all of these regulatory mechanisms. How much time is spent foraging, versus courting, versus sleeping? All are necessary activities for survival and reproduction, but knowing how appetite, sex drive, parental behavior, and foraging are each independently regulated does not help much with understanding how relative priorities among the behavioral possibilities are determined. This section focuses on the practical aspects of time budgets: How is a time budget constructed? How does it reflect behavioral trade-offs for animals? How can time budgets be used to ask interesting questions about animal behavior?

The time budget for an animal is a simple documentation of the proportion of time spent in each available activity by that animal. Time budgets are deceptively simple and low-tech; construction of a time budget gives investigators a powerful tool for testing hypotheses. For example, does disturbance affect an animal's time budget? How much time does it spend in each possible behavioral mode? How frequently does it switch from sleep to foraging? Time budgets provide a critical tool for analyzing an animal's internal state and for comparing animals across developmental stages, among seasons, or between sexes.

A good time budget relies first on the construction of an accurate ethogram of the animal's behavior. An ethogram is usually based on many hours of observation, including analysis of video recordings of a species' behavior. The finished ethogram will have captured the full range of behavioral possibilities for a species and provided a list, with definitions, of the

Playful observation   Play balancing

Play climbing
on partner

FIGURE 4.16
Behaviors that form the basis for a small portion of a sample ethogram. The ethogram was designed to compare the behavior of five species of monkey, including the Hanuman langur. Playful observation is defined as: "A playful monkey performs play intention movements...while watching its play partner". Play balancing involves jumping on top of another monkey and attempting to balance. Play climbing occurs when one monkey climbs on another. These simple definitions allow observers to easily differentiate among behaviors. The descriptions must be precise enough so that another investigator can replicate the analysis. *Adapted from Petru, M., Spinka, M., Charvatova, V., & Lhota, S. 2009, Revisiting play elements and self-handicapping in play: A comparative ethogram of five old world monkey species.* Journal of Comparative Psychology, *Vol. 123, pp. 250–263.*

119

behaviors (see Figure 4.16). Photos or drawings are very helpful accompaniments. When coupled with records of the frequency of each behavior, an ethogram becomes a time budget.

Other interesting approaches to time budgets deal with developmental and seasonal changes in behavior, and comparisons of males and females during mating and nesting. Clearly, there will be differences between juvenile and adult organisms in time budgets. How does the frequency of play behavior change with maturation? Are there specific hormonal correlates with changes in play? A time budget is a quantitative tool for investigating these questions. In many eusocial insects—bees, ants, wasps, and termites that live in colonies—the time budgets of workers change as they age; age-related specialization allows workers to be efficient in their use of time. Typical seasonal changes in time budgets in birds and mammals reflect shifts associated with mating, nesting, and preparing to migrate or overwinter. Finally, a question such as "Do males contribute equally in nest care in birds?" is easily answered when male and female time budgets are compared. Time budgets and need to allocate time for tasks may even influence the geographic distribution of species.[27]

Occasionally, time budget analyses can seem a little obvious. For example, mallard ducklings (see Figure 4.17) shift their time budgets depending on how resource-rich or poor their foraging area is.[28] Ducklings in resource-poor lakes spend more time moving, presumably searching for food . It is not surprising that hungry animals look for food, but consider this question from a different point of view: by observing the behavior of the ducklings, ecologists can infer much about the

FIGURE 4.17
Mallard ducklings adjust their activity budgets to reflect, among other things, foraging success. *Photo: Michael Breed.*

environmental conditions under which they live. The time budget of ducklings whose time budget is biased to feeding tells us that their environment is resource poor, relative to populations that invest proportionally less time in feeding. Likewise, seals adjust their time budgets during breeding season (see Figure 4.18).[29]

Recall that ethograms represent the behavior of a single animal, or what that single animal does when it is interacting with other animals. Two related approaches focus on transitions, or sequences of behavior. When the objective is to study single animals, the probability of shifting from one behavior to other behavioral possibilities is recorded. An interaction diagram captures the sequence of events when two or more animals communicate.

FIGURE 4.18

Habitat characteristics influence the amount of time that grey seals spend in the water during breeding season. *Photo: Michael Breed.*

Behavioral interactions can also be quantified in a *transition matrix*. In the matrix, the ethogram for the first animal in the interaction is listed in the descending column and the reaction of the second animal runs along the top of the matrix. The matrix shows how often each behavior of the first animal stimulates each possible response of the second animal. Through the use of transition matrices, comparisons can be made, for example, between sib and nonsib playmates, or between male–male encounters and male–female encounters. A more statistically formal way of analyzing the sequences captured in transition matrices uses *Markov chains*, which allow investigators to look at each behavioral transition in a long sequence.

KEY TERM A transition matrix quantifies the actions of one animal in an interaction and the resulting responses of the second animal.

120

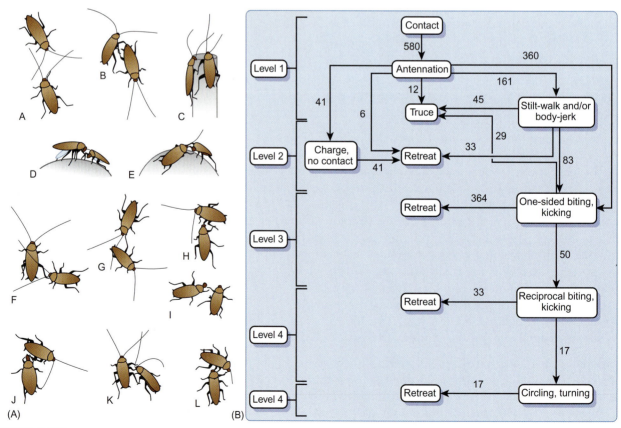

FIGURE 4.19

(A) These diagrams illustrate a variety of interactions that occur during aggressive interactions between cockroaches. Note that these interactions can be placed into a kind of flow chart (B) that shows the frequency with which one such interaction leads to another. *Adapted from Bell, W.J., Sams, G.R. 1973. Agressiveness in the cockroach* Periplaneta americana *(Orthoptera, Blattidae). Behav. Biol. Vol. 9, pp. 581–593.*

A special type of matrix captures the result of dominant/subordinate interactions between animals (see Figure 4.19). Animals A, B, C, and D are represented in the left column and also across the top of the matrix. The number of times each animal dominated another animal is then recorded. In a perfectly linear dominance hierarchy, A always dominates all of the others, B dominates C and D, and C dominates D. *Linear* means that the animals are lined up in hierarchical order, with strict rules of priority access to a resource. (Different resources may produce different hierarchies; thus, an animal that always competes successfully for water may not be as successful for nesting sites.) In the real world of animal behavior, dominance interactions are not quite so predictable, and non-linear relationships (A dominates B, B dominates C, but C dominates A) can occur. A somewhat different approach to this same type of question is to construct a sociogram, which is a diagram of how frequently animals are located near one another. Sociograms have recently come into the forefront in analyses of social networks in animals (see Chapter 13 for more information on social networks).

OF SPECIAL INTEREST: MAINTAINING SKIN, FUR, AND FEATHERS

Many animal species devote time and effort to keeping their body surface free of parasites, waterproof, or as good insulation. Simple behavior, such as licking or scratching, can be an efficient means of removing parasites, dead skin, and old fur or feathers (see Figures 4.20A and B). Bathing is also a common means of skin, fur, and feather maintenance (see Figure 4.20C). Animals with disease conditions or which are suffering from age often lose their ability to maintain themselves. This results in poor-appearing pelage or feathers, may increase external parasites, and may affect an animal's ability to court mates. Some animals, including many aquatic birds, groom oils to their feathers to improve their waterproofing and allow them to float on the water surface without absorbing water into their feathers (see Figure 4.20D). Oils on mammalian hair, such as lanolin on sheep, may have similar waterproofing functions.

121

FIGURE 4.20
Maintaining skin, fur, and feathers. *Photos: (A, B) Ben Hart; (C) Michael Breed; (D) Ben Pless.*

Like humans, animals have 24 hours in a day. Time budgeting by animals inevitably involves behavioral trade-offs. An animal that spends time feeding is doing that at the expense of another activity, such as courting potential mates. From an evolutionary perspective, the single critical result is having offspring and ensuring their survival—establishing genetic fitness. Natural selection directs all of an animal's activities to this ultimate goal. Self-maintenance and longevity are usually critical to reproduction, and keeping this point in mind helps us to understand the decisions that animals make as they choose among possible activities.

SUMMARY

This chapter has been an intellectual journey. The journey began by asking how animals make decisions about their behavior at any one moment. It continued through important regulators of behavior such as hunger, sleep, fear, and pain, and it finished with some practical tools for field and lab analyses of how animals spend their time.

Homeostasis is an umbrella term for the behavioral and physiological processes that animals use to maintain appropriate internal and external environments. Motivation and drive theory help us to understand homeostatic decisions by animals; although drive theory has been replaced by more sophisticated ways of looking at the physiology behind behavioral decisions, it remains a useful concept, especially in the area of animal welfare. Behavior plays a crucial role in the maintenance of animal homeostasis, and homeostatic needs often motivate behavioral choice.

Much of the discussion, such as that about self-directed behaviors and measures of pain, gives students of animal behavior objective, quantifiable tools for assessing animal welfare. Displacement behavior and redirected behavior, in the form of self-directed behaviors, are tools from drive theory that provide powerful insights for assessing animal welfare. Time budget analyses allow quantification and interpretation of behavioral decisions that maintain homeostasis. Time budgets are among the simplest, yet most powerful, tools available to biologists studying animal behavior in field settings.

STUDY QUESTIONS

1. What is the value of fear in an animal's life? What are the costs of fear?
2. At the zoo, you observe a primate pacing repetitively, plucking its hair, and rubbing against the cage bars until it wounds itself. What is your interpretation of this behavior and why? What would you suggest to the zookeeper in terms of changes in the animal's habitat?
3. Spend an hour watching an animal—any animal you can find will do as long as it remains in view for a while—and construct a time budget for that animal. When the animal switches between activities, is the motivation for the shift always apparent? Does the animal's allocation of time seem well adapted to its needs?
4. What is pain? How can pain be measured in a nonhuman animal?
5. What is a zeitgeber? What happens to an animal's circadian clock when the animal is kept in continuous light or dark?
6. What is satiety and how is it regulated?

Further Reading

Bouchard, T.J., 1994. Genes, environment, and personality. Science 264 (5166), 1700–1701.

Giebultowicz, J., 2010. Circadian window of opportunity: What have we learned from insects? J. Exp. Biol. 213, 185–186.

Harlow, H.F., Dodsworth, R.O., Harlow, M.K., 1965. Total social isolation in monkeys. Proc. Natl. Acad. Sci. USA 54, 90–97.

Harlow, H.F., Suomi, S.J., 1971. Social recovery by isolation-reared monkeys. Proc. Natl. Acad. Sci. USA 68, 1534–1538.

Hinde, R.A., 1959. Unitary drives. Anim. Behav. 7, 130–141.

Hughes, B.O., Duncan, I.J.H., 1988. The notion of ethological need, models of motivation and animal-welfare. Anim. Behav. 36, 1696–1707.

McFarland, D.J., Sibly, R.M., 1972. 'Unitary drives' revisited. Anim. Behav. 20, 548–563.

Toates, F.M., 1986. Motivational Systems. Cambridge University Press, Cambridge.

Notes

1. Blumstein, D.T., 2002. Moving to suburbia: Ontogenetic and evolutionary consequences of life on predator-free islands. J. Biogeogr. 29, 685–692.
2. Berger, S., Wikelski, M., Romero, L.M., Kalko, E.K., Roedl, T., 2007. Behavioral and physiological adjustments to new predators in an endemic island species, the Galapagos marine iguana. Horm. Beha. 52, 653–663.
3. Bouchard, T.J., 1994. Genes, environment, and personality. Science 264 (5166), 1700–1701.
4. Weiss, A., King, J.E., Figueredo, A.J., 2000. The heritability of personality factors in chimpanzees (*Pan troglodytes*). Behav. Genet. 30 (3), 213–221.
5. Svartberg, K., Forkman, B., 2002. Personality traits in the domestic dog (*Canis familiaris*). Appl. Anim. Behav. Sci. 79 (2), 133–155.
6. Hinde, R.A., 1959. Unitary drives. Anim. Behav. 7, 130–141.
7. McFarland, D.J., Sibly, R.M., 1972. 'Unitary drives' revisited. Anim. Behav. 20, 548–563.
8. Toates, F.M., 1986. Motivational Systems. Cambridge University Press, Cambridge.
9. Hughes, B.O., Duncan, I.J.H., 1988. The notion of ethological need, models of motivation and animal-welfare. Anim. Behav. 36, 1696-1707.
10. Troisi, A., Belsanti, S., Bucci, A.R., Mosco, C., Sinti, F., Verucci, M., 2000. Affect regulation in alexithymia—An ethological study of displacement behavior during psychiatric interviews. J. Nerv. Ment. Dis. 188 (1), 13–18.
11. Manson, J.H., Perry, S., 2000. Correlates of self-directed behaviour in wild white faced capuchins. Ethology 106 (4), 301–317.
12. McFarland, D.J., Sibly, R.M., 1972. 'Unitary drives' revisited. Anim. Behav. 20, 548–563.
13. Baker, K.C., Aureli, F., 1997. Behavioural indicators of anxiety: An empirical test in chimpanzees. Behaviour 134, 1031–1050.
14. McGinnis, M.Y., 2004. Anabolic androgenic steroids and aggression—Studies using animal models. Youth Violence: Scientific Approaches to Prevention Ann. N. Y. Acad. Sci. 1036, 399–415.
15. Castles, D.L., Whiten, A., Aureli, F., 1999. Social anxiety, relationships and self-directed behaviour among wild female olive baboons. Anim. Behav. 58, 1207–1215.
16. Maestripieri, D., Schino, G., Aureli, F., Troisi, A., 1992. A modest proposal—displacement activities as an indicator of emotions in primates. Anim. Behav. 44 (5), 967–979.
17. Leavens, D.A., Aureli, F., Hopkins, W.D., Hyatt, C.W., 2001. Effects of cognitive challenge on self-directed behaviors by chimpanzees (*Pan troglodytes*). Am. J. Primatol. 55 (1), 1–14.
18. Harlow, H.F., Dodsworth, R.O., Harlow, M.K., 1965. Total social isolation in monkeys. Proc. Natl. Acad. Sci. USA. 54, 90–97.
19. Harlow, H.F., Suomi, S.J., 1971. Social recovery by isolation-reared monkeys. Proc. Natl. Acad. Sci. USA. 68, 1534–1538.
20. Konopka, R. J., Benzer, S., 1971. Clock Mutants of *Drosophila melanogaster*. PNAS, 68, 2112–2116.
21. Giebultowicz, J., 2010. Circadian window of opportunity: What have we learned from insects? J. Exp. Biol. 213, 185–186.
22. Maerz, J.C., Panebianco, N.L., Madison, D.M., 2001. Effects of predator chemical cues and behavioral biorhythms on foraging, activity of terrestrial salamanders. J. Chem. Ecol. 27, 1333–1344.
23. Meschini, E.A., Gagliardo, F.P., 2008. Lunar orientation in sandhoppers is affected by shifting both the moon phase and the daily clock. Anim. Behav. 76, 25–35.
24. Yokogawa, T., 2007. Characterization of sleep in zebrafish and insomnia in hypocretin receptor mutants. PLOS Biology 5 (10), 2379–2397.
25. Chabot, C.C., Menaker, M., 2004. Effects of physiological cycles of infused melatonin on circadian rhythmicity in pigeons. J. Comp. Physiol. A: Neuroethol. Sens. Neural. Behav. Physiol. 170, 615–622.
26. Cuthill, I.C., Maddocks, S.A., Weall, C.V., Jones, E.K.M., 2000. Body mass regulation in response to changes in feeding predictability and overnight energy expenditure. Behav. Ecol. 11 (2), 189–195.
27. Korstjens, A.H., Lehmann, J., Dunbar, R.I.M., 2010. Resting time as an ecological constraint on primate biogeography. Anim. Behav. 79, 361–374.
28. Nummi, P., Sjoberg, K., Poysa, H., Elmberg, J., 2000. Individual foraging behaviour indicates resource limitation: An experiment with mallard ducklings. Can. J. Zool. 78 (11), 1891–1895.
29. Caudron, A.K., Joiris, C.R., Ruwet, J.-C., 2001. Comparative activity budget among grey seal (*Halichoerus grypus*) breeding colonies—the importance of marginal populations. Mammalia, 65, 373–382.

123

Learning

125

LEARNING OBJECTIVES

Studying this chapter should provide you with the knowledge to:

- Discover how learning allows animals to cope with unpredictable elements in their environment.

- Know that cellular changes, including protein synthesis and remodeling of synapses, underlie memory formation.

- Understand how the simple mechanisms of habituation and sensitization help animals sort important from unimportant information.

- See that the concept of conditioning provides important models for understanding how animals respond to repeated experiences.

- Know that in social learning, animals observe one another and gain critical information from the experiences of other animals.

- Appreciate the concept that play is important in learning and development for many animals, and provides experiences that enhance the development of both physical and social skills.

5.1 INTRODUCTION

Learning is a change in behavior as a result of experience. Genetic information is molded across generations by evolution; learned information is molded during each animal's lifetime. Learning is useful for animals with relatively long lives, because experience takes time. Recalling a previously experienced situation and using that experience to efficiently solve a current problem provide animals with key tools for survival and reproductive success.

> **KEY TERM** Learning is the modification of behavior due to stored information from previous experience.

CASE STUDY

Learning and Vampire Bats

The academic study of learning is frequently imagined as something restricted to white rats in mazes or chickens that press levers in strange boxes. It extends, however, to animals as exotic as vampire bats. With some eloquence, bats inform scientists about a unique kind of learning and the nervous system attributes that might be inferred from what the bats do and do not learn.

Taste-aversion learning is a special kind of learning that will be covered in detail later in this chapter. It is exactly what its name implies: it is a way that animals learn about taste and how to avoid some tastes. Animals that have varied diets benefit from such learning; as they sample different diet items, they can remember what tastes are likely to have unpleasant consequences. But what about animals that have restricted diets?

Consider the vampire bat (see Figure 5.1). (You will encounter this animal later in this chapter as well.) To say that the diet of a vampire bat is restricted would be a considerable understatement. Vampire bats consume the blood of living animals—usually cattle or horses. Because of this, their food is not variable and not likely to produce ill effects. John Ratcliffe

and associates wondered if vampire bats could nonetheless learn to avoid foods that made them ill.[1]

To be able to answer this question, they compared taste-aversion learning in three generalist bats with that in vampire bats (*Desmodus rotundus*). The generalist bats were the big brown bat (*Eptesicus fuscus*), the Antillean fruit-eating bat (*Brachyphylla cavernarum*), and the Jamaican fruit bat (*Artibeus jamaicensis*). The study took place in two phases. First, the investigators had to demonstrate that they could give the bat a cue about the blood quality, that is, alter the taste of the blood in a way that the bat could perceive. Vampire bats reacted to blood flavored with citric acid; they consumed less of it than when fed unflavored blood, indicating that they could taste citric acid. Because the other bats encounter citric acid in their natural foods and do not become averse to that flavor, they were given cinnamon as a novel flavor cue.

The second phase of the experiment was slightly more complex than what is reported here, but for our purposes, the following description is sufficient: for 10 days, the bats were given unflavored foods (blood for the vampire bats and fruit or proteinaceous substances for the other bats). On the 11th day, the bats were given flavored foods. Experimental animals (generalists and vampires) were injected with lithium chloride (LiCl), a treatment that induces extreme nausea. Controls were injected with physiological saline. For 3 days thereafter, all of the generalist bats exhibited a "profound" aversion to cinnamon-flavored food. Vampire bats, on the other hand, did not react to their citric acid-flavored food, even though it had made them ill.

Ratcliffe and coworkers concluded that vampire bats have lost the neural pathways involved in taste-aversion learning. After all, the bats' normal diet cannot possibly include spoiled or toxic food, and natural selection would not favor the retention of taste-aversion learning once vampire bat diets became specialized. This raises intriguing possibilities about the evolution of taste-aversion learning and the cost of its maintenance. In another comparative study, Paradis and Cabanac looked for taste-aversion learning in four species of reptiles and two species of amphibians.[2] They found strong avoidance of the trial food in the reptiles, but not in the amphibians, and hypothesized that

FIGURE 5.1
Vampire bat. *Photo: J. Scott Altenbach.*

(Continued)

within the vertebrate lineage, the origin of reptiles may have loosely coincided with the evolution of taste-aversion learning.

This case study asks Tinbergen's questions to greater or lesser extents. Questions of causation and survival value are clearly addressed; we understand how the animals learn to avoid tainted food, and we understand that in the case of generalists, this has adaptive value. It does not have adaptive value for the vampire bat, and this addresses evolutionary questions about how traits may be lost. This study was performed on adult animals, so little is known of ontogenetic aspects of taste-aversion learning from this experiment, although there is evidence from rodents that age has little influence on the ability to avoid toxic foods.

The optimal environments for behavior based on learned information and that based on genetic information are different. Chapter 3 focused on the behavioral value of inherited information when an animal's environment is predictable from generation to generation. One advantage of learned information is that it allows more tolerance for unpredictability across generations. Within any one animal's lifetime, however, predictability is what makes learning advantageous; there is no point in learning if the situation arises only once per lifetime. Predictability makes learned behaviors applicable to future events.

Chapter 1 discussed the uneasy relationship between comparative psychology and ethology in the study of learning and memory. Some students may have been taught that there is a fundamental gulf between laboratory studies on learning in animals such as mice, rats, and pigeons and field-based studies of learning in animals in natural environments. In fact, field studies of learning should be informed by the laboratory findings, and finding the bridges between the two areas of inquiry is often not too difficult.

The important questions that scientists ask about learning and memory can all be explored in detail, under controlled laboratory conditions, and they all have applications to field studies of behavior. In the field, the following questions are frequently asked: (1) How long is memory retained? (2) Is developmental stage important; for example, do nestlings show special proclivities for learning certain information? (3) Do gender differences in learning reflect the differences in male and female activities in a given species? (4) Does context affect what is learned or how well it is learned? The last question in particular reveals the critical relationship between studies of learning and *umwelt*. Learning ability in most animals is quite context-specific. For example, a bird species may be a champion at learning and remembering where it has cached food, but be rather dull at other tasks. This difference in abilities is reasonable, from an evolutionary standpoint, but reflects the complexity addressed in Chapter 2: How do we define intelligence?

This chapter first discusses some standard topics in the study of learning. After the basic structure of memory, habituation, and sensitization are covered, they are followed by imprinting and conditioning. These topics provide important background for studying the fascinating area of cognition, covered in Chapter 6.

5.2 LEARNING AND MEMORY

Steep learning curve is a term with ominous overtones. No one seems to want to experience a steep learning curve. What is this curve? **Learning curves** are an important tool in studies of learning because they allow learning, especially its rate, to be quantified. They are the visual representation of learning as defined here: a change in behavior as a result of experience, which takes time. Figure 5.2 shows a sample learning curve. The x-axis represents the number of trials or time; a measure of learning performance, such as percentage of correct attempts, is on the y-axis.

127

FIGURE 5.2
A sample learning curve. Learning curves are usually a graphical expression of change in performance as time/experience increases.

Thus, a steep learning curve means that behavior changes rapidly over time; in other words, tasks with steep learning curves can be challenging because they offer a lot to learn! What is undesirable, however, is a flat learning curve, which indicates an inability to modify behavior with experience.

The reverse of the learning process is captured in *extinction* and *forgetting curves*.[3,4,5,6] The concepts of extinction and forgetting will be considered in greater detail later in this section.

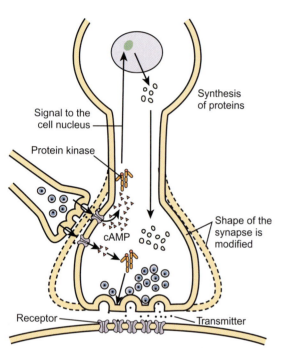

FIGURE 5.3
The cellular pathways of short-term memory. The key elements are the activities of protein kinase and changes in the shape of the synapse. *Adapted from: http://nobelprize. org/nobel_prizes/ medicine/laureates/2000/ med-5.jpg.*

Humans can create memories through imagination, that is, forming a mental image. Imagination is certainly linked to sensations—touch, smell, and vision— experienced in the past, but it also represents new and sometimes startling combinations of what is already known. Whether nonhuman animals can imagine is an interesting question, one that is revisited in Chapter 6 on cognition.

Memory generally starts with a sensory input. This results in sensory or "electrical" memory, which is transient; it just lasts a few milliseconds as the sensation is being formed and transmitted. Sensory input that has immediate importance is stored in short-term memory. *Short-term*, or working, *memory* is stored and available for relatively short periods of time— seconds or minutes. It has limited capacity; in most animals six or seven elements, or chunks, can be remembered at a time. Short-term memory is useful in remembering the sequence of a series of tasks. For example, a bee probably does not need to revisit flowers on the same plant within a short span of time. It can store previously visited locations in short-term memory and consequently avoid wasted effort.

At a cellular level, cyclic AMP, an important messenger within cells, and protein kinase, an enzyme involved in synthesizing proteins, are essential elements of the formation of short-term memories (see Figure 5.3). If protein synthesis is biochemically prevented, short-term memory fails. As expected, the proteins associated with short-term memory do not remain intact for long.

Of course, many memories last longer than short-term memories, and are used over extended time, up to weeks or months. These are called *long-term memories*. The gateway between short- and long-term memory is **selective attention**. Based on prior strong experiences or genetic programming, the brain recognizes certain types of information as being important in the longer term. An interesting example of selective attention is the *cocktail party effect*, which is the ability to sort sounds of interest, such as your name, from a very noisy background; more generally, it is the ability to sort the important from the unimportant, and store the latter information for future use.

Long-term memory also relies on kinases, but in addition involves transcription (RNA synthesis) and the growth of synaptic connections among neurons. The changes in the synapses appear to be particularly important for the formation of long-term memories. Maintenance of synaptic connections allows continued ability to access long-term memories.

Why Have Short- and Long-Term Storage of Information?

Assume that learning is a costly act for an animal. The costs include the time it takes to learn something, storage space for information in the brain, and in the neurochemicals required to store the information. If this assumption is true, then animals should learn only those parts of their experience that may be relevant and useful in the future.

In general, animals divide learned information into two types. The first is information that may be useful in the next few minutes, but which, if remembered later, just clutters thinking. A coyote might spy a mouse running under a specific bush. This information is useful while the coyote hunts for the mouse, but later (considering the large number of bushes in the coyote's world), it is probably not helpful. The same coyote may remember the location of a stream where it drinks; the water remains in the same location and a long-term memory of that location could be quite useful.

This logic predicts that short- and long-term memory should be handled somewhat differently in animals' brains and that the persistence of a memory should be tied to the length of time over which it might be useful. An excellent example of the importance of long-term memory in animals is the retrieval of cached food, a topic explored in greater detail later in this chapter.[7]

Where Is Memory?

If there is a holy grail of neurobiology and behavior, it is the *engram*—the internal record of a remembered event.[8] The basic principle underlying long-term memory is that synapses are "plastic," that their responses to the synapses of neighboring cells can change, and that this change becomes the record of a remembered event. The big question is where in the vertebrate brain this occurs. The most likely locations are the *neocortex*, the *amygdala*, and the *hippocampus* (see Figure 5.4).[9,10,11] Substantial debate over the past 40 years has failed to completely resolve this question. In addition, not all memory is formed in the same way. Human long-term memory is thought to be divisible into several kinds of memory (episodic, working, conditioning, and skill); it is entirely possible that different pathways and storage locales are invoked for each kind of memory.[12] In insects, long-term memory resides in the Kenyon cells of the *mushroom bodies*, which exhibit the same sort of synaptic plasticity that has been postulated for the parts of the vertebrate brain that are involved in memory.[13,14,15,16]

Reinforcement, Consolidation, Strength of Memory, and Forgetting

Some investigators have hypothesized that in addition to short-term and long-term memory, there is a third type of memory called *long-lasting memory*. While long-term memory lasts from hours to months, long-lasting memory stays with the animal for time spans in the range of months to an entire lifetime.[17] Long-lasting memory seems to differ from long-term memory in the strength of the associations formed between the remembered item and other memories.

KEY TERM The engram is the encoded storage of memory in the brain. Finding the exact nature of the engram has presented a deep scientific challenge.

KEY TERM The neocortex is the outer layer (gray matter) of the vertebrate brain.

KEY TERM The amygdalae are small nuclei (concentrations of neurons) within the temporal lobe of the vertebrate brain.

KEY TERM The hippocampus is a nucleus (concentration of neurons) within the temporal lobe of the vertebrate brain.

129

KEY TERM The mushroom bodies are concentrations of neurons in the upper part of the arthropod brain. So named because of their mushroom-like shape when viewed in cross-section, the mushroom bodies are thought to be the center of higher functions in arthropod brains.

FIGURE 5.4
The vertebrate brain, highlighting the hippocampus, which has been strongly implicated in memory.

Hippocampus

Understanding these strong associations involves a process called *reinforcement*, a term that occurs repeatedly in the study of learning and memory. Reinforcement is anything that alters the probability of a behavior. At the structural level, one hypothesis is that periodic reinforcement of an item in long-term memory causes the synaptic connections associated with that item to persist in time, thus strengthening the memory. *Periodic reinforcement* is a critical component of the process of *memory consolidation*, in which long-term memory is converted into long-lasting memory.

> **KEY TERM** Periodic reinforcement occurs when a behavior is sometimes rewarded, usually at unpredictable intervals. At other times, the behavior is unrewarded. Periodic reinforcement can be a powerful shaper of behavior. Humans playing a slot machine are periodically reinforced with jackpots; they continue to play even though the inevitable outcome is the loss of all their money.

As with selective attention, consolidation is an important process in separating valuable from expendable information. Access to experiences that are not reinforced and consolidated is, after a time, lost. It may in fact be that the memory itself is not lost, but that its synaptic connections disappear, so there is no route, within the brain, to find the information. In humans, this would explain why memories can be "dredged up" when a person is presented with an appropriate set of associations.

The strength of a memory may also be linked to specific contexts; the brain may be set up to facilitate strong, long-lasting memory of important items. Specific aversions, for example, are easily learned and difficult to forget. Humans and animals probably are primed in similar ways to place important information about social interactions, such as information used to recognize close relatives, into long-lasting memory quickly and with little or no reinforcement. Another example of strong memories comes in the context of imprinted information, discussed later; both the narrow time period in which imprinting occurs and the subsequent accurate recall point to the involvement of neural priming in the easy creation of strong, long-lasting memories.

130

BRINGING ANIMAL BEHAVIOR HOME: WHY ARE CATS PICKY EATERS?

Learning to avoid poisonous foods is clearly important for animals, and specific aversions help animals to learn quickly so they do not become ill a second time from the same source. Domestic cats are known for being finicky, a likely result of a mix of aversions and preferences for foods experienced when young; this behavior is notoriously challenging for the human caretakers of cats.[18] No one is willing to test whether his or her cat would rather starve to death than eat a particular brand of food. Exposure of cats to a wide range of foods during and immediately after weaning makes them more accepting of foods later in life. Could advertising this fact to the general public result in a change of feeding practices for cats?

Of course, memories do not always get stronger; sometimes they weaken. When items are lost from memory, they seem to disappear at an exponential rate.[19] *Forgetting* is largely a function of failure to reinforce a memory through repeated experiences; as indicated previously, reinforcement is critical in the transition of memory from short-term to long-term and from long-term to long-lasting. Forgetting is measured by observing whether the animal expresses the learned behavior over time, using periodic assessment. While forgetting is not exactly a mirror image of learning, in some animals, the rate of forgetting may be more easily studied than the rate of learning. For instance, it is difficult to study the rate at which caching animals learn their cache locations, but the rate at which they forget them can be more easily documented.

> **KEY TERM** Forgetting is measured by waiting periods of time and then reassessing whether the animal expresses the learned behavior. Likely, the behavior has not been totally lost, but the means within the brain of connecting with the behavior have weakened. Periodic reinforcement of a learned behavior can be quite important in maintaining access to the behavior.

Extinction, on the other hand, occurs when the expression of a learned response is suppressed.

> **KEY TERM** Extinction is the suppression of a learned response. The behavior is not forgotten, but its expression is gone.

Memory Capacity

Memory imposes costs on an animal; both the brain size and the metabolic energy used to support memory can be expensive.[20,21] Discarding unneeded memory reduces this cost. While even relatively small-brained animals, such as the honeybee, can have impressive capacity for memory,[22] storage selectivity leads to efficient use of neural capacity. Honeybees are very good at learning landmarks, which are essential for finding the way home, but rather less adept in learning experiments irrelevant to their natural setting.

Animals foraging on flowers learn how to "handle" the flower to get at the nectar or pollen. Many plants present complicated flowers that serve as puzzles for the forager to solve. A study of cabbage butterflies showed that these butterflies could learn a limited number of flower types; when presented with novel flowers, they "forgot" previously learned flowers. Subsequent studies have shown some ability of butterflies to generalize handling learned on one flower type to another flower.[23] They may even use what they have learned about places to lay eggs in their choice of flowers. In other cases, the inability of many foragers on flowers to learn how to handle multiple flower types helps to ensure fidelity to one or a few plant species; this, in turn, improves pollination, as pollen is carried to flowers of the appropriate species.

Less is known about memory and learning capacities in birds and mammals. While some investigators have postulated that learning capacity may be roughly linked to brain size, this assertion has been controversial. Across primate species, innovative behavior, social learning, and tool use are correlated with the size of the part of the brain devoted to "executive function" (neocortex and striatum). It is probably true that across a wide range of species, brain size is a meaningful measure of brain capacity. However, within species or when making comparisons between brains with relatively small differences in size, factors other than size have a greater influence on the animal's comparative abilities to learn and remember.

Of all the questions that Tinbergen asked, causation is perhaps the most difficult to answer when considering the physiology of learning and memory. Causation can also be addressed with models for learning—the ways in which animals learn—and these process-oriented approaches hold more promise. Because models for learning involve external events and experiences that are, in some cases, time- or stage-sensitive, they can also address Tinbergen's question about ontogeny.

5.3 BASIC MODELS FOR LEARNING

This section addresses the major ways that animals learn, that is, alter behavior in response to experience. Although that sounds simple enough, learning takes many forms. It occurs across animal taxa, and includes behaviors ranging from *Paramecium* avoiding shock to dogs sitting for treats, from goslings recognizing mothers to squirrels locating their winter hoard. Learning happens in many ways.

Imprinting

Imprinting was first identified as an important form of learning by Konrad Lorenz (see Chapter 1). Lorenz is best known among biologists for his pioneering work on imprinting in young animals. During a *critical period* early in their lives, many young animals learn the identity of their mother and father. Once learned, this information is firmly fixed and may be used later in life in identifying mates, in forming flocks, and in other social interactions. Lorenz found that by substituting himself for the mother during this critical period, he could induce

> **KEY TERM** Imprinting is the learning of a critical feature in the environment, such as the identity of a parent, at a young age and the retention of this knowledge for later use.

131

> **KEY TERM** A critical period is a short, defined period of time during which imprinting may take place.

young geese to imprint on him. Famous photographs of Lorenz show him being followed by geese imprinted in this way. Indeed, the word *imprint* refers to the seal of an important person placed on molten wax. The wax hardened and the image of the seal remained—permanent evidence of the authoritative origin of the document. Examples of imprinting include the mother's identity in geese and ducks, species-specific characteristics of birdsong, and chemical characteristics of water in natal streams for salmon.

Habituation and Sensitization

Much of the sensory input an animal receives is ignored—filtered out because it is unimportant. *Habituation* is one way this filtration can happen (see Table 5.1 and Figure 5.5).

> **KEY TERM** Habituation is the loss of response to a stimulus when the stimulus is given repeatedly.

It is an extremely simple form of learning, in which an animal stops responding to a stimulus after repeated exposure. Habituation is a key mechanism in filtering important from irrelevant information. It can function at the level of the receptor, or within the central nervous system.[24,25,26]

TABLE 5.1 Animals Are Most Likely to Habituate in Response to:
Weak stimuli
Less important (relevant) stimuli
Repeated stimuli
Frequent stimuli

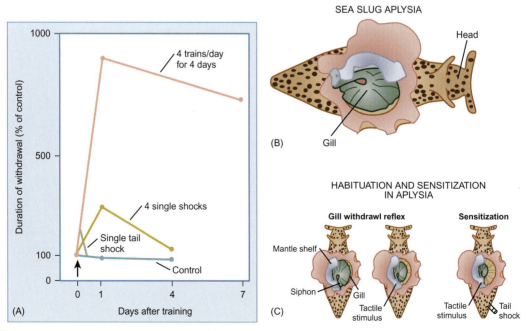

FIGURE 5.5

Habituation and sensitization in the sea slug, *Aplysia californica* (A). The gill and siphon are delicate areas and tactile stimulation results in a quick withdrawal of these structures (B). If stimulated 10–15 times in rapid succession, the gastropod soon habituates and the withdrawal response diminishes. The withdrawal response can be restored and the animal sensitized by one strong stimulus (e.g., shock) to the region. Sensitization can also occur independently of habituation, but the sea slug offers a clear juxtaposition of both (C). *Adapted from Nobel Prize Lecture, Eric Kandel, 2000.*

How does habituation occur at different levels in the nervous system? After a while, sensory systems may stop sending signals to the brain in response to a continuous or repetitive stimulus.[27] Lack of continued response to strong odors is a common example of sensory habitation. Habituation to complex stimuli may also occur at the level of the brain; the stimulus is still perceived, but the animal has simply "decided" to no longer pay attention.

Habituation is important in filtering the large amounts of environmental information bombarding an animal. By habituating to less-important signals, an animal can focus its attention on the most important features of its environment. A good example of this is species that rely on alarm calls to convey information about predators. Alarm calls cease when animals become familiar with other species in their environment that turn out not to be predators. Habituation is an important component of "not crying wolf" when nonthreatening animals come close.

Prairie dogs give alarm calls when mammals, large birds, or snakes approach. Individual prairie dogs are particularly susceptible to becoming food for a coyote, hawk, or rattlesnake, but collectively they are quite well defended because their alarm calls facilitate escape in burrows (see Figure 5.6). When prairie dog towns are located near trails used by humans, giving alarm calls every time a person walks by is a waste of time and energy for the group. Habituation to humans is beneficial in this context. By habituating and not responding to animals that have proven not to be a threat, the prairie dogs save time for foraging and display defensive behavior only when needed.

In animal behavior studies in the field, investigators often rely on the study animals becoming habituated to the presence of the investigator. Jane Goodall's famous studies of chimpanzees, for example, depended on the chimpanzees learning to tolerate her presence. This complex level of habituation differs greatly from learning to ignore an odor, for instance, but has a similar role in helping the animal ignore irrelevant stimuli.[28,29]

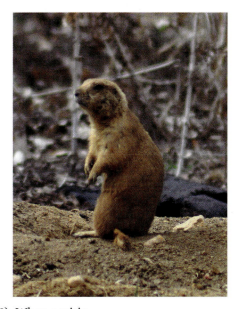

FIGURE 5.6
In prairie dog towns near busy trails, the rodents may habituate to the presence of people and their pets. *Photo: Michael Breed.*

133

KEY TERM Sensitization is an increase in responsiveness to a stimulus due to experience with that stimulus.

Sensitization looks like the opposite of habituation: as a result of experience, a sensitized animal becomes more responsive to a stimulus. In the case of sensitization, the stimulus is usually relevant and has potential consequences for the animal. Although sensitization and habituation seem like opposites, the neural pathways differ.

Conditioning (Associative Learning)

Much of animal learning is captured by the *conditioning* paradigm. Conditioning is the building of a learned association between a stimulus and a response. In its simplest form (probably what actually happens under field conditions), an association is formed between an action and a reward. *Associative learning*—that is, associating certain conditions or actions with certain outcomes— allows birds to efficiently find bugs under rocks and bees to find nectar in specific flowers. This simple type of learning increases opportunities for animals to behave efficiently, seeking resources where they have been found before, or collecting them in ways that have worked previously. It is reasonable to hypothesize that animals may also associate certain locations or conditions with unpleasant events and thus hone their predator avoidance abilities. This seems to be true for honeybees, which will avoid feeding at stations if they drink a nonlethal (but possibly unpleasant) pesticide there.[30]

KEY TERM Conditioning is a form of learning in which an association is made between an action and a reward for performing that action.

Much insight into animal learning and memory has been gained from the study of conditioning. Ethologists and behavioral ecologists often dismiss laboratory tests of conditioning as being so

FIGURE 5.7
Although bells are normally irrelevant to a feeding dog, the dog in Pavlov's laboratory associated the bell with the food. When this type of association is made, the bell alone, with no food present, can stimulate salivation. The formation of the association—a result of conditioning—transforms the neutral stimulus (the bell) into the conditioned stimulus (CS) and salivation becomes the conditioned response (CR).

134

far removed from the animal's "natural" biology as to be irrelevant. Observations of animals in the field, though, suggest the trial-and-error learning (see section 5.3) that often is used to gain experience with the environment is identical to associative learning. There are two major types of associative learning: classical conditioning and operant conditioning.

CLASSICAL CONDITIONING

Classical conditioning occurs when an animal associates a relevant stimulus, such as food, with a stimulus that would be irrelevant, were it not for the food. The animal is learning that two things are often associated; the animal has no control over either of them.

The most famous example of classical conditioning is the case of Pavlov's dog (see Figure 5.7). Nobel Laureate Ivan Pavlov was a gastrointestinal researcher studying insulin production in dogs. When a helper entered the kennel with food for the dog, a bell was sounded. After a while, the dog began to salivate at the sound of the bell, whether or not food was available. What happened to cause the dog to salivate at something as irrelevant as the sound of a bell? To begin with, consider a normal stimulus, such as the odor of food; this is likely to elicit salivation in dogs. Food (or its odor) is a highly relevant stimulus; a bell, of course, is not, unless something happens that causes the animal to associate the bell with food. In this situation, as classical conditioning begins, the odor is termed the *unconditioned stimulus* (UCS) and the response (salivation) is called the *unconditioned response* (UCR). No learning is required for the dog to salivate at the odor of food. When Pavlov rang the bell (a *neutral stimulus*—NS) at the same time he presented food (a UCS), and did so repeatedly and consistently, the dog formed an association between the bell and the food and began to salivate when it heard the bell. At that point, the bell became what is called a *conditioned stimulus* (CS) and salivation became the *conditioned response* (CR).

OPERANT CONDITIONING

In classical conditioning, the animal has no control over its environment; it is simply making associations between things that happen at the same time in that environment. In operant conditioning, the animal learns to operate an environmental feature to produce a consequence, often a reward. For example, people often think of rats and pigeons that press levers in boxes (the "Skinner box") when they think of operant conditioning; a hand-shaking dog has probably been operantly conditioned as well.

In laboratory studies of operant conditioning, an animal is typically allowed to explore its environment and to perform an action, such as pressing a lever. This can be viewed as a trial-and-error phase of learning. If the animal chooses the "correct" operation, it is rewarded with a food item. Thus, its action has a conseq   uence. Repetition of the operation and the resulting reward cause the animal to associate the two. More complex series of actions can be chained together by requiring the animal to add additional actions to what it has already learned. For example, two lever presses might be required before the reward is delivered.[31]

Many animals can associate a reward with a seemingly irrelevant action if the reward is received immediately after the action. Dogs can be taught to "shake" hands based on their expectation of receiving a food reward after shaking. Once the dog associates the hand signal and the extension of its paw, the food reward is no longer necessary. Such behavior is built on positive reinforcement—association of the desired action with the receipt of a reward.

Not all animals can learn something as irrelevant to their biology as shaking hands. Ease of conditioned learning is often dictated by the importance or relevance of the events to the animal species' evolutionary history and ecological conditions. Animals may be *prepared* to learn a task that is relevant to their evolutionary history (e.g., food choices in rats), they may be *unprepared* to learn a task that has no relevance, and they may be *contraprepared* to learn a task if their evolutionary history inclines them in the opposite direction.[32,33,34] By the same token, although food rewards are frequently invoked in this discussion both for convenience and because many animals will work for food, in fact, there is much individual variation in what animals value. For some domestic animals, a favorite toy is more important than food and thus may yield better results in training.

OF SPECIAL INTEREST: PREPAREDNESS

Much like *umwelt*, the concept of preparedness tells us that we must take the evolutionary history and inherent tendencies of the animal into account when designing learning studies. Not every question is suitable for every animal. When we fail to pair appropriate questions with appropriate animals, we, not the animal, fail to learn.

An additional stimulus can be delivered at the same time as the food reward. With dogs, horses, and other domestic animals, a *clicker* is often used in training; this approach uses a combination of operant and classical conditioning (see Figure 5.8). The sound of the clicker is delivered at the same time as the food reward. After a while, the food can be omitted from the sequence, and the animal accepts the "click" as the reward in place of the food. Clicking can then be used to modify the animal's behavior as much as the initial reward might have been. In clicker training, as with most associative learning, timing is everything: if the click is not delivered at the right instant, that is, nearly simultaneously with the food, the animal will not associate the click with the food reward. Try *shaping* the behavior of a friend or roommate by using only a clicker. It can be a simple behavior, such as touching a light switch. The difficulty and frustration that can be encountered in this game teaches the importance of appropriate timing, an essential aspect of animal training. Clickers also implement another important aspect of conditioning: consistency. Unlike a human voice, the sound that a clicker makes does not vary. Consistency and timing are crucial to successful conditioning of animals. Clickers are especially useful in conditioning large animals and aquatic animals.

> **KEY TERM** A clicker is a device used in animal training. The trainer builds an association between reward and the sound of the clicker so that the click itself becomes rewarding. The clicker can then be used to help the trainer get the animal to understand what behavior is desired.

> **KEY TERM** Shaping is the incremental training of an animal to perform a specified set of complex movements. Training a dog to roll over or do a head stand are examples of shaping.

135

FIGURE 5.8
Clicker training of dogs (and dolphins and chickens and just about anything that can hear!) combines both classical and operant conditioning. First the dog is taught to associate something very good (e.g., a treat reward) with the sound of a clicker (inset); this depends on classical conditioning. Eventually, the association between "reward" and the click is strong enough that the click can be used instead of the reward in operant conditioning, reinforcing desired behaviors by "rewarding" them with clicking. This is thought to be more effective than verbal praise, for instance, because it is more consistently delivered (the click always sounds the same) and delivered more quickly. Timing and consistency are essential in effective training. *Photos: Janice Moore.*

TABLE 5.2	The Terminology of Punishment and Reinforcement
Punishment	• Always decreases the likelihood of a behavior
	• Is positive if it causes a decrease in the probability of the target behavior because of the addition of an aversive stimulus immediately following the target behavior
	• Is negative if it results in a decrease in the probability of the target behavior by removing a desired stimulus immediately following the target behavior
Reinforcement	• Always increases the likelihood of a behavior
	• Is positive if it causes an increase in the probability of the target behavior when a stimulus is added immediately following the target behavior
	• Is negative if it causes an increase in the probability of the target behavior when an aversive stimulus is removed immediately following the target behavior

Training by Positive and Negative Reinforcement

Much learning results in a change in the probability of a behavior occurring. Behaviorists use a specific vocabulary for the discussion of this sort of learning, summarized in Table 5.2. Thus, giving a treat to a dog when it sits on command is *positive reinforcement*: it increases the probability of the "sit" response (that is, it reinforces the behavior) and does so by adding a treat. The addition of the treat (or anything else) is deemed positive. Turning away from a dog that is obnoxiously demanding attention is *negative punishment*; it is punishment because the behavior is less likely to occur in the future, and it is negative punishment because something (attention) has been removed.

These distinctions do not trip off the tongue and can degenerate into semantic squabbles. As a result, many behaviorists are often now content to use *reinforcement* and *punishment* without adding qualifying terms such as *positive* or *negative*.

Negative reinforcement might seem to be as effective as positive reinforcement. Indeed, many animals can be taught boundaries or limits on behavior based on painful experiences, such as electric fences or shock collars. Most training using negative reinforcement, though, is less effective than positive reinforcement. The reason is that when positive reinforcement is used, an animal is encouraged to associate an act with the reward; in addition, the trainer is planning in advance to elicit the desired response. Thus, the trainer is prepared, and the reinforcement occurs in a timely fashion—almost immediately. Returning to the example of dog shaking hands, the training is likely to begin by showing food to the dog. This gets the dog's interest and helps it to associate the desired behavior and subsequent reward.

Positive punishment, on the other hand, generally occurs after an act that the trainer cannot predict and therefore is unprepared to condition. There is often a significant lapse between the behavior and the consequence, making association difficult. Thus, if a dog urinates on the floor and then gets hit, will it associate the punishment with the bad behavior? Probably

BRINGING ANIMAL BEHAVIOR HOME: NEGATIVE REINFORCEMENT AND COMPANION ANIMALS—DOES IT WORK?

The plastic sheet shown in Figure 5.9 is an example of a negative reinforcement device used in training dogs to stay off furniture. It has embedded wires, which deliver a slight electric shock when the plastic is touched. While not harmful to the animal, the shocks are annoying and cause the animal to avoid the plastic; it is not a comfortable place to nap. As a result, the dog learns to avoid the plastic, but not the furniture. This pattern of learning is very typical of punishment training; the punishment is learned, but the animal fails to form an association between the punishment and the act that the trainer is trying to discourage and has no opportunity to learn a behavior that the trainer might want to encourage. The association is between the plastic and the shock, not between the furniture and the shock. Consequently, when the plastic is not present, the animal uses the furniture as a bed. Long-term experience with the shocking device does not lead the dog to generalize from the punishment to the principle that the trainer is trying to achieve. In the same manner, dogs that are trained with shock collars often must wear a "dummy" shock collar when not in training because they quickly associate the collar with the shock and the absence of the collar may not result in the desired behavior.

When a dog is presented with an alternative bedding location, in addition to experiencing negative reinforcement associated with bedding on furniture, it learns to use the alternative location (as in Figure 5.9). The positive reinforcement—a warm, comfortable location—presented as an alternative to the punishment enhances the animal's ability to learn appropriate behavior. In general, much training of domestic animals has moved toward a "pure positive" stance, based on the reality that animals work harder and faster to gain rewards than they do to avoid unpleasantness. The agility competitions that you may have seen on television follow this method. The dog must do the course correctly but must also do it at great speed. To encourage that speed, trainers do not use punishment.

FIGURE 5.9
An electrical punishment mat can be used to prevent animals from making themselves comfortable on furniture. The soft bedding on the right serves as a positive reinforcement. *Photos: Michael Breed.*

FIGURE 5.10
A Steller's Jay (*Cyanocitta stelleri*), a common bird in mountain habitats of the western United States. The Steller's Jay can foil ingenious devices meant to keep it out of camp food, and it is an aggressive predator on the eggs of other birds. Its ability to obtain food in the face of strong defenses has gained this species a reputation for cleverness and an ability to use trial and error learning in solving problems. *Photo: Michael Breed.*

137

not, because unlike the attention the dog might devote to a treat, there is no anticipated punishment, nor is the trainer prepared to deal with the dog in a timely fashion. Even if the association is made, how does the dog know what alternative behavior will receive a positive reward? Most dogs that intend to urinate become restless and sniff around their environments. Moving such a dog to an appropriate location, combined with positive reinforcement when urination occurs in the desired place, is much more likely to allow the dog to associate urination with an approved location and a job well done. Understanding these critical differences between reinforcement and punishment (both positive and negative) is an important step in learning how to train animals.

Trial-and-Error Learning

Trial-and-error learning allows learning of optimal responses for conditions by attempting a solution to a problem, and if that solution is incorrect, then another is tried until the correct solution is discovered (see Figure 5.10). The operant conditioning model fits well for this type of learning. Note that trial-and-error learning can be costly, depending on the nature of the "error." In the next chapter, we will explore the possibility that some animals can do trial-and-error learning mentally, without taking risks. This is what humans do when they think through consequences for actions before performing the actions.

KEY TERM Trial-and-error learning occurs when an animal attempts a series of solutions of a problem, eliminating possible solutions that don't work.

Taste-Aversion Learning

By far the most important context in which an animal needs to learn quickly occurs when a behavioral choice has put that animal's life in danger. Eating a poisonous or tainted meal can lead to the formation of a specific aversion to that type of food based on a single experience. *Emesis*, the process of vomiting a poisonous food item, costs an animal not only that specific meal, but other food items and often the ability to eat for a period of time. For animals that are on tight energy budgets, this loss can be costly; thus, selection has favored a route for quickly learning to avoid foods that result in illness and for retaining this information as strong, long-lasting memory.

Taste-aversion learning is widespread among animals. It is a learned pattern of aversion to a specific food. The ability to learn food aversion has been favored by natural selection and helps animals avoid poisonous foods. While mice vary in their ability to taste and respond to aversive stimuli, some species demonstrate an ability to form strong specific aversions.[35] Poisonous marine larvae can elicit taste aversion in their predators.[36] Interestingly, vampire bats, which feed on blood and would not normally encounter toxic food, do not demonstrate an ability for taste aversive learning to unpalatable food, although other bats have that ability (see the case study at the beginning of this chapter for more detail).[37] For humans, eating a food and experiencing strong nausea and vomiting shortly thereafter result in an aversion to that food that often lasts for decades. Rats have excellent abilities to form longstanding food aversions, a facility that allows them to avoid human attempts at poisoning them. Birds that eat insects quickly learn to avoid the poisonous ones, and that information is retained over the life of the bird.

Taste-aversion learning facilitates the evolution of chemical defense by plants and animals. A plant or animal that can produce or obtain a toxin that causes emesis has an excellent chance to avoid being eaten because its potential consumers will develop specific aversions to the food type (see Figure 5.11). Perhaps the best-studied examples of this are those remarkable omnivores, the rats, as well as birds that encounter toxic butterflies.[38,39,40,41,42,43]

Taste-aversion learning may appear identical to associative learning, but it is remarkably different. First, the aversive substance needs to be experienced only once for the aversion to be formed; in most associative learning tasks, repeated exposure is needed. Second, the taste-aversive substance can cause illness up to 24 hours after ingestion and still form the aversion; in classical conditioning, a gap of only seconds between the NS and the US will slow the learning process. In fact, taste-aversion learning can even occur under anesthesia. To summarize, taste-aversion learning and associative learning are vastly different, and there is a high likelihood that they use different neural mechanisms. This may be why vampire bats, given their restricted diet, have lost the unnecessary ability to do taste-aversion learning while retaining other forms of learning.

FIGURE 5.11
A corvid, exhibiting taste-aversion learning upon beginning to consume a monarch butterfly.

Cache Retrieval

Many animals *cache* (store) food items, leaving them for later retrieval. Some animals, such as mountain lions and ravens, leave spoilable food, such as

KEY TERM A cache is a location where food items are stored.

138

deer carcasses, for relatively short periods (hours or days) before retrieval and consumption. Others, particularly rodents (squirrels and chipmunks; see Figure 5.12) and birds (chickadees, jays) store seeds for weeks or even months before using them later. Seeds are particularly appropriate for long-term caching, and many seed-eating ants, birds, and mammals cache seeds as they drop from plants.[44,45,46,47,48,49]

Caching animals often *scatter hoard*—hide food items throughout their home range. Dispersing food items in this manner has the advantage of not creating a single large food resource in one location, which might be quite attractive to other animals, but has the disadvantage of requiring retrieval of food from dozens, or even hundreds, of locations. The alternative to scatter hoarding is to assemble the food items in a central place, such as a hollow tree or a cavity in the ground. This is called a *larder hoard*. Such a cache doesn't present great challenges in finding the food at a later time but may require defensive efforts.

Obviously, then, cached food is valuable only if it can be retrieved later. The most commonly hypothesized mechanism for efficient cache retrieval invokes learning cache locations and returning later to those spots. Some animals actually do this, but two other hypotheses must also be considered: reforaging and searching by rule. Each of these hypotheses can reasonably explain cache retrieval.

FIGURE 5.12
Tree squirrels cache food for use during the winter.
Photo: Jeff Mitton.

LEARNED CACHE RETRIEVAL

Psychologists refer to the type of memory used in learned cache retrieval as *episodic memory*. This means that the memory is attached, or associated, with a specific experience in an animal's life. In this model, each caching event is an episode that is stored in the caching animal's memory. Birds in the families Paridae (chickadees and their relatives) and Corvidae (jays, crows, and their relatives) are champions at retrieving cached food from remembered locations (involving spatial memory). Rodents, such as squirrels, may rely partially on memory but have other strategies that are discussed later.[50]

KEY TERM Scatter-hoarding animals distribute their cached food throughout their home range, so no one cache contains a large proportion of their food store.

KEY TERM A larder hoard is a large cache of food, often all or a large fraction of an animal's stored food placed in one location. The honey stored by a honeybee colony is a larder hoard.

The easiest experimental approach to determine if learning is involved in cache retrieval is to allow an animal to cache food in an observation area. (This could be a natural home range, if the behavior can be seen throughout the area, or an enclosure.) The rate of cache recovery is then observed and compared with that of a second animal that is allowed to search the area. To prevent observational learning of cache locations by the second animal (see following sections), the animals are allowed to forage for cached items separately. In this type of experiment, which has been conducted using both birds and mammals, the caching animal has a significant advantage over the other animal in finding food if it learned cache locations, and these data are interpreted as support for the learned retrieval hypothesis. Whenever an experiment tests the hypothesis of learned cache retrieval, two alternative hypotheses, *reforaging* and *searching by rule*, must always be considered.

KEY TERM Reforaging is finding cached food using cues produced by the cached items. Decomposing animal flesh may produce odors that are used in reforaging for cached items.

KEY TERM Searching by rule is using a simple rule to guide a search for cached items. For example, if food is always cached under rocks of a certain size, then the animal looking for cached food may be able to find the food by turning up rocks of that size. In that case, no learning of cache locations is necessary.

REFORAGING

Simply put, reforaging is a strategy in which an animal places food items in its home range, where they may be easily found, but it does not actually learn the locations of the food items. Instead, it searches them out. While not as efficient as learned retrieval, reforaging has the advantage of not requiring elaborate mental capabilities. In rodents, reforaging may supersede learned retrieval under moist conditions, when seeds can be retrieved by their scent.

SEARCH BY RULE

Another strategy for finding hidden food stores is to have rules by which they are cached and then to use those rules as clues in a retrieval search. For example, the rule may be "always cache food under rocks." The caching animal then knows to look under rocks for food, even if it does not remember locations of the food. Probably all cache retrieval involves search by rule at some level; "reasonable" (in the animal's view) cache locations are considered when looking for a cached item, whereas "unreasonable" locations need not be bothered with in the search. The role of search by rule in discovery of cached items is difficult to quantify, however, and it is simply a factor that should be considered in design of cache retrieval experiments. Human filing systems—at least ones that work—are based on search by rule.

PILFERAGE

Pilfering of caches by other animals—both of the same species and of other species—is an important problem for caching animals. Even for scatter hoarders, *pilferage* may substantially reduce the cache. Pilfering animals have two strategies. One is to search for food in another animal's home range. Interestingly, weather conditions greatly affect the effectiveness of pilferage. When it is dry, yellow pine chipmunks and deer mice use spatial memory to find their stored food, and a caching animal has a substantial advantage over a pilferer in finding food. Moist conditions bring out the scents from cached seeds, enabling pilferers to sniff them out. Under such conditions, the cacher has little or no advantage over the pilferer in discovering seeds.[51]

> **KEY TERM** Pilferage is stealing food from another animal's cache.

The second strategy employed by a pilfering animal involves observational learning. Scrub Jays, when observed caching by another bird, are likely to recache the food at a later time. Similarly, Pinyon Jays and Mexican Jays may watch other birds and learn their cache sites so that they can pilfer the sites later. Ravens display elaborate avoidance behavior to keep other ravens from observing their cache sites. These observations invite consideration of the next topic, observational learning.[52,53]

5.4 SOCIAL LEARNING: TRADITIONS AND "CULTURAL" TRANSMISSION OF INFORMATION IN ANIMALS

Observational learning occurs when one animal watches the actions of another and learns from those actions. This may be as simple as learning the location of a food source or as complicated as learning a sequence of actions that leads to a reward. Many animals can do this; good experimental evidence is available for the common octopus, quail, rats, and a variety of primates. Sometimes social learning can help an animal to avoid a risky location.[54] This chapter focuses on some examples of observational learning in which a task or preference is learned. In later chapters dealing with cognition (6) and social behavior (13), observational learning is placed into the contexts of social groups and animal culture.[55,56,57]

Observational Learning in Octopi

The common octopus provides a classic example in the study of observational learning (see Figure 5.13). When one octopus is trained to perform a task with a reward, another

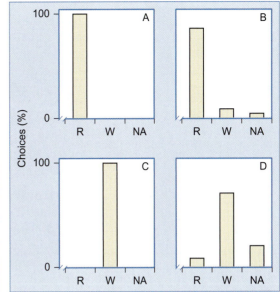

FIGURE 5.13
Observational learning in octopi. (Left) In this experimental set-up, two octopi have shelters in enclosures that are separated by a glass divider. One octopus watches while the other chooses between retrieving a red or white ball and is rewarded for retrieving a ball of a specified color. (Right) A. When a demonstrating animal has been trained to retrieve the red ball, it always does so. B. The observing octopus almost always follows the behavior of the demonstrator and retrieves the red ball. C. The demonstrator retrieves the white ball. D. The observer almost always does likewise. *Adapted from: Fiorito, G., Scotto, P., 1992. Observational learning in Octopus vulgaris, Science 256 (5056), 545–547.*

observing octopus can duplicate the task. This example demands attention on several levels. First, it shows that a passive observer can note the details of another animal's behavior, see that the behavior is rewarded, and then perform the behavior in an attempt to receive the same reward. Second, the 1992 discovery of this ability in a mollusk stretched the limits of how scientists imagined the operation of nonhuman animal minds. This study was on the leading edge of a remarkable scientific expansion in considering learning and memory and how they might function in a broad range of animals.[58] It helped open the door to an increased appreciation for complexity of learning and memory in a broad range of animals.

Food Preferences and Bait Shyness in Rats

"Humans try to poison rats, and rats learn to avoid the poisons." This simple statement captures the essence of the relationship between humans and a persistent and unwelcome animal associate. Rats easily form specific food aversions to baits that are laced with poison: if they survive one poisoning, they are unlikely to eat the same thing again. But, can rat pups use observational learning to shape food choices? Careful experiments show that rat's food preferences can be influenced by chemicals that are transmitted from the mother to the pups when they are still in the uterus. Also, after birth, when pups begin foraging, they feed near adults, using the same foods, and food preferences are learned via shared feeding experiences. Observational learning of food preferences is probably widespread in vertebrates, but this hypothesis has been rigorously tested in only a few species.

Birds and Milk Bottles

The most famous putative example of observational learning was the spread of the ability to open milk bottles among Blue Tits and European robins in Great Britain (see Figure 5.14). These birds learned to rob cream from the tops of milk bottles during the early part of the twentieth century, and the blue tits later adapted to the use of aluminum foil seals on the bottles, learning to tear them to access the cream. There are actually two hypotheses that could explain the increase in these behaviors: First, birds might observe other birds feeding in this manner and adopt the behavior. Second, each bird might, independently, discover this feeding option. This second possibility is particularly likely if a bottle opened by one bird serves as a clue to other birds that the bottles are food resources. David Sherry and Jeff Galef, leading scientists in the study of observational learning and cognition, argued for just this interpretation of this classic example.[59]

Although Sherry and Galef's reinterpretation is controversial, this example illustrates the need for carefully controlled experiments and an open mind to considering alternative explanations. Indeed, this alternate hypothesis is the nemesis of many who would conclude that observational learning occurs in a species or that cultural transmission of a behavior has taken place. If one animal has performed the novel behavior, then it is obviously possible for such a novelty to appear in the population. At that point, how does a scientist determine—especially under field conditions—that it has not appeared multiple times? The Blue Tits offer a mystery, made all the more frustrating because this intriguing behavior has disappeared from birds in the United Kingdom, perhaps because people now rely more on buying milk in stores rather than having it delivered.

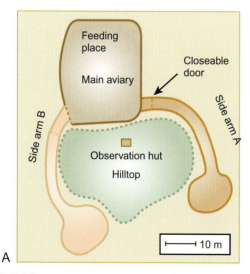

FIGURE 5.14
A Blue Tit robbing cream from a milk bottle. The short lives of these small birds combined with the increasing dependence on supermarket purchase of milk—and lowfat milk, at that—mean that the behavior is rarely observed in modern Britain.

142

Cache Raiding

Ravens are very clever birds and are highly capable of observational learning. One context in which they demonstrate this is in pilferage of each other's caches; they keenly observe each other's cache locations and pilfer if given the opportunity (see Figure 5.15).[60] Observational learning can play a role in exploiting other species' food caches as well.

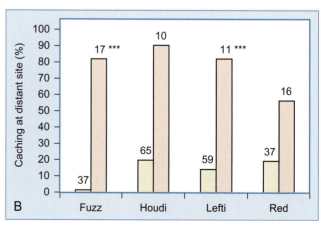

FIGURE 5.15
When ravens kept in a group in an aviary cache food, they respond to the presence of other, potentially raiding, ravens. Each of four ravens (names across the bottom of B) flew much further to cache food when other ravens were present (pink bars) than when other ravens were absent. This demonstrates the potential of observational learning by ravens stealing other birds' caches, and suggests that the presence of potential pilferers causes ravens to be more devious in their selection of cache locations. *Adapted from Heinrich and Pepper, 1998. Animal Behavior 56, 1083–1090.*

OF SPECIAL INTEREST: TOOL USE IN WOODPECKER FINCHES: OBSERVATIONAL OR TRIAL AND ERROR LEARNING?

One of the classic examples of tool use in animals is the Woodpecker Finch (see Figure 5.16). This bird, which lives in the Galapagos Islands, can use sticks or thorns as tools to remove insects from their hiding places on trees. Many biologists had assumed that tool use was transmitted among birds by observational learning. But, in fact, a recent study showed that birds without demonstrators were as likely to use tools as birds who observed others using tools. This experiment suggests that each bird "invents" the behavior of tool use through trial and error.[61] This example reinforces the cautionary note sounded by the Blue Tit example.

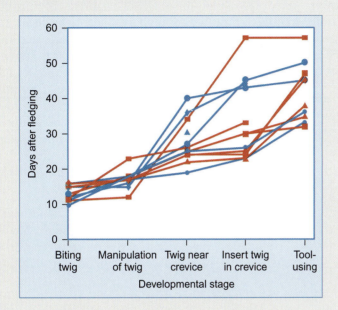

FIGURE 5.16
Each line in this graph represents an individual woodpecker finch using tools to collect food. Woodpecker finches progress through the developmental stages of tool use, regardless of whether they observe a tool-using bird (red lines) or not (blue lines). *Adapted from: Tebbich, S., Taborsky, M., Fessl, B., Blomqvist, D., 2001. Do Woodpecker Finches acquire tool-use by social learning? Proc Biol Sci. 268, 2189–2193.*

143

Survival Value of Learning

Tinbergen asks about the survival value of behavior, and aspects of learning clearly address that question. An animal that learns can avoid dangerous situations and can use a variety of cues, ranging from environmental correlates to the behavior of other animals, as a way to find food and shelter.

Tinbergen's question about evolution of behavior is a bit trickier in the world of learning, but it can nonetheless be addressed. Some workers have examined fossil skulls in an effort to determine brain size or shape. Others look for phylogenetic patterns in the distribution of learning abilities. There may also be life history correlations: learning takes time, so longer-lived animals are more likely to make use of learning. Social animals also have a propensity for learning; they have many social rules and relationships to understand and many group members to remember.

5.5 PLAY, LEARNING, AND DEVELOPMENT

Play is not only about learning, and because of that, the topic of play could easily be placed in at least three other chapters. Play is the engagement of an animal in seemingly purposeless activity that has no immediately apparent survival value. Play may help to develop reasoning skills, so a discussion of play could be considered in Chapter 6 on cognition. Some kinds of play may relate to developing skills for hunting, and might go into Chapter 9 on foraging. Still other types of play seem essential for developing social skills, and could easily find a home in Chapters 13 and 14, which deal with social behavior. Play is discussed in this chapter on learning, but notice the links between play and other topics and refer to this chapter as needed.[62]

Play is much more common in mammals than other kinds of animals, but arguments have been made for the existence of play in birds and in some invertebrates. Perhaps play is more easily recognized in mammals because there are striking similarities among mammals in play behavior; rat or monkey play exhibits the same basic patterns. These play patterns in mammals are mirrored in human play. Playing animals appear to be deeply engaged in an activity without immediate survival value. They may use an object, such as a stick or a fragment of a dead animal, in their play. Oftentimes play involves social signals of aggression, courtship, copulation, or parenting, but the signals are incomplete, muted, or delivered by animals that are physically incapable of performing the complete behavior because they are too young or not fully developed. At other times, play resembles hunting, but prey are absent. Signals such as play bows (see Figure 5.17), invitations to chase, and solicitations for tug of war games are seen in many species.

Play serves critical roles in the behavioral development in many animals. Play is often divided into two distinct categories: *autoplay*, or playing alone, and *social play*. Autoplay and social

> **KEY TERM** Autoplay is a behavior in which an animal plays alone, sometimes using an object to play with. A kitten with a ball of string is engaging in autoplay.

> **KEY TERM** Social play is a behavior in which animals play in pairs or groups; interaction with other animals is a component of the play behavior.

144

A

B

C

FIGURE 5.17
A. A dog's playbow. Dogs preface (and punctuate!) their play with this unmistakable signal.
B. Dogs at play.
C. Dogs seem to play "fair," at least where puppies are concerned. In this tug of war, the adult dog could have easily won, but didn't.
Photos: (B) Michael Breed; (C) Janice Moore.

play can both serve as platforms on which an animal's brain development is stimulated. An animal that plays gains motor skills, fosters muscular development and coordination, learns important details about its environment, and can experiment while being protected by parents or other group members. Social play can have the added functions of helping animals to learn social skills, to form bonds with other group members, and to establish social relationships that may persist into adulthood.[63]

In chimpanzees, play appears to allow an individual to assess the potential for cooperative behavior in itself and other members of its social group.[64] Self-assessment may be an important motivation for play, although it is difficult observationally to separate self-assessment from self-improvement (getting better at a motor skill by practicing it).

Play can be risky: animals may be injured or exposed to predation while playing. The cost of play implies significant benefits, even if they are difficult to demonstrate with any rigor. Rough physical play is common in carnivores and primates. Social play is often repressed in habitats in which risk is high.[65,66] Interestingly enough, animals whose play is restricted in natural habitats appear to develop "normally," but highly restrictive captive environments result in adults with abnormal social behavior.[67] Recently, play has been proposed as a measure of animal welfare; animals maintained in good conditions are thought to play more than animals under stressful conditions.[68]

Do animals play fair? The root of fair play lies in the hypothesis that a "social contract" keeps play within bounds so that injury is rare. In fair play, boundaries with respect to potential injury, turn-taking, and resource sharing are respected. Observations of play suggest that, indeed, young animals do tend to play fair. Why might this be? A simplistic explanation is that animals have a "theory of mind," which allows them to imagine the emotions or feelings of other animals and act in a sympathetic manner. This type of hypothesis is discussed in more detail in Chapter 6. Alternatively, does play generally occur with close relatives? Kin selection could put bounds on rough play. There is strong evidence in primates that animals "self-handicap" in play, evening the playing field among young of different sizes, ages, and abilities.[69]

DISCUSSION POINT: PLEASURE IN ANIMALS

The concept of "pleasure" is subjective and difficult to define for a human, much less for an animal. Do you think your dog or cat derives pleasure from play? Is it a reasonable hypothesis that a sensation of pleasure serves as an internal reinforcer for behavior that is important for survival but that is not otherwise directly reinforced? In other words, because play does not result in immediately receiving food, mating, shelter, or the like, has "pleasure" evolved as a means to ensure that play occurs? How comfortable are you with attributing the subjective state of "pleasure" to a dog or cat?

Play bouts involving more than one species of animal are intriguing. Sometimes interspecific play is pretty one-sided; it amounts to harassment. Magpies appear to be playing when they pluck hairs from the tail of a tethered dog, and certainly cats "playing" with mice before finishing them off is not play for the mouse. Other cases seem more mutual. Squirrels descend to lower branches on trees and chatter at dogs, which respond with jumping and barking. Neither animal seems to take these bouts seriously; the squirrels are safely out of the reach of the dog, and after some experience the dog may learn that the squirrel is uncatchable. Yet both animals persist in their contest, engaging in occasional bouts for weeks or months when a dog is fenced near trees. It is not clear why they do this, but perhaps the close but safe encounter gives each a pleasurable adrenalin rush, or maybe they both benefit from the experience gained in these encounters when a more risky situation arises.

OF SPECIAL INTEREST: THE RAVEN AND THE COYOTE

Ravens can use observational learning to exploit the caches of other animal species, such as coyotes (see Figure 5.18). We have just such a story for you. The actors in this scenario are a coyote and a raven that have been eating a deer. Birds in the family Corvidae—crows, ravens, magpies, jays, and their relatives—have a reputation for high intelligence. Humans have this impression because these birds are adept problem solvers, good at caching and finding food, and they even seem to have a sense of humor. There is much to admire about corvid behavior; consider this natural history description:

> On the morning of January 15, 1938 I saw a coyote trotting along the base of Mount Everts on the margin of a wide flat. Across the flat a raven was standing on a snow-drift. When the coyote had trotted to a point opposite the raven, about 200 yards away, it turned its course directly toward the bird. By that time, a second raven had alighted beside the first one to feed on a tiny food morsel it had been carrying. When the coyote was somewhat less than ten yards from the feeding raven, it made a quick dash. The raven easily escaped and lit again a few yards to one side. The coyote sniffed the spot where the raven had been feeding and then made another dash for it. These tactics were repeated six or seven times before the bird flew off about 250 yards. After peering at the departing bird, and seeming to hesitate whether or not to follow, the coyote trotted after it. When the coyote had covered half the distance, the raven circled back, wheeling fifteen or twenty feet over the coyote, which looked up. The raven lit on the snow again to feed on its morsel, and the coyote trotted along as if to pass it, but suddenly turned to make another quick charge. The rushes, as before, were repeated five or six times. Once the coyote leaped high in the air toward the raven and rolled over twice when it hit the snow. The raven finally flew away along the river and the coyote disappeared in a draw. It appeared that both animals were enjoying the fun, for the raven could easily have flown away to escape if it were annoyed, and it would seem that the coyote, which was probably well fed by the abundant carrion, would not have been so persistent unless he were enjoying the play.[70]

The behavior is so human-like, on the part of both the raven and the coyote, that it is hard not to conclude that these are thinking, intelligent animals. As scientists, we admit that this account lacks rigorous evidence. As humans, we believe that few days could have a better beginning than this one day in January did. Whatever the case may actually be, the account is charming.

FIGURE 5.18
A coyote and a raven, frequent competitors for food. *Photo: Lynne K. Stone.*

SUMMARY

Learning provides animals means for coping with unpredictable elements in their environments. Learning provides animals with the means for acquiring information about their environment and for updating that information as conditions change. This flexibility is both the great advantage of learning and its downfall. Learning allows instantaneous adaptation, but an inexperienced animal may be at real risk because it has not had the opportunity to learn key facts about its environment. In this chapter our approach has been to distill a complex and difficult academic subject into basic principles that apply to the study of animal behavior in field settings.

Is the study of learning and memory in the domain of psychology or biology? Like the nature–nurture issue discussed in Chapter 3, the question of which academic discipline "owns" these topics is an interesting part of the history of the scientific study of animal behavior. This book adopts a more biological approach, and this is reflected in somewhat limited coverage of the topic of conditioning and the use of laboratory rats, mice, and pigeons in testing learning paradigms. For the student who is interested in highly structured laboratory studies or the physiological and molecular bases for learning, psychology departments offer entire courses on these topics. Human learning and the adaptive behavior of humans find their way into the curricula of psychology, sociology, economics, and education departments.

The biology or psychology student whose primary interest is animal behavior, though, should not underestimate the value of laboratory studies of learning. Such studies provide the basic framework for understanding learning and memory, and they are included in this chapter because they are clearly relevant to understanding the behavior of animals in the field. Other behaviors, such as learned cache retrieval and observational learning of food preferences, illustrate the importance of learning as a topic in behavioral biology. Learning is one of the areas of animal behavior where laboratory and field are wonderfully complementary.

This chapter began with the cellular changes, including protein synthesis and remodeling of synapses, that underlie memory formation. This then led into a discussion of habituation and sensitization, which are simple mechanisms for sorting important from unimportant information. From there, the discussion moved into conditioning, a set of learning phenomena that have been developed in laboratory studies but have broad application in the interpretation of learning in field studies. Social learning, in which animals learn by observing one another, is an important and multifaceted topic. Finally, play is considered in the context of learning. Play is a component of learning and development, and understanding play sets the stage for discussions of a variety of topics, including cognition, foraging, and social behavior.

147

STUDY QUESTIONS

1. Members of a bird species learn their songs (which vary among individuals) at an early age and do not change their songs later in life. How would you test to determine whether this species has a critical period for song learning?
2. Killer whales "play" with their sea lion food, much in the same manner that domestic cats "play" with small rodents or birds. Propose one or two reasonable hypotheses for the existence of this behavior, and briefly describe how you would test your hypothesis or hypotheses.
3. What is the difference between classical and operant conditioning? How might both play roles in the life of an animal in a natural environment?
4. What is the role of forming specific aversions in shaping the diet of animals? Why is this type of learning an important mechanism?
5. What is a "critical period" for learning? If a male duck is raised by a female of another species, what happens to his courtship behavior when he is an adult, and why?
6. A squirrel buries nuts in a scattered pattern throughout its home range. It could either find the nuts by reforaging (randomly digging or smelling them) or by learning and

remembering the locations of the cached nuts. How would you design an experiment to test between the hypotheses of reforaging and learning?

7. The cream-thieving birds of Britain set up an intriguing problem: have they learned from each other, or have they independently invented this foraging tactic? If they were still involved in cream-thievery, how would you determine the relative amount of support for these two hypotheses?

8. Survey your friends and discover the methods they use for remembering various passwords. (We are not advising that you poach their passwords!) How do these methods compare to cache retrieval methods? Do they use rules? Do they use obvious passwords (reforaging)?

9. Why is "fair play" and suppression of the use of weaponry important in social play?

Further Reading

McGaugh, J.L., 2000. Neuroscience–Memory—a century of consolidation. Science 287 (5451), 248–251.

Pearce, J.M., 2008. Animal Learning and Cognition, third ed. Psychology Press, Florence, KY, 432 pp.

Rudy, J.W., 2008. The Neurobiology of Learning and Memory. Sinauer, Sunderland, MA, 380 pp.

Notes

1. Ratcliffe, J.M., Fenton, M.B., Galef, B.G., 2003. An exception to the rule: common vampire bats do not learn taste aversions. Anim. Behav. 65, 385–389.
2. Paradis, S., Cabanac, M., 2004. Flavor aversion learning induced by lithium chloride in reptiles but not in amphibians. Behav. Processes 67, 11–18.
4. White, K.G., 2001. Forgetting functions. Anim. Learn. Behav. 29 (3), 193–207.
3. Capaldi, E.J., Neath, I., 1995. Remembering and forgetting as context discrimination. Learn. Mem. 2 (3–4), 107–132.
5. Sangha, S., McComb, C., Lukowiak, K., 2003. Forgetting and the extension of memory in *Lymnaea*. J. Exp. Biol. 206 (1), 71–77.
6. Sargisson, R.J., White, K.G., 2003. The effect of reinforcer delays on the form of the forgetting function. J. Exp. Anal. Behav. 80 (1), 77–94.
7. Dukas, R., 1999. Costs of memory: ideas and predictions. J. Theor. Biol. 197 (1), 41–50.
8. Zhang, W., Linden, D.J., 2003. The other side of the engram: Experience-driven changes in neuronal intrinsic excitability. Nat. Rev. Neurosci. 4 (11), 885–900.
9. Holland, P.C., Bouton, M.E., 1999. Hippocampus and context in classical condition. Curr. Opin. Neurobiol. 9, 195–202.
10. Brodin, A., Lundborg, K., 2003. Is hippocampal volume affected by specialization for food hoarding in birds? Proc. R. Soc. Lond. B 270 (1524), 1555–1563.
11. Bliss, T.V.P., 2003. A journey from neocortex to hippocampus. Philos. T. Roy. Soc. B. 358, 621–623.
12. Griffiths, D.P., Clayton, N.S., 2001. Testing episodic memory in animals: a new approach. Physiol. Behav. 73 (5), 755–762 (special issue).
13. Menzel, R., Mueller, U., 1996. Learning and memory in honeybees: From behavior to neural substrates. Annu. Rev. Neurosci. 19, 379–404.
14. Rosenkranz, J.A., Grace, A.A., 2002. Dopamine-mediated modulation of odour-evoked amygdala potentials during pavlovian conditioning. Nature 417 (6886), 282–287.
15. Pare, D., 2002. Mechanisms of Pavlovian fear conditioning: has the engram been located? Trends Neurosci. 25 (9), 436–437.
16. Daly, K.C., Christensen, T.A., Lei, H., Smith, B.H., Hildebrand, J.G., 2004. Learning modulates the ensemble representations for odors in primary olfactory networks. PNAS 101, 10476–10481.
17. McGaugh, J.L., 2000. Neuroscience–Memory—a century of consolidation. Science 287 (5451), 248–251.
18. Bradshaw, J.W.S., Goodwin, D., LegrandDefretin, V., Nott, H.M., 1996. Food selection by the domestic cat, an obligate carnivore. Comp. Biochem. Physiol. A. Physiol. 114, 205–209.
19. White, D.J., Galef, B.G., 1998. Social influence on avoidance of dangerous stimuli by rats. Anim. Learn. Behav. 26 (4), 433–438.
20. Dukas, R., 1999. Costs of memory: ideas and predictions. J. Theor. Biol. 197 (1), 41–50.
21. Weiss, M.R., Papaj, D.R., 2003. Colour learning in two behavioural contexts: how much can a butterfly keep in mind? Anim. Behav. 65, 425–434.
22. Menzel, R., Mueller, U., 1996. Learning and memory in honeybees: from behavior to neural substrates. Annu. Rev. Neurosci. 19, 379–404.
23. Lewis, A. C., 1986. Memory constraints and flower choice in *Pieris rapae*. Science 232, 863–865.
24. Rose, J.K., Rankin, C.H., 2001. Analyses of habituation in *Caenorhabditis elegans*. Learn. Mem. 8 (2), 63–69.

25. Deshmukh, S.S., Bhalla, U.S., 2003. Representation of odor habituation and timing in the hippocampus. J. Neurosci. 23 (5), 1903–1915.

26. Burrell, B.D., Sahley, C.L., 2001. Learning in simple systems. Curr. Opin. Neurobiol. 11 (6), 757–764.

27. Cohen, T.E., Kaplan, S.W., Kandel, E.R., Hawkins, R.D., 1997. A simplified preparation for relating cellular events to behavior: mechanisms contributing to habituation, dishabituation, and sensitization of the Aplysia gill-withdrawal reflex. J. Neurosci. 17 (8), 2886–2899.

28. Werdenich, D., Dupain, J., Arnheim, E., Julve, C., Deblauwe, I., van Elsacker, L., 2003. Reactions of chimpanzees and gorillas to human observers in a non-protected area in South-Eastern Cameroon. Folia Primatol. 74 (2), 97–100.

29. Van Krunkelsven, E., Dupain, J., Van Elsacker, L., Verheyen, R., 1999. Habituation of bonobos (*Pan paniscus*): first reactions to the presence of observers and the evolution of response over time. Folia Primatol. 70 (6), 365–368.

30. Abramson, C.I., Singleton, J.B., Wilson, M.K., Wanderley, P.A., Ramalho, F.S., Michaluk, L.M., 2006. The effect of an organic pesticide on mortality and learning in Africanized honey bees (*Apis mellifera L.*) in Brasil. Am. J. Environ. Sci. 2, 33–40.

31. Turnbough, P.D., Lloyd, K.E., 1973. Operant responding in Siamese fighting fish (*Betta splendens*) as a function of schedule of reinforcement and visual reinforcers. J Exp Anim Behav. 20, 355–362.

32. Domjan, M., Cusato, B., Krause, M., 2004. Learning with arbitrary versus ecological conditioned stimuli: evidence from sexual conditioning. Psychon. Bull. Rev. 11, 232–246.

33. Johanson, I.B., Hall, W.G., 1982. Appetitive conditioning in neonatal rats: conditioned orientation to a novel odor. Dev. Psychol. 15, 379–397.

34. Cleaveland, J.M, Jäger, R., Rößner, P., Delius, J.D., 2003. Ontogeny has a phylogeny: background to adjunctive behaviors in pigeons and budgerigars. Behav. Processes 61, 143–158.

35. Glendinning, J.I., 1992. Effectiveness of cardenolides as feeding deterrents to *Peromyscus* mice. J. Chem. Ecol. 18 (9), 1559–1575.

36. Lindquist, N., Hay, M.E., 1995. Can small rare prey be chemically defended—the case for marine larvae. Ecology 76 (4), 1347–1358.

37. Ratcliffe, J.M., Fenton, M.B., Galef, B.G., 2003. An exception to the rule: common vampire bats do not learn taste aversions. Anim. Behav. 65, 385–389.

38. Mikulka, P., Klein, S., 1980. Resistance to extinction of a taste aversion: effect of level of training and procedures used in acquisition and extinction. Am. J. Psych. 93, 631–641.

39. Rusiniak, K.W., Palmerino, C.C., Rice, A.G., Forthman, D.L., Garcia, J., 1982. Flavor-illness aversion: potentiation of odor by taste with toxin but not shock in rats. J. Comp. Physiol. Psychol. 96, 527–539.

40. Brower, L.P., Fink, L.S., 1985. A natural toxic defense system—cardenolides in butterflies versus birds. Ann. N. Y. Acad. Sci. 443, 171–188.

41. Malcolm, S.B., Brower, L.P., 1989. Evolutionary and ecological implications of cardenolide sequestration in the monarch butterfly. Experientia 45 (3), 284–295.

42. Ritland, D.B., 1995. Comparative unpalatability of mimetic viceroy butterflies (*Limenitis archippus*) from 4 southeastern United States populations, Oecologia 103 (3), 327–336.

43. Galef, B.G., 1989. Socially mediated attenuation of taste-aversion learning in Norway rats—preventing development of food phobias. Anim. Learn. Behav. 17 (4), 468–474.

44. HadjChikh, L.Z., Steele, M.A., Smallwood, P.D., 1996. Caching decisions by grey squirrels: a test of the handling time and perishability hypotheses. Anim. Behav. 52, 941–948.

45. Moller, A., Pavlick, B., Hile, A.G., Balda, R.P., 2001. Clark's nutcrackers *Nucifraga columbiana* remember the size of their cached seeds. Ethology 107 (5), 451–461.

46. Tomback, D.F., 1980. How nutcrackers find their seed stores. Condor, 82 (1), 10–19.

47. Vander Wall, S.B., 2000. The influence of environmental conditions on cache recovery and cache pilferage by yellow pine chipmunks (*Tamias amoenus*) and deer mice (*Peromyscus maniculatus*). Behav. Ecol. 11 (5), 544–549

48. Cristol, D., 2005. Walnut-caching behavior of American Crow. J. Field Ornithol. 76, 27–32.

49. Devenport, J.A., Luna, L.D., Devenport, L.D., 2000. Placement, retrieval and memory of caches by thirteen-lined ground squirrels. Ethology 106, 171–183.

50. Balda, R.P., Kamil, A.C., 1992. Long-term spatial memory in Clark's nutcracker, *Nucifraga columbiana*. Anim Behav 44, 761–769.

51. Leaver, L.A., Daly, M., 2001. Food caching and differential cache pilferage: a field study of coexistence of sympatric kangaroo rats and pocket mice. Oecologia 128 (4), 577–584.

52. Emery, N.J., Clayton, N.S., 2001. Effects of experience and social context on prospective caching strategies by Scrub Jays. Nature 414 (6862), 443–446.

53. Heinrich, B., Pepper, J.W., 1998. Influence of competitors on caching behaviour in the common raven, *Corvus corax*. Anim. Behav. 56, 1083–1090.

54. Griffin, A.S., Boyce, H.M., MacFarlane, G.R., 2010. Social learning about places: observers may need to detect both social alarm and its cause to learn. Anim. Behav. 79, 459–465.

55. Altshuler, D.L., Nunn, A.M., 2001. Observational learning in hummingbirds. AUK 118 (3), 795–799.

Bednekoff, P.A., Balda, R.P., 1996. Observational spatial memory in Clark's nutcrackers and Mexican Jays. Anim. Behav. 52, 833–839.

56. Lindberg, A.C., Kelland, A., Nicol, C.J., 1999. Effects of observational learning on acquisition of an operant response in horses. Appl. Anim. Behav. Sci. 61 (3), 187–199.

57. Sasvari, L., Hegyi, Z., 1998. How mixed-species foraging flocks develop in response to benefits from observational learning. Anim. Behav. 55, 1461–1469.

149

58. Fiorito, G., Scotto, P., 1992. Observational learning in Octopus vulgaris. Science 256 (5056), 545–547.

59. Sherry, D.F., Galef, B.G., 1984. Cultural transmission without imitation—milk bottle opening by birds. Anim. Behav. 32, 937–938.

60. Bugnyar, T., Kotrschal, K., 2002. Observational learning and the raiding of food caches in ravens, *Corvus corax*: is it 'tactical' deception? Anim. Behav. 64, 185–195.

61. Tebbich, S., Taborsky, M., Fessl, B., Blomqvist, D., 2001. Do woodpecker finches acquire tool-use by social learning? Proc. Biol. Sci. 268 (1482), 2189–2193

62. de Oliveira, C.R., Ruiz-Miranda C.R., Kleiman, D.G., Beck, B.B., 2003. Play behavior in juvenile golden lion tamarins (Callitrichidae: Primates): organization in relation to costs. Ethology 109, 593–612.

63. Spinka, M., Newberry, R.C., Bekoff, M., 2001. Mammalian play: training for the unexpected. Q. Rev. Biol. 76, 141–168.

64. Palagi, E., Cordoni, G., Tarli, S.M.B., 2004. Immediate and delayed benefits of play behaviour: new evidence from chimpanzees (*Pan troglodytes*). Ethology 110, 949–962.

65. Harcourt, R., 1991. Survivorship costs of play in the South American fur seal. Anim. Behav. 42, 509–511.

66. Caro, T.M., 1995. Short-term costs and correlates of play in cheetahs. Anim. Behav. 49, 333–345.

67. Boissy, A., Manteuffel, G., Jensen, M.B., Moe, R.O., Spruijt, B., Keeling, L.J., et al., 2007. Assessment of positive emotions in animals to improve their welfare. Physiol. Behav. 92, 375–397.

68. Marquez-Arias, A., Santillan-Doherty, A.M., Arenas-Rosas, R.V., Gasca-Matias, M.P., Munoz-Delgado, J., 2010. Environmental enrichment for captive stumptail macaques (*Macaca arctoides*). J. Med. Primatol. 39, 32–40.

69. Petru, M., Spinka, M., Charvatova, V., Lhota, S., 2009. Revisiting play elements and self-handicapping in play: a comparative ethogram of five old world monkey species. J. Comp. Psychol. 123, 250–263.

70. Murie, A., 1940. Ecology of the Coyote of the Yellowstone, US Government Printing Office, public domain, pp. 33–34.

Cognition

151

LEARNING OBJECTIVES

Studying this chapter should provide you with the knowledge to:

- Understand that "brain" and "mind" are different. While the brain is the physical structure that coordinates behavior, the mind is where thought occurs.

- Define cognition as the ability to visualize the self in the context of the larger world, to forecast the results of actions and emotional states, to solve complex problems, and to interpret the emotional states of other animals.

- See how cognition reduces impulsive behavior, and allows reflection, planning, and foresight—in other words, nonimpulsive behavior.

- Consider the historical unwillingness of many biologists to credit nonhumans with cognition and how modern insights have broken down this cognitive/noncognitive divide.

- Know that cognition intertwines with perceptions of intelligence, problem solving, and insight.

- Understand personality as a noncognitive expression of an animal's strategy for survival. When animals interpret each other's personalities, this becomes a part of the cognitive picture. Behavioral syndromes are tools for understanding and measuring variation in animal personalities.

- Appreciate cognition as a mechanism by which an animal assesses its own emotional state and the feelings of surrounding animals; it is not necessary for emotional life or personality.

6.1 INTRODUCTION: WHAT IS COGNITION?

Of all the topics covered in animal behavior, cognition is the trickiest. Scientists struggle with defining *cognition*, much less measuring it. Cognition involves the concept of self as a separate identity from the surrounding environment, and at its highest levels involves the imaginative ability to bring seemingly unrelated facts or ideas together to create a novel solution to a problem or a unique plan for future action. A consistent thread through this chapter is the question of how can animals be tested to determine if they use cognitive processes like those that are so obvious from human perspectives?

> **KEY TERM** Cognition is the ability of an animal to separate itself from the moment in which it is living and to contemplate the past, predict the future, and act accordingly.

Cognition is particularly tricky because of its history in behavioral studies. In the study of cognition—a vibrant, edgy science that asks what animals think and how they think it—history of science itself is rich with ideas and examples that extend far beyond a creaky collection of factoids and dates. Victorian ideas of clever foxes and diligent insects, displays of early twentieth century horses that do arithmetic (for example, Clever Hans, see Chapter 1), and early psychologists' fondness for rats in boxes equipped with levers (and not much else) have a strong effect on scientists today, who are only now emerging from those shadows. The reaction of the first modern behavioral scientists such as Morgan (of Morgan's Canon; see Chapter 1) to the *anthropomorphism* of earlier generations was in some ways accurate: anthropomorphic imaginings have no place in science. As is true with many human endeavors, however, the rejection of anthropomorphism led to overreaction. In their zeal to avoid further anthropomorphism, many behavioral biologists rejected any and every suggestion of cognition in the animal world. A review of Chapter 1 will reveal that only courageous resistance to this overreaction by scientists such as Donald Griffin has led to a recent increase in acceptance of cognition as an honored and fruitful part of animal behavior.

152

> **KEY TERM** Anthropomorphism is the uncritical attribution of human characteristics to nonhuman animals.

We will introduce this topic with a discussion of some simple ways to assess cognition, including an examination of the importance of language in such assessments. This will be followed by the concepts of self and *self-awareness*, and how they fit into the cognitive framework. From there, the subject expands to include foresight and planning, including developing the idea of *mental time travel*. Because cognition is often interpreted as intelligence, difficulties of cross-species comparisons of intellectual capacity are important. The next sections deal with the cognitive aspects of variation among animals; personality, impulse control, and emotion all play important roles in cognition and behavioral variation within animal groups.

> **KEY TERM** Self-awareness is the ability of an animal to assess its own condition and to contrast that condition with the states of other animals in its population.

> **KEY TERM** Mental time travel is the ability to use past experiences to forecast the future.

In addition to more routine topics, this chapter includes *intelligence*, *insight*, *personality* and *emotion*. These attributes are difficult to separate from cognition because human perceptions of cognition are intertwined with how smart, engaging, and emotional the public thinks an animal is. In addition, there are reasonable arguments that the emotional lives of animals show that they have cognitive lives. In fact, cognition itself is closely associated with some of the material about learning and memory covered in Chapter 5. Finally, while reading about animal cognition, consider whether cognitive abilities elevate humans above other animals and how explorations of animal cognition should inform the ongoing discussion of the relationships between humans and the rest of the natural world.

Cognition is a reasonable hypothesis that explains how at least some animal species mentally operate. That reasonable hypothesis yields a difficult question: If animals are cognitive, how should this affect human behavior toward animals? Humans have, at times, awarded themselves a privileged position in the natural

KEY TERM Intelligence is a measure of the ability to learn, to remember, and to solve problems.

KEY TERM Insight is the ability to consider a set of information and to derive from that information a novel solution to a problem.

KEY TERM Personality is the sum of behavioral tendencies of an individual that make it unique among animals of its species.

KEY TERM Emotions are mental states, or "feelings," that humans and perhaps other animals experience. Examples include joy, sadness, anger, and love.

CASE STUDY

153

Cognitive Elephants?

The ability to form complex categories based on experience could be an example of cognition. For instance, some potential predators may be more dangerous than others, and it is advantageous to distinguish between them. Many aquatic animals can sense predator chemicals. Prairie dogs and some primates can do this, distinguishing more or less dangerous species. African elephants, however, have taken this a step further than any other animal that that has been studied. They categorize members of one predator species—humans—into subgroups based on the threat that they pose. In Kenya, Maasai men spear elephants as proof of their virility; Kamba men, with a more agricultural heritage, do not do this. Elephants reacted more fearfully to the smell of clothes worn by Maasai men than to clothes worn by Kamba men or to clean clothing presented as a control. (All of the clothing was red.) In addition, because Maasai men wear red clothes and Kamba men tend to wear white, when shown red material, elephants behaved aggressively, with stamping and head shaking; they did not show such behavior when shown white clothes (see Figure 6.1).[1]

FIGURE 6.1
(A) Elephants react to the scent of a shuka—a garment worn by the Maasai men who hunt them—and sniff the air excitedly, prior to (B) running away. Photos: *(A) K. E. Slocombe, (B) R. W. Byrne.*

ideas. This is a "last resort" because other perceived distinctions have evaporated as scientific knowledge of animal behavior has expanded. Thus, humans are no longer the only animals that use tools, they are no longer the only animals that care for nonrelatives.… Indeed, the more that is learned about animals, the more some standard-bearers for human superiority focus on redefining the defining traits of humans! These arguments stretch the thread of scientific fact to near, or even past, the breaking point.

OF SPECIAL INTEREST: TOOL-MAKING BIRDS

Rooks are corvids, a group of birds known for intelligence and creativity that also includes jays, crows, and magpies. In contrast to New Caledonian crows, which can make tools in the wild and in the laboratory, rooks do not seem to make tools in the wild. Therefore, it surprised scientists to see them do it in the laboratory (see Figure 6.5).[7] The rooks were presented with a delicious waxworm (a caterpillar) in a difficult-to-access location in the laboratory. When given a variety of choices among potential tools, they selected the correct tool, they used the wrong tool to acquire the correct tool, and they made tools for retrieving the caterpillars. Why rooks show this ability in captivity, but apparently not in the wild, is a mystery. The investigators argue that their findings demonstrate that the animals show insight—the ability to put together the pieces of a mental puzzle.[8,9,10,11,12,13]

FIGURE 6.5
A rook using a piece of metal, curved into a hook, to retrieve food. *Photo: Nathan Emery.*

DISCUSSION POINT: DO HUMANS STAND ABOVE ANIMALS?

This discussion of the relative status of humans and other animals raises a bothersome question: Why would a person want to elevate humans above other animals? If we are simply looking for distinctive traits, then every species, by definition, is distinct from every other. Moreover, as superior as we would like to feel, it is difficult to imagine a human being that matches the filter-feeding talents of, say, a clam or the persistent architectural genius of an orb-weaving spider. Why might humans be compelled to feel superior? Do humans have a special place in the universe? Are some species better than others? Are humans co-equals in a web of life that defies imagination? What do you think? Why?

In addition to asking if other species have language, there are equally puzzling questions about cognition that reside squarely within the confines of nonhuman animal behavior. For instance, if a nonhuman animal is thinking, in a cognitive way, what system of internal representation does it use? It is certainly possible that an animal would have a fully developed internal symbolic language that is never expressed externally, but how could this be discovered? Nonhuman primates exhibit human processes in mirror tests and gaze following, topics that are discussed in following sections; it is very likely that similarities to human language also are found in nonhuman primates. The hypothesis that language is used entirely internally—as a facilitator of thought, rather than as a communicatory mechanism—cannot be refuted.

> **KEY TERM** Universal grammar is the hypothetical grammatical structure of human language that is based on the nature of the human brain. Universal grammar, if it exists, would be identical among all human languages.

Following this idea about human language, consider for a moment its nature. The linguist Noam Chomsky developed the concept of *universal grammar*, which hypothesizes that underlying neural features of the human brain drive similarities in structure among all human languages.[14,15] He reasoned that human language must function within the structure of the human brain (see Figure 6.6). As human brain structures are derived from the brain structures of primate ancestors, this suggests that the basis for language may be shared with at least some other primates. In the words of Chater et al., "language has evolved to fit the human brain."[16]

Returning to the larger issue of language and cognition, the definition of cognition presented in this chapter involves planning and a sense of self. Neither of these requires language, and it is up to the creative mind of humans to imagine whether nonhuman animals have internal language-based reasoning or if their thinking is entirely nonlinguistic. Discovering the extent of language use by animals is one of the great challenges for the next generation of animal behaviorists.

6.2 THE CONCEPT OF SELF
Minds and Bodies

The *theory of mind* features prominently in studies of animal cognition and intelligence. An animal with a theory of mind can form hypotheses about the thoughts of surrounding animals. This means it can discern another individual's intentions, reflect on the other's desires, and understand its beliefs. People hope that others in their group have theories about their minds and that they act on these theories in beneficial ways. But when a dog looks at a person, is it forming a theory about that person's mind? Crows can observe people and remember which person is a threat and should be avoided;[17] this is an example of making predictions about behavior based on a "theory" of another animal's mind.

FIGURE 6.6
Can Chomsky's idea of "universal grammar" for humans be extended to other species?

159

Consciousness of self is a key characteristic of humans as a cognitive species. A person separates him- or herself, mentally, from other humans, from the larger living world, and from the physical environment by self-recognition—that is, recognizing him- or herself as a unique entity. People examine themselves, mentally, and reflect on their individual experiences. This mental life is distinct, in many ways, from human physical life, and includes conjectures, dreams, and fantasies that merge the past, present, and future. Indeed, humanity has at times been defined by the fact that humans are thinking animals, with a mental reality that separates mind from brain and body. It is possible to imagine that a broad range of animals live cognitive lives much as humans do. Because human communication by gesture, expression, and voice is deeply rooted in the

> **KEY TERM** Theory of mind is the ability to form hypotheses about the thoughts of surrounding animals.

evolution of vertebrates, much of the human range of communication can be observed among other animals, particularly mammals. This does not, however, reveal what is going on in the animal's brain while it is communicating. Does it experience the same mind/body dichotomy that humans do? Philosophical thought on the difference between the mind and the brain extends back at least to Plato, and remains a point of fascination for us. As might be imagined, considering this dichotomy in other species becomes more difficult as phylogenetic distance from humans increases. In perhaps a related vein, that distance also reduces the risk of anthropomorphism.

Self-Awareness and Mirror Tests

Some general tests for cognition can be found in the introduction to this chapter. Other tests address self-awareness in particular, and they are covered here. For instance, a simple form of cognition is the ability to differentiate "self" from others. Some of the earliest explorations in cognition used mirror tests to determine if animals showed evidence of self-awareness, that is, an ability to separate their concepts of their own bodies (self) from the bodies of others (see Figure 6.7). When a dog or cat walks in front of a mirror, it may show some interest in the image in the mirror and in some cases become aggressive or territorial toward it, as if the image were an interloper in the home. Ultimately, the dog or cat habituates to the image (see Chapter 5), perhaps because it does not respond in interesting ways, perhaps because it has no smell. There is nothing about the dog's or cat's behavior that says "that's me in the mirror." How might a scientist determine if an animal has some perception of self when it looks at its image in a mirror? One way to do this is to put a mark on the animal or otherwise change its physical appearance.

FIGURE 6.7

Do elephants have a theory of mind? They do if mirror tests are indicators of such a state. Although previous mirror tests had not revealed such behavior on the part of elephants, when the setting was made more elephant-friendly, elephants made it clear that THEY were the elephants in the mirror! *Photo: Richard Byrne.*

The basic procedure for a mirror test is to provide a mirror in a focal animal's habitat, and to allow it to observe and explore the image in the mirror. Given some time, the animal will either figure out that it is looking at itself, or it will continue to perceive the image as something other than itself. After a defined interval, the observer changes the appearance of the *focal animal*, perhaps by dyeing a patch of fur on its head a contrasting color that the animal can see. After this manipulation, if the focal animal sees the image as its own reflection, it may respond by touching the dyed patch, or otherwise acknowledging that its own appearance has been changed. This acknowledgment supports the hypothesis that the animal has a sense of self and therefore is cognitive.

To date, dolphins, magpies, elephants and nonhuman apes such as chimpanzees have all shown positive responses in mirror tests.[18] The ability of the great apes to perform well in mirror tests is not surprising, given their evolutionary proximity to humans; the expectation is that minds of animals that are closely related to humans work similarly to human minds. While very young human infants do not respond positively in mirror tests, the response develops by the time a child is 18–30 months old. The major drawback to these tests is that a negative result cannot be interpreted as a lack of cognition. An animal may perfectly well recognize, internally, that it is seeing itself in the mirror and that its appearance has changed, but it may not be primed to display any overt response. This is a good example of why scientists are cautious when they get negative results from an experiment; in the long run, more refined tests of the same hypothesis may give a positive result.

KEY TERM The focal animal is the animal under observation. Focal animal sampling involves extended observation of one animal, in contrast to scan sampling, which is intermittent and repeated observation of many animals.

The importance of design and its effect on results can be seen especially clearly in the case of elephant mirror self-recognition. Initial mirror tests with elephants showed no self-recognition; the mirrors were relatively small and were kept out of range of the animals' trunks. When these two design features were changed, and when large (elephant-proof) mirrors were placed close enough for elephants to touch and explore the mirrors, results

160

changed, and elephants displayed self-recognition. The three female elephants began by trying to look behind the mirror and continued with a variety of behaviors, including ones that they did not perform in any other location, such as exploring their oral cavities in front of the mirror and moving food to a location immediately in front of the mirror, where they consumed it. One of the three showed a dramatic increase in head-touching (where the mark was located) in the mark test.

Does this concept of self as revealed in mirror tests really demonstrate that an animal has a theory of mind—for example, does it suggest a cognitive ability to predict how the animal's own actions might then translate into subsequent events within a social group? This issue is certainly open to debate; an alternative hypothesis is that mirror tests measure problem-solving ability and insight learning, rather than cognition that is driven by a theory of mind. This uncertainty plagues research on animal cognition because it is always difficult to separate the result of associative thinking from actual cognition. For now, mirror test results are intriguing and suggestive but do not conclusively support arguments for cognition. Other tests of theory of mind hypotheses have yielded equal uncertainty about how the results should be interpreted.[19]

Gaze Following

As described previously, gaze following is another basic measure of an animal's ability to separate self from others. Humans are very sensitive to gaze direction; a person can easily observe the direction that another person is looking and identify the object of that person's attention. Human infants develop the ability to follow gazes at 6–8 months old, much younger than their development of responses in mirror tests. For many years, investigators thought that only humans could follow gazes; newer studies demonstrate gaze following in a variety of primates, suggesting, that, like numerical assessments, gaze following is deeply rooted in primate evolution. Rhesus monkeys can follow gazes by looking at the orientation of another monkey's head, rather than focusing on the eyes. Most dog owners believe that their dogs can follow their gaze, but this ability has only recently been established scientifically for dogs;[20] studies of wolves suggest that gaze following may be a unique result of the domestication of dogs, along with artificial selection that favors dogs who can follow human instructions (see Figure 6.8).[21]

161

FIGURE 6.8
Gaze following experiments may be used as evidence—albeit controversial—of theory of mind. These animals are looking intently at something, and humans can follow their gaze, but can they follow each other's gazes? This is an interesting, and difficult to test, question in animal cognition. Likely the chimpanzee (right) can follow other chimpanzee's gazes, while the hypothesis is untested in ring-tailed lemurs (left). *Photos: Michael Breed (left) and John Mitani (right).*

Like mirror tests, positive results of gaze following studies can be interpreted to support a concept of self for an animal that follows a gaze; animals that do not follow a gaze may not have a theory of mind—the ability to project the thoughts of other animals. Returning to the definition of cognition presented earlier in this chapter, humans use gaze following to assess intent and forecast the future. But do other animals, even chimpanzees, use gaze in the same way? Perhaps gaze following is merely a functional adaptation that allows an organism to compete for food, spot potential predators, and perform similar simple and not necessarily cognitive operations. As with the mirror tests, the interpretation of gaze following as evidence of cognition remains controversial.

Self and Self-Consciousness

A level of self-consciousness is implicit in a discussion of self and self-awareness. Self-consciousness involves an ability to judge one's own actions in the context of values or traditions within a social community. Self-conscious judgments show a knowledge of normative or standard behavior and a realization of how others would view an individual's behavior.

How would a scientist know if an animal is self-conscious? The central argument for self-consciousness in animals derives from behavior that humans interpret as representing embarrassment or remorse. When a person leaves the room and his or her dog filches food from a table or visits the kitchen garbage, the person may later see behavior—tail down, eyes drooping, ears low, slinking away—that could be interpreted as embarrassment or remorse. This interpretation carries a high risk of anthropomorphism (see Chapter 1) and is not viewed as strong evidence for cognition.

OF SPECIAL INTEREST: EMPATHY

Theory of mind is seen as a prerequisite for *empathy*. Are animals empathetic? How can this be tested? Some workers have suggested that self-recognition (see the earlier discussion of mirror tests) is related to empathetic behavior; they point out that species that can recognize themselves in mirrors have also displayed behaviors that can be interpreted as empathetic. In more nearly direct tests, other scientists have examined facial mimicry, a behavior that is closely related to "emotional contagion," the human tendency to share the emotions of others. Facial mimicry (smiling when smiled at, yawning in the presence of yawning people) can be voluntary or involuntary; the latter can occur within a second or so of the initial event. In orangutans, the open mouth face is an affiliative expression and is mimicked within one second (see Figure 6.9). Scientists concluded that this building block of empathy is exhibited by a great ape that last shared an ancestor with humans at least 12 million years ago.[22]

FIGURE 6.9
The open mouth face of orangutans (shown here) is mimicked within one second by another orangutan; facial mimicry is one facet of empathetic behavior.

KEY TERM Empathy is the ability to project or "feel" the emotions of another animal.

6.3 THOUGHT, FORESIGHT, AND PROBLEM SOLVING

Another way of considering cognition and a theory of the mind is by using the concept of mental time travel, that is, thinking about the past and using that information to form plans for the future. Humans reflect on the past and hope that knowledge gained from those experiences helps in coping with the future. Mental time travel is more complex than simply benefiting from trial-and-error learning (see Chapter 5 on learning); an important component of mental time travel involves bringing together seemingly unrelated pieces of information, from experiences that perhaps occurred at different times, to anticipate the solution to a problem. *Chronesthesia* is a term related to mental time travel. A chronesthetic animal is aware of past and future, and its life in the present is shaped by that awareness. As with the theory of mind concept, testing animals for chronesthesia is a challenge; experiments must ask animals to respond to complex tasks in a way that could be possible only if the animal has chronesthetic awareness.

> **KEY TERM** Chronesthesia is the awareness of the past and the future, and the use of that information in the present.

Memory and Cognition

The differences among semantic, procedural, and episodic memories are important to understanding mental time travel. *Semantic memory* is the abstract mental representation of concrete identification or concept. The ability to learn and remember language is semantic memory. *Procedural memory* is the ability to remember and apply a series of steps to a task. *Episodic memory* places learning into the contexts of what, when, and where, allowing cognitive reflection and forecasting.[23]

SEMANTIC MEMORY

The average dog can learn and react correctly to 100 verbal/commands, such as "walk," "sit," "stay," "doggie," "squirrel," and so on, but can also respond to the specific names of the people with whom it interacts. Rico, a border collie, learned over 200 items and could associate each item with a spoken word.[24] This suggests a degree of semantic memory that extends beyond the range variation in canine vocalizations. Alex, an African grey parrot (see Figure 6.10) who worked with Irene Pepperberg, knew 150 words and could identify 50 objects. While 150 or 200 words is impressive, it pales in comparison with the 60,000-word vocabulary of a human high school graduate. But perhaps animals such as dogs and parrots have much larger internally held "vocabularies" that are not represented in the tests that humans have devised. Semantic memory is not considered cognitive, because even though the animal may draw an abstraction (such as "doggie" as a general term for dogs), no mental time travel or creative use of combined facts is necessary for semantic memory.

163

FIGURE 6.10
An African Grey Parrot. Parrots have gained considerable fame for their cognitive abilities. Alex, shown here, could count up to six objects and was learning to count seven and eight when he passed away.
Photos: James D. Liberatos.

Experiments that feature a single animal, such as Rico or Alex, are subject to considerable suspicion. Are the experimenters' claims really supported, or is there a design flaw? Of particular concern is the Clever Hans effect. This effect, detailed in Chapter 1, essentially says that it is possible for animals to produce the "right" answer by reading unintentional cues from their human trainers. In the experiments with Rico, the dog was given the command to retrieve an item, and then went into an adjacent room to select the object from an array of objects; no one was in the room with him, so a Clever Hans effect was impossible. With Alex, his ability to interact and respond to a variety of humans argues against a Clever Hans effect in his use of vocabulary and object identification. The assessments of the semantic abilities of Rico and Alex are well verified.

PROCEDURAL MEMORY

Procedural memory is memory of how to perform a task. The steps of a task, such as a bee working a complicated flower to obtain nectar or a monkey opening a difficult fruit, are remembered in detail and in order. Procedural memories are formed by trial and error and can be modified and improved over time. Despite the impressive accomplishments that can result from procedural memory, it is not cognition because it does not require reflection and integration of information.

EPISODIC MEMORY

Episodic memory captures individual experiences of particular times and places. Episodic memory associates what, when, and where, and is the basis for cognition, because the animal that remembers episodes can use them to mentally travel back in time, or to project forward, predicting future events. For example, if an animal has previously forded a rapidly running stream, it might reflect on that experience and form a plan for fording a newly encountered stream. Sometimes reference to an occurrence at a specific time is useful, as in time–place learning (see the following section on time–place learning). While humans seem to refer to specific times during the circadian cycle (the time of day that an event happened), rats measure the time elapsed since the event (that is, how long ago the happened).[25] Either approach is useful in cognitive processes. Experiments designed to test for cognition in animals focus on episodic memory.

Time–Place Learning

Many ecological events occur at predictable times and places. Often, predators appear, flowers produce nectar, and potential mates display at predictable times and places. *Time–place learning* is the ability to remember both the location of a resource and the time at which the resource is present. The honeybee was the first well-studied example of time–place learning; foragers can learn to return to feeding stations at the same time on a daily schedule. Studies of time–place learning usually involve training the animal to a feeding location where food is provided for a limited interval, such as 15 minutes, once a day. After a few days—usually two or three are sufficient—the food is withheld. If time–place learning has occurred, the animals will show up at the appointed time and place, even though the food is not there to attract them. Control measurements, taken through the day, will show few, if any, visits to the feeding location at nonfeeding times, with the exception of times just prior to feeding. The appearance of trained animals prior to the exact feeding time, termed *anticipation*, is typical of time–place learning, as anyone who feeds a pet on a schedule can verify (see Figure 6.11).

> **KEY TERM** Time-place learning is the ability to associate a reward or behavioral consequence with a specific location and time, and to use that information to return to the location at the specified time in the future.

Time–place learning brings together two remarkably different pieces of neural machinery: the animal's internal clock and the animal's spatial memory of its landscape. To the best of

164

FIGURE 6.11

Time–place learning in a tropical stingless bee (*Trigona fulviventris*). Both of these graphs show the number of bees responding over time. (A) For several days prior to the day that these data were collected, sugar water was provided at time "0." As a result, some bees are showing up as early as 20 minutes before feeding time. This is called "anticipation." (B) In this case, no food was offered. Bees showed up early, but gave up quickly after no sugar water appeared at time zero. n = number of bees. *Adapted from Murphy, C.M., Breed, M.D., 2008.*[26]

human knowledge, all organisms have a circadian clock—even fungi have one!—and many animals can learn key aspects of their environments and navigate repeatedly to specific locations. This is particularly true of central place foragers; these are animals such as social insects, birds, and mammals, that leave their fixed nesting locations to forage and then return to the nest. (See Chapter 9 on foraging for more information about central place foraging.) Time–place learning can be very much a part of foraging for these animals. Fish, however, tend not to be central place foragers. Nonetheless, some fish (e.g., rainbow trout) can also time–place learn.[27] Genetic studies of mice show that mutations of circadian clock genes disrupt time–place learning; this is not surprising, but confirms the importance of the circadian clock in time–place learning.[28]

Time–place learning can be remarkably complex; honeybees can learn up to nine different time–place associations and can hold this memory for about a week. For a bee, the ability to retain the memory for several days is important because rain or cold may cause bees to skip a day or two of foraging.

Is time–place learning episodic (mental time travel) or semantic memory? The ability to associate time and place is complex only in drawing together two separate neural processes involving time and space; the actual act of associating the time, place, and resource is simple. But if episodic memory is defined as the ability to remember what, when, and where, and to use that information to shape future behavior, time–place learning merges with episodic memory. What is lacking in time–place learning, and makes it fall a bit short of the definition of cognition, is any necessity for mental reflection. This illustrates how tricky it is to separate examples of what–when–where learning from truly cognitive behavior.

Caching and Thievery

Collecting more food than can be immediately consumed leads to an interesting set of problems. The food can be protected by active defense (fighting off competitors), or it can be hidden to prevent discovery by competitors. The general term for hiding food is *caching*, a topic introduced in Chapter 5 and dealt with in more detail in Chapter 9 on foraging. Hiding (caching) presents two problems of its own; the first is finding the food at a later time, and the second is reacting appropriately to the presence of potential thieves.

165

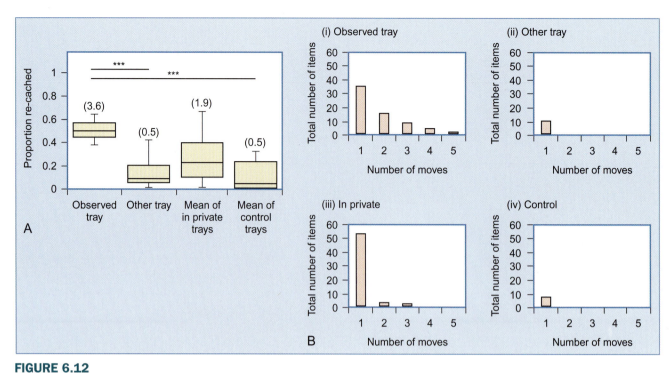

FIGURE 6.12

Cache-hiding in Scrub jays (*Aphelocoma californica*). In A, focal birds have foraged on food while being observed by other birds in an enclosure. These birds frequently move their caches, suggesting an awareness of their competitors and the use of a caching strategy to avoid loss of food. B is a condition in which the focal birds foraged in private; in this circumstance they rarely moved their cache. *Adapted from Daily, J.M., Emery, N.J., Clayton, N.S., 2006. Food-caching Western Scrub-Jays keep track of who was watching when. Science Vol. 312, 1662–1665.*

The classic example of cognitive processes is found in a bird, the Western Scrub-Jay.[29,30] Focusing on the problem of cache robbers and using the logic of "it takes a thief to know a thief," the experiment compared nonplunderers with experienced cache plunderers and asked if experienced birds behaved in ways that better protected their own cache. Indeed, birds with experience as thieves displayed greater alertness when a potential robber observed their caching behavior, and they responded by later recaching the food (see Figure 6.12). This is consistent with the idea that recaching happens more frequently in birds that rob caches because thieves foresee that other birds may behave like thieves, and use this mental forecast to shape their own behavior.[31]

Cognitive Maps

Some investigators have argued that the use of *cognitive maps* is evidence for cognition in honeybees.[32] A cognitive map is a mental representation of landscape that an animal can use to calculate optimal routes between locations. Because calculating a route from an internally held map involves reflection, problem solving, and foresight, an animal that plans its routes in this way could be considered cognitive. (See Chapter 8 on navigation for more detailed information about the experiments that test for cognitive maps in bees and other animals.) While the use of cognitive maps would reflect sophisticated thinking and a certain type of mental time travel, it does not require a theory of mind or involve personality, emotion, or empathy, which are other features often used to characterize cognitive animals. As a result, skeptics feel that even if bees use cognitive maps, this does not establish them as cognitive animals in the same sense as has been argued for birds and mammals.[33]

> **KEY TERM** A cognitive map is a mental representation of an animal's landscape used for calculating optimal routes.

DISCUSSION POINT: COGNITIVE MAPS

I have a yellow labrador retriever named Clyde. My house has front-facing windows on the first and second floors. When the doorbell rings, Clyde looks out the ground-level windows to see who is at the door. If the shutters are closed, he runs upstairs and looks out the second floor window. There are two possible explanations for his "knowledge" that if he runs upstairs, he will see the same view from a different perspective. He could be using a complex cognitive map of his world to forecast that both perspectives will allow him to see what he wants. Or, he could be using simple associative learning so that he identifies the two views as essentially the same, without referring to anything resembling a cognitive map. This simple example illustrates the conundrum of cognition: How can an experiment be designed to separate simple association from more complex cognitive processes?

Experimental Approaches to Mental Time Travel and Foresight

A study of bonobo chimpanzees and orangutans investigated tool use, intelligence, and ability to plan for the future.[34,35] The animals first learned to use an object as a tool to obtain a reward. They then were allowed to choose among suitable and unsuitable tools for this task and to take one into a waiting room. An hour later, they were allowed back into the test room. In almost half of the trials, subjects selected a suitable tool, transported it out to the waiting room, and then brought it back into the test room to get the reward. Subsequent experiments showed the same ability to plan ahead when the time interval between tool choice and reward was much longer. The experiment also demonstrated that animals vary in cognitive ability. The star performer was the orangutan Dokana, who in the first study succeeded 15 out of 16 times—more than twice as often as the second best performing animal.

Like many of the questions raised in this text, intraspecific differences among animals pose problems that reach far beyond cognition itself. Because selective pressures vary in space and time, many traits should exhibit this kind of intraspecific variation. As Dokana shows us, care must be taken when speaking of a "typical" orangutan...or a "typical" anything else!

Another approach to the detection of mental time travel involves asking an animal an "unexpected" question that requires it to reflect on its experience and, if it can, to generate a novel answer. Zentall and his colleagues took this approach with pigeons. Facing an unexpected question calls on episodic memory, the type of memory used to make time associations and engage in mental time travel.[36] Pigeons were trained to peck at locations to receive rewards and then had to refer to that previous experience when confronting a new task. The behavior of the pigeons suggested that they do, indeed, use episodic memory to inform answers to unexpected questions.

Problem Solving

There are numerous examples of clever problem solving by animals;[37] the example discussed here is based on the behavior of dogs. If a handler asks a dog to find an object, many dogs (but sadly, not all of them) have adequate semantic memory to associate the name of an object with the object itself and to find a ball or other toy based on its name. Interactions between humans and dogs in this process typically involve other cues from the human (pointing or selective encouragement when the dog is close to the object) that help the dog in its search. Dogs can be trained to find a toy hidden under one of two identical containers following cues from a human (pointing at the correct container). But what happens when the human is not present to deliver these cues? In the absence of a human, dogs may be able to use inferential reasoning to find a toy.[38] The dog does this by figuring out how to find the toy in the absence of a human handler, using strategies gained from previous experience in finding toys.

167

Counting

Animals often need to determine how many items are present. Counting and the ability to make calculations using numbers are strongly related to the issue of problem solving. While the ability to count is not necessarily evidence of cognition, the ability to make comparisons, such as "more than" or "less than," and the ability to predict future rewards based on current numerical values do suggest cognitive processes.

Determining whether one quantity is more or less than another quantity can also provide critical information. Counting the number of eggs or offspring being tended can provide much needed information to a parent. Assessing the number of food items, and comparing relative numbers of food items between patches of food, is elemental to efficient foraging. A reasonable hypothesis is that counting improves survival in many ways, and that an ability to count, when it occurs, should be favored by natural selection. Is counting a cognitive process? Counting often allows animals to gather information that shapes their future decisions. In the sense that counting is part of forecasting the future, counting is a cognitive process.[39,40]

Testing the hypothesis that an animal can count is nonetheless difficult, and adequate tests have been performed on only a handful of animals. Tests for counting are similar to tests for color vision, in that they usually involve a two-step process: first giving the test animal the opportunity to learn (count) objects and then giving the test animal a choice between sets, one of which has the same number of objects as the learned set.

In honeybees (*Apis mellifera*), Chittka and Geiger trained workers to feed at a station along a 300 meter route from their hive.[41] Small tents, which the bees could use as landmarks to guide their flight, were set up along the route. Bees that learned the route using a fixed number of tent landmarks changed their foraging distance—the location where they searched for the feeding station—if the investigators changed the number of landmarks. Inserting landmarks between the hive and the feeding station caused bees to fly a shorter distance before searching for food, while removing landmarks caused them to fly farther. This suggests that flying bees could count the number of landmarks and that they extended their flight in search of additional landmarks when they did not find the expected number.

Coots face the challenge of preventing other birds from laying eggs in their nest. Female American coots (*Filuca americana*) often attempt to lay eggs in the nests of unsuspecting conspecifics. If they succeed, their young will be reared by another bird, a clear benefit to the cheating female. Coot eggs vary in the amount of speckling on their surface, a feature that sometimes allows coot females to tell their own eggs from eggs that have been imposed upon them by other females. Coot females who reject the eggs of cheaters must do so based on maintaining a count of their own eggs in the nest.[42]

Tests of monkeys (Rhesus macaques, *Macaca mulatta*) and chimpanzees (*Pan troglodytes*) suggest not only an ability to count, but also abilities to add and subtract. In Hauser's (a well known scientist in the field of cognition at Harvard University) tests of counting and numerical representations in macaques, the monkey was shown a small number of objects that were then hidden with a screen.[43] While the monkey could not see the objects, the investigator could leave the same number (this is the control) or change the number by adding or subtracting objects (this is the experimental treatment). The screen was then removed, and the amount of time the monkey spent looking at the objects was recorded. If the number did not change (control), the monkey spent only a second or two looking at the objects, but if the number did change, the monkey stared at the objects for 3 or 4 seconds. This and other tests suggest that macaques and chimpanzees can count at least small numbers of objects and perform simple addition and subtraction (see Figure 6.13).

Rats and pigeons, when tested using conditioned responses, showed similar counting abilities as those of rhesus macaques. They could also generalize; rats trained to respond

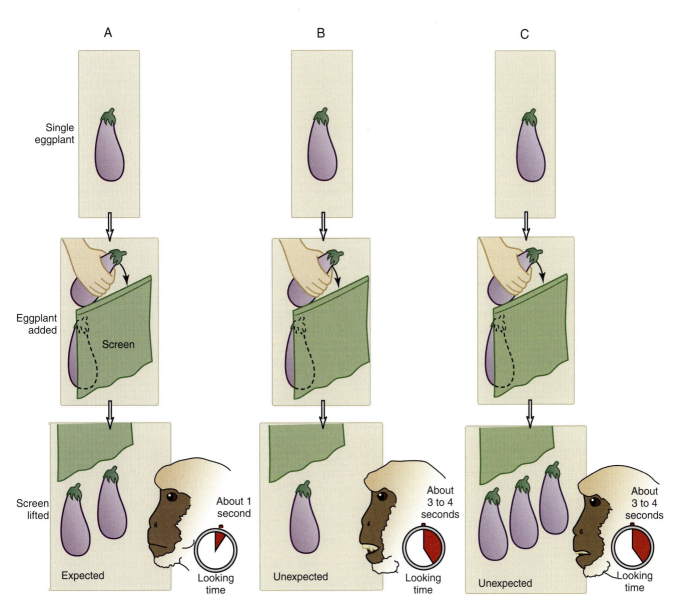

FIGURE 6.13
Some non-human primates may be able to count. In tests where a monkey observed two eggplants being placed behind a screen, the monkey stared for a longer period of time if an unexpected number of eggplants (one or three) was revealed. *Adapted from: Hauser, M.D., 2000. What do animals think about numbers? Am. Sci. 88, 144–151.*

to a certain number of light flashes responded to the same number of bell rings. Clever experiments are almost certain to reveal numerical abilities in a wide variety of animals.

Researchers at Kyoto University's Primate Research Institute showed sequences of numerals (1–9) to three mother-offspring pairs of chimpanzees.[44] The chimpanzees learned the routine sequence (1-2-3, etc) and transferred this knowledge to sequences that were missing numbers (e.g., 2-3-5-8-9). The researchers then showed the chimpanzees touchscreens with numbers appearing in a variety of locations. For each screen, when the first number was touched, the remaining numbers were masked. The chimps had to recall the location of each number, and touch each masked location in numerical order. The results of these experiments show that chimpanzees can learn—at a glance—the location of nine numerals on a screen. The best young chimpanzee and the best mother chimpanzee were then compared to human subjects (nine university students). The university students did slightly better than the adult chimpanzee, with performance in both declining as duration of exposure to the screen decreased. The young chimpanzee, on the other hand, did better than either university students or adult chimpanzee, in terms of both speed and accuracy, and its performance did not decline with duration of exposure to the screen. This may suggest that young chimpanzees have stronger attentiveness than the other test groups.

6.4 INTELLIGENCE AND SOCIAL COGNITION

The definition of intelligence is fraught, as it influences the content of intelligence tests and therefore influences attempts at measuring a valued trait among humans. Humans usually think of intelligence as a combination of quickness to learn, ability to retain information, and facility in solving problems. An animal can have all of these and not be particularly cognitive, but intelligence certainly helps an animal to be cognitive. Intelligence tests for humans are notoriously unreliable because they are subject to cultural and socioeconomic biases, and often favor test-takers who have strong learned skills in multiple-choice questions.

Human intelligence tests are certainly fraught with difficulty, but it is more challenging, by far, to design intelligence tests for nonhuman animals. As in humans, if an investigator presents an animal with an intelligence test, the test should (1) give the same result if it is taken again by the same animal; (2) in a demonstrable way, measure a combination of quickness, retention, and problem solving; and (3) have good value for predicting future performance in tasks. In addition, given the principles outlined in the initial discussion of *umwelt* in Chapter 1, problems or challenges in animal intelligence must also include measures that are relevant to that species' ecological context and evolutionary history.

What kinds of animals are generally regarded as "smart"? The list includes primates, mammalian carnivores (members of the Order Carnivora, such as canids, felids, hyenas, raccoons, seals, and their relatives), corvids (members of the bird family Corvidae, such as jays, magpies, crows, and ravens[45]), dolphins, and cephalopod mollusks (particularly octopuses). What do these animals have in common that might predict intelligence? They are only distantly related evolutionarily, so the trait called intelligence cannot be interpreted as being monophyletic, that is, as having evolved only once among the animals on a single branch of the evolutionary tree. Intelligence must have evolved more than once (see Figure 6.14).

What conditions favor this evolution? Two commonalities stand out: predatory life styles and living in social groups. Some species in this list are both predatory and social; others are one or the other. British psychologist Nicholas Humphrey receives credit for the social intelligence hypothesis—the idea that animal societies are highly shaped by previous social interactions, and that animals have much to gain if they can predict the outcomes of future

FIGURE 6.14

Animals that are credited with "intelligence" because of their problem solving abilities and the possibility that they may show social cognition. (A) Steller's jay, (B) Magpie, (C) Orangutan. The assertion of intelligence in a non-human animal can stimulate interesting discussion about how intelligence is defined and measured. *Photos: Jeff Mitton (A), Michael Breed (B, C).*

OF SPECIAL INTEREST: SMART RATS

Rats can reflect on their own mental processes—and can tell if they are likely to perform well (or not) on a duration–discrimination test. They were asked to decide if a sound that they recently heard was long or short. Short tones lasted from 2 to 3.6 seconds; long ones, from 4.4 to 8 seconds. (Note that 3.6 seconds is more difficult to discern from 4.4 seconds than 2 seconds is to discern from 8 seconds. Rats understand this, apparently.) After hearing the sounds, a rat had two choices: it could bail on the test by sticking its nose into one hole and receiving a small reward, or it could opt to take the test about the difference in duration by sticking its nose into a different hole and receiving a big reward if it made the correct choice (registered by pressing a lever). An incorrect choice resulted in no reward. Rats were more likely to decline the test (and receive the smaller reward) the more difficult the test was, that is, the more similar in duration the two sounds were.[46] In other words, rats can assess their own cognitive states.

interactions and recall previous ones, that is, if they can engage in mental time travel.[47] The social intelligence hypothesis has had strong backing, although some investigators, such as Kay Holekamp of Michigan State University,[48] have questioned the validity of sociality as an explanatory factor in intelligence. Successful carnivory, independent of sociality, can also require quickness, retention, and problem solving; many carnivorous animals are social and gain advantages over larger prey when they hunt in groups. This makes it difficult to disentangle the roles of sociality and carnivory in the evolution of intelligence.

Interestingly, the social insects—the termites, ants, bees, and wasps—have been omitted from the list of smart animals. These insects can be impressively clever, but their problem-solving abilities are limited to highly specific contexts, such as learning how to get nectar from a complicated flower or remembering how to travel from the nest to food and back, and any social politics are worked out over evolutionary time, rather than within the moment-to-moment culture of the colony. (See Chapter 13 on social behavior for a more detailed discussion of the evolution of cooperation.)

The junction of social cognition and apparent intelligence provides some interesting examples.[49] Because of the unique and widespread nature of human interactions with dogs, people probably think more about dog intelligence than that of any other nonhuman species. Dogs can seem hugely clever and act in ways that appear to express social cognition, both with humans and with other dogs. Nonetheless, attempts to design actual intelligence tests for dogs are confounded by scientific inability to distinguish activity and inquisitiveness from problem solving, by human tendencies to project feelings and abilities onto canine companions, and by a desire to see dogs solve problems that are outside the range of conditions their species would have experienced over evolutionary time.

Punishment in a social context is sometimes called *vengeance* and should not to be confused with punishment in the context of learning (see Chapter 5). It provides an excellent final example of how social cognition, requiring some level of intelligence, plays out. To exact punishment, an animal must understand that another animal did something to it, it must project that there are ways of punishing that animal, and it probably needs to consider the possible response if the punishment is meted out (counter-punishment). Punishment is the social opposite of reciprocal altruism, discussed in Chapter 13. (*Spite*, by the way, is behavior that damages both the recipient and the conveyor of the spiteful behavior; spite is unlikely to evolve because there is no evolutionary benefit to the spiteful animal.)

> **KEY TERM** Punishment in a social context is retaliation by one animal as a result of the action of another animal. Punishment benefits the animal delivering the punishment, to the detriment of the recipient.

> **KEY TERM** Spite is an action which is detrimental to both the animal that acts and the animal that receives the act.

In a test of punishment in chimpanzees, one chimp was given the opportunity to take food from another one that was feeding.[50,51] The chimpanzee whose food was taken could then punish the other chimpanzee by causing the food to drop out of reach. Chimpanzees did not make the food vanish it if was removed by an unseen manipulator or by a human, but they clearly punished other chimpanzees for removing the food. This experiment contained an interesting special circumstance: there was no cost to the punisher for making the food vanish. This design probably maximized the chances that theft would be followed by punishment; nonetheless, the punished animal did not escalate the conflict. Conflict and reconciliation are normal parts of wild chimpanzee behavior.[52]

6.5 PERSONALITY AND BEHAVIORAL SYNDROMES

Knowing the personality of a friend or enemy helps shape interactions with that friend (or enemy) to be able to achieve goals. This brings the assessment of personality clearly within the scope of social cognition.

What is personality? *Personality* is a collection of traits that characterize an individual's social responses.[53] These collections vary across populations and make individuals, if not unique, at least behaviorally distinguishable from most other individuals, which have their own collections of social response traits. It is this variation that makes personality so intriguing. Given experience with an individual, personality can be a predictive measure of future behavior.

Nonhuman animals can definitely have personality. Humans think of the animals with which they interact—dogs, cats, horses, and so on—as having personality. This is part of what makes the company of a pet so interesting. Scientists working on social behavior of birds or mammals are often struck by differences in personality among their study animals. Primates, canids, parrots and their relatives, crows and their relatives, and dolphins are known for the variety of personalities they present, but such variation can be found in a broad, and sometimes surprising, range of animals, including, for example, insects.

In recent years some scientists have come to prefer the term **behavioral syndrome** over **personality**; *behavioral syndrome* does not have as much of a sociocultural overlay from human biology and can be used a bit more broadly than *personality*. A behavioral syndrome is a set of correlated responses that is relatively stable over time for individual animals.[54]

> **KEY TERM** Behavioral syndrome and personality refer to consistently expressed behavioral tendencies of an individual.

For example, in a recent study, aquatic spiders within a population were classified along a continuum of shyness-boldness.[55] These spiders skim on the surface of the water, supported by surface tension. The shy spiders tended to hide for long periods of time when disturbed, whereas the bold spiders quickly returned to the water surface. Shy spiders may be better protected from predation, but miss opportunities for feeding and mating that more bold spiders can exploit. Often, differences within populations in behavioral syndromes reflect behavioral options that are each successful, allowing animals to use different pathways to survival and reproduction. The behavioral syndrome concept includes personality but puts the focus on how behavioral differences among animals affect strategies in a broader range of contexts than that of social behavior. It also emphasizes that behavioral syndromes can be constant within a population, but differ among populations depending on evolutionary history and ecological conditions.

Behavioral syndromes often have high heritabilities. This suggests that offspring will resemble their parents in personality. There is also evidence that in at least some animals, learning plays a role in shaping the expression of personality.

The discussion of variation in personality leads to a consideration of how variation can appear in a population. There are two major ways to achieve variation in a trait within a population, and this applies to personality as well as to other traits: (1) individuals within a population are "plastic" with respect to the trait (that is, any individual can express any one of a number of phenotypes), or (2) individuals themselves are not variable (that is, they each express a limited phenotype), but because they differ from one another, the population as a whole exhibits variation.

In the first case, experience can shape personality. For instance, success as a dominant animal may require a different personality than that of a subordinate; expression of personality depends on status within the social group. In this type of system, an animal's personality varies, depending on the circumstances. This also opens the door for early developmental experiences to shape the animal's personality later in life.

In the second case, personality may be genetically influenced for a given animal but vary among individuals. Different personality phenotypes (and their underlying genotypes) continue to coexist because several strategies can be successful; for example, both aggressive and passive animals may succeed in finding mates, but the strategies they use in mate finding will differ greatly. (These two sources of variation in personality—experience and genetics—are reminiscent of the nature–nurture controversy discussed in Chapter 3; it may be useful to review that material now.)

If personality varies among animals but is strongly influenced by genetic make-up for an individual, then the study of personality lies within the realm of behavioral genetics (refer to Chapter 3). The relevance of personality for this chapter, no matter what the genetic basis, is that personality creates the opportunity for multiple strategies within an animal population. This is critically important in social groups where, if all animals had exactly the same strategy, there could be no effective cooperation, the group might then descend into chaos, and the individual reproduction of group members would be affected.

173

Among nonhuman animals, personality is best known in chimpanzees and domestic dogs. In chimpanzees, personality is described by these variables:[56]

- Dominance
- Extraversion
- Dependability
- Emotional Stability
- Agreeableness
- Openness

The last five of these dimensions are thought to describe human personality; their presence in both chimpanzees and humans can be thought of as representing shared evolutionary history. Interestingly enough, in populations of both humans and chimpanzees, these personality traits have relatively high heritabilities and show virtually no effect of rearing environment. In a study of chimpanzees, there was a particularly strong heritability for social dominance, and weak heritabilities for the other dimensions of chimpanzee personality. As in human studies of personality, the same study found little effect of environment (in this case, different zoos) on personality.[56] Human twins who are separated at birth and reared in very different environments show startling similarities in personality. These findings support the hypothesis that for both humans and chimpanzees, genes strongly influence personality, and variability of personality in a population may result from many different personalities, each of which tends to be fixed, rather than a population full of highly variable individuals.

BRINGING ANIMAL BEHAVIOR HOME: COMPANION ANIMALS: PERSONALITY IN DOGS

In dogs, Svartberg and Forkman identified the main variables describing personality[57]:

- Playfulness
- Curiosity/fearlessness
- Desire to chase
- Sociability
- Aggressiveness

They suggest that the first four factors are all influenced by a single "broad" personality dimension, with aggressiveness working separately. This is, interestingly, parallel to the role of dominance in chimpanzee personality. Dog personalities have substantial heritable components.[58] Personality tests designed for dogs attempt to rate dogs for each of these characteristics. Animal shelters often use such tests to determine the suitability of a dog for adoption. Prospective adopters or purchasers of a dog should consider these personality traits when choosing a dog for their household.

In summary, personality is a set of attributes, such as sociability, aggressiveness, and willingness to please, that come together to form both the individual and social behavior of animals. It is real, measurable, and seems to be strongly influenced by genes. Variability in personality reflects, in a sense, genetic variability. This suggests that, like other traits for which there exists variation, different personalities can be successful and persist in evolutionary time; if only one personality type were successful, natural selection would eliminate this variation.

174

6.6 THE FRONTAL LOBE AND IMPULSE CONTROL

The ability to control impulses is a key aspect of personality in social groups, particularly when uninhibited behavioral expression might damage either the animal's future in the group or the survival of the group. *Social restraint* and related behaviors such as submission and appeasement are most commonly exhibited by the lower-ranking animals in a social group. One hypothesis is that the enlarged frontal lobes of humans evolved partly to enhance social restraint (impulse control) in human societies. The frontal lobes have a large number of other critical functions, such as problem solving, memory, and language, so judgment and impulse control are not the only selective factors. The primate frontal lobe is larger than that of other mammals and is larger in apes than in monkeys. The volume of the human frontal lobe is about three times larger than what would be expected from looking at the brains of other apes.[59,60]

> **KEY TERM** Social restraint is the ability of an animal to pause and consider the social consequences of their response to the actions of another member of the social group. Because social restraint involves impulse control and foresight, it can reflect cognitive processes.

Social restraint, submission, and appeasement have costs for animals in lower social ranks. In vertebrates, stressful conditions result in lowered levels of the neurotransmitter serotonin and other neurotransmitters in the brain. (See Chapter 2 for a more detailed discussion of neurotransmitters and mood.) Low serotonin stimulates the hypothalamus of the brain to release Corticotrophin Releasing Factor (CRF) and vasopressin into the circulation. These hormones act on the adrenal glands (small glands associated with the kidneys) so that stress hormones, called *corticosteroids* (also called *glucocorticosteriods*), are released. The steroid hormone cortisol is particularly important in stress responses. Under normal circumstances, corticosteroids prepare the animal physiologically to cope with stressful conditions by increasing blood flow and mobilizing the body's energy resources. This can be part of the "fight-or-flight" syndrome that is key to survival under dangerous conditions.

But what happens to personality if an animal is chronically stressed, such as when a low-ranking animal must continuously restrain its behavior? In a wide range of animals, cortisol levels are much higher in low ranking animals.[61] Conveniently, cortisol levels can be measured from fecal samples, meaning that the social stress can be measured without the added stress of capture and blood sampling. Hypothetically, an animal's cognition of its low rank may be mediated by feelings that are analogous to depression and anxiety, although this can be inferred only from external attributes, such as self-directed behaviors (see Chapter 4). The effect of stress on personality leads into the next section, which focuses on emotions in animals.

6.7 ANIMAL EMOTIONS

Can a dog love a person? A discussion of emotion logically follows a consideration of personality. Emotion is at the core of human cognitive life. Human self-perception of emotions is one of the roots of the human separation of mind from body. Human theory of mind depends on the ability to perceive the emotions of other humans, the ability to project the effect of those emotions, and the ability to understand how the consequences of human actions depend on the emotional state of the recipient. This leads many humans, scientists and nonscientists alike, to conclude that emotion is a fundamental aspect of cognition. Many humans long to believe that nonhuman animals have emotional lives parallel to human emotional lives. Is this true? Or is it simply a projection of human emotional lives onto animals that in reality have no emotional lives of their own? (For that matter, many people need to believe that animals have no emotional lives.)

> ### BRINGING ANIMAL BEHAVIOR HOME: LOVE, AFFECTION, AND PETS
>
> What if the "love and affection" that a dog seems to shower upon a person is not really that at all? Imagine that deep in dog and human evolutionary history, dogs were attracted to food scraps and other leavings around human campsites. To the humans, dogs probably appeared as just another potential food item. Under these circumstances, natural selection would favor any sort of dog behavior that averted being eaten. What behaviors did the dog already have in its repertoire? Wild canid social behavior includes affiliative and submissive behaviors that are nicely congruent with human expectations for social bonding. In this scenario, natural selection should favor the expression of affiliation and submission by the dog, whether the dog felt any "emotional" attachment or not (see Figure 6.15).[62]
>
>
>
> **FIGURE 6.15**
> Hundreds (and thousands) of years of natural and artificial selection have resulted in dogs that people love and that seem to love people. Who can know the mind of a dog? *Photo: Ross Madden.*

In reality, the core question is whether emotions such as joy, sorrow, love, and sadness can be ascribed to a dog (or any animal) as feelings that underlie behavioral actions. Beyond the person's own emotional response—"Yes, of course, my dog loves me"—can the hypothesis that animals feel emotion be tested? A direct test is probably impossible; exactly what a dog thinks may never be known, although brain scans like those discussed at the beginning of this chapter may offer a window into other species' thought processes. However, there is a line of evidence that can be considered as support for the hypothesis that animals have emotions of the human sort.

First, the neural architecture that regulates behaviors and their underlying emotions in humans did not arise *de novo* when humans evolved. Humans share brain structure, neurotransmitters, and hormones with the entire vertebrate lineage and beyond. Vertebrate homologies, that is, attributes that are shared as the result of descent from a common ancestor, extend beyond having four limbs and similar bone structures. For instance, human emotions are greatly affected by neurotransmitters such as serotonin; likewise, treatment of animals such as dogs with drugs such as Prozac (fluoxetine) has behavioral effects on dogs consistent with the behavioral effects that Prozac has on humans. Even mice exhibit behavior that reflects "anxiety."[63] This supports the argument that emotions could have deep evolutionary roots within the vertebrates, if not deeper.

In addition, consider the logic of how emotions compel human behavior. In "normal" behavior, human emotions serve as motivators to express evolutionarily appropriate behavioral responses. (See the discussion of motivation in Chapter 4 for more on this issue.) An excellent example of this is the role of emotion in stimulating parental care of infants. If emotion is the great motivator of human behavior, would it be surprising that animals with similar brains are motivated in the same way? If emotion does not drive the behavior of animals with physiology that is similar to humans, what does?

BRINGING ANIMAL BEHAVIOR HOME: "GUILTY" DOGS

Many dog owners think their dog looks "guilty" if the animal violates a household rule. "Guilty" looks in dogs include submissive facial expressions, lowered tails, and cowering. Common violations include taking food from a counter or rummaging in the garbage when no one is in the room. Alexandra Horowitz tested this perception using a simple but clever experimental design.[64] The experiment focused on dogs in their familiar household environment. The dog's owner left a room, and then for half the dogs, the experimenter gave an illicit treat, while half the dogs received nothing. The owners were then either told that their dog had violated the family rule or that the dog had complied. Half of the owners received the truth, while half were told a lie.

In two types of dog-owner pairs, the owner knew the truth:
1. the dog received a treat and the owner knew it, and
2. the dog did not receive a treat and the owner knew that. In the other two types, the owner had false information:
3. the dog received a treat but the owner thought it had complied with the rule, or
4. the dog did not receive a treat and the owner thought it had broken the rule. In conditions 1 and 4, the owners perceived that their dog was a violator, even though only half of these dogs had actually violated the rule. In conditions 2 and 3, the owner thought the dog had followed the rules, when in fact in half of the cases it had not.

Owners judged their dogs as looking equally guilty in conditions 1 and 4, whether or not the dog had actually followed the rule. Owners judged their dogs as looking not-guilty in conditions 2 and 3, even though half of these dogs were guilty. The driving factor was the owners' punitive demeanor when told that their dog had violated the rules. This illustrates how sensitive dogs are to human moods, and argues against dogs feeling a moral sense of regret when they take an opportunity to violate a rule.

Ultimately, emotionality in animals may be an untestable hypothesis. Descriptors of emotion, such as fear, sadness, loneliness, and affection, are subjective representations of a human state. They are difficult enough to quantify in humans and to compare among humans, much less to document in animals. While joy and sorrow are qualitatively different, there is no way of knowing whether everyone's experience of an emotion is the same.

Indeed, some rather incendiary miscommunication can occur if one assumes that human emotional lives are the same. If emotionality is difficult to document in animals that are similar to us, how much more difficult is it to document in, say, an octopus?

In some ways, this chapter has come full circle, in a return to the challenge of anthropomorphism. For the past century, biologists have been trained to avoid anthropomorphism, probably rightly so, because ascribing human emotions and motivations to animals interferes with objective hypothesis testing. It is certainly unscientific to thoughtlessly apply subjective descriptors to the behavior under observation. It is equally unscientific to ignore reasonable hypotheses because of a well-established taboo among scientists. Enough is known now about the neurochemical bases of behavior that the brain is no longer a black box; an extensive fund of knowledge concerning the roles of hormones and neurotransmitters in behavioral modulation has replaced pure imagination.

What should students and researchers of animal behavior do? When appropriate, emotion should be considered as a hypothetical shaper of behavior. In manipulative experiments and clinical treatment of animal behavior, matching subjective descriptions of human emotion with an animal's state gives excellent clues about how pharmacological treatment might affect the animal's behavior. This approach, though, needs to be tempered with the recognition that such subjective representations can lead to gross misinterpretations or oversimplifications of animal behavior and social systems. As with any hypothesis, emotion should not be thoughtlessly denied or affirmed, but should be seen as a potentially testable statement, rather than as an assumption.

6.8 ARE COGNITIVE ABILITIES UNDER- OR OVER-ATTRIBUTED TO ANIMALS?

Human history is littered with attempts to draw a distinguishing boundary between animals and humans. This is an impossible task because, of course, humans are animals, and whatever makes humans feel special, relative to the rest of the animal world, is simply a set of adaptations for the peculiarities of human life. In a sense, these adaptations create "human-ness," but any other species under observation has its own unique set of adaptations; other species are different, but are not lesser forms of life. In a recent book, Clive Wynne argued: "the psychological abilities that make human culture possible— enthusiasm to imitate others, language, and the ability to place oneself imaginatively into another's perspective on events—are almost entirely lacking in any other species."[65] This is a common point of view, and the contrast between this conclusion and the arguments of some scientists in favor of more widespread cognition in animals is part of what makes this such an interesting field of study. In grade schools in the 1960s, students were taught that specific characteristics—language, tool use, upright posture, and government— separate humans from nonhuman animals. Scientific experience over the intervening 50 years seems to have dissolved all of these barriers. Humans may be better at performing some tricks than nonhuman species, but other species have capabilities humans can only dream about.

Human-centric thinkers tend to give animal species less credit for cognition than they deserve. In contrast, other humans claim to have special empathy for animals, thereby feeling that they are elevated on a "moral" platform above most of their fellow humans. Over evolutionary time, human interactions with animals have passed from coexistence and relationships in which sometimes humans were prey and sometimes were predators, to the domestication and employment of animals as companions, food, and beasts of burden. In the process, humans have become the greatest threat to the continued existence of many animal species. As these relationships have changed over time, humans, as thinking, contemplative animals, have been challenged to reshape ethical and moral views of their interactions with animals.

One of the major points of contention raised by scientific findings about animal cognition is whether humans should then use the discovery of the cognitive lives of animals to inform choices about treatment of those species. Bluntly put, does knowing an animal is cognitive of how it is being treated make humans, as a society, want to treat it differently? If so, how should it be treated? This question is as difficult to answer as others in this chapter because there is good evidence that (not surprisingly) different species are not identical in the ways that they experience "comfort." There are no good, definitive, answers to those questions, but they certainly will be the subject of lively debate for the foreseeable future.

Humans can be quite calculating when it comes to assessing the ability of an animal to perform a task, but that facility at predicting animal behavior is not the same as a window into the animal's actual thought processes. In other words, even though humans can project their interpretations onto animal behavior, that does not mean that animals think the same way that humans do.

OF SPECIAL INTEREST: ANECDOTE AS EVIDENCE IN STUDIES OF COGNITION

One of the biggest scientific barriers to acceptance of animal cognition, particularly of personality and emotion as components of the cognitive life of nonhuman animals, is that much of the evidence comes from anecdotes. Typical of this type of information are stories about dolphins rescuing drowning humans by supporting them in the water. If true, this could be argued as evidence for dolphins living an empathic life. But how strong is the evidence? Were the observers reliable? Was the story changed or amplified through multiple retellings? Could the circumstances be replicated? What about all the times when no dolphin appeared? Reliability and replicability are hallmarks of scientific veracity.

Donald Griffin (see p. 21), argued that accumulated anecdotal evidence for cognition could develop into compelling scientific evidence. The current scientific community is divided on this issue, with probably the majority of animal behaviorists having some healthy skepticism about anecdotal evidence for cognition. Accepted scientific standards require that results be repeatable and adequately controlled, a set of criteria not often met by anecdotal accounts supporting cognitive events in animals' lives.

What if the argument that cognitive abilities are widespread among other animal species is accepted? Do scientists also accept the corollaries that personality is a reflection of cognition and that emotions are part of the lives of animals? The question of whether this informs/changes in human interactions with animals is still open. The debate over animal cognition reaches into human religion, spirituality, and legal systems. An interesting approach taken in Boulder, Colorado, and other cities has been to legislate that humans are "guardians" rather than "owners" of animals. *Companion animal* becomes the preferred term over *pet*. Some people argue that this change in terminology can modify, in a meaningful way, the relationship between humans and nonhuman animals, whereas others view this as semantic silliness. (This calls up the previous discussion of language and mental symbols!)

Following this line of reasoning, critics of the maintenance of animals in zoos sometimes argue that part of the cruelty of animals' captivity stems from animal cognition of surroundings and the conditions of confinement. Chapter 4 presents methods of measuring the stress level of animals and strategies for changing the conditions of captivity to at least eliminate the behavioral manifestations of stress. But the argument against captivity based on hypothesized cognition reaches more deeply, suggesting that it is immoral to maintain an animal whose cognitive abilities allow it to understand that it is captive. For some people, this argument outweighs the positive roles of zoos in education and conservation. Others argue that hypotheses about cognition are nowhere near as well supported as the risk of extinction. Is it better to allow extinction than to use captive breeding? Can animals acclimate to captivity? What about captive-born animals?

SUMMARY

What is cognition? This proves to be a difficult concept to define, and it is fair to say that cognition means different things to different scientists. The separation of the brain as the physical structure that coordinates behavior and the mind as a philosophical concept is central because cognition is a property of the mind. This then leads to questions about tests for cognition. One key element of many definitions of cognition is the ability to visualize oneself in the context of the larger world. Following this line of reasoning, simple behaviors such as gaze following and mirror tests offer some guidance about the cognitive abilities of an animal. Harder to test but also important is how animals forecast social relationships and solve complex problems. Time–place learning, caching, and thievery may also require foresight and planning, making them cognitive processes.

Intelligence, personality, and emotion also play central roles in discussions of cognition. Intelligence is as hard to define as cognition, but it is clear that phenomena that can be interpreted as reflecting intelligence often are cognitive processes. When animals interpret each other's personalities, personality becomes a part of the cognitive picture. The concept of the behavioral syndrome has yielded a somewhat broader tool than personality to understand variation in the behavioral tendencies of animals. Emotionality is, in itself, not a cognitive process, but empathy for the emotions of others is such a process, and the discussion of emotions in this chapter helps in understanding impulsivity, impulse control, and empathy. Cognition separates an animal from being a purely impulsive creature, in that cognition allows reflection, planning, and foresight—in other words, nonimpulsive behavior.

Do humans under- or overattribute cognitive abilities to animals? People seem to be pulled in both directions. It is tempting to project thought and emotional processes onto animals without critically testing hypotheses of cognition. On the other hand, as more data have been collected, particularly over the past decade, scientific arguments credit animals with much greater levels of cognition than previously suspected. Modern insights have broken down the cognitive/noncognitive divide that many humans have constructed between animals and humans.

What is the endpoint for this evolving topic? Developing an understanding of animal cognition is a blossoming field in animal behavior, and of all the topics covered in this book, it is the one for which scientific understanding is mostly likely to dramatically change over the next one or two decades. For students of animal behavior, the topic of cognition should prove a rich area for debate and discussion. Scientific knowledge in this area is far from definitive, and the potential ethical ramifications seem to have no end.

STUDY QUESTIONS

1. What is cognition?
2. How might the roles of sociality and carnivory in the evolution of intelligence be disentangled?
3. If animals are cognitive, how should this affect human behavior toward animals?
4. How could a test be devised for the hypothesis that a nonhuman animal has a theory of mind?
5. In the laboratory manual that accompanies this text, students are challenged to develop reliable intelligence tests for dogs (Chapter 15 in the laboratory manual[66]). This is harder than it might seem at first glance!
6. Dog owners often boast of the intelligence of their dog and sometimes show a startling tendency to view their dog's emotional state as a reflection of their own emotions. Dogs can appear quite empathic, but are they truly so? Can empathy be formulated as a testable hypothesis and be subjected to the same sorts of rigorous tests that are required

in other areas of animal behavior? How can humans separate their projection of human behavior from what the animal is actually doing?

7. Recall Tinbergen's questions. Although cognition is a new and unsettled field of animal behavior, some topics nonetheless might claim to answer all of his questions. Review this chapter and find a topic or two that you think might be fruitful in that regard.

Further Reading

Balda, R.P., Pepperberg, I.M., Kamil, A.C. (Eds.), 1998. Animal Cognition in Nature. Academic Press, New York, 456 pp.

Bekoff, M., 2002. Minding Animals: Awareness, Emotions and Heart. Oxford University Press, Oxford, 256 pp.

Bekoff, M., Allen, C., Burghardt, G., 2002. The Cognitive Animal. MIT Press, Boston, 504 pp.

Emery, N., Clayton, N., Frith, C., 2008. Social Intelligence: From Brain to Culture. Oxford University Press, Oxford, 432 pp.

Griffin, D., 1985. Animal Thinking. Harvard University Press, Boston, 265 pp.

Hauser, M., 2001. Wild Minds: What Animals Really Think. Holt, 336 pp.

Hess, E., 2008. Nim Chimpsky: The Chimp Who Would Be Human. Bantam, 384 pp.

Stamp Dawkins, M., 1998. Through Our Eyes Only?: The Search for Animal Consciousness. Oxford University Press, Oxford, 206 pp.

Wasserman, E., Zentall, T., 2009. Comparative Cognition. Oxford University Press, Oxford, 720 pp.

Wright, W., 1999. Born That Way: Genes, Behavior Personality. Routledge, 304 pp.

Wynne, C., 2006. Do Animals Think? Princeton University Press, Princeton, NJ, 288 pp.

Wynne, C., 2002. Animal Cognition: The Mental Lives of Animals. Palgrave Macmillan, 231 pp.

Notes

1. Bates, L.A., Sayialel, K.N., Njiraini, N.W., Moss, C.J., Poole, J.H., Byrne, R.W., 2007. Elephants classify human ethnic groups by odor and garment color. Curr. Biol. 17, 1938–1942.

2. Bahlig-Pieren, Z., Turner, D.C., 1999. Anthropomorphic interpretations and ethological descriptions of dog and cat behavior by lay people. Anthrozoos 12 (4), 205–210.

3. Darwin, C., 1872. The Expression of the Emotions in Man and Animals, John Murray, London, 350 pp.

4. Kamphuis, S., Dicke, P.W., Thier, P., 2009. Neuronal substrates of gaze following in monkeys. Eur. J. Neurosci. 29, 1732–1738.

5. Struhsaker, T.T., 1977. Infanticide and social organization in the redtail monkey (Cercopithecua ascaniua schmidti) in the Kibale Forest Uganda. Z. Tierpsychol. 45, 75–84.

6. Owren, M.J., 1990. Acoustic classification of alarm calls by vervet monkeys (Cercopithecus aethiops) and humans (Homo sapiens): II. Synthetic calls. J. Comp. Psychol. 104, 29–40.

7. Emery, N.J., Seed, A.M., von Byern, A.M.P., Clayton, N.S., 2007. Cognitive adaptions of social bonding in birds. Phil. Trans. R. Soc. B. 362, 489–505.

8. Bird, C.D., Emery, N.J., 2009. Insightful problem solving and creative tool modification by captive nontool-using rooks. PNAS 106, 10370–10375.

9. Visalberghi, E., Addessi, E., Truppa, V., Spanoletti, N., Ottoni, E., Izar, P., et al., 2009. Selection of effective stone tools by wild bearded capuchin monkeys. Curr. Biol. 19, 213–217.

10. Sanz, C.M., Shoning, C., Morgan, D.B., 2009. Chimpanzees prey on army ants with specialized tool set. Am. J. Primatol. 72, 17–24.

11. Pruetz, J.D., Bertolani, P., 2007. Savanna Chimpanzees, Pan troglodytes verus, hunt with tools. Curr. Biol. 17, 1–6.

12. Mercader, J., Barton, H., Gillespie, J., Harris, J., Kuhn, S., Tyler, R., et al., 2007. 4,300-year-old chimpanzee sites and the origin of percussive stone technology. PNAS 104, 3043–3048.

13. Kenward, B., Weir, A.A.S., Rutz, C., Kacelnik, A., 2005. Tool manufacture by naïve juvenile crows. Nature 433, 121.

14. Nowak, M.A., Komarova, N.L., Niyogi, P., 2001. Evolution of universal grammar. Science 291, 114–118.

15. Henry, C., 1995. Universal grammar. In: L. Rocha (Ed.). Communication and cognition—artificial intelligence. 12 (1–2), 45–61. Special Issue. Self-Reference in Biological and Cognitive Systems.

16. Chater, N., Reali, F., Christiansen, M.H., 2009. Restrictions on biological adaptation in language evolution. PNAS 106, 1015–1020.

17. Marzluff, J.M., Walls, J., Cornell, H.N., Withey, J.C., Craig, D.P., 2010. Lasting recognition of threatening people by wild American crows. Anim. Behav. 79, 699–707.

18. Plotnik, J.M., de Waal, F.B.M., Reiss, D., 2006. Self-recognition in an Asian elephant. PNAS 103, 17053–17057.

19. Penn, D.C., Povinelli, D.J., 2007. On the lack of evidence that non-human animals possess anything remotely resembling a 'theory of mind'. Philos. Trans. R. Soc. B. 362, 731–744.

20. Byrne, R.W., 2003. Animal communication: what makes a dog able to understand its master? Curr. Biol. 13, R347–R348.

21. Wynne, C.D.L., Dorey, N.R., Udell, M., 2008. Wolves outperform dogs in following human social cues. Anim. Behav. 76, 1767–1773.

22. Marina, D., Menzier, S., Zimmerman, E., 2008. Rapid facial mimicry in orangutan play. Biol. Lett. 4, 27–30.

23. Brown, M.F., Farley, R.F., Lorek, E.J., 2007. Remembrance of places you passed: social spatial working memory in rats. J. Exper. Psych. 33, 213–224.

24. Kaminski, J., Call, J., Fischer, J., 2004. Word learning in a domestic dog: evidence for "fast mapping". Science 304, 1682–1683.

25. Roberts, W.A., Feeney, M.C., MacPherson, K., Petter, M., McMillan, N., Musolino, E., 2008. Episodic-like memory in rats: is it based on when or how long ago? Science 320, 113–115.

26. Murphy, C.M., Breed, M.D., 2008. Time-place learning in the stingless bee, *Trigona fulviventris*. Journal of the Kansas Entomological Society. 81, 73–76.

27. Heydarnejad, M.S., Purser, J., 2008. Specific individuals of rainbow trout (*Oncorhynchus mykiss*) are able to show time–place learning. Turk. J. Biol. 32, 209–229.

28. Van der Zee, E.A., Havekes, R., Barf, R.P., Hut, R.A., Nijholt, I.M., Jacobs, E.H., et al., 2008. Circadian time–place learning in mice depends on *Cry* genes. Curr. Biol. 18, 844–848.

29. Daily, J.M., Emery, N.J., Clayton, N.S., 2006. Food-caching Western Scrub-Jays keep track of who was watching when. Science 312, 1662–1665.

30. Raby, C.R., Alexis, D.M., Dickinson, A., Clayton, N.S., 2007. Planning for the future by Western Scrub-Jays. Nature 445, 919–921.

31. Clayton, N.S., Dally, J.M., Emery, N.J., 2007. Social cognition by food-caching corvids. The Western Scrub-Jay as a natural psychologist. Philos. Trans. R. Soc. B, Biol. Sci. 362, 507–522.

32. Gould, J.L., 1986. The locale map of honey-bees—do insects have cognitive maps. Science 232, 861–863.

33. Collett, M., Collett, T.S., 2006. Insect navigation: No map at the end of the trail? Curr. Biol. 16, R48–R51.

34. Mulcahy, N., Call, J., 2006. Apes save tools for future use. Science 312, 1038–1040.

35. Fragaszy, D.M., Pickering, T., Liu, Q., Izar, P., Ottoni, E., Visalberghi, E., 2010. Bearded capuchin monkeys' and a human's efficiency at cracking palm nuts with stone tools: field experiments. Anim. Behav. 79, 321–332.

36. Zentall, T.R., Singer, R.A., Stagner, J.P., 2008. Episodic-like memory: Pigeons can report location pecked when unexpectedly asked. Behav. Processes 79, 93–98.

37. Gruber, T., Muller, M.N., Strimling, P., Wrangham, R., Zuberbuhler, K., 2009. Wild chimpanzees rely on knowledge to solve experimental honey acquisition task. Curr. Biol. 19, 1806–1810.

38. Erdohegyi, A., Topal, J., Viranyi, Z., Miklosi, A., 2007. Dog-logic: inferential reasoning in a two-way choice task and its restricted use. Anim. Behav. 74, 725–737.

39. Hyde, D.C., Spelke, E.S., 2009. All numbers are not equal: an electrophysiological investigation of small and large number representations. J. Cogn. Neurosci. 21, 1039–1053.

40. Gallistel, C.R., Gelman, R., 2005. Non-verbal numerical cognition: from reals to integers. Trends Cogn. Sci. 4, 59–65.

41. Chittka, L., Geiger, K., 1995. Can honeybees count landmarks? Anim. Behav. 49, 159–164.

42. Andersson, M., 2003. Behavioural ecology: Coots count. Nature 422, 483–485.
Lyon, B.E., 2003. Egg recognition and counting reduce costs of avian conspecific brood parasitism. Nature 422, 495–499.

43. Hauser, M.D., 2000. What do animals think about numbers? Am. Sci. 88, 144–151.
Hauser, M.D., MacNeilage, P., Ware, M., 1996. Numerical representations in primates. Proc. Natl. Acad. Sci. USA 93, 1514–1517

44. Inoue, S., Matsuzawa, T., 2007. Working memory of numerals in chimpanzees. Curr. Biol. 17, R1004–R1005.

45. Seed, A., Emery, N., Clayton, N., 2009. Intelligence in Corvids and Apes: A Case of Convergent Evolution? Ethology 115, 401–420.

46. Foote, A.L., Crystal, J.D., 2007. Metacognition in the rat. Curr. Biol. 17, 551–555.

47. Humphrey, N.K., 1976. The social function of intellect. In: Byrne, R.W., Whiten, A. (Eds.), Machiavellian Intelligence: Social Expertise and the Evolution of Intellect in Monkeys, Apes and Humans. Clarendon Press, Oxford, UK, pp. 285–305.

48. Holekamp, K.E., 2007. Questioning the social intelligence hypothesis. Trends Cogn. Sci. 11, 65–69.

49. Hare, B., Brown, M., Williamson, C., Tomasello, M., 2002. The domestication of social cognition in dogs. Science 298, 1634–1636.

50. Silk, J., 2007. Chimps don't just get mad, they get even. Proc. Natl. Acad. Sci. USA 104, 13537–13538.

51. Jensen, K., Call, J., Tomasello, M., 2007. Chimpanzees are vengeful but not spiteful. Proc. Natl. Acad. Sci. USA 104, 13046–13050.

52. Subiaul, F., Vonk, J., Okamoto-Barth, S., Barth, J., 2008. Do chimpanzees learn reputation by observation? Evidence from direct and indirect experience with generous and selfish strangers. Anim. Cogn. 11, 611–623.

53. Bouchard, T.J., 1994. Genes, environment, and personality. Science 264, 1700–1701.

54. Sih, A., Bell, A., Johnson, C.J., 2004. Behavioral syndromes: an ecological and evolutionary overview. Trends Ecol. Evol. 19, 372–378.

55. Johnson and Sih 2007.

56. Weiss, A., King, J.E., Figueredo, A.J., 2000. The heritability of personality factors in chimpanzees (*Pan troglodytes*). Behav. Genet. 30, 213–221.

57. Svartberg, K., Forkman, B., 2002. Personality traits in the domestic dog (*Canis familiaris*). Appl. Anim. Behav. Sci. 79 (2), 133–155.

58. Wilsson, E., Sundgren, P.E., 1997. The use of a behaviour test for selection of dogs for service and breeding 2. Heritability for tested parameters and effect of selection based on service dog characteristics. Appl. Anim. Behav. Sci. 54, 235–241.

59. Semendeferi, K., Lu, A., Schenker, N., Damasio, H., 2002. Humans and great apes share a large frontal cortex. Nat. Neurosci. 5, 272–276.

60. Passingham, R.E., 2002. The frontal cortex: does size matter? Nat. Neurosci. 5, 190–192.

61. Ostner, J., Heistermann, M., Schulke, O., 2008. Dominance, aggression and physiological stress in wild male Assamese macaques (*Macaca assamensis*). Horm. Behav. 54, 613–619.

62. Range, F., Viranyi, Z., Huber, L., 2007. Selective imitation in domestic dogs. Curr. Biol. 17, 868–872.

63. Nyberg, J., Vekovischeva, M.O., Sandnabba, N.K., 2003. Anxiety profiles of mice selectively bred for intermale aggression. Behav. Genet. 33, 503–511.

64. Horowitz, A., 2009. Disambiguating the "guilty look": salient prompts to a familiar dog behavior. Behav. Processes 81, 447–452.

65. Wynne, C.D.L., 2006. Do Animals Think? Princeton University Press, 7 pp, 288 pp.

66. Tillberg, C.V., Breed, M.D., Hinners, S.J., 2007. Field and Laboratory Exercises in Animal Behavior. Academic Press, 232 pp.

182

Communication

183

LEARNING OBJECTIVES

Studying this chapter should provide you with the knowledge to:

- Understand that communication involves a signal produced by a sending animal and interpreted by a receiving animal.

- Be able to explain the co-evolution between the sender (or signaler) and receiver as a necessary part of the communication process. Evolution refines both the sender's ability to send a meaningful signal and the recipient's ability to perceive and understand the signal.

- Know how signals used in communication evolve following the basic principles of co-option, ritualization, stereotypy, and redundancy.

- Identify the circumstances favoring honest and deceitful signaling in animal populations.

- See that signals become public information that is available both to the intended recipient and to other animals, which might take advantage of the signal.

- Understand how noise influences the evolution of communication.
- Apply the concept that, in interspecific signaling, signals evolve to influence the behavior of other species.

7.1 INTRODUCTION: COMMUNICATION THEORY

Communication can be as simple as a female moth producing an odor that attracts potential mates or as complicated as honeybee dances conveying information about food locations.[1,2] *Communication* always involves the transmission of information from one animal to another. This information is not merely random; communication differs from other forms of information because it involves *signals* sent from one animal to another.

> **KEY TERM** Communication is the transmission of signals from a sender to a receiver.

> **KEY TERM** A signal is a product of an animal that has evolved to carry a specific message to another animal.

Humans are highly communicative organisms, with lives so embedded in communication that we rarely think about it as a phenomenon; nonetheless, communication is also foundational for much of the behavior that occurs among animals in the natural world. It is difficult to imagine the existence of social behavior, for instance, without communication.

CASE STUDY

184

Making Their Own Light

Light production by an animal can be a powerful signal. Animal species that make their own light include fireflies (lightning bugs, beetles in the Family Lampyridae), larvae of fireflies and some other insects, some species of click beetle (Family Elateridae), and a large number of deep sea animals. In bioluminescent beetles (e.g., fireflies), light production clearly has a signal function in attracting mates. Deep-sea organisms may use luminescence as a signal, as a lure for prey, or as illumination so they can see their underwater world.

Male fireflies in the genus *Photinus* fly above the ground and flash signals of their mating availability; the signals vary so that the intensity of the flash and delays between flashes identify the males' species. Females on the ground respond with a single flash; the male and female then signal back and forth, a prelude to the female choosing a male. Better-quality males produce longer flashes, so male signaling advertises more about the male than just his species.[3] The male gives the female a "nuptial gift" of a nutritious glandular secretion (see Chapter 11 on mating) that nourishes the female, so finding better males affects a female's reproductive ability. Fireflies in the genus *Photuris* prey on male *Photinus* fireflies by mimicking *Photinus* signals; they draw in unsuspecting males using this form of aggressive mimicry and then eat them (see Figure 7.1).

Bioluminescent marine animals often use their light as a defense, startling or confusing predators, or as an offensive weapon, illuminating or attracting prey.[4] In fact, a vast preponderance of deep sea animals use bioluminescence in some offensive or defensive aspect of their lives. In some cases, though,

bioluminescence in the ocean serves as a signal to attract mates. A good example of this is a small crustacean, *Photeros annecohenae* (Class Ostracoda). (An ostracod looks like a tiny clam with a shrimp inside.) *Photeros annecohenae* males use bioluminescence as a signal to attract females for mating.[5] Males of this species swim together, with many males flashing at the same time, producing a spectacular miniature lighting display. Some males "sneak," swimming near flashing males while not using their own signal. Bioluminesence works well as a communication mode in dark environments, opening the door for these interesting interactions.

© Joyce Gross

FIGURE 7.1

In this example of aggressive mimicry, a female *Photuris* firefly has mimicked the flashing signal of a female *Photinus* firefly and has lured a male *Photinus* to his death. *Photo: Joyce Gross.*

What Is the Adaptive Significance of Communication?

This is a question straight out of Niko Tinbergen's playbook: Why has animal communication evolved? Thinking broadly, perhaps finding a mate is the most elemental and universal reason that communication exists. Finding a mate means being able to identify another animal's location, its species, and sometimes its quality. Such broad need for information favors the evolution of communication.

As if the advantages that communication confers upon mating behavior were not beneficial enough, parents may communicate key survival information to offspring. Within social groups, animals sometimes communicate status, needs for care or desire to play, alarm when predators or competitors are present, and information about the location and quality of food or shelter. For individuals or social groups, communication plays key roles in establishing and maintaining territories.

Many of the studies in the current animal behavior literature ask if signals are honest or dishonest.[6] Honesty and deception in communication revolve around whether the animals share interests in the outcome of their interaction, or if their interests differ. Signals are expected to be honest in the case of shared interests; if interests differ, signals might evolve toward dishonesty.

What Is a Signal?

Simply put, a signal is an animal product that has evolved to carry a specific meaning to another animal. Signals can be external attributes of the animal, such as color; actions, such as waving an arm or producing a sound; a chemical compound, such as an alarm pheromone; or an energetic output, such as an electrical discharge or a magnetic field. The scientific literature offers somewhat more complex definitions. Maynard Smith and Harper define a signal as "any act or structure which alters the behaviour of other organisms, which evolved because of that effect, and which is effective because the receiver's response has also evolved."[7] If we step back from the apparent complexity of these definitions, the key feature is that signals have evolved specifically for communication.

This feature, then—that signals are the result of natural selection that favors information content—is what differentiates a signal from other types of information. Virtually everything an animal does creates information in the environment. Even as signals have been honed by natural selection to carry information, so it is that animals able to receive the signals have also been favored by natural selection. What point would there be in producing a signal that cannot be received or being primed to receive a signal that never arrives? The definitions are complicated only because there are so many types of signals. Some structural signals are produced almost continuously, whereas other behavioral signals are produced at certain times, when they are likely to be received. Thus, a male Canyon Wren (see Figure 7.2A) is brown and speckled all of the time, even though that signal is not always needed. In contrast, it sings when it "wants" to be heard (see Figure 7.2B). Both the color and the song are signals, but one is a persistent part of the physical nature of the animal and the other is a product that is periodically broadcast into the world around the animal.

DISCUSSION POINT: WHAT IS A SIGNAL?

It is sometimes difficult to test the hypothesis that a given attribute is a signal, that is, that a structure or behavior has evolved for communication. This creates some uncertainty about whether or not information is a signal. To put this into an easily understood framework, consider that you can probably recognize your friends from a great distance by how they walk; in fact, recognizing someone's gait may be your first clue as to his or her identity. Clearly, there is information in the gait, but is it a signal? It would only be a signal if gait has evolved for the specific communicative function of identifying a person. Can you design an experiment to test whether gait is a signal?

FIGURE 7.2
(A) A male Canyon Wren and (B) a sonogram of its song.

Senders and Receivers

Most of the animals we study in this book have, over evolutionary time, modified receptors that originated early in the history of life (see Chapter 2). Regardless of size and complexity, the sender has a structure that constitutes the signal (for example, the peacock's tail) or that produces the signal (such as vocal cords, scent glands, and electric organs). The receiver has receptors, which are usually modified nerve cells. These cells, such as rods and cones in the eyes, hair cells in ears, and the olfactory epithelium, *transduce* the signal into nerve impulses that are processed by the nervous system. Decoding sensory information is a critical task of the nervous system; noise must be separated from meaningful signals. In some cases, an animal communicates with itself; bats do this when they echolocate. Communication with oneself is called *autocommunication*. In the study of communication, *noise* refers to a disturbance or variation in a stimulus that interferes with a signal. Noise is not limited to auditory signals but refers to stimuli of any sort.

In sum, these are the building blocks of communication: The sender creates a signal by converting energy into something—a molecule, a movement, an electric field, for example—that can be transmitted into the environment. The receiver uses receptors to transduce the signal into a form that can be used inside the receiver. The information content of a signal can be measured by its effect on the behavior of the receiver. (See Chapter 2 for more information about transduction.)

Public Information and Eavesdropping

When an animal broadcasts a signal, there are "intended" recipients, or targets. A female emitting sex pheromone is luring males of her own species. A bull elk bugling in the fall is signaling his dominance to elk males and his attractiveness to females. When the information is broadcast, however, it becomes *public information*. For instance, eavesdropping female cowbirds can listen to the interactions between male and female cowbird pairs. Male cowbirds produce calls to attract mates, and when a female chooses a male (presumably based on his quality as a mate), she responds to his vocalizations with "chattering" calls. Eavesdropping females learn to prefer male calls that have already elicited chattering by other females. Thus, the eavesdropper could potentially take advantage of public information to find a high-quality mate without having to go through the process of assessing a large number of males (see Figure 7.3).[8]

7.2 THE EVOLUTION OF COMMUNICATION
The Evolution of Signals

How does communication evolve? That is a reasonable question; after all, if communication requires a sender and a receiver, how likely is it that two such organisms would accidentally be able to share information? At first glance, this seems to be an unlikely coincidence, but it makes good sense to return to basic principles: All living organisms require energy sources, and the ability to locate those energy sources has always been strongly favored by natural selection. Thus, photosynthetic organisms must locate light sources, chemosynthetic organisms must locate sources of appropriate chemicals, and heterotrophic animals must locate other organisms for their nutrition. Because all living things emit metabolic waste and byproducts, an organism that could respond to such chemicals in the environment would be at an evolutionary advantage—it could find more energy and do so more quickly than organisms lacking such ability. The ability to sense and respond to energy sources—light, chemicals, prey—means that corresponding receptors exist for those sources—photoreceptors, chemoreceptors, and the like.

> **KEY TERM** Autocommunication, or communication with oneself, most commonly occurs in two contexts. The first is echolocation. The second is the use of territorial or trail markers for orientation; an ant might leave a pheromone trail, which it then uses later to return to a feeding location. If no other ants use the trail, then this is autocommunication.

> **KEY TERM** Noise is unwanted information that interferes with signals. Acoustic noise, for example, from other animals or from human sources, such as traffic, may interfere with sounds produced as signals.

> **KEY TERM** Public information is signals or attributes of an animal that are generally available to other animals in its environment. Public information can be exploited by other animals, sometimes to the detriment of the animal that has produced the information.

187

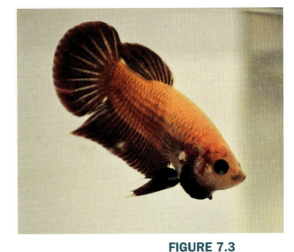

FIGURE 7.3
Bettas, *Betta splendens*, are known to eavesdrop on dominance interactions among other bettas and to use this information in their own interactions. *Photo: Michael Breed.*

FIGURE 7.4

This flicker reveals redundant information to potential mates through two communication channels: its drumming ability (auditory) and its colorful patterning (visual). *Photo: Michael Breed.*

However, this correspondence is not yet communication. A rabbit in hiding that has to sneeze is not communicating with the predator; she is just having a bad day. What is important here is that efficient receptors have been favored by natural selection since the early days of living things—as evidenced by the chemical receptors of protists, for instance. These efficient receptors, combined with the ability of a receiver to glean information about a living information source, are the necessary ingredients for the evolution of communication. Indeed, receptors on one-celled organisms also paved the way for intercellular communication in multicellular organisms; as a result, the evolution of signaling systems between organisms and the evolution of signaling systems within organisms are intertwined. Given the right variation, animals can evolve receptors for any form of energy in the environment, and once these receptors have evolved, they become fair evolutionary game for use in communication systems.

Now consider an animal that is ready to reproduce. Mate-finding can be hazardous and time consuming. If a reproductively receptive organism has an attribute that can be perceived by a receiver that is ready to mate and that then responds to the attribute, both animals benefit from the increased reproductive efficiency. To generalize, if selection favors changes or variation in the sender's attributes, and favors the receiver's ability to ascribe meaning to this variation, then communication is taking place. At this point, the sender's attribute can be called a *signal*.

188

Co-option, Ritualization, and Stereotypy

The evolutionary course of *co-opting* an attribute to form the basis of a signal is called *ritualization*. Ritualization is adaptive because in communication, both parties stand to benefit from increased efficiency and accuracy. Therefore, in many cases, selection favors reduction in variation in the signal; this is the evolutionary process of *stereotypy*. Reduced variation helps minimize uncertainty about the meaning of the signal by reducing possible overlap with other signals; it allows the evolution of both finely tuned receptors and specific neural mechanisms for recognizing the signal. For example, birds that use different calls for mating and territoriality gain an advantage in clarity of communication, which comes from having each type of call highly stereotyped and distinctive.[9]

> **KEY TERM** Co-option is the evolutionary adoption of something the animal already does, or has, for use in a different form of communication.

> **KEY TERM** Ritualization is the association, through evolution, of a meaning with a signal.

> **KEY TERM** Stereotypy is the evolutionary reduction of the variation of a signal so that its meaning is more easily understood.

> **KEY TERM** Redundancy is the use of multiple signals that have the same meaning; this increases the clarity of a message and reduces possible confusion.

Redundancy

The final key element in the evolution of communication is *redundancy*. This is the use of multiple signals, often in different modes, to communicate the same message. Dogs mark their territories with odors from urine but also use visual displays such as leaving scratch marks on the ground near their scent mark. Redundant signals such as these help to ensure that the recipients will get the message (see Figure 7.4).

BRINGING ANIMAL BEHAVIOR HOME: SCENT MARKING IN DOGS AND CATS

The wild ancestors of dogs and cats used territorial markings to announce their presence in an area, and this behavior has carried over into domestic dogs and cats. The territorial markings of dogs are a composite signal, using both olfactory and visual cues. Dogs distribute olfactory cues by urinating small amounts on objects in their home range. This behavior is most marked in male dogs, which lift one hind leg—a behavior that develops as the male matures—and spray urine on an object. Males urinate frequently and do so more often in areas that are less familiar to them. Females urinate less frequently, but also direct their urine at objects and increase the frequency of urination when in less familiar territory; females may also lift their legs when placing urine on objects. Some objects—the bases of certain trees, lampposts, or fire hydrants—become olfactory signposts that attract much sniffing and marking. In habitats with winter snows, the urination sites are recognizable as "yellow snow." For students of animal behavior, these sites provide an opportunity for experimentation on canine territoriality because patches of yellow snow can easily be moved using a shovel.[10]

The development of raised leg urination is slowed and may never develop fully in castrated males. The younger the age of castration, the less likely the behavior will develop. Dogs, particularly males, sometimes also scratch the ground with their hind paws after urinating; this happens more often when other dogs are in view and may serve as a visual signal of presence of the scent mark. Ground scratching may also leave a lasting visual signal on the ground.[11,12] In the case of giant pandas, males even do handstands in order to leave urine in high places, a marking strategy that seems to intimidate subadult males. Is this deceit? Could only a REALLY large panda mark that high? Or does it communicate competitive determination?

Cats have a similar system of territorial marking, in which males, and sometimes females, spray urine on vertical surfaces. Cats differ in a critical way from dogs, though; cats concentrate their urine spraying at the core of their territory. For domestic cats, this means spraying on surfaces, such as walls, doorways, and furniture, within their household. About 10% of castrated male cats spray and about 5% of female cats spray. Compounding the problem, cat urine has a much stronger odor to humans than does dog urine. The most effective prevention of spraying is castration of males at a young age. Sometimes spraying results from behavioral problems related to stress or anxiety, and veterinarians occasionally suggest behavioral therapy, perhaps in combination with antianxiety medication, for cats that spray excessively.

189

7.3 MODES OF COMMUNICATION
How the Type of Signal Is Chosen

The evolution of communication allowed for a wide variety of communication *modes*, ranging from ones familiar to humans (e.g., vision and sound) to those that are far outside the human experience (e.g., electrical discharges, seismic waves). The goal in this section is to develop an understanding of the circumstances that cause evolution to favor a particular mode of communication over the other possible modes and to show how each mode presents special opportunities for communication, while at the same time presenting limitations.

> **KEY TERM** Mode refers to sensory mode—the type of sensory receptor used to receive the signal. The primary signaling modes are visual, auditory (sound), touch, and olfactory (taste and smell). Other modes, such as electricity, are also possible routes for communication.

These opportunities and limitations arise because different environments and different messages are best communicated with different types of signals. Disparate modes are favored by natural selection, driven by divergences in how directional and specific a signal might be, how far it might travel, and how fast it might travel. In some cases, signals that travel rapidly over long distances are favored, but distance and speed are not always important or desirable attributes of a signal; some signals are best restricted to close-by individuals and may actually attract predators or parasites if they travel too far. The function and context of the signal have a strong influence on the mode that is used.

Scientists have sometimes been fairly anthropocentric when assessing how animals use signals; they have assumed that the animal world is much like the human world when it comes to communication. It does not take deep thinking, though, to realize that dogs rely more on olfaction than humans do, and that fish swimming in muddy water face very different communication challenges than either dogs or humans. This fits with the overall theme of *umwelt*, and the unique perceptual world of each animal species (see Chapter 2). The general lesson in sensory perception and communication is that any form of energy in the environment has the potential to be perceived and that anything that is perceived has the potential for use as a signal. The challenge to human scientists is to consider channels for which we, as a species, are receptor-poor. To have even a small chance at filling such a tall order, we must again address first principles: What abiotic and biotic environments favor what modes of communication? The following sections address major communication modes, including signal production in that mode, limitations on use and perception of those signals, and examples of signals using that particular mode. Sensory perception is discussed in detail in Chapter 2; refer to that chapter for a basic review of how each signal mode is perceived.

Chemical Signals: Messages by Smell and Taste

Chemical signals may have been the first signals to evolve, if signaling began with increasing sensitivity to chemicals in the environment that indicated resources. Chemical signals are certainly ubiquitous, used by many animals. Chemical signals work well in dark environments and are not impeded by obstacles. Depending on their composition, they may linger in the environment or travel large distances. They are usually inexpensive to produce.

If the production of a molecule has been shaped by evolution so that it has a signaling function within a species, then it is a *pheromone*. Animals also produce defensive molecules; this use is discussed in Chapter 10. Humans tend to differentiate between smell (molecules carried in the air and perceived on the olfactory epithelium of the nose) and taste (molecules dissolved in liquid and perceived on the tongue), but both of these senses fall into the broad category of chemoreception.

190

> **KEY TERM** A pheromone is a chemical signal used in transmitting information within a species.

Virtually all pheromones are molecules with carbon chain backbones. The major differences among pheromone molecules are size and polarity; these two features are matched with the function of the pheromone. Small molecules and less polar molecules are relatively volatile; an airborne pheromone, such as a moth sex attractant, is likely to be a smaller molecule. A pheromone that does well to persist in the environment, such as scent that marks a territory, is more likely to be higher molecular weight and relatively more polar. The reason is that higher molecular weight molecules require more energy (heat) to evaporate, and polarity makes molecules "stick" (due to hydrogen bonding) to each other. Thus, the main alarm pheromone of the honeybee, isopentyl acetate, has only seven carbons and evaporates almost immediately after release. Rapid evaporation enhances quick defensive responses by bee colonies. The main component in honeybee queen pheromone is a larger, more polar molecule, (E)-9-oxodec-2-enoic acid, which is passed from the queen to workers, and among workers, via contact (see Figure 7.5). This lower volatility causes it to persist within the nest and retains the signal among the bees.

Pheromones provide valuable channels for communication when audible or visual signals might fail; thus, a nocturnal animal might use chemicals to mark its territory. In addition to functioning well in the dark, pheromones can be deposited on surfaces where they remain after the animal has left; the signal retains its information value over time and in the absence of the sender. Potential disadvantages of chemical signals are their slow and

FIGURE 7.5
Structures of honeybee alarm pheromone and honeybee queen pheromone. These chemical structures of honeybee pheromones are instructive not only in their basic chemistry but also what they tell us about the pheromones' functions. The alarm pheromone is a small molecule, meant to disperse rapidly and warn of danger. The queen pheromone is larger and used for information within the nest.

uncertain dispersal relative to visual or auditory signals (no faster or more predictable than the air currents that carry them) and the difficulty of conveying complex information with a molecule. That difficulty is especially apparent when considering directional information. In a situation in which consistent wind or water currents carry the signaling molecules, the direction the signal is coming from may be obvious to the receiver. In other cases, gusts or sidedrafts can break up the odor stream, forcing the receiver to use more complex search strategies to discover the directional information (see Chapter 2). Pheromone trails on surfaces, such as those laid by ants from the nest to food, do not convey information about direction; an ant encountering the trail must gain directional information from direct contact with other ants on the trail or by using landmarks that establish the direction of the nest.

In contrast to visual or vibrational communication, noise is not a major issue in olfactory communication because natural selection favors unique pheromones (and receptors) not otherwise found in the environment. The most likely sources of interference with olfactory communication are similar species in the same environment; in such instances, sex pheromones are commonly blends of several compounds that together convey the full message. Differences in these blends distinguish sex pheromone messages of moths such as the tobacco budworm (*Heliothis virescens*) and its close relative *Heliothis subflexa*.[13]

Tactile Signals

Tactile signals (touching) can be remarkably efficient when animals are close together and do not require ambient light or air currents. Their major limitation is the requirement of proximity; they are ineffective for distant communication and, for that matter, do not work well in the presence of obstacles.

Tactile senses are an important source of information for animals. Touch provides information when light is unavailable, when noise interference obscures sound, or when vocalizations might attract predators. In mammals, touch is a significant aspect of affiliative behaviors in social groups. Touch also provides important information about proximity of food, predators, and other environmental features.

Touch functions in both aggressive and affiliative interactions. In any aggressive encounter, direct physical force is the most escalated and threatening form of communication; bites, slashes, and kicks carry powerful messages. On the other end of the social spectrum, grooming is a powerful affiliative behavior for many animals. Allogrooming (mutual grooming) is widespread in mammals, birds, and social insects. Social groups are a fertile breeding ground for external parasites, and one driving force, evolutionarily, for all grooming is the removal of parasites. Might unusually frequent grooming be an (inadvertent) signal of high parasite loads?

OF SPECIAL INTEREST: TANDEM RUNNING IN ANTS

Some ants lead nestmates to food. In tandem running, the leading ant moves to the goal while the next ant follows, maintaining antennal contact with the leader. Following may be stimulated by a pheromone of the leader, as well as by contact. Sometimes many ants form chains (see Figure 7.6). This example illustrates a number of principles in communication. First, the signaling is **honest** because the sender and receiver have shared interests: they both wish to find the food or nesting site. Second, there is some redundancy as the ants use both tactile and chemical signals. Because the requirement of contact between ants makes the communication somewhat inefficient, many ants have evolved the ability to use chemical trails that can be left by one ant and later followed by another without direct contact between the two.

> **KEY TERM** Honest signals accurately reflect the meaning and intent of the signaler. Honest signals reflect shared interests of the sender and receiver.

FIGURE 7.6
These two ants are engaged in tandem running. Both contact (tactile communication) and pheromones are used to lead nestmates to food sources. Tandem running can result in long chains of moving ants.

Audible Signals

Audible signals work well in the dark and are not impeded by obstacles such as rocks and trees. They do not linger in the environment; the distance that they travel depends in part on the medium and the type of sound (wavelength) involved. The production of audible signals tends to be energy intensive.[14]

WHAT IS SOUND?

Sound is vibration. Some sound is audible to a given species and some is not, depending on the frequency of the vibration. Humans usually sense some vibrations in air or water as sound, but other species sense vibrations that people do not perceive, including vibrations in soil and rock. The physical characteristics of the medium through which the sound travels have a major effect on how sound can be used. Density of the medium is particularly important and affects both the energy required to cause the vibration and the speed at which the vibration travels. For instance, it requires less energy to make air vibrate than water, and making the ground vibrate takes substantial energy. Thus, the use of vibrations in communication depends on the ability of an animal to make a material vibrate, so not surprisingly, only very large animals, such as elephants, can use the soil as a transmitting medium for vibrational signals. Density of medium also affects speed. In air, sound travels at about 334 m/sec (at 20°C). This seems fast, but the rate of travel in surface ocean water is nearly five times faster, 1,520 m/sec, and in rock and soil it can even be faster, up to 13,000 m/sec. Again, all of these speeds depend on the density of the medium, which changes with temperature, salinity of water, and type of rock or soil. In some ocean depths, cold temperatures and moderate pressures yield ideal conditions for transmission, resulting in a SOund Fixing And Ranging (SOFAR) channel.

OF SPECIAL INTEREST: PSYCHOPHYSICS AND NONLINEAR SENSORY RESPONSES

Our discussion of auditory communication merits a small detour into psychophysics. *Psychophysics* is the study of how the physical properties of a stimulus affect the interpretation of the stimulus by the nervous system. Psychophysics adds several important elements to our understanding of how perception works in communication. For instance, psychophysics tells us about the threshold of perception for a stimulus, the animal's ability to judge differences in intensity between stimuli, and the energy needed to produce a stimulus that will result in a given type of receptor response. A basic principle is that receptor response is not linearly related to stimulus strength.

Sound is a great example of this *nonlinearity*. To us, 0 decibels is the threshold for hearing, a 10-decibel sound is barely perceptible, the tiniest of rustles; whereas a 130-decibel sound is immense, deafening. If you ask people to rate intensities of sound, most would tell you the sound roughly doubles with every 10-decibel increase. (The relationship is actually more complex and is captured by the sone scale; see Figure 7.7). Yet the decibel scale is logarithmic, so each 10-decibel interval corresponds to a 10-fold increase in the intensity (energy) of the sound. From a 10 db rustle to a 20 db whisper is a 10-fold increase in sound energy; from 10 to 30 db is 100-fold; from 10 to 40 db is 1000-fold. Normal conversation, 60 db, is 100,000 times more intense that the 10 db rustle. In communication, this means that at the low end of the loudness range, relatively small differences in sound energy can be perceived, but as loudness increases, only major differences in intensity can be perceived (see Figure 7.7). In other words, the perception of loudness may double with every 10 db increase, but the additional energy required for this increase is far greater than a mere doubling.

As we move through our discussion of communication, keep in mind that the nonlinear nature of receptor systems means that the cost of producing more intense signals may not pay off, in terms of perceived increases in the intensity of the signal. For a male frog, high-*amplitude* (loud) calls may be better, but at the upper end of his ability to make high-amplitude sounds, the female's perception of loudness may not allow her to tell the difference.

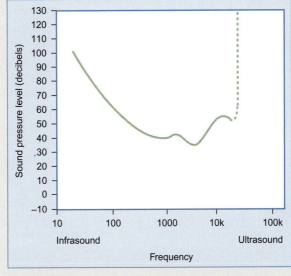

FIGURE 7.7
The line represents perception of loudness across a range of sound frequencies (x-axis). This curve, developed from human data, is representative of how bird and mammal ears function. Sounds at low (infrasound) and high (ultrasound) end of the hearing range require more energy (dB) to be heard.

> **KEY TERM** Nonlinearity occurs in relationships between two variables when one variable increases at a faster rate than the other. This results in something other than a straight line when the two variables are graphed. Nonlinear relationships are often described as exponential, logarithmic, or power curves.

> **KEY TERM** Amplitude is the intensity, or loudness, or a sound.

193

Sounds can travel thousands of kilometers in this SOFAR channel, and marine mammals can communicate across oceans in this way. The frighteningly rapid travel of tidal waves (tsunamis) and even faster rate of travel of earthquake tremors through the earth's crust reflect the speed at which any vibration will travel in these dense media.

> **KEY TERM** Dissipation is the loss of intensity as sound travels.

> **KEY TERM** Pitch is the tone of a sound. Bass notes are low pitched, treble notes are high pitched. Pitch is also termed *frequency* because it reflects the number of sound oscillations during a given period of time. Human hearing responds to pitches from about 50 cycles per second to about 20,000 cycles per second.

> **KEY TERM** Reflection is the bouncing of sound or light waves off dense objects. Echoes result from reflection and may confuse communication using sound.

> **KEY TERM** Refraction is the change in frequency of waves when they move into a medium of different density. Refraction of sound changes its pitch, which sometimes can affect its signal value.

194

There are a few other relevant points about sound as a mode of communication. *Dissipation* is the loss of intensity as sound travels. One important contributor to this loss is the effect of spreading; as the sound spreads out over a larger area, less energy is present at any one point. Mathematically, the inverse-square law describes dissipation: sound energy decreases exponentially over distance. Another cause of loss of sound is absorption and scattering by the medium in which the sound is traveling. Low-*pitched* signals have considerable energy and carry well, making them suitable for territorial, group-cohesion, and other long-distance calls; "famous" low-pitched signals include the vocalizations of whales[15,16,17,18] and the seismic-wave communication of elephants.[19,20,21] Higher-pitched signals are employed in echolocation systems.

Sound is also subject to *reflection* and *refraction*. When sound hits a denser medium, some of the energy bounces, or echoes, off the surface of the denser medium. Echoes can be useful, as in the echolocation system of bats, or can result in confusing noise. When sound moves from a medium of one density to a medium of a different density, it refracts, or changes wavelength as well. This results in distortions of the sound.

SOUND PRODUCTION

There are four common sound production mechanisms. Essentially, any movement or action that causes molecules to vibrate or makes pressure waves produces sound. Most animals use special structures to make sound, but a few simply take advantage of a resonant substrate by slapping, tapping, or drumming on it. Specific sound production mechanisms include

- Vibrating a drum-like membrane. Vibrating membranes are used by cicadas to produce their mate attraction calls. The membrane, or tymbal, is held in place by rigid exoskeletal structures and moved by a muscle attached directly to the membrane.
- Stridulating using a file and scraper. In crickets, grasshoppers, some ants, and a variety of other insects, one body part is equipped with a scraper (imagine something like a guitar pick slightly protruding from the animal), and an adjacent body part has a rough or file-like structure. The animal makes a chirp or whir by moving the two body parts against each other, dragging the scraper across the file. The entire structure (file and scraper) is called a *stridulatory organ*. The frequency (pitch) changes depending on how fast the two structures are moved. Because insects are ectotherms, their stridulations may vary in pitch depending on the temperature.
- Vibrating a membrane in an air flow. Like the reed in a saxophone or clarinet, the larynx of frogs, toads, and mammals and the syrinx of birds feature a membrane that vibrates as air from the lungs is pushed past it. The rate of vibration, and hence the frequency (pitch) of the sound, can be regulated by muscles that tighten or loosen the membrane.

- Hitting a substrate. Beavers slap their tails on the surface of the water when alarmed. Because sound travels much faster in water than air, this is an effective warning signal. Some termites, such as the Formosan subterranean termite (*Coptotermes formosanus*), bash their heads against substrate when alarmed, producing sound and substrate vibrations. Literally millions of termites in a colony may do this synchronously, producing a sound that is audible for several meters around the nest. Woodpeckers, such as the flicker (*Caleptes cafer*), knock on hollow trees (or buildings that resonate like hollow trees) as a way of attracting mates.

OF SPECIAL INTEREST: FRT SIGNALS IN HERRING

Animal communication has given us the Ig Nobel Prize in Biology (2004) (this is a prize given for research that is both humorous and thought-provoking) with a study of burst pulse sounds in Pacific and Atlantic herring (*Clupea pallasii* and *C. harengus*), respectively. Although these fish are commercially important, their sound production has been poorly studied. Wild-caught Pacific herring produced a train of pulses (1.7–22 kHz) that lasted up to 7.6 seconds. This occurred mostly at night, and video analysis of Atlantic herring showed that it was associated with air release from the anal duct region. These are termed *Fast Repetitive Tick* (FRT) (think about it—FRT ...) sounds. Because the *per capita* rates of sound production increased with fish density, Wilson et al. hypothesized that these sounds are involved in social behavior in these fish.[22]

REGULATING PITCH AND AMPLITUDE

Each of the first three mechanisms (tymbal, stridulatory organ, larynx/syrinx) offer the animal a way to modulate pitch, making it easier to produce calls that are distinctive for a species, sex, or individual. In these cases, amplitude (loudness) can be varied as well, by changing the amount of energy with which the sound is made. Because of the relationship between energy and amplitude, body size limits sound amplitude. In an absolute sense, strength is a function of body size, so the loudest sounds are reserved for the largest animals.

Special adaptations help some organisms surmount the size–strength–sound limitation. One such adaptation involves using a resonant structure to amplify the sound. This can be external to the animal, such as a cricket calling from inside a hollow stem, or internal to the animal, such as the hyoid cartilage of the howler monkey. With vocal sacs, animals like howler monkeys and many frogs can force larger volumes of air through the larynx. While these alone do not amplify the sound, they allow the production of longer duration sounds than would be produced by air from the lungs alone.

INFRASOUND AND COMMUNICATION

Infrasound and ultrasound are not audible to humans. Respectively, they are below and above the normal range of human hearing. The scientific exploration of infrasound communication is just beginning, and the realization that some animals use this channel has alerted us to potential (and previously unsuspected) social complexity in those animals. Elephants and killer whales are two such species.

Infrasound has the special characteristic of traveling well in ground or water; in fact, the waves of an earthquake can be thought of as a form of infrasound. Because sound travels much faster in ground than in air (or water), ground-borne vibrations, if perceived, can serve as an early-warning system, arriving well before airborne sound from the same source arrives. Infrasound dissipates less rapidly in air, making it ideal for long-distance communication. Perception of infrasound, however, presents some specific problems. An object smaller than the distance between waves is a poor receiver for infrasound waves. Thus, infrasonic receivers need to be large and tend to be found on the large animals able to generate infrasound. This is probably the reason that infrasonic communication is used by only a few animals, and the best-understood infrasonic communication system is that of the African elephant (see Figure 7.8). The elephant's large pinnae (external portion of the ears) may play an important role in perception of low-frequency sounds, which are significant in elephant communication.

Infrasonic calls, because they travel great distances, provide an important mechanism for communication and affiliation in these wide-ranging animals. By remembering the identity of different callers, elephants can maintain contact with specific individuals. Female African elephants use "contact calls" to communicate with other elephants in their bands (usually a family group). These infrasonic calls, with a frequency of about 21 Hz and a normal duration of 4–5 seconds, carry for several kilometers, and help elephants determine the location of other individuals. Calls vary among individual elephants, so responses to familiar and unfamiliar calls differ; perhaps elephants can recognize the identity of the caller.

FIGURE 7.8
African elephants have large pinnae (external ears) that assist in perception of infrasound, which is important in elephant communication. *Photo: Pen-Yuan Hsing.*

The social use of individual identification is not known. Nonetheless, as little as we understand about elephant communication, we suspect that it functions at least in part in elephant social behavior. African elephants have a social structure best described as fluid; animals move freely over wide areas, sometimes affiliating with other animals. Female members of a family tend to stay together, and of course their juveniles travel with them. These female-centered groups may merge with other such groups periodically. Adult males are less likely to join groups.

Killer whales (*Orcinus orca*) also live in matrilineal (mother plus her offspring) social groups (see Figure 7.9). In the summer, these groups sometimes come together with up to seven other matrilineal groups to form a *pod*. Unlike many other mammals, male killer whale offspring tend not to disperse; they remain with their mothers. Female offspring ultimately have offspring of their own and eventually split from their maternal group to form their own group. Killer whales fitting this social pattern are termed *residents*. Transient killer whales show fewer group affiliation tendencies; their social behavior is not as well understood.

FIGURE 7.9
Killer whales use infrasound for communication and group cohesion. Some calls are stereotyped while others are highly variable. *Photo: Jennifer Marsh.*

Killer whales make calls and echolocation clicks. Calls are repeated (pulsed) sounds in the 1 to 10 KHz range; between 7 and 17 calls can be recognized by scientists, depending on the matrilineal group under study. Scientists recognize calls by listening to them, or by examining a graphical representation of the physical characteristics of the call, called a *sonograph*. The number of different calls that the whales (as opposed to scientists) actually perceive is more difficult to determine, and this remains an area of active scientific investigation.

Calls may be stereotyped so that a call is quite similar each time it is given and varies little among individuals who give it. Stereotyped calls can be interpreted based on genetically coded information in the receiver, and are thus consistent with part of the model for innate releasing mechanisms of classical ethology (see Chapter 1). Call stereotypy is particularly valuable when the communicatory context is critical and brooks no misinterpretation. Alarm calls are especially suitable for call stereotypy.

In contrast, other calls may vary among individuals or social groups. Call variation can encode information about age, sex, social group membership, or other aspects of social status. To take advantage of variable calls, animals must be able to perceive call differences, associate a behavioral context with the call, learn that association, and remember it for future use. Variable calls are particularly useful in more complex aspects of social interactions. Killer whale calls were more similar within matrilineal groups than between groups, and during this study, groups actually became more distinctive in their calls.[23] Thus, killer whale calls contain information that can be used to identify social groups and may play a function in maintaining kin groups over time.

ULTRASOUND AND COMMUNICATION

Ultrasound presents two challenges for the recipient animal. First, recall that high frequency translates to short wavelengths; the hearing organ must be miniaturized to match the wavelength. Second, high-frequency sounds tend to be supported by little energy. They not only dissipate rapidly as the sound travels, making them relatively faint even close to the source, but also may be absorbed by the hearing organ without being transduced into a neuronal signal. To accommodate the lower energy of ultrasound, the hearing membrane, or tympanum, is typically thinner in animals that rely on ultrasound for communication or navigation. The outer ears (pinnae) of mammals that can perceive high-frequency sound may be quite complex; bat ears are characterized by grooves and channels that help carry sounds to the tympanum, as well as maintain small differences in frequency (pitch) and amplitude (volume) which can be used to localize sound sources.

Ultrasonic signals are produced in two major contexts. First, several species use ultrasound in social communication. Bats, murid rodents (rats and mice), and various sorts of moths use ultrasounds to communicate with mates, and rodent pups use it to call their mothers if they become isolated. The second context—echolocation—is more famously associated with bats producing high-pitched sounds that are reflected off objects in the bat's flight path.[24,25,26,27,28,29] Because bats prey on insects, many insect species are attuned to bat echolocation calls and take evasive measures if they hear a bat call (see Figure 7.10). Males produce a calling song to attract females in greater wax moths (*Galleria mellonia*) and lesser wax moths (*Achroia grisella*); they stop calling if a calling bat approaches; presumably the bat can orient to the mating call of the moth.

197

FIGURE 7.10
Bats are famous for their ability to catch aerial prey such as insects with echolocation. Echolocation can be used to locate other types of prey as well. *Noctilio*, a fishing bat, uses echolocation to sense ripples in the water that indicate a fish prey is near the surface. *Photo: J. Scott Altenbach.*

High-pitched sounds (ultrasound) have several advantages in echolocation: (1) They take relatively little energy to produce; (2) because of their short wavelength, they are more likely to bounce back to the bat, which is essential for echolocation; and (3) they dissipate rapidly, reducing confusion from "old" sounds that could still be bouncing around an area.

CHORUSES

What do many frogs, insects, and birds have in common? They use acoustical signals—sound—to attract mates. Generally, the male sings mightily, sometimes all day or all night, in hopes of attracting an interested female. Males may sing to demonstrate their strength, compared to other males, or to attract females to a territory. If intrasexual competition pushes males to perform to their maximum capability, then these songs become an honest signal of male condition (see Chapter 11 on mating).

Insect and frog "choruses" are common features of summer nights in many locations; all of the animals seem to sing synchronously (see Figure 7.11).[30] Why might they do this? Greenfield and his research group have led the way in answering this question.[31] Perhaps the simplest model comes from species in which females have a preference for the first male in a group to sing. This "leading" male receives the female's attention. Other males are then

FIGURE 7.11
During calling, the frog's vocal sac inflates and then forces air through the larynx to produce sound. *Photos: Matthew Bolek.*

pressured to sing ahead of the "leader." Ultimately, the males sing in chorus, because each is trying to be first. The amount of time between songs is governed by the males' recovery time and by the female's response to repeated songs. If the lag between a pair of songs is too short, then the female regards the second song as a "following" song. Once enough time has elapsed, the female is receptive to a "leading" song. These factors combine to push the males into synchronous singing.

BRINGING ANIMAL BEHAVIOR HOME: VOCAL MIMICRY IN BIRDS

Most birds must sing the songs they inherited, or songs they learned from their parents at a young age. Others—15–20% of songbird species—are remarkable vocal imitators, able to listen to sounds, learn them, and reproduce complex vocal patterns.[32] This ability stays with the bird through its life. Parrots, parakeets, and mynahs are impressive mimics and are often kept in human households. (Mynahs are more commonly kept in households in South Asia.) Good imitators not usually kept in households are starlings and catbirds. The evolutionary and functional reasons for vocal mimicry in birds are not well understood. These birds respond well to repetition and reward when imitating human voices; parrots are most adept at this. One of the more disappointing possibilities for a child is to bring home a female parakeet and to try to teach it to talk. Only male parakeets are vocal mimics!

Chorusing (synchronous singing) has its disadvantages. If dozens, hundreds, or even thousands of males participate in a chorus, auditory confusion may result for females. Consequently, in at least in some species, females have evolved mechanisms for selectively listening to a nearby subsample of the males.[33] This ability allows a female to evaluate the signals of individual males and use the information in choosing and locating her mate. Other animals that chorus include katydids,[34] humpback whales,[35] and chimpanzees.[36]

Male crickets such as *Acheta domesticus,* the common household cricket, also chorus to attract females. Producing the monotonous chirps for long periods of time must be taxing to the male crickets, making persistent calling an honest signal of male energetic capabilities. Females can discriminate male condition based on the calls and are preferentially attracted to certain males.[37] Other males listen in and are attracted as well, preferring the same, generally larger, males.[38] Why might males be attracted to the songs of other males, particularly those males preferred by females? Occasional chirps—advertisement calls—are cheaper for the males than walking around searching for mates, but trilling—continuous calling used in courtship—is twice as expensive as chirping.[39] Meanwhile, by increasing proximity to a trilling male, a chirping male may intercept a female. Many other insects, such as periodical cicadas (*Magicicada*), also engage in acoustical contests to attract females.[40,41]

AUDITORY NOISE

Most environments have lots of sound noise. The wind whistling, rain dripping, trees rustling, and water splashing can all obscure airborne animal-produced sounds. Rain, waves, and currents produce equally omnipresent underwater noise. Add all the birds and whales singing, frogs croaking, lions roaring, and so on, and the auditory world is noisy, indeed. Animals cope with this noise using three basic strategies. One strategy is to produce signals in a limited range of pitches and to hear them selectively with ears that respond only in the acoustic range of the signal. Another strategy is the neural ability to sort meaningful sound from background noise. Finally, acoustic signaling can be partitioned by time of day so that species evolve to use certain times and not others for signaling. In this way, they avoid each other's signals. Hearing is difficult when the signal-to-noise ratio is low, as when ocean surf overwhelms shorebird calls. Evolution is a powerful tool for overcoming problems like noise, and each of these solutions help animals to communicate within life's cacophony.

The "cocktail party effect" is a good example of how animals can evolve to deal with noise using the second strategy. This phenomenon gained its name from the ability of humans to isolate a single voice in a crowded, noisy room, and to focus on the signal content of that one voice. Birds[42] and frogs[43] are known to have the same abilities to separate relevant sounds from background chatter (see Figure 7.12).

Of greater concern is anthropogenic noise—sounds produced by humans. In urban environments and in areas around roadways, noise pollution can overwhelm animal sounds. Rustling leaves measure about 20 decibels, whereas cars, motorcycles, and trucks can produce 80–100 decibels within 25 meters of a roadway. Remember that the decibel scale is logarithmic, so motorized vehicles are producing approximately 10^6 more energetic sounds than soft natural noises. Recent studies have shown that noise pollution does actually affect how birds and frogs communicate.[44,45,46] Similarly, noise from ships, SONAR, and other types of location systems have the potential to disrupt auditory communication in the oceans. In addition, recently developed underwater communication devices have the potential for disrupting animal communication in the SOFAR channel. The effects of noise on animal communication present a major conservation concern.

Vibrational Communication

Many animals lack ears—that is, they do not have a membrane that vibrates in sympathy with the surrounding air—but they are still able to gather vibrational information.[47] Infrasound is a common type of vibrational information carried in rock and soil instead of air. (Signals carried in the ground are sometimes called *seismic signals.*) Predators can detect movements of large herding animals by using vibrational information carried in the soil, much as a

199

FIGURE 7.12
A colony of pelicans can be cacophonous. Birds in such environments are able to screen sounds and pay attention to relevant information, much as humans can at loud cocktail parties. *Photo: David Hall/US Fish and Wildlife Service.*

person might detect a passing car radio that has the bass turned up. Most examples of direct communication via vibration come from the invertebrate world. Some termites communicate alarm by whacking their heads on the substrate in unison; the unified effort of many termites causes the substrate to vibrate. Wolf spiders, which do not build webs, drum the ground during courtship; signals reach their potential mate through a combination of vibrations in the ground and in the air. Orb-weaving spiders use vibrations in their web to alert them to the presence and location of prey, and in some species, male spiders intentionally vibrate the web to gain the attention of their potential mate on the web.[48,49] Vibrational information also plays an important role in communication in the honeybee dance.

Visual Signals

VISUAL SIGNAL PRODUCTION

Visual signals work well in the presence of light and in the absence of objects (such as vegetation) that do not block transmission. Visual signals reach their destination quickly—at the speed of light. Visual signals can be fairly inexpensive to produce, but they do not linger in the environment in the absence of the action or structure that produces them.

FIGURE 7.13
Lizards use the dewlap, a flap of skin that can be extended and retracted, as a visual signal in territorial and sexual displays. *Photo: Jonathan Losos.*

Light moves through water, as well as air, but visual signals in water face three related challenges. First, water absorbs light, so with increasing depth less light is available for animals to see signals. Some waters are "crystal clear," and good illumination is present to depths of 20 meters or more, but other waters have suspended particles of clay, dirt, or sand (called *turbidity*) that greatly reduce visibility even near the surface. Second, water absorbs orange, red, and yellow wavelengths, so as depth increases, these colors become much less useful. Colorful fish and other aquatic animals are usually found in clear waters at shallow depths; fish from deeper waters are often more drab. Finally, light entering water is refracted, so interpretation of cues and movements at the interface between water and air can require experience; this is probably more important for predators than it is in visual signaling (see Figure 7.13). (See Chapter 9 on foraging.)

Visual signals can be assigned to one of three categories. First, visual signals can be patterns on the surface of the animal, or patterns combined with color.[50,51] Coral reef fish present many outstanding examples of combinations of color and pattern to produce signals (see Figure 7.14). Second, visual signals can involve movement, such as a fiddler crab waving its large claw. Not surprisingly, many visual signals combine the two elements of pattern/color and movement, exemplified when a peacock spreads its tail (movement) to reveal the magnificent pattern and color. Signals in these first two categories require sunlight or moonlight to be functional. The third category does not require an external light source because it involves animals that produce their own light for visual signals and do not need sunlight to illuminate a visual display. A few animals, such as fireflies and deep-sea organisms, fall into this category.

Conceptually, visual signals are not difficult to understand. As covered in Chapter 2, receptive cells that transduce light (generally considered to range from long-wave infrared radiation to shortwave ultraviolet radiation) are widespread in animals. In many animals, light-receptive cells are concentrated in a light-receptive surface, the retina, and light is focused on the retinal surface by a lens or lenses. The processing of light information is complex and involves filtering and enhancement of the image in several stages, beginning at the level of individual receptor cells and proceeding to interactions among adjacent receptors. The central nervous system integrates the information about image formation and pattern recognition. (See Chapter 2 for more details.)

Recall that in communication, the key element is that the signal and the receptor have co-evolved, so the intended target can perceive the signal. This may seem obvious, but

(A)

(B)

(C)

(D)

FIGURE 7.14
Visual signals in the animal world range across all major taxa that have image-forming eyes. (A) Colorful reef fish advertise sexual and dominance status. (B) Fiddler crabs (so named for their large, signaling claw) position themselves against backdrops that increase their visibility and wave their major claw in the air. (C) Peacocks are legendary for using their colorful, large tail to signal to peahens. (D) Some fish have bioluminescent extensions that lure other fish near their mouths.

FIGURE 7.15
Loosejaw "stoplight" fish produce and perceive red bioluminescence, an unusual signaling adaptation. The red light is produced by a bioluminescent organ beneath the eye.

evolution can yield elegant variations on this theme. Many deep-sea organisms are bioluminescent, for example, but most produce blue-green light. Loosejaw fish (Family Malacosteidae; see Figure 7.15) make red light, instead. This means they can signal to each other without interference by predators, but to accomplish this, both the light production system and the eyes had to evolve to use red wavelengths.

As indicated previously, pattern, color, and bioluminescence are three categories of visual signals. Color will be discussed in more depth here (see Figure 7.16). There are two major ways to produce color that do not involve bioluminescence; color can be produced with pigments and with physical structure. Of these, pigments are more commonly used. Most colors result from pigments stored on or underneath the animal's surface. A pigment that absorbs all visible (to that species) wavelengths is black; a pigment that reflects all wavelengths is white. Pigments give animals color by reflecting specific wavelengths. Generally, pigments are chemicals with double-bonding structures among carbons or between carbon and nitrogen; the longer the series of such double bonds within a molecule, the darker the pigment. Most animal pigments are pterins (white, yellow, or red), quinones (yellow, red, or orange), melanins (generally yellowish-brown, brown, or black) and carotenoids (yellow, orange, or red). Animals usually produce pterins, quinones, and melanins themselves, but carotenoids often come from their diet, so the quality of food available to an animal, or its ability to compete for food, can affect its coloration. The bright red of male house finches in mating season depends on having carotenoid pigments in their diets.

Unlike the blues and greens commonly produced by bioluminescence, blue and green have been scarce in this discussion of pigments. A few blues, such as bird eggshell blues, are pigmentary, but blues and greens are often nonpigmentary colors, such as Tyndall blues, which are discussed later.

FIGURE 7.16
Blue occurs less commonly among animals, and is most likely to be found in birds. *Photos: (A, C) Jeff Mitton; (B, D) Michael Breed.*

202

Sometimes light is absorbed by a pigment and then retransmitted (by fluorescence) at a different wavelength; this is fluorescence and is known to provide color in some coral reef fish. Fluorescent pigments are interesting because they allow an animal to produce a color that may not be present in the ambient light; for instance, by the time light has traveled a few meters through water, the red wavelengths have been absorbed, resulting in a blue-green underwater world. Fluorescence allows fish living at greater depths to produce red coloration by absorbing the predominantly blue and green wavelengths and emitting the energy as red wavelength light.[52]

OF SPECIAL INTEREST: IS RED REALLY A WARNING COLOR?

Red is used as an exclamation point signal; the color red stands out against the green, gray, brown, and black backgrounds that are common in nature (see Figure 7.17). Sometimes red is produced as a signal within a species, such as the red chevrons of the male red-winged blackbird. From a human perspective, however, we often think of red as a generic warning color, a thought that is culturally reinforced by red stop signs, poison labels, and the like. Is red generically used as an interspecific warning signal?[53] This could clearly fall into our definition of communication, because a red signal might evolve specifically to warn away potential predators, and selection could easily favor receivers who use this information. The existence of red-striped coral snakes, red poison frogs, and red-marked black widow spiders all seems to support this thought. But apples, cherries, and strawberries are all also red and are hardly poisonous; in these instances, the red seems to have evolved as an attractive signal. In some cases, such as birds feeding on insects, the bird has the opportunity to sample and build an association among color, pattern, and palatability. One could then hypothesize that red is a color that is easily learned, perhaps. In contrast, a bird bitten by a coral snake often does not get a second chance; this suggests that sometimes learning may not be important in recognizing warning colors. There is at least some evidence that in birds color-based aversions are inherited, then reinforced by experience. In chicks, red-colored fruit shapes are attractive, but red-colored insect shapes are repellent;[54,55] thus, context (shape) interacts in an important way with color.[56] To answer our original question, there is a least some evidence for innate avoidance of red by some animals, but there is more evidence that aversion to red is learned and in some cases is cultural. The general topic of warning colors is treated in more depth in Chapter 10 on self-defense.

FIGURE 7.17
The color red is ubiquitous among major groups of living things, from berries to ladybeetles, birds to grasshoppers. As a signal, it communicates many things. *Photos: (A) Pine grosbeak, Jeff Mitton; (B) chili peppers, Michael Breed; (C) lubber grasshopper, Jeff Mitton; (D) poison dart frog, Michael Breed.*

FIGURE 7.18
The iridescence of butterfly wings, a peacock's tail, and a damselfly results from structural aspects of the wings. *Photos: (A, C) Jeff Mitton; (B) Michael Breed.*

203

Colors can result from the structure of an animal's surface rather than pigments. Blues and greens in animals are almost always structural colors. Iridescent colors of butterfly wings are the product of thin layers within the scales on the wings that selectively reflect wavelengths in the blue or green color range (Figure 7.18). The blue of the iris of some human and animal eyes is a Tyndall blue, which is produced by the scattering of blue wavelengths of light by small particles in liquid in the iris. The blue of the sky is also a Tyndall blue, produced by the scattering of light from molecules.

Note
In 1859, John Tyndall realized that small particles in suspension scattered short wavelengths more than longer ones, thus explaining why the sky is blue.

VISUAL NOISE AND THE DISRUPTION OF VISUAL SIGNALS

Visual noise includes irrelevant movements, patterns, and colors in an animal's environment. In natural habitats, animals can separate visual noise from signal, often through natural selection acting to produce novel combinations of pattern and color; birds, coral reef fish, and butterflies all present examples of signal divergence to overcome problems of visual noise (see Figure 7.19). Camouflage (see Chapter 10 on self-defense) can use visual noise to hide an animal against its background.

FIGURE 7.19
These cabbage butterflies (family Pieridae) occur in the same habitat but have divergent wing coloration. *Photos: Michael Breed.*

Human activities complicate the visual environment and can result in various types of visual noise that disrupt animal behavior. Animals may mistake colors on signs, automobiles, and the like for signals. Tinbergen's famous observation of three-spined stickleback fish responding to red British mail vans in the distance is a classic example of this. Reflective surfaces, such as windows or mirrors, may cause an animal to attempt to court or fight its image, which it mistakes for the visual signals of another animal. Streetlights disrupt the commuting routes of bats between perches and feeding areas.[57] Porchlights and streetlights at night may be mistaken for the moon or stars in navigation (see Chapter 8); while navigation suffers the direct effect of light pollution, these lights may also disrupt visual or pheromonal communication among potential mates.

Electrical Signals

Electrical signals are largely limited to fish swimming in muddy waters, where visual signals are not useful. Sharks may also use sensitivity to electrical impulses to detect prey. Like chemical cues inadvertently released as a result of metabolic activity (discussed earlier in this chapter), all multicellular animals emit some form of electrical signal. The reason is that nerves and muscles use changes in membrane polarity to function. A few organisms (e.g., electric fish) have enhanced this ability with the evolution of electric organs that can produce well-defined electric signals.[58,59]

Receiving electrical signals can give an animal any of three valuable inputs:

- Electrolocation information
- Receipt of communication from other animals
- Location of other animals, particularly potential prey, which unintentionally produce electrical fields

Electroreception is common in aquatic organisms such as fish and amphibia, but unknown in terrestrial vertebrates and invertebrates. Because air is a poor conductor of electricity, such electrical discharges are limited to animals that live in water; they would be of little use to a terrestrial organism. Usually, there are enough ions in water to make the medium a good electrical conductor, and electrical currents and fields are excellent sources of information. Additionally, the interaction of water currents and the earth's magnetic field produces electrical fields that can be used in navigation.

Electrical fish come in strongly electric forms, which use their electricity primarily for predation and defense, and weakly electric forms, which use electrical signals for communication and electrolocation, similar to the use of sound in echolocation (see Figure 7.20). Electroreceptors of fish include ampullary receptors, canals opening from the surface of the fish into cavities

FIGURE 7.20
The geographic distributions of strongly and weakly electric fish illustrates the broad use that such electric signals may serve. Strongly electric fish use their electric emanations for predation; weakly electric fish use them for communication. *Adapted from: Moller, P., 1995. Electric Fishes: History and Behavior. Chapman & Hall, 583 pp.*

lined with nerve cells, and tuberous receptors, found on weakly electric fish (mormyriforms and gymnotoforms). The tuberous receptors are structurally similar to the ampullary receptors but are designed to reduce the loss of the incoming signal.[60,61]

7.4 MULTIMODAL SIGNALING AND ENCODING COMPLEX MESSAGES

A *multimodal signal* is exactly what the term implies: a signal that uses more than one mode to convey its message. Sometimes the multiple modes establish signal redundancy, which involves repeating the same message in more than one mode. Primates, for example, recognize each other using both visual and olfactory signals. In other cases of multimodal signaling, each mode contributes a unique element to the overall meaning. In a video clip, the soundtrack and the video itself each contribute uniquely to the information gained.

> **KEY TERM** A multimodal signal is composed using two or more signaling modalities. When a male grasshopper jumps to attract females, it makes a clicking sound and flashes its colorful wings; this is a multimodal signal.

The more complex the message, the more likely an animal is to use multimodal signaling. The dance language of the bees is a good example of multimodal signaling. In their dance language, bees combine olfaction, vibration, touch, and sometimes visual signals to *encode* detailed navigational information.

> **KEY TERM** Encoding is the translation of information into a signal that can be decoded by the recipient.

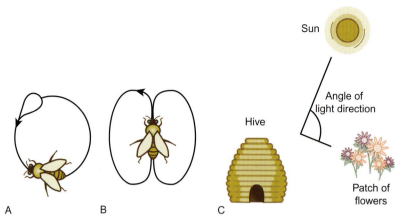

FIGURE 7.21
(A) Diagram of the round dance. This alerts bees to food near the hive but does not convey directional information.
(B) Diagram of the waggle dance. The tempo of the dance tells recruits how far to fly (the slower the dance, the greater the distance) and the angle of the straight part of the dance tells them the direction to fly. (C) A hive forms the base of an angle between the sun and the food source. Because the inside of the hive is dark and the comb is vertical, bees convert the angle of the dance on the vertical comb to the angle formed by the feeding station, the hive, and the sun. A dance straight up the comb, as illustrated in B, tells recruits to fly toward the sun. A straight-down dance tells them to fly directly away from the sun. Dances at angles to the vertical indicate intermediate flight directions. The bees' circadian clocks allow them to correct their dances for the movement of the sun in the sky. *Adapted from* von Frisch, K 1993, *The Dance Language and Orientation of Bees,* Harvard University Press.

Von Frisch observed that once one honeybee finds a feeding station, many others soon appear at the same station. This suggests that the first bee recruits other bees to the food. How might honeybees recruit help in collecting food? Von Frisch's discovery of the dance language of the honeybee required careful determination of the correlations between movements of bees inside the hive and the locations of feeding stations. He found two types of dance. The round dance (see Figure 7.21A) causes bees to look for food a short distance (up to about 50 meters) from the hive. The waggle dance (see Figure 7.21B) tells bees the direction and distance to fly to find more distant food sources. Scout bees use these dances to recruit assistance in collecting food.[62] Similar dances are used to help find a new home when bees swarm. In this case, scouts dance to direct bees in the swarm to hollow trees, caves, or other likely nesting sites. After a number of bees have visited each nesting site, a "voting" process takes place, until one site (generally the best available location) wins out by having more bees dance for it.

OF SPECIAL INTEREST: THE DANCE LANGUAGE CONTROVERSY

Some controversy has surrounded the dance language hypothesis. Adrian Wenner and his colleagues argued that the correlations between bee dances and information conveyed about floral locations is spurious and that the bees are really cueing on odors associated with landscape and food.[63] Wenner felt that von Frisch had not adequately controlled for the possible influence of information other than that from the dance. As a result of Wenner's critique, two types of experiments were conducted to once again test the dance language hypothesis. The first type was misdirection experiments, that is, experiments that asked if the bee dance or landscape odors mattered more to bees experiencing the dance.[64] In these experiments, bees were induced, by shifting their circadian clocks, to dance to indicate locations different from the location predicted by the worker bees attending the dance. The workers went to the location predicted by the dances, rather than landscape odors. The second type of experiment used mechanical robot bees to simulate dances; again, the workers attending the dances responded as predicted by the information in the dances.[65] The overwhelming scientific evidence supports the dance language hypothesis (Figure 7.22). Wenner and colleagues' hypothesis was a reasonable alternative and warranted testing because olfaction is important in the behavioral biology of honeybees.

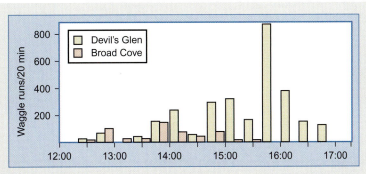

FIGURE 7.22
Honeybees in a swarm "vote," choosing between two potential nest sites. At the beginning of the process, scouts dance equally for two sites, Devil's Glen and Broad Cove, but later in the day one of the sites, Devil's Glen, wins out and the swarm departs for that location. *Adapted from: Seeley, T.D., Visscher, P.K., 2003. Choosing a home: how the scouts in a honey bee swarm perceive the completion of their group decision making. Behav Ecol Sociobiol 54, 511–520.*

OF SPECIAL INTEREST: INFORMATION THEORY

How can we know the extent to which communication actually occurs between a pair of interacting animals? This is a tricky business, because the only scientific approach to this question is to measure the effects of one of the animal's signals on the subsequent behavior of the second animal. In more formal terms, a *bit* is the amount of information needed to answer a single yes/no or on/off question. If an animal is making a complex decision, then it may need more than one bit of information; the minimum number of bits to make a decision (*H*) among *M* possibilities is

$$H = \log_2 M$$

Another way of looking at this is to ask how much information is contained in an array of sender behaviors by looking at the correlation between sender behaviors and receiver responses. This one aspect of communication—sender effect on receiver behavior—can reveal the extent of communication that is occurring. For instance, if the sender's action results in a high diversity of receiver responses with no evidence of pattern, then little or no communication has taken place. On the other hand, if the sender's action restricts the receiver's response to one or two behaviors from a large list of possibilities, then there is evidence for influential communication.

Information theory provides an equation for measuring the diversity of responses to sender signals:

$$H(X) = -\sum_{i=1}^{n} p(x_i) \log_b p(x_i)$$

In this formula, there are *i* possible behaviors. The probability (p) of each behavior is calculated and multiplied by the log of that probability. In ecology, this same equation is often used as a measure of species diversity in communities. The next step is to look at sender–receiver interactions as Markov chains, in which the dependence of receiver response on each signal is measured.

Information theory measures of communication are particularly useful when the communication value of a potential signal is unknown. If a bird, for example, does something that seems odd, does the odd behavior have a communicatory function? This hypothesis can be tested by creating a list of receiver behaviors (an ethogram; see Chapter 14). One then compares the diversity of behavior in the receiver in response to a variety of sender behaviors, including the "odd" behavior in question. Highly diverse responses suggest no communication; restricted responses suggest a strong signal value. Clearly, this approach is very helpful in analyzing behavioral interactions. Information theory analyses can be coupled with the construction of a sociogram.

You will recognize the human game "Twenty Questions" in this discussion. How much information is needed to make a decision? For humans, any arbitrarily selected object can be identified with yes or no answers to 20 well-chosen questions. This approach is also similar to the concept of "Six Degrees of Separation," in which one human can find any another human based on six social linkages. The reason there are fewer than 20 degrees of separation is that each question can narrow more than half the field of people. If you are looking for social linkages to someone in Moscow, you might ask, "Do you know anyone in Moscow?" If the response is positive, you have narrowed the earth's population considerably with a single social link.

7.5 RUNAWAY SEXUAL SELECTION AND SIGNALING

Most signals are optimized for their purpose. They are produced to balance cost with benefit, so they are strong enough to convey the message over the needed distance without wasting effort by broadcasting information beyond the distance at which the signal is useful. The value of communication is also balanced against the risk of attracting a predator. Evolution, as a process, is outstanding at finding cost-efficient ways of conveying information.

In the light of such efficiency, mate choice has a peculiar effect on signaling. If one sex (generally the female) prefers mates to have a strongly expressed signal, selection will favor making that signal bigger, longer, or stronger, with the upper limits well beyond what would be minimally necessary to convey the message. Hence, animals have bright colors, huge horns and antlers, and impressive tails (see Figure 7.23). This is sometimes called *runaway sexual selection*, and occurs when traits in one sex are magnified beyond any reasonable scale as a result of preferences of the other sex for very large, loud, or bright signals. Such preferences may have their roots in animals' heightened responses to supernormal stimuli; these are stimuli that are greater in magnitude than a normal stimulus and also more attractive (see Figure 7.24).

208

FIGURE 7.23
As Tinbergen observed, a gull will react more strongly to an oversized egg than to one of its own, even though the oversized egg cannot possibly be its own egg. A "larger than life" stimulus that elicits such a response is called a *supernormal stimulus.*

7.6 DECEIT VERSUS HONEST SIGNALING

Natural selection should favor any signal that enhances an animal's likelihood of survival and reproduction. Honest signals in communication are given when both the sender and

FIGURE 7.24
Signals that may have resulted from runaway sexual selection. Contrast the large horns of the male bighorn sheep (on the left) with the smaller horns of the female. The male mallard (above) is brightly adorned while the female (below) is drab. *Photos: bighorn sheep, Jeff Mitton; mallards, Michael Breed.*

receiver have the same interest in the result.[66] *Deceit* occurs when one animal can exploit another in order to improve its fitness.[67] The most obvious arena for deceit is mate choice. Natural selection strongly favors any device that increases an animal's reproduction; if a dishonest or deceitful signal does that, then genes for the signal will spread rapidly in a

> **KEY TERM** Deceitful signals have evolved to mislead. A male attempting to attract a mate, for example, may attempt to mislead a female about his parental abilities.

population. Another context in which deceit is often found is in the area of parental care, in which young may use signals that exaggerate their actual needs so that they can extract more than necessary care from a parent.

DISCUSSION POINT: DECEIT

While deceit is most common in mate choice, complex social systems, such as those found in apes, provide opportunities for deceitful signaling within the social group. Certainly deceit is a noticeable part of human social behavior. Can you find examples of deceit in our great ape relatives in the literature or on the Web? How credible, scientifically, are these reports? How is deceit signaled, or how is concealment of signals used in deceit?

Given the apparent advantages of deceptive behavior, studies that demonstrate intraspecific deceit are surprisingly rare, with the exception of extrapair copulations (EPCs). This rarity may be due to the fact that the receiver has the upper hand in choosing whether or not to respond to a signal, and consequently can enforce honesty. Scientists have only recently recognized the high frequency of extrapair copulations in seemingly monogamous species of birds and mammals. These EPCs often result in extrapair paternity (EPP) of offspring. In species that have apparently monogamous mating systems, so that a male and female are pair-bonded, the individuals may nonetheless seek copulations with other partners, doing so at times or locations that make detection by their pair-bonded mate unlikely. This phenomenon is best studied in apparently monogamous birds; averaged over a large number of species, 13% of the offspring were sired by nonpair males. Behavioral deception (secretively copulating) plays an in important role in facilitating this behavior.

Costly Signals as a Way of Ensuring Honesty

When a mate attraction signal is costly to produce, as are cricket trills or frog calls, males that are unable to produce attractive, yet costly, signals sometimes adopt a satellite strategy. Satellite males lurk near calling males, intercept females approaching the attractive male, and attempt, sometimes successfully, to mate. This type of deceit takes advantage of a calling males' inability to guard females as they approach and was mentioned briefly in the discussion of auditory signals.

Why produce a costly mating attraction signal? As previously indicated, deceit is uncommon in part because the receiver can impose stringent requirements upon the sender; natural selection favors a receiver that is not easily deceived. Thus, communication can be an evolutionary conflict between the sender and the receiver, in which one is trying to take advantage of the other. If the signal is costly to the sender and if the receiver responds only to costly signals, then the communication evolves to reflect the sender's ability to pay the cost; that is to say, senders that can produce costly signals will be favored by natural selection. This can lead to examples of the *handicap principle*[68], which is the second major explanation for expensive signals and is especially relevant to mate choice and parental care communication. The ability of a receiver to impose a cost on the signals of the sender by preferentially

FIGURE 7.25
Male moose may be examples of the handicap principle. The heavy antlers may be an honest representation of the male's ability to carry weight and maneuver. *Photo: Jeff Mitton.*

210

responding to expensive signals is a key factor in the evolution of signaling; the response of the receiver (e.g., mating receptivity in this case) must be worth the cost to the sender.

Does the fact that communication often involves individuals who share a common goal, such as mating, feeding, or caring for their young, mean that communication evolves as simple sharing of information? In some cases, this is correct. For instance, this sharing of information would certainly enhance the pair's ability to reach their commonly held goal. In other cases, however, there are underlying conflicts of interest and asymmetries. Knowledge of animal behavior is not sophisticated enough to allow recognition of all the costs and benefits involved in communication; in some circumstances, deceit may be more common than honest signaling. Interacting animals generally act in their own best interest and convey the information, honest or not, needed to enhance their fitness. Figure 7.25 captures this concept; from the sender's point of view signals should be advantageous (whether they are honest or not) and from the receiver's point of view the challenge is to get as much information as possible, while parsing honesty and dishonesty in the signal.

From the sender's point of view, communication evolves

- To minimize the time and effort spent on communication
- To present a message (signal) advantageous to the sender

From the receiver's point of view, communication evolves

- To enhance acuity so that the maximum information can be obtained from the signal
- To separate honest from deceitful signals

If a signal is used intraspecifically, then it is vulnerable to exploitation by another sender. The sender can be a different species, provided that sender can produce the signal. Because of this, deceit is common in signals that function between species,. Carnivorous fireflies mimic the mate attraction signals (light flashes) of other firefly species, attracting males looking for would-be mates; the encounter ends in death instead.[69] Bolas spiders use moth sex pheromones in a similar manner.[70] "Guests" in ant nests acquire the chemical signature used by the ants to discriminate nest members from nonmembers.[71] Signals such as these are aggressive mimicry; they take advantage of the victim's need to communicate.

Studies of Deceit

An anecdotal account reports that a female marsh harrier courted a male to be able to obtain access to food he had stored.[72] She then took this food and fed it to chicks that had been fathered by another male. More extensive studies focus on possibly deceitful behavior in the pied flycatcher (*Ficedula hypoleuca*), a species in which males may possess more than one territory simultaneously. In this species, females gain from mating with a male that has no other mates; males may attempt to deceive females about their mating status (mated or unmated). Females assess whether a male has already mated; if he is alone on a territory during repeated visits by the female, then he is probably unmated. Mated males will be absent from the territory (presumably because they are at another territory with their mate). By repeated sampling of male behavior, females are usually able to avoid mating with previously mated males.[73,74,75]

7.7 GAME THEORY AND COMMUNICATION

Game theory uses mathematical or logical models to predict how animals with mutual interests should behave.[76] Game theory is particularly appropriate for modeling and analyzing interactions involving communication, because the exchange of information between the two animals can be manipulated in the model and provides a very useful framework for analyzing communication. Remember that when an animal communicates, it has a goal, which may be in conflict with the goals of other animals.

To be successful within a game, an animal needs an overall strategy. The strategy defines the goal and a general plan for achieving the goal. In a Prisoner's Dilemma game, the strategy may be to induce the other animal to cooperate while benefiting more than the other animal from the cooperation.[77] Within the overall strategy, tactics are employed to achieve the goal. In Prisoner's Dilemma,[78] there are various tactics, such as "always defect," "alternate defect and cooperate," and "tit for tat." This makes it sound as though animals that engage in games must be very cognitive—able to remember the responses of the other animals in the game and to calculate and adjust their responses as the game proceeds. This is not necessarily true, because evolution can take the place of cognition in producing this or most other types of behavior. Natural selection will shape strategy and tactics in communicatory games if the reproductive success of the animal is influenced by the outcome of the game. Clearly, strategy and tactics play important roles in animal communication.

An important consideration in these games in nature is the potential for death or serious injury when communication escalates into outright fights. The misconception that animals do not fight to death is widely held. This is simply not true; when the stakes are high for both individuals, such as two males competing for breeding rights with a harem of females, death or serious injury are common outcomes. We may also underestimate the cost of injuries that result only in veterinary care for domestic animals; these may be much more costly in wild animals.

DISCUSSION POINT: THE PRISONER'S DILEMMA

The Prisoner's Dilemma is a two-player game in which the object is to gain more points than your opponent by inducing your opponent to cooperate with you. The players take turns; in each turn a player chooses to either cooperate or defect and the other player responds. The payoff depends on how the response matches the initial play. Table 7.1 shows the payoffs for the two players, given the choice of cooperating or defecting for each player. If both cooperate, each gets 3 points. If both defect, each gets 1 point. The highest payoff, 5, comes if your opponent cooperates and you defect. In this case, your opponent gets the "sucker's payoff" of 0. The number of turns in the game is not known to the players, although you do know that at some point the game will be stopped.

Think about playing the game. Can you devise a strategy (a combination of cooperation and defection) that will induce your opponent to cooperate? After you have thought about your strategy for a while, check these strategies.

TABLE 7.1 Prisoner's Dilemma Payoff Matrix

Prisoner's Dilemma

	Cooperate	Defect
Cooperate	3 Reward	0 Sucker
Defect	5 Cheat	1 Punishment

What is your best strategy if you know the next turn is the last turn of the game?

Try the game a few times, and see which of your startegies gives you the best result. To see how various strategies fare in computer simulations, search the Web for "Prisoner's Dilemma"; you will find several fun sites, such as http://www.iterated-prisoners-dilemma.net/.

Game theory can be taken one step further. Given the variety of tactics available to an animal in a game and the variety of tactics that can be used in response by the other animal, some tactical choices will be poor under any conditions, whereas in the right conditions, others will perform optimally. These optimal choices are called *evolutionarily stable strategies* or ESSs. An ESS is a strategy that cannot be displaced in a population by a new or mutant strategy. If Prisoner's Dilemma is played repeatedly between the same two players with no definite endpoint (this is called *reiterated Prisoner's Dilemma*), then tit for tat is an ESS, and no new strategy in the population will succeed over tit for tat.

To apply game theory, a behavioral biologist goes through the following steps: (1) Identify possible strategies. They may be based on observation (what do the animals actually do?) or logic (what alternatives can the scientist think of?). (2) Determine the outcomes when different tactics are employed by each animal. For instance, in the Prisoner's Dilemma, there were payoffs for cooperating and defecting, depending on the behavior of the opponent (3) Through repetitions of the game, determine whether one strategy is always superior (an evolutionarily stable strategy). (4) Observe animals in the field to see if their communicatory strategies fit game theory predictions.

In addition to the Prisoner's Dilemma, Hawk versus Dove is another game that is often used in the analysis of signaling in animal behavior.[79] This game is used primarily to model whether animals should fight or not in disputes such as those over territory. In a simple Hawk versus Dove game, two tactics are available to the participants. A hawk always attacks (aggressive signaling), and a dove always retreats or yields (submissive or passive signaling).[80]

Try this with a classmate: Put out a number of food items. (Ten candy bars is a good starting point.) In the first round of the game, put all the food in one pile. Assign one person to play a hawk strategy and the other to play a dove strategy, but both are to try to capture the food. In this instance, the hawk should quickly come into possession of the food and will continue to possess it because the hawk can efficiently signal aggression and drive the dove away. Now spread the 10 pieces of food so that they are a meter or two apart. When the hawk tries to signal defense at all of these food locations, it will end up spending all of its time running after the dove, which should have an easy time collecting the food. This is an excellent example of the application of game theory in the analysis of communication because it reveals that a strategy that works well under one set environmental conditions may not fare well in a different environment. Thus, animal species that have evolved in variable environments are expected to have alternate tactics that they can employ, depending on the circumstances. Again, using these tactics does not require cognition or deep thinking on the part of the animal; alternative courses of action can be hard-wired, and the appropriate action can be triggered by environmental conditions.

DISCUSSION POINT: HAWKS VERSUS DOVES

One interesting sidelight of the Hawk versus Dove game is that in many species, a size or strength advantage does not determine the winner of a battle for possession of a territory or key environmental feature, such as a shell employed as a home by a fiddler crab. Numerous studies in a wide variety of animals show that the possessor of the resource has an advantage in the encounter, even if it is smaller or weaker than the potential usurper. Why might this be?

7.8 INTERSPECIFIC SIGNALING

Most signals evolve for intraspecific perception. Some signals, though, have clear interspecific functions. All of the principles of signaler/receiver coevolution that apply in intraspecific signals are relevant to interspecific signals as well.

Signals between species can be divided into three basic groups. First are attractive devices, used by one species to lure members of another species. Examples of attractive devices include floral attractants (odor, color, and shape) for pollinators and baits or lures used by predators to attract prey, such as angler fish. In some cases, attraction can be related to shared interests, such as interspecific mutualisms, but clearly luring prey is a selfish, one-sided piece of communication.

Cleaner fish represent a very neat example of signaling in an interspecific mutualism.[81] Cleaning has evolved several times; the most commonly mentioned cleaners are wrasses and gobies (see Figure 7.26). Cleaner species eat ectoparasites and dead skin from the surface of other, larger species of fish. Both species benefit; cleaners obtain food and clients suffer from few parasites. For this relationship to work, the cleaner must signal in a way that the client recognizes the cleaner as a mutualist, rather than a prey item. The client must also signal its willingness to be cleaned so that the cleaner is assured it will not be the victim of predation or a defensive action. There is plenty of room for cheating in this relationship—cleaners may take a bite out of the client, or the client may eat the cleaner after it has been cleaned—and signaling helps to keep the participants' intentions clear. Cleaner species may engage in dancing displays and tend to have distinctive colors and stripes, which make them stand out; this is exaggerated by the tendency of cleaners to group at "cleaning stations." The stripes

FIGURE 7.26
Cleaner wrasses and gobies, with clients. *Photos: Alexandra Grutter.*

tend to be blue and yellow;[82] through the convergence of many cleaner species on the same type of signal, the effectiveness of the signal is increased.

The second group of interspecific signals is repellent or defensive signals, such as distinctive sounds (rattlesnakes, for example), colors (red, in some cases; see "Of Special Interest: Is Red Really a Warning Color?"), and patterns (the striped abdomen of bees and wasps, the striped body of coral snakes) that serve to cue potential predators about the dangerous prospects. In an odd way, warning signals used as repellants serve a mutual goal between the two species: the sender and receiver each avoid damage. Finally, many interspecific cues are given off inadvertently, or are eavesdropped by animals wishing to exploit the sender. While not signals in the sense of having evolved for the resulting interspecific communication, these merit mention here because eavesdropping can place strong selective pressure on how signals are used.[83]

SUMMARY

Communication is the foundation for mating behavior (Chapter 11) and social behavior (Chapters 13 and 14)—in fact, any behavior routinely involving more than one animal. Communication involves a signal that is produced by a sending animal and received by another animal that interprets the signal. Senders (or signalers) and receivers co-evolve in the communication process. Evolution refines both the sender's ability to send a meaningful signal and the recipient's ability to perceive and understand the signal. Signals used in communication evolve following the basic principles of co-option, ritualization, stereotypy, and redundancy.

Modes, or channels, such as sight, sound, or touch are used in animal communication. Each mode has advantages and disadvantages, depending on the environmental conditions and the signaling and receiving capacities of the species. Signals are often honest, but sometimes deceitful signaling is advantageous, which is especially important in mate choice. Once broadcast, signals become public information, available both to the intended recipient and to other animals, which might take advantage of the signal.

Noise interferes with signaling and influences the evolution of communication by causing selection to favor sensory mechanisms that filter noise or behavior that avoids noise. The potential for human-caused noise to disrupt animal communication is a major conservation concern. Interspecific signaling occurs when a signal has evolved to influence the behavior of other species; examples include attractive signals between mutualists and warning signals produced to avoid predation. Communication is the keystone in the arch of behavior from the individual to its participation in behavioral interactions with other animals.

STUDY QUESTIONS

1. What are ritualization and stereotypy? Why are these concepts important in understanding the evolution of communication?
2. Why might animals signal honestly? When and why might dishonest signals be expected?
3. Under what circumstances should animals present honest signals about their own condition (gender, strength, ovulatory status, and so on)? Are there circumstances under which it might be to an animals' advantage to deceive other animals? Why might deception be advantageous? Design an experiment to test the hypothesis that males attempt to deceive females when females are choosing mates.
4. Watch a honeybee dancing. It is moving very slowly and the straight part of the dance is horizontal, from left to right. What information about the food location can be gleaned from the dance?

Further Reading

Axelrod, R.M., 1984. The Evolution of Cooperation. Basic Books x, New York, 241 pp.

Davis, M.D., 1997. Game Theory: A Nontechnical Introduction. Dover Publications, Mineola, NY, 252 pp.

Dugatkin, L.A., Reeve, H.K., 1998. Game Theory & Animal Behavior. Oxford University Press, New York. xiv, 320 pp.

Lucas, J., Freeberg, T.M., 2010. Communication: An Overview, In: Michael D. Breed and Janice Moore (Eds.), Encyclopedia of Animal Behavior, Academic Press, Oxford, Vol 2. pp. 337–339 (http://www.sciencedirect.com/science/referenceworks/9780080453378).

Maynard Smith, J., Harper, D., 2003. Animal Signals. Oxford University Press, Oxford, Oxford Series on Ecology and Evolution.

Ord, T.J. 2010. Evolution and Phylogeny of Communication, In: Michael D. Breed and Janice Moore (Eds.), Encyclopedia of Animal Behavior, Academic Press, Oxford, Vol. 2 pp. 652–660 (http://www.sciencedirect.com/science/referenceworks/9780080453378).

Ryan, M.J., 1998. Sexual selection, receiver biases, and the evolution of sex differences. Science 281, 1999–2003.

Searcy, W.A., Nowicki, S., 2005. The Evolution of Animal Communication: Reliability and Deception in Signaling Systems. Princeton University Press, Princeton. cxii, 270 pp.

von Frisch, K., 1956. Bees: Their Vision, Chemical Senses, and Language. Cornell University Press, Ithaca, NY, 157 pp.

von Frisch, K., 1993. The Dance Language and Orientation of Bees. Harvard University Press, Boston, 592 pp.

Wenner, A.M., Wells, P.H., 1990. Anatomy of a Controversy: The Question of a "Language" among Bees. Columbia University Press, New York, xiv, 399 pp.

Notes

1. von Frisch, K., 1993. The Dance Language and Orientation of Bees. Harvard University Press. Esch, H.E., Zhang, S., Srinivasan, M.V., Tautz, J., 2001. Honeybee dances communicate distances measured by optic flow. Nature 411, 581–583.
2. Nieh, J.C., 2004. Recruitment communication in stingless bees (Hymenoptera, Apidae, Meliponini). Apidologie 35, 159–182.
3. Cratsley, C.K., Lewis, S.M., 2003. Female preference for male courtship flashes in *Photinus ignitus* fireflies. Behav. Ecol. 14 (1), 135–140.
4. Haddock, S.H.D., Moline, M.A., Case, J.F., 2010. Bioluminescence in the sea. Ann. Rev. Mar. Sci. 2, 293–343.
5. Rivers, T.J., Morin, J.G., 2009. Plasticity of male mating behaviour in a marine bioluminescent ostracod in both time and space. Anim. Behav. 78, 723–734.
6. Maynard Smith, J., Harper, D., 2003. Animal Signals. Oxford University Press, Oxford Series on Ecology and Evolution. Searcy, W.A., Nowicki, S., 2005. The Evolution of Animal Communication: Reliability and Deception in Signaling Systems. Princeton University Press, Princeton, cxii, 270 pp.
7. Maynard Smith, J., Harper, D., 2003. Animal Signals. Oxford University Press, Oxford Series on Ecology and Evolution, 3 pp.
8. Freed-Brown, G., White, D.J., 2009. Acoustic mate copying: female cowbirds attend to other females' vocalizations to modify their song preferences. Proc. R. Soc. B–Biol. Sci. 276, 3319–3325.
9. Stoddard, P.K., Beecher, M.D., Horning, C.L., Campbell, S.E., 1991. Recognition of individual neighbors by song in the song sparrow, a species with song repertoires. Behav. Ecol. Sociobiol. 29, 211–215.
10. Bekoff, M., 2001. Observations of scent-marking and discriminating self from others by a domestic dog (*Canis familiaris*): tales of displaced yellow snow. Behav. Processes 55, 75–79.
11. Bekoff, M., Ground scratching by male domestic dogs: a composite signal? J. Mammal. 60, 847–848.
12. White, A.M., Swaisgood, R.R., Zhang, H., 2002. The highs and lows of chemical communication in giant pandas (*Ailuropoda melanoleuca*): effect of scent deposition height on signal discrimination. Behav. Ecol. Sociobiol 51, 519–529.
13. Groot, A.T., Ward, C., Wang, J., Pokrzywa, A., O'Brien, J., Bennett, J., et al., 2004. Introgressing pheromone QTL between species: towards an evolutionary understanding of differentiation in sexual communication. J. Chem. Ecol. 30, 2495–2514.
14. Gammon, D.E., Baker, M.C., 2004. Song repertoire evolution and acoustic divergency in a population of black-capped chickadees, *Poecile atricapillus*. Anim. Behav. 68, 903–913.
15. Au, W.W.L., Mobley, J., Burgess, W.C., Lammers, M.O., Nachtigall, P.E., 2000. Seasonal and diurnal trends of chorusing humpback whales wintering in waters off western Maui. Mar. Mammal. Sci. 16, 530–544.
16. Miller, P.J.O., Bain, D.E., 2000. Within-pod variation in the sound production of a pod of killer whales, *Orcinus orca*. Anim. Behav. 60, 617–628.
17. Ford, J.K.B., 1989. Acoustic behavior of resident killer whales (*Orcinus orca*) in coastal waters of British Columbia. Can. J. Zool. 67, 727–745.

215

18. Ford, J.K.B., 1991. Vocal traditions among resident killer whales (*Orcinus orca*) in coastal waters of British Columbia. Can. J. Zool. 69, 1454–1483.

19. Moss, C.J., Poole, J.H., 1983. Relationships and social structure of African elephants. In: Hinde, R.A. (Ed.), Primate Social Relationships: An Integrated Approach. Blackwell Scientific, Oxford.

20. Poole, K., Payne, K., Langauer Jr., W.R., Moss, C.J., 1988. The social contexts of some very low frequency calls of African elephants. Behav. Ecol. Sociobiol. 22, 385–392.

21. McComb, K., Moss, C., Sayialel, S., Baker, L., 2000. Unusually extensive networks of vocal recognition in African elephants. Anim. Behav. 59, 1103–1109.

22. Wilson, B., Batty, R.S., Dill, L.M., 2004. Pacific and Atlantic herring produce burst pulse sounds. Proc R. Soc. Lond. B 271 (Suppl.), S95–S97.

23. Miller, P.J.O., Bain, D.E., 2000. Within-pod variation in the sound production of a pod of killer whales, *Orcinus orca*. Anim. Behav. 60, 617–628.

24. Blake, B.H., 2002. Ultrasonic calling in isolated infant prairie voles (*Microtus ochrogaster*) and montane voles (*M. montanus*). J. Mammal. 83 (2), 536–545.

25. Greenfield, M.D., Baker, M., 2003. Bat avoidance in non-aerial insects: the silence response of signaling males in an acoustic moth. Ethology 109 (5), 427–442.
Greenfield, M.D., Tourtellot, M.K., Tillberg, C., Bell, W.J., Prins, N., 2002. Acoustic orientation via sequential comparison in an ultrasonic moth. Naturwissenschaften 89 (8), 376–380.

26. Jones, G., Barabas, A., Elliott, W., Parsons, S., 2002. Female greater wax moths reduce sexual display behavior in relation to the potential risk of predation by echolocating bats. Behav. Ecol. 13 (3), 375–380.

27. Moles, A., D'Amato, F.R., 2000. Ultrasonic vocalization by female mice in the presence of a conspecific carrying food cues. Anim. Behav. 60, 689–694.

28. Surlykke, A., Yack, J.E., Spence, A.J., Hasenfuss, I., 2003. Hearing in hooktip moths (Drepanidae: Lepidoptera). J. Exp. Biol. 206 (15), 2653–2663.

29. Thornton, L.M., Hahn, M.E., Schanz, N., 2003. Genetic and developmental influences on infant mouse ultrasonic calling: Patterns of inheritance of call characteristics. Behav. Genet. 33, 721–772.

30. Gerhardt, H.C., Klump, G.M., 1988. Masking of acoustic signals by the chorus background noise in the green tree frog: a limitation on mate choice. Anim. Behav. 36, 1247–1249.

31. Greenfield, M.D., 1994. Synchronous and alternating choruses in insects and anurans—common mechanisms and diverse functions. Am. Zool. 34, 605–615.
Snedden, W.A., Greenfield, M.D., 1998. Females prefer leading males: relative call timing and sexual selection in katydid choruses. Anim. Behav. 56, 1091–1098.
Minckley, R.L., Greenfield, M.D., Tourtellot, M.K., 1995. Chorus structure in tarbush grasshoppers—inhibition, selective phonoresponse and signal competition. Anim. Behav. 50, 579–594.
Snedden, W.A., Greenfield, M.D., Jang, Y.W., 1998. Mechanisms of selective attention in grasshopper choruses: who listens to whom? Behav. Ecol. Sociobiol. 43, 59–66.

32. Hindmarsh, A.M., 1984. Vocal mimicry in starlings. Behaviour 90, 302–324.

33. Snedden, W.A., Greenfield, M.D., 1998. Females prefer leading males: relative call timing and sexual selection in katydid choruses. Anim. Behav. 56, 1091–1098.

34. Tauber, E., 2001. Bidirectional communication system in katydids: the effect on chorus structure. Behav. Ecol. 12, 308–312.

35. Au, W.W.L., Mobley, J., Burgess, W.C., Lammers, M.O., Nachtigall, P.E., 2000. Seasonal and diurnal trends of chorusing humpback whales wintering in waters off western Maui. Mar. Mamm. Sci. 16, 530–544.

36. Mitani, J.C., Gros-Louis, J., 1998. Chorusing and call convergence in chimpanzees: tests of three hypotheses. Behaviour 135, 1041–1064.

37. Gray, D.A., 1999. Intrinsic factors affecting female choice in house crickets: Time cost, female age, nutritional condition, body size, and size-relative reproductive investment. J. Insect Behav. 12 (5), 691–700.
Gray, D.A., 1997. Female house crickets, *Acheta domesticus*, prefer the chirps of large males. Anim. Behav. 54, 1553–1562.

38. Kiflawi, M., Gray, D.A., 2000. Size-dependent response to conspecific mating calls by male crickets. Proc. Roy. Soc. London (B) 267, 2157–2161.

39. Hack, M.A., 1998. The energetics of male mating strategies in field crickets (Orthoptera: Gryllinae: Gryllidae). J. Insect Behav. 11, 853–867.

40. Cooley, J.R., Marshall, D.C., 2001. Sexual signaling in periodical cicadas, *Magicicada spp.* (Hemiptera: Cicadidae). Behaviour 138, 827–855.

41. Wells, M.M., Henry, C.S., 1992. The role of courtship songs in reproductive isolation among populations of green lacewings of the genus *Chrysoperla* (Neuroptera: Chrysopidae). Evolution 46, 31–42.

42. Aubin, T., Jouventin, P., 1998. Cocktail-party effect in king penguin colonies. Proc. R. Soc. Biol. Sci. 265, 1665–1673.

43. Bee, M.A., Michey, C., 2008. The "Cocktail Party Problem": What Is It? How Can It Be Solved? And Why Should Animal Behaviorists Study It? J. Comp. Psychol. 122, 235–251.

44. Bee, M.A., Swanson, E.M., 2007. Auditory masking of anuran advertisement calls by road traffic noise. Anim. Behav. 74, 1765–1776.

45. Pohl, N.U., Slabbekoorn, H., Klump, G.M., Langemann, U., 2009. Effects of signal features and environmental noise on signal detection in the great tit, *Parus major*. Anim. Behav. 78, 1293–1300.

46. Swaddle, J.P., Page, L.C., 2007. High levels of environmental noise erode pair preferences in zebra finches: implications for noise pollution. Anim. Behav. 74, 363–368.

47. Hill, P.S.M., 2009. How do animals use substrate-borne vibrations as an information source? Naturwissenschaften 96, 1355–1371.

48. Maklakov, A.A., Bilde, T., Lubin, Y., 2003. Vibratory courtship in a web-building spider: signalling quality or stimulating the female? Anim. Behav. 66, 623–630.

49. Hebets, E.A., 2008. Seismic signal dominance in the multimodal courtship display of the wolf spider *Schizocosa stridulans* Stratton 1991. Behav. Ecol. 19, 1250–1257.

50. Cuthill, I.C., Partridge, J.C., Bennett, A.T.D., 2000. Avian UV vision and sexual selection. In: Espmark, Y., Amundson, T., Rosenquist, G. (Eds.), Animal Signals. Tapir Academic Press, Trondheim, pp. 61–82.

51. Senviratne, S., Jones, I., 2010. Origin and maintenance of mechanosensory feather ornaments. Anim. Behav. DOI 10.1016.

52. Michiels, N.K., Anthes, N., Hart, N.S., Herler, J., Meixner, A.J., Schleifenbaum, F., et al., 2008. Red fluorescence in reef fish: a novel signalling mechanism? BMC Ecol. 8, 16.

53. Pryke, S.R., 2009. Is red an innate or learned signal of aggression and intimidation? Anim. Behav. 78, 393–398.

54. Gamberale-Stille, G., Tullberg, B.S., 2001. Fruit or aposematic insect? Context-dependent colour preferences in domestic chicks. Proc. R. Soc. Lond., B, Biol. Sci. 268, 2525.

55. Aronsson, M., Gamberale-Stille, G., 2009. Importance of internal pattern contrast and contrast against the background in aposematic signals. Behav. Ecol. 20, 1356–1362.

56. Skelhorn, J., Griksaitis, D., Rowe, C., 2008. Colour biases are more than a question of taste. Anim. Behav. 75, 827–835.

57. Stone, E.L., Jones, G., Harris, S., 2009. Street lighting disturbs commuting bats. Curr. Biol. 19, 1123–1127.

58. Hopkins, C.D., 1999. Design features for electric communication. J. Exp. Biol. 202, 1217–1228.

59. Dunlap, K.D., Zakon, H.H., 1998. Behavioral actions of androgens and androgen receptor expression in the electrocommunication system of an electric fish, *Eigenmannia irescens*. Horm. Behav. 34 (1), 30–38.

60. Hagedorn, M., Heiligenberg, W., 1985. Court and spark: electric signals in the courtship and mating of gynotid fish. Anim. Behav. 33, 254–265.
   Hagedorn, M., Zelick, R., 1989. Relative dominance among males is expressed in the electric organ discharge characteristics of a weakly electric fish. Anim. Behav. 38, 520–525.

61. Heiligenberg, W., 1990. Electrosensory systems in fish. Synapse 6, 196–206.

62. Matilla, H., Burke, K., Seeley, T., 2010. Genetic diversity within honeybee colonies increases signal production by waggle dancing foragers. PNAS 275, 809–816.

63. Wenner, A.M., 1967. Honey bees: do they use the distance information contained in their dance maneuver? Science 155, 847–849.

64. Gould, J.L., 1975. Honey bee recruitment dance language. Science 189, 685–693.

65. Michelsen, A., Andersen, B.B., Kirchner, W.H., Lindauer, M., 1989. Honeybees can be recruited by a mechanical model of a dancing bee. Naturwissenschaften 76, 277–280.

66. Smith, J.M., 1991. Honest signaling—the Philip Sidney game. Anim. Behav. 42, 1034–1035.

67. Dawkins, M.S., Guilford, T., 1991 The corruption of honest signaling. Anim. Behav. 41, 865–873.

68. Zahavi, A., 1975. Mate selection—selection for a handicap. J. Theor. Biol. 53 (1), 205–214.
   Zahavi, A., 1977 Cost of honesty (further remarks on handicap principle). J. Theor. Biol. 67 (3), 603–605.

69. Lloyd, J.E., 1975. Aggressive mimicry in *Photuris* fireflies: signal repertoires by femmes fatales. Science 197, 452–453.

70. Eberhard, W.G., 1977. Aggressive chemical mimicry by a bolas spider. Science 198 (4322), 1173–1175.

71. Vandermeer, R.K., Wojcik, D.P., 1982. Chemical mimicry in the myrmecophilous beetle *Myrmecaphodius excavaticollis*. Science 218 (4574), 806–808.

72. Simmons, R., 1992. Brood adoption and deceit among African marsh harriers *Circus ranivorus*. Ibis 134, 32–34.

73. Searcy, W.A., Eriksson, D., Lundberg, A., 1991. Deceptive behavior in pied flycatchers. Behav. Ecol. Sociobiol. 29 (3), 167–175.

74. Getty, T., 1996. Mate selection by repeated inspection: more on pied flycatchers. Anim. Behav. 51, 739–745.

75. Davis, M.D., 1997. Game Theory: A Nontechnical Introduction. Dover Publications.
   Dugatkin, L.A., Reeve, H.K., 1998. Game Theory & Animal Behavior. Oxford University Press, New York, xiv, 320 pp.

76. Lomborg, B., 1996. The Structure of Solutions in the Iterated Prisoner's Dilemma. (Center for International Relations Series). University of California LA.

77. Axelrod, R.M., 1984. The Evolution of Cooperation. Basic Books, New York, x, 241 p.

78. Dale, S., Slagsvold, T., 1994. Polygyny and deception in the pied flycatcher—can females determine male mating status. Anim. Behav. 48 (5), 1207–1217.

79. Crowley, P.H., 2001. Dangerous games and the emergence of social structure: evolving memory-based strategies for the generalized hawk-dove game. Behav. Ecol. 12 (6). 753–760.

80. Enquist, M., 1985. Communication during aggressive interactions with particular reference to variation in choice of behavior. Anim. Behav. 33, 1152–1161.

81. Lettieri, L., Cheney, K.L., Mazel, C.H., Boothe, D., Marshall, N.J., Streelman, J.T., 2009. Cleaner gobies evolve advertising stripes of higher contrast. J. Exp. Biol. 212, 2194–2203.

82. Cheney, K.L., Grutter, A.S., Blomberg, S.P., Marshall, N.J., 2009. Blue and yellow signal cleaning behavior in coral reef fishes. Curr. Biol. 19, 1283–1287.

83. Stowe, M.K., Turlings, T.C.J., Loughrin, J.H., Lewis, W.J., Tumlinson, J.H., 1995. The chemistry of eavesdropping, alarm, and deceit. Proc. Natl. Acad. Sci. U.S.A. 92, 23–28.

CHAPTER 8

Movement: Search, Navigation, Migration, and Dispersal

CHAPTER OUTLINE

219

LEARNING OBJECTIVES

Studying this chapter should provide you with the knowledge to:

- Understand how animals integrate navigational information from genetic and environmental sources.

- Know that basic navigational mechanisms, including kineses (undirected movements) and taxes (directed movements) allow movement to resources.

- Learn that directional information comes from sequential or simultaneous comparisons of environmental stimuli.

- See that search behavior often starts with undirected or looping movements, shifting to goal-directed movements as cues are gained from the environment.

- Understand the concept that animals can use complex mechanisms, including path integration and compasses, to set their course.
- Know that homing behavior allows animals to return to a central place, such as a nest or den.
- Understand that migration gives animals the opportunity to exploit more than one habitat during their life cycles; knowledge of migratory routes may be innate, learned, or a mixture of the two.
- Know how dispersal allows animals to avoid competition and to colonize new habitats.

8.1 INTRODUCTION

Animals move to find better conditions. Movement is a prelude to life-sustaining behaviors such as feeding, avoiding predators,[1] and mating; indeed, a variety of motivations prompt animals to move. The ability to change locations at almost any time during their lives distinguishes most animals from fungi and plants, which, once established, are more or less stationary in their environments.

Movement is one of the most fascinating topics in animal behavior because it requires an exquisite integration of sensory inputs, neural calculation, and execution of behavior. This chapter begins with a look at the tools animals use to shape their movements. It then follows the progression that an animal might use in its own daily and seasonal routines. This progression starts with *searching* movements[2] and then moves on to goal-directed *navigation*, which includes *homing* and migration.[3] The chapter concludes with a discussion of dispersal—movements that take an animal from its birthplace to a different population or that lead to colonization of new habitats—and how dispersal can prevent inbreeding and reduce competition within populations.

Scientific explorations of animal movement have generated clever and intriguing experimental designs that provide windows into the use of unexpected sensory inputs, such as geomagnetic fields,[4,5] and navigational abilities of seemingly simple animals that far exceed human capabilities. Complex movements, such as homing and migration, are the culmination of interactions between sensory inputs and innate abilities to calculate how to reach a goal.

> **KEY TERM** Search is movement to find a goal. Search can be entirely random, can be patterned to cover space efficiently, or it can be shaped by new information as it comes in.

> **KEY TERM** Navigation is guided movement from one location to another, typically using a compass or landmarks.

> **KEY TERM** Homing is movement with the goal of reaching a known resource, such as a shelter, that is used as central place within an animal's home range.

220

CASE STUDY

Navigation and Whooping Crane Conservation
Studies of navigation and migration often form critical elements in conservation plans. If a migratory animal is endangered, then wildlife managers need to understand its migratory behavior.

This includes the migratory route, the locations of its breeding and overwintering grounds, and stopover places on the migratory route. A conservation crisis exists if any of the key habitats—breeding, overwintering, and stopover—are threatened

(Continued)

CASE STUDY (CONTINUED)

by destruction. If this is the case, then wildlife managers need to know much more about the species' migratory behavior; they must understand whether the animals' knowledge of the route is learned or innate, what triggers migration, and what cues the animals use to set their route as they travel. (You may wish to contrast this with our discussion of a failed whooping crane conservation program in Chapter 15. That program relied on cross-fostering whooping crane eggs into sandhill crane nests.)

The highly endangered whooping crane is a classic example of how conservation concerns and knowledge of migratory behavior can come together. This large, attractive bird overwinters at only one site, on the Texas coast, and breeds at only one site, in northern Canada. Habitat disturbance at either location could doom this species.

Scientists are attempting to establish a second population, with its winter home in Florida and summer home in Wisconsin. The direction of the new migratory path matches the established population's route, so genetic information about the migratory path works in the new population. However, there was one major behavioral hurdle remaining before a new population could be established: distance, stopovers, and strategies for completing the route are learned by young birds when they follow experienced birds on their first migration.[6] How would the second population learn that route?

The solution lay in using ultralight planes to lead the birds on the new route, teaching them the endpoints and stopovers

FIGURE 8.1

These whooping cranes are following an ultralight as conservation biologists, taking advantage of the fact that migration routes can be learned, attempt to establish a second wintering population of the endangered birds. *Photo: U.S. Fish and Wildlife Service, Operation Migration.*

(see Figure 8.1). In succeeding generations this information will be culturally transmitted among the birds. This approach is complex and requires considerable planning and funding, but the outcome may be the establishment of a second population of cranes that provides critical insurance for the survival of this elegant species.

221

Another way of looking at this situation is to think of a progression of types of movements. This navigation starts with actions that have no set direction. Sometimes ***random movement*** accomplishes the goals of finding food or shelter. However, in most instances animals use ***directional movements*** to travel from one place to another. Finally, navigation is not just directional movement, but it is movement with a specific destination. Movement directed from one point to another is essential to survival in most animal species. The roots of the word *navigation* can illuminate this concept. For ancient people, travel over long distances often involved the sea; the Latin roots of *navigation* refer to ships (*navis*) and guidance (*agere*). For animals, navigation can occur at vastly different scales, ranging from a movement of a few body lengths that takes the animal to food or shelter to the thousands of kilometers (40,000, in fact) that an Arctic Tern spans on its nearly pole-to-pole migration (see Figure 8.2). Navigation seems like an obvious enough behavior—a simple task—but what does going from one place to another entail? How do animals gather the information they need for navigation, and what do they do with it? These are the central questions for this chapter, which covers the full range of possible travels, from short trips around nests or dens to long journeys across oceans and continents.

> **KEY TERM** Random movements cannot be predicted by either the animal's previous movement or by assessing the surrounding environment.

> **KEY TERM** Directional movements are oriented with respect to the surrounding environment.

To be able to navigate, an organism needs to know at least two things: its current location and the location of its goal. Thus, navigation is not the same as simply moving in a

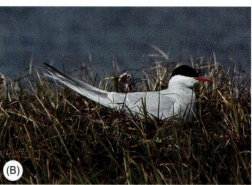

FIGURE 8.2
The Arctic Tern accomplishes the longest migration of any organism as it flies pole to pole. *Photos: (A) Ben Pless; (B) Tim Bowman/ US Fish and Wildlife Service.*

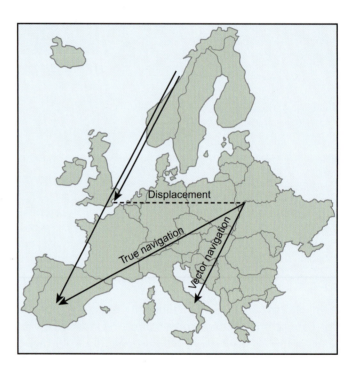

FIGURE 8.3
If an organism is *orienting* toward, say, southwest, it will move in a southwesterly direction; if displaced hundreds of kilometers to the east, it will continue to move southwest, hundreds of kilometers east of the path it would have taken prior to displacement. If an animal is *navigating* toward a goal in the southwest and is displaced hundreds of kilometers to the east, it will compensate for the displacement, shift its path west, and continue to move toward the goal. *Adapted from Berthold, P., 2001. Bird Migration: A General Survey, Second ed. Oxford University Press, 272p.*

randomly chosen direction, even if that movement is in a straight line. In this regard, it is useful to note the distinction between *navigation* and *orientation*. *Orientation* is a word that indicates movement in a given direction—classically, in a compass direction, although a variety of stimuli that have nothing to do with compass direction may also elicit orientation. (*Orient* itself is from a French word meaning "east.") However, moving in a direction is not the same as moving toward a goal or destination. If an orienting animal is moving north and it is displaced to the east, it will continue to move north. If a navigating animal is moving north and is displaced to the east, it will correct its path, moving to the northwest; in so doing, it will reach the same destination that it had prior to being displaced (see Figure 8.3).

Thus, navigational movements often involve an environmental cue, such as a landmark, the sun, or the earth's geomagnetic field. In some cases navigation involves only internal calculation of the distance and direction to be traveled; this interesting process is called **path integration**. Studies of navigation often focus on discovering how animals obtain the

FIGURE 8.4
A bird's search for appropriate habitat or food shifts among levels, including a very local level (upper left), an intermediate landscape level (perched on a tree, center) and a larger landscape level (lower right).

information needed to direct their movements (see Figure 8.4); such investigations have led to spectacular findings of homing pigeons using odors to navigate home from distances of hundreds of kilometers and of sea turtles using magnetic fields in the ocean to orient their long-distance migrations.

> **KEY TERM** Path integration is the ability to track an outward path and to calculate the shortest possible path back to the starting location.

As noted previously, navigation is central to survival in many species. Probably because of this, there are reinforcing linkages among many of the concepts discussed in this chapter, such as search and navigation to and from a central place, and elements of Chapter 9 on foraging. The discussion of cognitive maps in Chapter 6 comes into play as well; refer to that discussion to assist with topics in this chapter, such as homing and migration.

The maintenance of homeostasis described in Chapter 4 and the nature–nurture collaboration explored in Chapter 3 on behavioral genetics are never far away. Continue to find and explore linkages while reading this chapter; this is how understanding of animal behavior grows.

Early ethologists took a highly mechanistic approach to animal movement, emphasizing the classification of movements and how sensory inputs and movements are integrated. This led to intriguing revelations about the relationships between sensory systems and navigation; animals without highly complex nervous systems, such as earthworms, for example, can make orientation decisions that bring them to appropriate habitat sites. More contemporary approaches to animal movement led scientists to ask questions such as the following ones: Are there hidden (to humans) sensory processes involved in collecting navigational information? How are multiple, sometimes contradictory, information inputs integrated to make navigational decisions? Can neural network models be used to analyze these decisions? What role do cognitive processes play in this decision making?

8.2 SOURCES OF NAVIGATIONAL INFORMATION

It seems obvious that most navigation is shaped by several sources of information. One form of information is a product of evolution—if the best shelter, such as hiding under rocks (dark, humid places for shelter in cockroaches, for example), remains in the same location generation after generation or if the same migratory route works well over evolutionary time, then if the appropriate genetic variation exists, natural selection may favor coding this information into a species' DNA. Another form of information is learned information about the animal's landscape. Squirrels, for example, may be genetically compelled to build nests, but a good nest location will be specific to an individual squirrel and likely changes with time, generation to generation. Thus, while the motivation to build a nest may have a large genetic component, the location of the nest itself will depend strongly on learned information. For instance, a squirrel needs to learn how to leave its nest, forage, and return. Learning provides a bridge between elements of an environment that may be predictable during an animal's lifetime and information that has been shaped over the species' evolutionary history. When making a decision about movement in "real time," an animal needs to integrate its genetically encoded information, what it has learned during its lifetime, and sensory inputs about what is happening at that moment. Those sensory inputs of the moment are determined by its condition and where it is located:

- Is it hungry?
- Does it need shelter?
- Is it fearful?
- Is it looking for a mate?
- Has it reached sexual maturity so that it is time to disperse from its parents' nest?
- Has a change of season triggered the initiation of migration?

As discussed in Chapter 4, changes in neurotransmitter and/or hormonal levels trigger these behavioral decisions that culminate in the impulse to move; refer to that chapter for a more detailed discussion of these changes. To repeat, information about movement can be either genetic, as in sea turtle migration, or learned, as in homing in pigeons.

DISCUSSION POINT: CLIMATE CHANGE, HOMING, DISPERSAL, AND MIGRATION

Based on what you have learned so far about animal behavior and information that you can glean from online sources, how might global climate change affect movements such as homing, dispersal, and migration? Form your own hypotheses at this point, as this will give you the basis for critically approaching the rest of this chapter.

Genetically determined habitat preferences play central roles in an animal's search for optimal habitat in its landscape. Studies of the genetics of habitat preferences are easier to conduct in insects than in birds or mammals because of insects' short generation times and the ease with which some species can be bred and tested for habitat preferences. The examples here are insects, but the same general principles apply to all animals.

Habitat preferences can involve simple information, such as the cockroach preference for dark places mentioned previously. More complex examples of stored information come from herbivorous insects that must find their host plant for themselves or their offspring. (In the case of host plants, "habitat" and "food" are near-synonymous.) Genetics provides information about host plant odors for some insects, so inexperienced individuals can use those odors to find the right plant. Pea aphids, for example, are widely distributed in North America and England (see Figure 8.5) and suck the juices from a number of plant species; populations of pea aphids are genetically differentiated in their preference for host plants such as clover and alfalfa.[7] Likewise, picture-winged flies (*Rhagoletis pomonella*, apple maggots) can live in a variety of fruits, including apples, hawthorns, and dogwoods. The adult flies find the appropriate fruit by its odor and lay their eggs on the fruit. Preferences among fruit species are genetically determined, and this fly species is divided into host plant races based on these preferences. Matings between host plant races of the fly result in offspring that are poor at locating fruit for egg laying.[8]

In other cases, host plant information comes from environmental sources. For instance, some insects learn host plant preferences by feeding on those plants as larvae. The common tobacco hornworm has a range of acceptable food plants, including members of the nightshade family like tobacco and tomato, but will accept a variety of other plants as food. Once a caterpillar has eaten a certain plant, it develops a preference for that plant type over other potentially acceptable plants and will choose that plant for oviposition (egg laying) when it is an adult (see Figure 8.6). This shows how learned information can shape orientation decisions that drive simple habitat preferences.

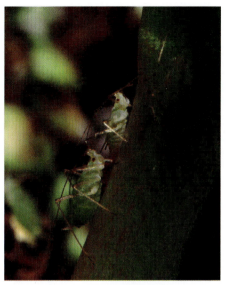

FIGURE 8.5
The geographic range of pea aphids is broad, covering continents, and their host plant preferences differ across populations. This translates into differing search strategies among populations. *Photo: Bernie Roitberg.*

FIGURE 8.6
(A) Changes in tobacco hornworm diet preferences during growth and development. *Adapted from de Boer, G., 2004. Temporal and developmental aspects of dietary-induced larval food preferences in the tobacco hornworm. Entomologia Experimentalis et Applicata 113, 197–204.* (B) Tobacco hornworm. *Photo: Shelly Adamo.*

Host plants are not the only source of environmental information. Many animals can use cues, learning critical *landmark* features or using *compass* information (from the sun, stars, patterns of polarization in the sky, or the earth's magnetic field) as referents so they can revisit locations;[9] this is particularly important for homing and migrating animals.[10] In addition, animals can use information about previous movements to set a route in the absence of environmental cues. This is the strategy used in path integration, a phenomenon in which an animal returns to a beginning point (perhaps its nest) by extrapolating the direct return path from the twists and turns of its outbound journey.[11]

8.3 SENSING THE ENVIRONMENT IN TIME AND SPACE

Unless an animal uses only idiothetic information to shape its navigational decisions, it needs directional information to get to the correct location in the environment. This is obvious to humans, who see the goal or another clue about the location of a destination and then move toward the goal. Depth perception and the ability to recognize shapes and objects in the environment make this visual task so easy that it requires little thought. Under total darkness, humans who are trying to find distant objects are reduced to using senses such as hearing and smell for which their cue-localization is much less well developed. This human analogy captures the central issue for animals in gaining sensory information from the environment; it is easy to perceive the presence and intensity of a cue such as a light, an odor, or a sound. The task of extrapolating the source of the cue is much more difficult, but if the animal is to organize its movements to reach the stimulus (or to use the stimulus as a landmark or compass), it must perform that extrapolation and find the source.

OF SPECIAL INTEREST: THE USE OF THERMAL DETECTION BY PIT VIPERS TO LOCATE PREY

Precise location of prey is extremely important for predators; predator strikes require precision navigation on a microcosmic level. Pit vipers, venomous snakes that include the copperhead, rattlesnakes, the bushmaster, and the fer-de-lance, have a pair of indentations, or pits, just below and slightly in front of their eyes (see Figure 8.7). The pits contain hundreds of infrared (temperature) sensitive cells, which form a thermal image of the area in front of them. This image is fairly low resolution, and the pit viper relies on choosing a hunting site with a cool background so warm-blooded prey will stand out. Because the pits are paired, there may be a stereo effect, helping localize the prey and increasing strike accuracy. For this reason, field biologists (and anyone else!) should avoid wearing open-toed shoes or sandals in areas where pit vipers are common.

FIGURE 8.7

The pits (for which pit vipers like this one are named) are small indentations just in front of the eyes. They are lined with infrared heat sensors that enable the animal to locate warm prey with amazing precision. *Photo: Jeff Mitton.*

How do animals locate the source of a stimulus? Generally, animals triangulate; that is, they measure the intensity of the stimulus from two different locations in some combination of time and/or space. In other words, to triangulate on something, an animal must perceive it from two different locations. This perception can be done simultaneously, as in stereoptic vision used by many animals, or sequentially, so one leg of the triangle is perceived and then the animal moves (or at least moves its head) to find the other leg of the triangle. For instance, humans locating a stimulus triangulate by simultaneously using two sensory organs (two eyes, two ears) to perceive the stimulus, and then using the difference in perception between the two organs to calculate distance/direction.[12,13,14]

Visual triangulation confers the ability to tell how far away an object is, although in reality people use additional strategies to refine this information. In the context of vision, triangulation is called *depth perception*, *stereopsis*, or *binocular stereopsis*. Stereopsis is the use of two simultaneous sensory inputs to triangulate on the distance of a stimulus. Binocular vision gives us depth perception because it allows stereopsis. Stereopsis is very important for predators that strike at their prey, because to strike accurately, they need to know the distance to the prey. Such animals tend to have forward-directed eyes that overlap considerably in visual field; the overlap facilitates the comparisons needed for stereopsis (see Figure 8.8).

FIGURE 8.8
A ladybird beetle larva moves its head from side to side, gathering sensory information sequentially. *Adapted from* Bell, WJ 1991, *Search Behaviour*, Chapman and Hall, London.

BRINGING ANIMAL BEHAVIOR HOME: FOLLOWING THE SCENT

While humans are good at visual triangulation, our ability to triangulate using olfactory information alone is unimpressive. The reason is that our nostrils lead to a single olfactory organ, the olfactory epithelium, and that eliminates the basis for triangulation using simultaneous input. By taking a step, or moving the head from side to side, and sniffing in different locations, we can use sequential comparisons of the odor strength to begin to discern the direction of the odor's source. With a single sense organ, sequential perception allows triangulation. By moving the sense organ from side to side, we create a triangle in time sequence. Watch a dog with its nose to the ground, following a scent. It stays on course by using small side-to-side movements of its nose, sequentially triangulating on the scent source. (Of course, you could move your head from side to side, sniffing, until your neck got very sore, and you would not be as successful at olfactory orientation as a dog, or many other mammals. The reason is the differences in the olfactory epithelium itself, including the number of receptors and their sensitivity.)

FIGURE 8.9

227

Lizards use triangulation in prey (or perch) location. This is the result of simultaneous input from both eyes (binocular vision) and is called simultaneous stereopsis. Lizards also use sequential stereopsis; they move their heads from side to side, thus increasing the angle of triangulation, and the accuracy of location. *Photo: Michael Breed.*

Many animals (grasshoppers and lizards are great examples) combine simultaneous with sequential sensory strategies to maximize their accuracy in stereopsis (see Figure 8.9). By moving their heads back and forth while triangulating on a prey item or a spot to jump toward, they increase the angle with which they triangulate, making their measurement more accurate, much as might happen if their eyes were further apart.[15]

In theory, animals with spatially separated odor detectors, such as the antennae of insects, could use that spatial separation to triangulate in a way that is analogous to binocular stereopsis. However, the common reaction of insects attempting to find an odor source, particularly a sex pheromone source, is to walk or fly upwind when they encounter an odor plume (see Figure 8.10).[16,17] Experimentally, it is difficult to know if this response is based on spatial odor comparisons between the antennae or sequential comparisons, but the generally held view is that walking or flying insects make sequential comparisons of odor concentrations and change their movements if they begin to find lower concentrations of the attractive odor.

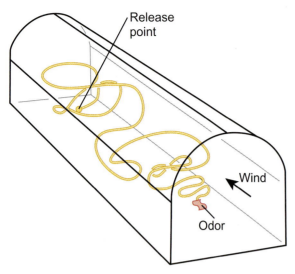

FIGURE 8.10

Insects are thought to move upwind when in an odor plume, and then crosswind when they lose the plume. This allows them to follow an odor, such as a sex pheromone, that may shift with the wind.

FIGURE 8.11

How a taxis keeps a an animal (or bacterium) in favorable habitat conditions. As long as the animal is in favorable conditions, it moves in a straight line or randomly, but when it passes a threshold (low humidity as indicated in the figure, or for a bacterium, low concentration of a nutrient), it turns more sharply, keeping it in the favorable habitat. *Adapted from http://www.rowland.harvard.edu/labs/bacteria/index_movies.html, http://jb.asm.org/cgi/content/full/182/24/6865.*

In contrast, bacteria (not exactly animals, but interesting enough to include in the discussion of navigation) are thought to sense and use micro-scale differences and gradients in concentrations of molecules in their surroundings. Many bacteria have a flagellum—a thread-like organ that can propel the bacterium in a more or less straight line, not to be confused with a eucaryotic flagellum. When a change of direction is needed, the flagellar motion changes to produce a tumbling movement that reverts to straightline swimming when the new direction is established (see Figure 8.11).[18,19]

When we are considering how animals sense the environment, the most important point to remember is that locating the source of an environmental cue requires a mechanism, either sequential or simultaneous, for comparing the strength of stimuli. Comparisons allow triangulation on the cue source. Whether organisms use sequential or simultaneous comparisons depends in part on the structure of their nervous and sensory systems.

OF SPECIAL INTEREST: NORTH-FACING COWS—INSIGHTS FROM GOOGLE EARTH™

On any given day, a scientist—or a farmer—might be able to observe a herd of cows and determine which direction they tended to face. This one observation would not allow us to determine the cause of their alignment—it could be a response to sun, wind, or a number of factors, depending on conditions—and would not give us numerous observations under a wide range of geographic and meteorological conditions. In other words, we could not generalize from such an observation, and the number of observations required for such generality would be daunting.

Sabine Begall and colleagues solved this problem by using three sources of data that addressed the orientation of cattle and deer.[20] First, they used satellite images from Google Earth that showed the body alignment of 8,510 cattle in 308 pastures on six continents; in this manner, they could *scan-sample* numerous herds, reducing the likelihood that the same weather events influenced them all at the time of the sample. Second, they observed alignment in over 2,000 red and roe deer under a variety of conditions and at various times and places in the Czech Republic. Third, they used data from the body impressions of red and roe deer in snow.

All of these data point to the fact that cattle and deer orient on a north–south axis in line with the magnetic poles of the earth. Because of the broad range of data collection, Begall and colleagues could eliminate many other hypothesized environmental causes of orientation, such as avoiding wind, sun-basking, and the like. Although the Google Earth images were too unresolved to allow anterior/posterior distinctions in the case of cattle, the field observations of deer indicated that they tended to face north, not south. Moreover, cattle and deer in the vicinity of power lines did not conform to this pattern, and indeed, displayed different alignment patterns, probably because power lines generate extremely low-frequency magnetic fields that can disrupt the animals' ability to sense the magnetic field of the earth.[21]

> **KEY TERM** Scan sampling is observing and recording the behavior of multiple animals repeatedly, at specified intervals for specified amounts of time. Thus, it is like a series of snapshots of a given animal, taken at known intervals.

8.4 HOW TO RESPOND TO SENSORY INFORMATION: A TOOLBOX FOR FINDING THE WAY

After the previous review of the ways in which animals gather and process information about movement, the next step is to learn how they use it and to learn about the tools they have to convert sensory information into locomotor decisions. These tools range from deceptively "simple" (e.g., "move" and "don't move") decisions to amazingly complex ones, involving mental maps and geomagnetic fields!

Kinesis

Early ethologists wanted to be able to describe animal behavior by classifying instinctive patterns of behavior. This approach was particularly common in ethological analyses of animal movements. The simplest locomotor responses to the environment, *kineses* (plural), involve changing velocity or altering turning rates depending on environmental stimuli. Indeed, the word *kinesis* (sing.) comes from a Greek word that means "motion." Kinesis does not include orientation to environmental stimuli and, consequently, requires only very simple sensory and neural systems. If an animal has a specific need, such as food, a more humid environment, or shelter from the sun, but lacks the sensory ability to gather information about the location of the needed resource, it may engage in an undirected search, or kinesis; that is, it will start moving (see Table 8.1). Such an instinctive response works very well for analyses of the behavior of many animals, including most invertebrates and cold-blooded vertebrates, such as amphibia and snakes.

TABLE 8.1 Definitions of Simple Movements and Orientation Patterns

Kineses	Changes in speed of movement without orientation to a stimulus source.
Orthokinesis	Motion is in a straight line but is undirected with respect to the source of the stimulus.
Klinokinesis	Turning movements that change in rate as stimulus strength changes. Examples: looping behavior of beetles during search, tumbling behavior of bacteria
Taxes	Changes in direction of movements that are oriented to a stimulus source.
Klinotaxis	Turning that is oriented to a stimulus. A classic example is the negatively phototactic (moving away from light) response of blowfly larvae.
Tropotaxis	If two stimuli are used, such as two lights—one perceived by each eye—then the animal averages the input and moves to a point between the two light sources. Many insects display this response.
Telotaxis	In this case, when two stimuli are presented, the animal chooses one and travels directly to that stimulus. A positively phototelotactic animal would choose the stronger of two lights.

Taxis

The integration of orientation (directional response to a stimulus) with kinesis (locomotor response to a stimulus) yields a *taxis* (pronounced *tax-sis*, not to be confused with cars for hire). The plural of *taxis* is *taxes* (*tax-eez*). In biology, a taxis is a movement that is directed with respect to a stimulus, such as an object, a light, or an odor source.

Kinesis and *taxis* are often paired with prefixes to indicate the nature of the stimulus that motivates the response. Attractive stimuli are termed positive, repulsive stimuli are negative. The nature of the stimulus is used as prefix for the type of movement (*anemo* = wind, *geo* = gravity, *photo* = light, and so on). Hence, positive phototaxis occurs as a result of attraction to light. This can be figured out by looking at the parts of the phrase. "Positive" refers to attraction, so that indicates movement toward. "Photo" refers to light, and "taxis" means oriented movement: so "positive phototaxis" means "movement toward light." In the same vein, negative anemotaxis means moving downwind, and so on. A moth that flies toward a light is positively phototactic. A cockroach moving away from light is negatively phototactic. Any environmental stimulus with a directional source can be used to orient a taxis.

Simultaneous or sequential comparisons of stimulus strength can be used to establish the direction of movement. While many organisms display simple positive and negative taxes, such as an insect moving directly to a light, more complex navigation like that which occurs in homing and migration can require the ability to keep the stimulus at a fixed angle somewhere between 0 degrees (toward the stimulus) and 180 degrees (away from the stimulus). Moving at an angle relative to the stimulus, rather than directly toward or away from the stimulus, is called *menotaxis,* which is useful because it can be applied to homing. An animal that leaves its nest and moves at an angle with respect to a landmark needs only to reverse that angle to return to its nest.

Counterturning

Straight movements are often thwarted by obstacles. (If the animal is an ant, these obstacles could be small pebbles; for larger animals, trees and boulders are effective barriers.) How can an animal continue to move in a relatively fixed direction when its path is frequently deflected? Counterturning is one simple mechanism used by many animals to maintain their course of movement.[22] In counterturning, each change of direction is balanced by a movement in the opposite (counter) direction when it becomes possible to do so. The classic test for counterturning forces a test animal to establish a movement direction, running or walking down the arm of a maze to a point where it must turn (see Figure 8.12). After following the second branch of the maze, it reaches a point at which it can choose to counterturn, thus re-establishing its original course, or choose to make a reverse turn. In an experimental context, a number of animals can be asked to move through the maze, and the frequency of counterturns then can be compared with random turning, which is characterized by a 1:1 ratio of counterturns and reverse turns. The distance of the second arm of the maze and the time taken to traverse it affect the likelihood of counterturning; as distance and time increase, the animal "forgets" the needed course correction and is more likely to make a random choice. Early work on counterturning in milkweed bugs set the stage for consideration of counterturning as a basic behavioral mechanism in animal navigation. Most of the recent literature that highlights counterturning as an important mechanism

FIGURE 8.12
In this classic test for counterturning, a test animal is forced to establish a movement direction, running or walking down the arm of a maze. It then reaches a point where it is forced to turn; this mimics obstacles the animal might encounter in the natural world. After traveling down the second branch of the maze, perpendicular to its original direction, it reaches a decision point where it can make a counter turn, which re-establishes the original direction of travel, or make a reverse turn.

Choice point - the animal can make a counterturn.

This arrow indicates the direction of travel if the animal makes a counterturn.

Forced turn (mimics obstacle)

Established direction of movement

deals with how moths find odors,[23] but counterturning is probably displayed by most animals, if they are given the opportunity (see Figure 8.13).

Landmarks

Animals can use prominent physical features in the landscape (a mountain, large trees, rocks, and so on) as landmarks in orientation. The classic demonstration of landmark orientation is an experiment by Niko Tinbergen on digger wasps. When a digger wasp leaves its nest, it makes several looping flight passes around the nest; this allows it to learn the landmarks it needs to know to return to the correct location.[24] To demonstrate this, Tinbergen first arranged a circle of pinecones around the wasp's nest entrance (a hole in the ground); by moving or altogether removing the cones, he discovered that he could change the return of the wasp. If the circle of cones was moved a few centimeters from the nest, the wasp searched, in vain, for the nest inside the relocated circle.

Many mammals, such as pika, deer, and members of the cat and dog families, create their own landmarks by scent marking. By placing glandular secretions on key locations, such as nest sites or territorial boundaries, the animal creates a guidepost for itself and a signal to other animals that the area is occupied. Scent marking was briefly discussed in Chapter 7, on communication, and is covered in more detail in Chapter 11, in the context of territoriality.

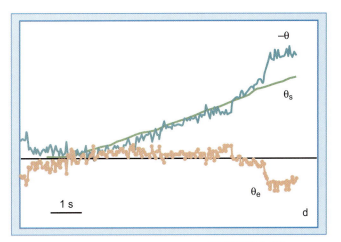

FIGURE 8.13
In an interesting experiment, crabs which home to a nest burrow were able to maintain a straight course of navigation, even when they were placed on a slowly rotating platform, as long as they had visual cues from the surrounding environment. The green line is the rotation of the platform, the red line is the course maintained by the crab relative to the external cues. *Adapted from Zeil, J., 1998. Homing in fiddler crabs (Uca lactea annulipes and Uca vomeris ocypodidae). J. Comp. Physiol. 183 (3), 367–377.*[77]

OF SPECIAL INTEREST: WHY DO MOTHS FLY TO A FLAME?

At night, a candle or light bulb can attract hundreds or thousands of moths, along with a few beetles, planthoppers, and occasionally less-common insects. Homeowners don't want a flock of moths hovering around their patio, landing in drinks and flying into guests' mouths and noses. Bats and

FIGURE 8.14
(A) In the natural, pre-electrified world, insects use light (always from distant sources) as a guide for straight-line movement. By flying at a constant angle relative to a very distant light source, they effectively move in a straight line. (B) The advent of electricity made nearby light sources abundant. (Candles, being expensive, were never so abundant.) When insects fly at a constant angle relative to a light bulb or candle flame, their flight path describes a spiral, ever closer to the light source, with no good effect for the insect.

spiders also move in to take advantage of feasting on the insects, adding to the homeowners' distress. A partial cure comes from employing a bug-zapper, usually a device that combines an ultraviolet light (which is more attractive to moths than a regular incandescent light) and an electric grid that shocks the insect. The bug-zapper is no respecter of insects, frying beneficial bugs along with nuisance and harmful ones, so the environmental effects of a bug-zapper must be balanced against the urge to enjoy an insect-free patio.

All this leaves open the question of why the insects are flying toward the light. The likely answer is that under natural conditions the insects use a distant light, such as the moon or a star, as a cue for their flight direction. Moving in a line at a constant angle relative to a distant landmark (menotaxis) results in more or less straight flight—the farther away the landmark, the straighter the movement. A closer landmark, for example, a porch light, has a different effect. Flight that maintains the same angle relative to the landmark results in an inward spiral, leading the moth to its death dance with the flame or hot light bulb. Because moths fly at night largely to find mates, the conservation effects of this type of light pollution—never mind bug-zappers—are likely immense, but few studies have addressed this question.

Meanwhile, one of us enjoyed the company of a cat-faced spider (*Araneus gemmoides*) who set up shop (and web) across the top of the patio door, near a light. It was good "hunting" for the spider, and good spider-watching for the biologist!

Snapshot Orientation

Snapshot orientation also uses landmarks but is more sophisticated than directional movement toward a stimulus. In snapshot orientation the animal remembers a visual image of the landscape, or at least of a composite of significant landscape features. It can then compare the landscape it sees at any given moment with a "library" of memorized landscape snapshots. If the current landscape matches the snapshot, then the animal is at the correct location. If the landmarks appear smaller than expected, this indicates to the animal that it needs to move closer to the goal, and if they appear too large, then the animal needs to move further away. If one landmark is at the wrong angle with respect to another landmark, then movements across the landscape can reduce this angular error, bringing the animal to the proper orientation. This is an interesting and simple solution to the problem of adjusting orientation that is based on the animal's perspective of the landscape.

The concept of snapshot orientation was first developed in studies of honeybees (see p. 236) and has been applied in bees and ants. Humans seem to be able to perform this sort of orientation, although differentiating between use of snapshot orientation and the use of a cognitive map (see p. 235) is difficult.

Path Integration

The ability to start at a fixed point, visit several locations, and return directly to the original point is fundamental to animal orientation. To accomplish path integration, the animal simply computes its return distance and direction vector from the vectors joining the locations on its route (see Figure 8.15).[25,26] This saves retracing steps. Honeybees, ants, dogs,[27] rats,[28] and humans are all known to have excellent path integration abilities. While a variety of animals can path integrate reasonably accurately in the absence of landmarks or other cues, long-distance orientation using path integration usually incorporates corrections from landmarks or celestial cues (see Figure 8.16).

FIGURE 8.15
Path integration over three stops. The animal changes direction at each stop, and distances among the stops vary. The distance and direction information is integrated to allow the animal to return directly to the starting place from the third stop.

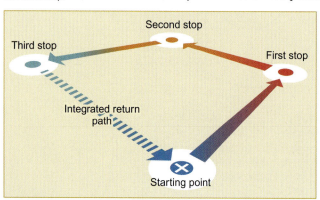

Second stop

Third stop

First stop

Integrated return path

Starting point

232

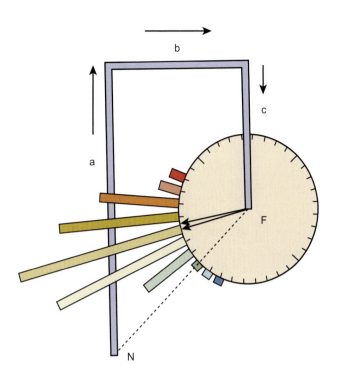

FIGURE 8.16

This diagram represents path integration in the desert ant *Cataglyphis fortis*. The ant begins at the nest (N), proceeds along path a, and is forced to turn twice, moving until it arrives at point F, a feeding station. The results are shown in this figure, where the length of the bars radiating away from the feeding station indicates the number of ants moving in that direction. The ants do not integrate precisely, but their approximation of the nest location is sufficient to return them to the nest, along with the use of landmarks and other environmental cues. *Adapted from* Müller, M., & Wehner, R., *Proceedings of the National Academy of Sciences of the United States of America*, 85 (1414), 5287–5290.

DISCUSSION POINT: CAN YOU PATH INTEGRATE?

Have a little fun with your friends and roommates by testing their abilities to path integrate. Blindfold your "victims" and take them over a route with two or three turns (walking 10–20 steps between turns is good). When you have them at the end of this route, turn them around in a circle two or three times, and then let them try to path integrate to find their starting point.

Compass Orientation

The position of the sun or the stars in the sky can be used as a compass. Because of the earth's rotation, the sun appears to move through the sky from the eastern to the western horizon. At the equator, this path travels through the vertical azimuth; away from the equator, the sun's path is not vertical, but the sun nonetheless remains a strong directional cue in most locations except those very near the poles.

To the earth-bound observer, the stars appear to be fixed on the inner surface of a rotating sphere, much like the view in a planetarium. It is similar to standing in the center of a rotating ball; the stars are points of light on the inside of the ball. The ends of the celestial axis of rotation (the southern and northern points in the ball at which there is no apparent movement of the stars) correspond to the north star (Polaris) and the Southern Cross. Sailors, of course, used these points to help determine their position prior to the invention of modern navigational methods, such as Global Positioning Systems (see Figure 8.17). Other animals are equally capable of using the sun or the stars as navigational reference points.

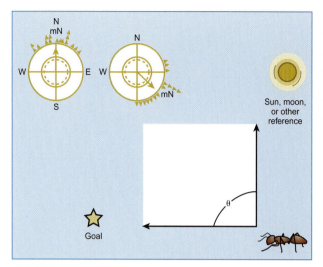

FIGURE 8.17
How an orientational compass works. An ant, lower right in the figure, may use the location of the sun or other reference and its goal (the nest or food item), to learn an orientation route. In the upper left of the figure, each dot at the margin of the compass circles indicates the direction of movement of an ant. Ants oriented to the north compass direction (lefthand circle), may be deceived by experimentally moving a landmark or using a mirror to make the sun appear in a different location (righthand circle, dots indicate a change in the orientation of the ants).

How do animals compensate for the apparent motion of the stars or the sun? The spin of the earth creates a problem; reference points may appear to move through the daily cycle of rotation. This apparent motion of the sun and stars is predictable on a daily basis, and animals can compensate for it by integrating a clock mechanism with their navigational system, something that is easily demonstrated by time-shifting animals in artificial light environments (habituating them to a day-night cycle that is a few hours displaced from the natural cycle) and then observing their orientation behavior when they are moved to a field situation.[29,30,31] Long-distance navigators, such as migrating birds, use geomagnetic cues to calibrate their compass. Through the use of multiple, interacting cues, a more accurate course can be set than could be obtained by using any one cue.

Odometers and Measuring the Distance Traveled

For an animal to find its way to a location, whether it is a bee returning to its hive or a salmon finding its stream after life at sea, it needs a measure of its desired travel distance. In a car, the *odometer* gauges distance traveled; ultimately, the car odometer is linked to the number of times the car wheels turn. Animals display a wide variety of strategies for gathering the same sort of information. Odometer mechanisms have the greatest importance when landmarks are unavailable (as in a fairly homogeneous desert environment) or when an animal is using genetic information to complete a migration.

Desert ants (*Cataglyphis cursor*) use something analogous to the car odometer.[32] They use step counting to assess distance traveled, much like a person counts paces to guess the dimensions of a field. In a clever experiment, scientists put stilt-like extensions on the legs of these ants before the ants' return to their nests. These ants had used path integration during an outward trip from the colony to "know" the distance to be traveled. Stilted ants moved the number of steps that would have been required to find their nest; because their legs were unexpectedly long, they ended up moving farther than their goal. In a reverse of the experiment, ants moving outward with stilts did not travel far enough to reach the nest after the stilts were moved. In other words, ants use a pedometer to gauge their progress.

Animals that fly across the landscape require different odometer mechanisms. Because of the effect of wind, effort alone is not a good measure of distance traveled. For instance, a headwind can result in much expenditure of effort but little progress, whereas a tailwind can greatly ease the difficulty of a trip. Eastbound passenger airplanes crossing North America fly with the jetstream as a tailwind, but the same air currents are headwinds for westbound planes; a cross-continent westbound trip takes about an hour longer than an eastbound trip because of these wind effects.

KEY TERM An odometer is any physiological or mechanical device that allows measurement of the distance traveled.

Migratory birds, because they encounter predictable upper air currents akin to the jet stream, may be able to use effort as an indicator of distance, with the expectation of quite different efforts for northbound and southbound trips. Migrating fish and sea turtles may encounter these difficulties, but their mechanisms for measuring distance traveled are less well understood. Honeybees gauge the rate of movement of landscape across their visual field; a bee flying into a headwind experiences a slow rate of change of visual images. This compensates for wind effects and results in an accurate odometer, a key component for accurately following dance information. Thus, animals that know the local landscape can also use landmarks to judge distances.

Cognitive Maps

A cognitive map is an internal neural representation of the landscape in which an animal travels. Animals that use cognitive maps can "visualize" the landscape and solve orientation problems by referring to these maps. While it is generally accepted that birds and mammals can form cognitive maps, and that the hippocampus is the most important part of the brain in their formation, considerable controversy has centered around whether other animals, such as honeybees, can form similar maps.[33] An animal with a cognitive map should be able to assess landmarks and compass information and then calculate its travel path to any location within its mapped area. In practice, though, it is difficult to design experiments that differentiate among landmark orientation, path integration, and use of a cognitive map. The strongest evidence for cognitive processes in orientation decisions comes from experiments that require animals to solve multiple problems, particularly to navigate to more than one point in the area around their home.[34] This topic is discussed in more detail in Chapter 6; remember that the use of landmarks and compasses alone does not establish the use of a cognitive map.

FIGURE 8.18
Examples of search patterns: (A) undirected search; (B) straight line search, covering large area; (C) spiral or looping search; (D) another type of straight line (large area) search.

235

8.5 SEARCH

Movement often starts with a search, which may look random or only slightly systematic to the casual observer.[35] Searching behavior is critical to resource location and is an example of a behavior that uses some of the simple responses to environmental information that are discussed in Section 8.4. In a search for food, shelter, or a mate, an animal may initially have little or no information about the location of its goal. The first step for the animal is to employ a strategy that will bring it that information, such as an odor, a visual sighting, a sound, or another cue coming from its goal. Search may be undirected, but even undirected search is usually not random; undirected search should be structured to maximize the amount of environment sampled (Figure 8.18, strategy A). A carnivore in search of prey typically walks in a fairly straight route, allowing it to sample as much of the environment as possible. (Turning or looping movements would likely cause it to cover less ground.) If the search is for an odor, visual, or sonic cue that can travel a considerable distance, then movement in a straight line allows sampling of a large habitat area (Figure 8.18, strategies B and D). Looping or spiral movements (Figure 8.18, strategy C) concentrate search and are effective when the target may be found only if it lies directly in the searcher's pathway.

Once the animal finds a cue, kinetic or tactic orientation mechanisms can kick in (see Figure 8.19). In many cases the first cue is not from the target itself, but rather is information about the environment that helps in directing the search. Animals searching for odor cues, such as a moth searching for sex pheromone, typically fly at right angles to the direction of the wind so that they sample the greatest number of odor plumes possible. If a cue is detected, but its location is uncertain, the animal may then change its direction of travel, turning more frequently (klinokinesis), so that it makes random loops in a fairly well-defined area; this behavior is likely to bring it back into contact with the cue. Some animals use a layered, sequential search, employing a hierarchy of cues. A parasite seeking a specific snail host will follow salinity gradients

FIGURE 8.19
This simple diagram illustrates the basics of an undirected search. The animal, traveling from left to right in the diagram, moves in a more or less straight line through unsuitable habitat. *Source: Bell, W.J., 1991. Search Behaviour, Chapman and Hall, London, Figure 6.1.*

to get to the salinity preferred by that species of snail; it will then use chemical cues from snail mucus to identify the correct species. Specialized types of searching behavior include sit-and-wait predation and the use of search images (both discussed in Chapter 9).

8.6 HOMING

The Concept of a Central Place

Many animals center their activities around a nest or shelter. As these animals range out from their central place, finding the way back becomes a critical problem. Animals "home" when they return to a central place, such as their nest or their territory. Homing often occurs after foraging bouts or other relatively local movements. Homing is therefore a repetitive act, and this distinguishes it from dispersal and migration, which are long-distance movements that usually occur seasonally in an animal's life. (These movements are discussed later in this chapter.) Understanding how animals glean the directional information needed for movement is critical to understanding homing behavior, and as will become apparent, homing may involve most of the toolbox covered so far.

Animals that have a central place are, in a sense, bound to that location; their movements are centered around a fixed spot, and their searches for food and mates are limited by the distance that they can travel from that central place before needing to return. Solving this problem is complex because animals usually search for food by using random circular movements or by traveling between a succession of spots where food is likely to be found (refer to Section 8.5 on search). A bird searching for insect larvae might fly from tree to tree, moving among a large number of trees before finding enough food to take back to its nest. Ultimately, its path probably has involved many loops and turns, and retracing its steps would be difficult indeed for the bird; path integration solves this problem. In contrast, most honeybee foraging trips involve traveling in a straight line (a beeline!) from the nest to the food source; in this case orientation using a compass or landmarks is the typical solution. In yet other examples, some animals leave a pheromone trail that guides the return to the nest. As with searching, animals frequently use such information hierarchically, so that one type of environmental information is important when they are distant from their home, and others become more important as they approach their goal.

Homing in Bee Foraging

Honeybees (*Apis mellifera*) are central-place foragers and may forage several kilometers from their hive, making them a good model species for studying navigation in central-place foragers. Honeybees primarily use path integration in making their way to and from foraging sites. Dance information (see Chapter 7) provides outgoing bees with a distance and direction to be traveled. Flight direction is set by sun compass orientation, and the distance of flight by an internal "odometer" that measures the rate at which visual images flow past the eyes.[36] Other inputs, such as odors, provide supplementary information. Once a route is learned, bees incorporate visual and olfactory landmarks when repeating visits to a foraging site. The return trip is governed by path integration as well, but also may be informed by landmarks.

How do bees incorporate landmarks into their orientation? Two basic models, snapshot memory and cognitive maps, have been proposed. The simplest, and probably correct, model is the snapshot model, which calls for the bee to remember a series of visual images ("snapshots") of the landscape as it passes. The bee also remembers images of particularly prominent landscape features. These images can then be compared with the actual landscape surrounding the bee at any given moment. Ultimately, the bee may be able to use the landmark snapshot information in an "allocentric" manner, projecting its position when it is displaced to an unfamiliar location but still in view of an array of familiar landmarks (i.e., viewing them from a novel angle).

The cognitive map model is more complex and requires the bee to construct a relatively complete neural representation of the landscape based on its experiences while flying. Tests of the cognitive map model demand that displaced bees calculate a novel route home, based on their memory of the landscape map (as humans might). Experimental support of the cognitive map explanation of bee navigation has been put forward, but critics have noted simpler explanations for these results. The allocentric model for snapshot use begins to converge with the cognitive map model but remains a simpler explanation for orientation.

Homing in Pigeons

Homing pigeons (*Columba livia*) carry almost the entire animal movement toolbox with them, and they may have tools yet to be discovered (see Figure 8.20). As will become apparent, the mechanisms with which they navigate are highly redundant. Redundancy is common in many essential biological processes, but pigeons are nonetheless impressive. Ever since ancient times, people have been fascinated by these birds, which find their way home from unfamiliar sites up to thousands of kilometers from their roost. Pigeon races may feature releases of birds from France, for example, which then successfully fly home to sites in England or the Netherlands. The extraordinary reliability of homing pigeons makes them excellent subjects for studies of navigation. This is central-place homing, albeit on a large scale, as it involves learning a location and returning to it. Genetic information is unlikely to be useful because the location of the home may change from generation to generation.

FIGURE 8.20
Homing pigeons exhibit large-scale central place homing, returning to home nests from hundreds of miles away.

237

How do pigeons find their way home when deposited in an unfamiliar location? To do this, they must have "map sense," that is, knowledge about their geographic location, and they must have "compass sense," that is, knowledge about the direction they need to fly from their new location in order to reach their home. In other words, they need to know where they are and what direction to fly. If either information source is disrupted, then homing fails or is delayed.

In familiar surroundings—locations from which pigeons have previously homed or landscapes through which they have flown—landmarks play a predominant role in homing. Pigeons learn visual features of the landscape and use these visual features to determine their current position (map location) relative to their roost.

Pigeons clearly use visual landmarks, based on the fact that pigeons orient better in familiar landscapes even when other sensory inputs, such as olfaction, are eliminated. Direct tests of landmark usage are difficult, however, because experiments manipulating visual landmarks are generally not feasible. One can hardly bulldoze mountains or cut forests as part of an experimental design, and interference with the eyes, such as using contact lenses, may be so much of a general disruption to the pigeon that it confounds tests of landmark perception in orientation.

Landmarks cannot be used in completely new locations—they are not landmarks under such novel conditions— and it turns out that in such situations, pigeons use olfaction to produce a map. In their roost, they associate odors with wind directions. When released, they assess the odor of their new location and extrapolate the map location from their roost-gained knowledge of winds and odors. Familiarity may enhance olfaction-based homing;

pigeons may home better if they have some time to use olfaction to experience their new surroundings prior to release.

Given this ability, it is not surprising that pigeons in visually unfamiliar territory whose sense of smell has been disrupted (by cutting olfactory nerves or treatment of the nasal passages with zinc sulfate solution) have a great deal of difficulty homing. Similarly, homing fails if the roost is blocked from winds and provided with filtered air.

Path integration is one possible orientation mechanism that seems to have been experimentally eliminated as a potential tool used in pigeon homing. There is no evidence that pigeons track their movement when being transported in cages and then use path integration to determine the distance and direction needed to reach home. Tests of this possibility involve eliminating sensory inputs during transport. Because this conclusion relies on a negative result (continued orientation ability in the absence of a sensory input), there is a small possibility that pigeons use an unknown (and consequently experimentally uncontrolled) sensory input for path integration during transport.

The primary compass information of pigeons comes from the position of the sun in the sky. By integrating their internal clock with the sun's position, they compensate for the apparent movement of the sun across the sky. Pigeons whose circadian rhythm is shifted by keeping them under an artificial light:dark cycle display incorrect orientations when released. For example, if artificial "sunrise" comes for the pigeons 6 hours prior to actual sunrise, then their orientation is shifted counterclockwise. If their "sunrise" is later than the actually sunrise, then their orientation shifts clockwise. Like the sun compass of migrating birds, the pigeon's sun compass interacts with a magnetic compass. Under some conditions, experimental modification of the magnetic field around pigeons causes problems in homing.

DISCUSSION POINT: MAGNETIC COMPASSES

The first scientists to argue that pigeons might be using a magnetic compass to find their way home encountered substantial resistance to their ideas. Can you form some hypotheses about why humans have difficulty crediting animals with sensory capacities that we lack? Do you think science is too conservative when new ideas meet with such scepticism, or does demanding tests of new ideas help to maintain scientific integrity?

Experiments with clock shifts and magnetic disruption often do not interfere as much as expected with homing. The reason is the redundancy in pigeon navigation systems; olfactory and landscape information used in establishing their map sense can also be used to correct for compass misinformation. Another possible source of landscape information is the pattern of reflecting sound from mountains. Ocean waves generate infrasound, which then reverberates through the atmosphere. However, the only evidence for use of infrasound by pigeons is the hypothesis that sonic booms from jets may have disrupted pigeon races by impairing the pigeons' hearing, causing them to be unable to use ultrasound in orientation. This is a far cry from the successful use of infrasound in homing, but then again, the pigeon's navigation toolbox probably holds many more surprises.

8.7 MIGRATION

Most migrations are movements of animal populations between seasonally appropriate habitats, but migration can occur between any two habitats and can occur more frequently than the typical annual cycle.[37,38,39] The fact that it is between habitats hints at one important difference between dispersal and migration: dispersal is a one-way ticket, whereas migratory behavior involves a return on the part of descendants, if not the migrants themselves. Migratory behavior is found in many kinds of animals, ranging from monarch butterflies to birds and mammals, and allows animals to take advantage of habitats that are favorable in different seasons, or that are favorable for different stages in an animal's life cycle.[40] Another major difference between dispersal and migration, then, is that the migrating animal must "know" the route.[41] Some animals rely entirely on inherited, genetic information; in migrants such as monarch butterflies and sea turtles, each young animal must find its own way to the next habitat, without guidance from experienced adults. In other species, young animals migrate alongside experienced migrators and can learn route details—distance, direction, stopover points—by following the choices of animals that have already completed the route one or more times. Investigations of migration typically begin with determining the extent to which learned and innate information is used to shape routes.

Traveling from point A to point B requires two basic pieces of knowledge: distance and direction. These topics were covered earlier in this chapter; review those sections for more detail. Direction is normally set by a compass; the sun, stars, or the earth's magnetic field are the typical referents for the compass.[42,43,44] To assess distance, one needs an "odometer"; typically, this is measured as physiological effort. Recall that bees measure the rate at which the landscape moves past their eyes; this suggests there are other types of odometers in migratory animals as well. Again, direction and distance knowledge can be a function of genetics, experience, or both. For longer-distance migrants, stopover points are often essential; they must have features such as food, water, and shelter.[45] Social learning of stopover points in group migrants is probably a key element for migratory success for many birds and mammals.

Not surprisingly, most migratory animals seek favorable conditions for travel—open, unobstructed routes for terrestrial animals, air currents for flying animals, and water currents for swimming animals. Routes that require less effort, even if they are longer, are typically preferred. Migrating birds must balance water loss with flight speed in their choice of flight altitude; higher, stronger wind currents provide a speed advantage but may expose the birds to higher rates of water loss. Generally, birds choose rapid transit over slower routes, even at the expense of water conservation; this means that migrating birds often fly quite high above the ground (see Figure 8.21).

Migratory animals present many of the world's most critical conservation problems. Migrants can be vulnerable in any or all of their three key habitats: their nesting or spawning areas, their transit routes, and their overwintering (or unfavorable season) habitats. Change in any one of the habitats can reduce migratory populations and affect the number of animals observed in all of the habitats, and conservation efforts in one habitat may be at least partially thwarted by failure to manage the species in another. Dams reduce populations of migrating salmon by affecting their transit routes, rainforest harvest decreases wintering habitat for songbirds, and beachfront development eliminates nesting habitat for sea turtles. Understanding the behavioral mechanisms behind migration helps in the development of solutions for the conservation of migratory animals.

Little is understood about the evolution of migration. How did Arctic Terns evolve to "know" to migrate from pole to pole, or how did sea turtles come to migrate from beach to open ocean

FIGURE 8.21
Migratory birds vary widely in the details of how and where they migrate, but the underlying physiology that makes migration possible is similar across avian species. These are sandhill cranes, which migrate from northern Canada to the Gulf of Mexico. *Photos: upper, Jeff Mitton, middle and lower, Ben Pless.*

239

and back? Perhaps exploratory behavior takes some animals to a new habitat, where they are successful, but then seasonal changes or physiological needs drive them back to their old habitat. If animals that carry out these movements have better reproductive success, then whatever genes predispose them to the movements would increase in frequency in succeeding generations. Some scientists have proposed hypotheses involving continental drift or species range expansion/contraction. These verbal models are a beginning, but the evolutionary roots of intricate long-distance migrations are still a mystery.

Bird Migration

We know more about bird migration than other animal migrations.[46,47,48,49,50] Some bird migrations are particularly spectacular because of the immensely long distances traveled. The migratory feats of very small birds, such as hummingbirds, capture the imagination as extraordinary behavioral and physiological feats.

Because bird species vary in the details of their migratory mechanisms, an in-depth analysis of migratory mechanisms in birds would be quite complicated. Fortunately, the basic mechanisms are the same in all birds, so a generalized picture of bird migration is easy to construct.

Seasonal migration in birds, which occurs twice yearly, is closely linked to their biological rhythm (see Chapter 4). There seem to be two distinct stages; there is a preparatory stage followed by the actual initiation of migration. Photoperiod is the most important external cue for the preparatory stage, while the initiation of migration that follows this preparation frequently depends on more variable cues, such as weather conditions. For temperate birds, photoperiod is a dependable cue for a major behavioral shift such as migration. Not only is it linked to more predictable long-term conditions than, say, daily temperature fluctuations, but it also allows the organism to "anticipate" annual change. The physiological shifts that accompany migration or breeding occur over several weeks, before the (average) weather permits such behavior. In the tropics, photoperiod stays much more constant: at the equator there are 12 hours of light and 12 hours of dark year-around, so birds that have migrated to the tropics must rely on cues other than photoperiod to time their annual return to their temperate breeding grounds. Much less is known about migration triggers that involve equatorial habitats.

Two crucial physiological events occur during the preparatory stage: fat deposition and *migratory restlessness*, or *zugunruhe*.[51] Some species fly nonstop (when possible) to their destination, and individuals of these species may lose up to 40% of their body weight during this arduous process; clearly, fat deposition is critical to successful completion of the journey. For species that stop along the way (see following text), it is perhaps less important. For any bird to be able to deposit so much fat, hormonal changes must occur. These are mediated by the pituitary gland (see Chapter 2), which controls metabolism and sex hormone production.

KEY TERM Zugunruhe is migratory restlessness, the behavioral manifestation of the physiological changes leading up to departure on a migration.

As preparation for migration continues, birds increase their activity levels. This has been monitored with electronic perches, and Figure 8.22 shows the dramatic difference after the onset of migratory restlessness. Note that restless birds not only hop more on their perches, but do this all day and all night, in seeming disregard for their usual circadian rhythm.

With the accumulation of sufficient fat reserves, good weather is the proximate cue for initiating migration. Indeed, migratory restlessness does not increase smoothly over time, but reflects weather conditions; restless birds are more restless in good (migration-friendly) weather.

While the image of bird migration is shaped by seeing waterfowl—ducks, geese, and cranes, in particular—migrate during the daytime, most avian migration is nocturnal. This fact is

FIGURE 8.22
(A) In male birds, an annual cycle of testicular development and regression is associated with the migratory cycle. (B) Immediately prior to migration, birds display migratory restlessness, or *zugunruhe*. Before the onset of *zugunruhe*, birds exhibit a typical circadian activity cycle, with much diurnal activity and no nocturnal activity. After the onset of *zugunruhe*, birds remain active throughout the night. *Adapted from* Farner, D.S., 1955, The animal stimulus for migration: experimental and physiologic aspects, in *Recent Studies in Avian Biology,* University of Illinois Press.

especially intriguing because when not migrating, most birds are diurnal. Why shift activity time during migration? Birds face two critical physiological problems in their migration: energy usage and loss of water. Flying at night reduces their thermal load (less heating from the sun, thus less dessication) and may provide some protection from predators at take-off and landing areas.

Although many of these night-flying birds stop on a daily basis during the daylight hours, there is another category of pausing during migration called a *stopover* that differs from these daily stops. For the many migrant birds that lack enough fat to fuel their entire migration, stopover locations become a necessary component of successful migration. Moreover, habitats that provide food, water, and shelter—essential for surviving migration—may be infrequent on the migratory route. Time spent at a stopover can vary greatly among birds within a species; those with lower fat reserves spend a longer time feeding, sometimes as long as several weeks, before moving on. In addition, depending on the migratory route, some species routinely employ stopovers; waterfowl migrating on the central flyway in

OF SPECIAL INTEREST: HAS GLOBAL CLIMATE CHANGE AFFECTED ARRIVAL AND DEPARTURE DATES FOR MIGRATING BIRDS?

For migration to work, the arrival of birds in their seasonal habitat needs to be synchronized with good weather and the availability of food. You can imagine that a bird which arrives too early or too late will be at a serious, perhaps fatal, disadvantage. For birds leaving a tropical habitat and flying to the temperate zone, the timing of their arrival is partly controlled by the timing of their departure and partly by their ability to wait along the route if they encounter unseasonal conditions as they make their way into the temperate zone. Ultimately, though, a predictable set of conditions needs to be present in their nesting habitat. What if global climate change causes flowers to emerge at the "wrong" time, or causes insect populations to peak too early? What if the birds themselves adapt to changing conditions and arrive early, only to find that the flora did not respond in the same way to the altered climate? There is much to be concerned about here, because migratory patterns may be disrupted and populations diminished. Although not all bird species are affected, clear patterns of early arrival have been demonstrated recently in some species (see Figure 8.23). How these changes in timing affect the survival and reproduction of bird species remains to be discovered.[52,53]

FIGURE 8.23

Changes over a 40-year time span in the arrival date of three species of bird at the Rybachy biological station on the Baltic Sea. (Left) Willow warblers; (center) chiffchaffs; (right) pied flycatchers. All three species show substantially earlier arrival dates as global climate conditions have changed. *Adapted from Sparks T.H., Bairlein F., Bojarinova J.G., Huppop O., Lehikoinen E.A., Rainio K., Sokolov L.V., & Walker D. 2005, Examining the total arrival distribution of migratory birds. Global Change Biology, Vol. 11, pp. 22–30.*

North America pass over vast expanses of dry upland and must take advantage of the relatively few low areas with water on their route, using areas such as the Platte River valley in western Nebraska for stopovers (see Figure 8.24). Stopovers also frequently occur prior to crossing important barriers, such as large bodies of water or mountain ranges.

FIGURE 8.24

Sandhill cranes stopping over in the Platte River Valley. *Photo: Ben Pless.*

Typically, information about migratory direction is innate (genetically coded). This is known from displacement experiments, in which naive birds (young birds that have never migrated) are transported a substantial distance, usually hundreds of kilometers, from their birthplace. When seasonal cues trigger migration, these birds follow the same flight direction as they would have had they not been displaced; the displacement does not change their flight direction. Cross-fostering of birds between populations with different migration routes yields the same result; the cross-fostered birds fly the direction inherited from their biological parents, not in the direction flown by their foster parents. The innate directional information is coordinated with a compass: the sun for diurnal migrants and a celestial compass for nocturnal ones. The moon is used as a reference point less commonly. The solar or celestial compass can be complemented by information from a geomagnetic compass.

A series of classic experiments used a planetarium to show that birds, indeed, can orient to the stars in the sky when they fly at night; they are genetically predisposed to learn the night sky that they see as nestlings and use that learned map to determine direction they should fly at migration time. For instance, in the northern hemisphere, the North Star is fixed; if the planetarium constellations rotated around a different star, the young birds used that star as an indication of north when migration time arrived. Nestlings that were not allowed to see the night sky did not learn the stellar map and were unable to identify north.[54]

Migration in Salmon

Salmon undergo a remarkable odyssey during their lives.[55,56,57,58] Young salmon leave the freshwater streams in which they hatched, move thousands of kilometers into the open ocean, and then return as adults to spawn in their natal stream. The question of how they find their way back after many months and across an enormous distance has intrigued scientists for decades.

Starting with an adult fish initiating its journey to freshwater, the migration can be broken down into two phases. The fish must first locate the coastline or general area of the freshwater source it is seeking. The second task is the specific identification of the stream from which it migrated earlier in life. The first task—finding the coastline—does not require the fish to experience the outward swim, but that experience is essential for the second task— locating the natal stream. Thus, it appears the fish learns some of the information about its course on its outward migration.

There are many possible tools that salmon might use to find the coastline. For instance, they contain magnetite and may be able to use geomagnetic information in orientation; such information might allow them to find a coastline in the absence of other information about the location of the coast. As in honeybees and pigeons flying home, salmon need a direction in which to swim, and perhaps an expected travel distance, to achieve the coastline. Salmon have several possible sources of compass information for orienting their travel (see Figure 8.25). These sources include polarized light, the sun, and geomagnetic information. Ocean temperatures and currents may also be informative. The salmon may use path integration, or the direction may simply be genetically encoded. There is little evidence to help sort out the possible mechanisms responsible for salmon migration.

Finding the natal stream requires that the salmon experience that natal stream early in life. When the salmon are close to their natal stream, they identify the appropriate freshwater

FIGURE 8.25
(A) A salmon ladder on the Columbia river. Salmon, migrating upstream, navigate these ladders to return to upriver spawning grounds.
(B) Spawning sockeye salmon in Alaska. *Photos: Michael Breed (A) and USFWS (B).*

source by olfaction. Young salmon, perhaps primed by the hormone thyroxine, imprint on a stream-specific chemical signature. This imprinting may take place during the transition from the parr to the smelt stage in development. (*Parr* and *smelt* are terms for immature salmon.) Understanding the imprinting period is critical to developing strategies for restocking salmon streams from hatchery stock; this is an area of current research interest.

Migration in Monarch Butterflies

Monarch butterflies are an excellent example of a long-distance migrator that uses genetically derived information to determine its migratory route. The butterflies in the southward-moving generation have had no contact with their parents; a different generation migrates each direction. Clearly, in this case there is no opportunity for learning the migratory route. Instead, ideothetic information, combined with orientation based on the sun compass and prevailing winds, must be sufficient to guide the butterflies thousands of kilometers to their overwintering sites in southern California and Mexico (see Figure 8.26).[59]

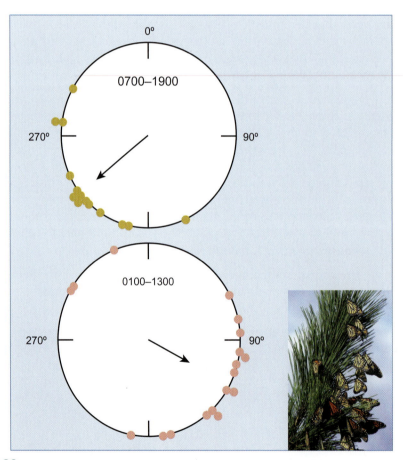

FIGURE 8.26

Animals like the monarch butterfly, which rely on a sun compass to orient their migration, can be fooled by shifting their circadian clocks. The circles in this diagram represent the horizon. The center of the circle is the point at which butterflies were released, and each dot represents the point at which a butterfly disappeared from view as it flew toward the horizon. This type of diagram is the standard way of representing flight direction of animals from a release point. The arrows show the mean flight direction for each treatment. When maintained in photophase from 7:00 am to 7:00 pm (upper circle), the butterflies flew in a southwesterly direction. Shifting the light part of their cycle from 1:00 am to 1:00 pm resulted in flight in a southeasterly direction. This effect of clock shifting suggests that the clock shifted animals misinterpret the position of the sun.[78] *Adapted from Froy, O., Gotter, A.L., Casselman, A.L., & Reppert, S.M. 2003, Illuminating the circadian clock in monarch butterfly migration, Science, Vol. 300:1303–1305. Inset photo: U.S. Fish and Wildlife Service.*

Migration in African Ungulates

In contrast, some of the most picturesque images of animal behavior come from the great migrations of hoofed mammals in Africa. Vast seas of wildebeest, zebra, and giraffes move in concert with the seasonal availability of water and forage (see Figure 8.27). Ungulate migrations tend to follow "traditional routes" along which there is much opportunity for learning, The group dynamic shapes the migratory route and from one year to the next, experienced animals remember routes of travel, locations of watering holes, and destinations. Young animals have the opportunity to learn routes and migratory cues from older animals, and movement in herds provides protection from the numerous predators that live along the migratory route or follow the migrating animals. North American ungulate populations, such as bison, elk, and pronghorn, engaged in similar migrations before they were decimated. Disruption of the African migration—"the last great migration"—by anthropogenic (human caused) changes in the landscape is a key conservation issue; changes in vegetation patterns in East Africa likely mean that the current generations of these species will be the last to engage in one of the most spectacular wildlife sights on the globe.

FIGURE 8.27
Like the great migration of mammals in Africa, some North American mammals also have impressive migrations. For example, Caribou migrate over 1000 km from summer to winter foraging ranges. *Photo: Aaron Collins, U.S. Fish and Wildlife Service.*

245

Turtle Migration

Sea turtles have a migration pattern somewhat similar to that of salmon, with a life cycle that takes them from hatching on beaches to feeding waters in the ocean and back to the beaches to lay the next generation of eggs.[60,61] Adult females migrate to their natal beach and lay eggs in the sand (see Figure 8.28). Hatchlings undertake a short journey to the shoreline (they orient to the horizon), fraught with danger from predators.[62] They then swim to the open ocean, initially orienting into the waves, which near the shore provide an indication of the direction of the open ocean. Their migration then continues for hundreds or thousands of kilometers into the ocean, where they probably use earth's magnetic field and the direction of the currents to find their way into favorable feeding grounds.

Even less is known about how sea turtles accomplish this feat than is known for salmon. Turtles may actually use the geomagnetic field to tell them their location (of course, the open ocean has no landmarks that humans know about); if this is the case, then their magnetoperception system seems to work on a more refined level than that of birds, which is thought to provide directional but not locational information. Some scientists think that sea turtles probably use the geomagnetic field to orient their swim in the deep ocean, but they would also have celestial cues available to them when close to the surface. One classic test for the use of magnetic fields in navigation and orientation consists of fitting magnets

FIGURE 8.28
(A) Sea turtle nest on Florida Atlantic coast marked as part of a conservation effort. (B) Emerging Kemp's Ridley sea turtle. *Photos: Michael Breed (A) and U.S. Fish and Wildlife Service (B).*

to an animal's head; the magnet will interfere with geomagnetic cues. However, just such an experiment with green turtles migrating between the open Atlantic and Brazil found that the turtles with magnets performed as well as their counterpart controls. Another (untested) hypothesis is that the directional vector of sea turtles is determined genetically. Also intriguing but unexplained is the ability of the turtles to return to their natal beach over their vast travel distance. Do they imprint, as salmon do, on olfactory features in the water? Or is the location pinpointed using geomagnetic information?

Migratory routes and distances vary among species of sea turtles. Green sea turtles migrate primarily along the coasts from nesting to feeding grounds. However, turtles from some populations will travel approximately 2,100 km (1,300 miles) across the Atlantic Ocean from the Ascension Island nesting grounds to the Brazilian coast feeding grounds. Loggerheads leave foraging areas and travel on breeding migrations that range in distance from a few to thousands of kilometers. Leatherbacks have the longest migration of all sea turtles. They have been found more than 4,800 km (3,000 miles) from their nesting beaches.

8.8 DISPERSAL

Dispersal is a topic in both behavior and ecology; many of the ecological principles concerning dispersal apply to both plants and animals, but of course here animal dispersal is of primary interest.[63,64,65,66,67,68] An ecologist is most likely concerned with the numbers of animals that disperse and how movement into and out of populations affects the growth of those populations. Animal behaviorists look at dispersal on a different scale, asking different questions: What causes an animal to decide to disperse? Is dispersal behavior linked to age, sex, or behavioral history? Within a population are there behavioral syndromes that make some animals more likely to disperse than others?

Dispersal allows animals to avoid competition, avoid inbreeding,[69] and to colonize new habitats. Animals disperse by leaving their natal area and finding new territories or home ranges. The dispersing animal, like the migrating one, is attempting to improve its lot in life by finding a suitable habitat. (As will become apparent, sometimes it is attempting to improve its lot in life by leaving increasingly hostile adult relatives!) Unlike migration, dispersal is usually a one-way trip to an unknown destination. Perhaps the greatest difficulty for a dispersing animal is finding unoccupied suitable habitat. In most species, all the best locations will already be occupied, and a disperser is faced with an uncertain future, including the possibility of starvation in an unsuitable habitat. For the vast majority of animal species, dispersal is the simple act of finding suitable habitat after hatching from an egg. Indeed, the choice of a location for oviposition is one of the most important decisions most mothers make.

Who Disperses?
PARENTS VERSUS OFFSPRING

Dispersers in birds and mammals are likely to be adolescents or young adults. Lacking the experience or strength to wrest a favored location from their parents, they leave, or are driven, from parental territories. Female dispersers are unlikely to remain without suitable habitat for long, as males with territories or harems will typically accept multiple females into their area. A striking counterexample to the view that younger animals always suffer the risks associated with dispersal is the honeybee, in which the reigning queen in a colony leaves with a swarm; this allows one of her daughters to inherit the nest, food, and the portion of the worker force that she leaves behind (see Figure 8.29).

FIGURE 8.29
Animals disperse at many scales, from monarch butterlies, whose migration disperses them over large georaphic areas, to elk, who may range hundreds of kilometers, to honeybees, which may disperse from a few meters to a few hundred meters.

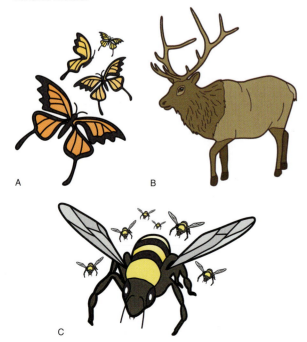

A   B

C

MALES VERSUS FEMALES

In most mammals, males are the dispersing sex,[70] whereas in most birds, females are more likely to disperse.[71] The general argument is that mammals tend to be polygynous—live in social groups in which one male is associated with many females. Wild horses are a familiar example of this type of system. (See Chapter 11 on mating systems for more details on polygyny.) Birds, on the other hand, tend to be monogamous and to have territorial males, leading to a situation in which females disperse to search out male territory holders. When exceptions occur, such as brown jays, in which males are more likely to disperse,[72] or hamadryas baboons, in which females are more likely to disperse,[73] complex social dynamics seem involved in the evolution of these exceptions to the general rules about gender-specific dispersal.

The example of prairie dogs is consistent with the conventional wisdom about mammalian dispersal. Prairie dogs live in dense colonies, which are subdivided into burrow systems occupied by families (coteries; see Figure 8.30). Two species of prairie dog, Gunnison's and the black-tailed prairie dog, have been well studied; the biology of the species differs, but the overall pattern of dispersal behavior and colony structure is similar between the species. Young males are the primary dispersers and can move a few kilometers to join a new colony; young females remain in their natal colony. This model of male dispersal and female philopatry (a tendency for an animal to remain close to, or return to, the animal's birthplace) facilitates inbreeding avoidance.[74] Prairie dogs tend to decimate the vegetation in their colony; because of this, sites on the edges of the colony may be favored because the forage is better. Perhaps for this reason, some dispersal of older animals also takes place, with older males and females equally likely to move.

Young dispersing male mammals sometimes form coalitions with other males in similar circumstances, forming "bachelor herds." Males in bachelor herds benefit from shared vigilance and shared information about feeding and watering locations, and also serve as foils in practice combat. Bachelor herds are common in ungulates, such as horses, mountain sheep, and deer, but are also found in marine mammals and in lions.

FIGURE 8.30
Young male prairie dogs disperse from their natal group; females remain, so prairie dog colonies are matrilineal. *Photo: Jeff Mitton.*

247

Mammalian females may leave social groups to find better locations for the birth and feeding of their young. This type of dispersal may protect the young from potential infanticide by adult males, and is well known in a variety of primates. Maternal dispersal is usually reversible, and once the juvenile is old enough, the mother and infant return to the social group. In some cases, though, dispersing pregnant females are responding to competitive conditions and, while pregnant, search for new suitable habitat.

Why Disperse?

COMPETITION DISPERSAL

Competition dispersal is probably a major cause of dispersal in plants; light, root space, and nutrients are depleted or dominated by the parental plant. In animals, it also plays a key role in many species and recurs as a theme in the discussion of dispersal. In competition dispersal, movement is typically an uneven affair, with gender and age biases. Typically, it is the young males that bear the brunt of the risk of dispersal. As a juvenile male approaches reproductive age, it may threaten the status of the dominant male in the group, or it may simply be an extra mouth to feed.

KIN COMPETITION AVOIDANCE

Kin competition avoidance is dispersal or other behavior that keeps sibs or parents and their offspring from competing with each other. Similarly, natural selection may favor animals that disperse and thus avoid competition with kin. This point is important to consider even though it is difficult to discriminate between avoidance of competition with kin and inbreeding avoidance (see later section). In both cases, the result is essentially the same: close kin end up not living near one another.

INFANTICIDE VERSUS DISPERSAL

As an aside, infanticide, the killing of a young animal by an older animal, is a common response to habitat depletion in rodents and can be thought of as the result of resource competition.[75,76] Infanticide is performed by females and males that are unrelated to or are unfamiliar with the pups (see the section on kin recognition in Chapter 13). In crowded populations of related mice and voles, infanticide is much more likely than it is in less-crowded circumstances. Infanticide is an alternative to dispersal, albeit an unpleasant one, which maintains resource conditions for the remaining animals in the population.

INBREEDING AVOIDANCE

Another result of dispersal is inbreeding avoidance. Inbreeding avoidance is any behavioral or physiological mechanism that prevents closely related animals from mating. If young of one sex remain near their birthplace and the other sex disperses fairly widely, the chances of inbred matings become low. For inbreeding avoidance, it does not matter which sex does the dispersing. Dispersal can help to avoid kin associations; meadow voles are more likely to disperse from plots occupied by siblings than from plots occupied by nonsiblings.

DISPERSAL FOR COLONIZATION

Colonizing species succeed by establishing populations in new habitats. Inseminated females are the most powerful dispersers for colonization; by mating before dispersing, they eliminate the need to find a mate in the new habitat (or to disperse in pairs). Invasive insects are particularly effective at exploiting this strategy to gain footholds in new locations; the red imported fire ant, which has successfully invaded ecosystems in the southeastern United States, Arizona, and California, is an excellent example of a species that combines strong dispersal and colonization abilities. Among mammals, Norway rats are highly effective dispersers and colonizers of new habitat. In the western United States, changing land-use patterns and perhaps climate change have opened habitats for house finches, which have proven to be good dispersers into previously unoccupied areas.

Recent experience with reintroductions of mammalian carnivores illustrates the complexity of thinking about dispersal. Wolves, which have been reintroduced into the greater Yellowstone ecosystem in Wyoming and Montana, have shown the ability to move hundreds of miles relatively rapidly. Thus, the occasional wolf ends up far from the introduction site and well away from the intended population area. Ultimately, this is promising for re-establishment of wolves in the western United States, but in the short term this behavior fuels the anxiety of some ranchers and residents of rural areas.

DISCUSSION POINT: BEHAVIORAL SYNDROMES AND DISPERSAL

In Chapter 6 on cognition, we talked about behavioral syndromes—sets of behavior that are relatively constant within an animal but vary among animals in a population. Many documented behavioral syndromes range along a continuum from shy to bold. Can you postulate adaptive values for both shyness and boldness in dispersal?

SUMMARY

The study of animal movements in the environment is a journey in itself, starting with the integration of simple sensory inputs and basic movements. For many animals, the ability to discover and orient to food, shelter, and mates is all that is needed for evolutionary success. Kineses, or undirected movements, and taxes, or directed movements, provide the basic mechanisms by which many animals find their appropriate habitats. The discovery of the direction or location of resources is often key to navigating to those resources. Search behavior typically begins with undirected or looping movements and, as cues are gained from the environment, shifts to goal-directed movements. Sensory systems, such as stereoptic vision, are needed to allow goal-oriented movements.

For some animals, navigational needs extend far beyond these abilities. Navigation sometimes relies primarily on genetically sourced information, the product of generations of evolutionary trial and error. In other cases, learning plays a larger part in navigation, giving the animal the flexibility to accommodate its current location and to exploit information obtained from other population members. Path integration and compasses are often used in homing. Finding home requires using sophisticated mechanisms such as path integration, landmarks, and snapshot memory.

Migration, moving seasonally between widely separated habitats, gives animals the opportunity to exploit more than one habitat during their life cycle. Migration means that animals must use highly intricate compass and odometer mechanisms. Migratory routes may be innate or learned; again, many animals rely on a mixture of these two types of information to direct their movements. The diversity of organisms that migrate and the vulnerability of the migratory lifestyle make migrating animals a particularly important conservation concern. Finally, dispersal is what allows animals to establish new populations and avoid the negative outcomes of inbreeding and competition. By dispersing, animals avoid competition and may colonize new habitats.

STUDY QUESTIONS

1. As part of an experiment, pigeons are kept in a room where the lights are controlled by a timer. For a week, the timer is adjusted so that the lights go on 3 hours before the natural sunrise at this location and go off 3 hours before the natural sunset. In this particular experiment, when released some distance from their roost, these "clock-shifted" pigeons fly in the same direction as control pigeons. Are they using a sun-compass? Why or why not?
2. A repeating theme in this book has been the interaction of genetic and environmental information in shaping behavior. As we discussed migration in this chapter, you saw examples of genetic shaping of migratory movements as well as learned migration information. Under what circumstances would you expect a species to rely on genetic information for migration and under what circumstances would learning be dominant?
3. What are the problems that animals face in orientation at night, and how have they overcome those problems? Subdivide your answer by types of orientation, such as search for food, homing, and migration. Why do most bird species migrate at night?
4. How would you test the hypothesis that humans use a magnetic compass in their orientation?
5. What circumstances would favor males as the dispersing sex? Females? Can you model a circumstance under which both sexes would be equally likely to disperse?

Further Reading

Åkesson, S., Hedenström, A., 2007. How migrants get there: migratory performance and orientation. Bioscience 57, 123–133.

Baker, R.R., 1978. The Evolutionary Ecology of Animal Migration. Hodder and Stoughton, London.

Bell, W.J., 1991. Search Behaviour. Chapman and Hall, London.

Clobert, J., Danchin, E., Dhondt, A.A., Nichols, J.D. (Eds.), 2001. Dispersal. Oxford University Press, Oxford, 480 pp.

Berthold, P., 2001. Bird Migration: A General Survey, second ed. Oxford University Press, Oxford, 272 pp.

Berthold, P, Gwinner, E., Sonnenschein, E. (Eds.), 2003. Avian Migration. Springer, Berlin

Dingle, H., 1996. Migration: The Biology of Life on the Move. Oxford University Press, New York.

Gauthreaux, S.A. Jr., 2010. Bird Migration, In: Michael D. Breed and Janice Moore (Eds.), Encyclopedia of Animal Behavior, Academic Press, Oxford, Vol. 2, pp. 211–219.

Lohmann, C.M.F., Lohmann, K.J., 2010. Sea Turtles: Navigation and Orientation, In: Michael D. Breed and Janice Moore (Eds.), Encyclopedia of Animal Behavior, Academic Press, Oxford, Vol. 2, pp. 101–107.

Pulido, F., 2007. The genetics and evolution of avian migration. Bioscience 57, 165–174.

Ramenofsky, M., Wingfield, J.C., 2007. Regulation of migration. Bioscience 57, 135–143.

Roff, D.A., Fairbairn, D.J., 2007. The evolution and genetics of migration in insects. Bioscience 57, 155–164.

Notes

1. Lindeyer, C.M., Reader, S.M., 2010. Social learning of escape routes in zebrafish and the stability of behavioural traditions. Anim. Behav. 79, 827–834

2. Bell, W.J., 1991. Search Behaviour. Chapman and Hall, London.

3. Bingman, V.P., Cheng, K., 2005. Mechanisms of animal global navigation: comparative perspectives and enduring challenges. Ethol. Ecol. Evol. 17 (4), 295–318.

4. Phillips, J.B., Borland, S.C., 1992. Behavioural evidence for use of light-dependent magnetoreception mechanism by a vertebrate. Nature 359, 142–144.

5. Sandberg, R., Backman, J., Moore, F.R., Lohmus, M., 2000. Magnetic information calibrates celestial cues during migration. Anim. Behav. 60, 453–462.

6. Sparling, D., Krapu, G., 1994. Communal roosting and foraging behavior of staging sandhill cranes. Wilson Bull. 106, 62–77.

7. Ferrari, J., Godfray, H.C.J., Faulconbridge, A.S., Prior, K., Via, S., 2006. Population differentiation and genetic variation in host choice among pea aphids from eight host plant genera. Evolution 60 (8), 1574–1584.

8. Dambrowski, H.R., Linn, C., Berlocher, S.H., Forbes, A.A., Roelofs, W., Feder, J.L., 2005. The genetic basis for fruit odor discrimination in *Rhagoletis* flies and its significance for sympatric host shifts. Evolution 59, 1953–1964.

9. Wiltschko, R., Wiltschko, W., 1999. The orientation system of birds—I. Compass mechanisms. J. Ornithol. 140 (1), 1–40.

10. Whishaw, I.Q., Brooks, B.L., 1999. Calibrating space: exploration is important for allothetic and idiothetic navigation. Hippocampus 9, 659–667.

11. Biegler, R., 2000. Possible uses of path integration in animal navigation. Anim. Learning Behav. 28 (3), 257–277.

12. Rossel, S., 1983. Binocular stereopsis in an insect. Nature 302 (5911), 821–822.

13. Kral, K., Vernik, M., Devetak, D., 2000. The visually controlled prey-capture behaviour of the European mantispid *Mantispa styriaca*. J. Exp. Biol. 203 (14), 2117–2123.
    Kral, K., 2003. Behavioural-analytical studies of the role of head movements in depth perception in insects, birds and mammals. Behav. Processes 64 (1), 1–12.

14. Steck, K., Markus, K., Hansson, B.S., 2010. Do desert ants smell the scenery in stereo? Anim. Behav. 79, 939–945

15. Kral, K., 2003. Behavioural-analytical studies of the role of head movements in depth perception in insects, birds and mammals. Behav. Processes 64 (1), 1–12.

16. Willis, M.A., Avondet, J.L., 2005. Odor-modulated orientation in walking male cockroaches *Periplaneta americana*, and the effects of odor plumes of different structure. J. Exp. Biol. 208 (4), 721–735.

17. Grasso, F., Basil, J., 2002. How lobsters, crayfishes , and crabs locate sources of odor: current perspectives and future directions. Curr. Opin. Neurobiol. 12, 721–727.

18. Thar, R., Kuhl, M., 2003. Bacteria are not too small for spatial sensing of chemical gradients: an experimental evidence. Proc. Natl. Acad. Sci. U.S.A. 100, 5748–5753.

19. Mitchell, J.G., Kogure, K., 2006. Bacterial motility: links to the environment and a driving force for microbial physics. FEMS Microbiol. Ecol. 55 (1), 3–16.

20. Begall, S., Cerveny, J., Neef, J., Vojtech, O., Burda, H., 2008. Magnetic alignment in grazing and resting cattle and deer. PNAS 105, 13451–13455.

21. Burda, H., Begall, S., Cerveny, J., Neef, J., Nemec, P., 2009. Extremely low-frequency electromagnetic fields disrupt magnetic alignment of ruminants. PNAS 106, 5708–5713.

22. Dingle, H., 1962. The occurrence of correcting behavior in various insects. Ecology 43, 727–728.

23. Vickers, N.J., 2000. Mechanisms of animal navigation in odor plumes. Biol. Bull. 198, 203–212.

24. Capaldi, E.A., Smith, A D., Osborne, J.L., Fahrbach, S.E., Farris, S.M., Reynolds, D.R., et al. 2000. Ontogeny of orientation flight in the honeybee revealed by harmonic radar. Nature 403, 537–540.

25. Menzel, R., Geiger, K., Joerges, J., Muller, U., Chittka, L., 1998. Bees travel novel homeward routes by integrating separately acquired vector memories. Anim. Behav. 55, 139–152.

26. Etienne, A.S., Maurer, R., Berlie, J., Reverdin, B., Rowe, T., Georgakopoulos, J., et al. 1998. Navigation through vector addition. Nature 396 (6707), 161–164.

27. Seguinot, V., Cattet, J., Benhamou, S., 1998. Path integration in dogs. Anim. Behav. 55, 787–797.

28. Benhamou, S., 1997. Path integration by swimming rats. Anim. Behav. 54, 321–327.

29. Dickerson, J., Dyer, F., 1996. How insects learn about the sun's course: alternative modelling approaches. In: Maes, P., Mafaric, J.J., Meyer, J., Pollock, J., Wilson, S. (Eds.), Animals to Animats. vol. 4. MIT Press, Cambridge, MA.

30. Dyer, F.C., 1991. Bees acquire route-based memories but not cognitive maps in a familiar landscape. Anim. Behav. 41, 239–246.
    Dyer, F.C., Berry N.A., Richard, A.S., 1993. Honeybee spatial memory—use of route-based memories after displacement. Anim. Behav. 45 (5), 1028–1030.

31. Wiltschko, R., Wiltschko, W., 1999. The orientation system of birds—I. Compass mechanisms. J. Ornithol. 140 (1), 1–40.

32. Collett, M., Collett, T.S., Wehner, R., 1999. Calibration of vector navigation in desert ants. Curr. Biol. 9 (18), 1031–1034.

33. Gould, J., 1986. The locale map of honey bees: Do insects have cognitive maps? Science 232, 861–863.

34. Menzel, R., De Marco, R.J., Greggers, U., 2006. Spatial memory, navigation and dance behaviour in *Apis mellifera*. J. Comp. Physiol. A Neuroethol. Sens. Neural. Behav. Physiol. 192 (9), 889–903.

35. Bell, W.J., 1991. Search Behaviour. Chapman and Hall, London.

36. Si, A, Srinivasan, M.V., Zhang, S., 2003. Honeybee navigation: properties of the visually driven 'odometer'. J. Exp. Biol. 206, 1265–1273.

37. Dingle, H., 1996. Migration: The Biology of Life on the Move. Oxford University Press, New York.

38. Dingle, H., Alistair Drake, V., 2007. What is migration? Bioscience 57, 113–121.

39. Fitzsimons, J.M., Parham, J.E., Nishimoto, R.T., 2004. Similarities in behavioral ecology among amphidromous and catadromous fishes on the oceanic islands of Hawai'I and Guam. Environ. Biol. Fishes 65, 123–129.

40. Baker, R.R., 1978. The Evolutionary Ecology of Animal Migration. Hodder and Stoughton, London.

41. Åkesson, S., Hedenström, A., 2007. How migrants get there: migratory performance and orientation. Bioscience 57, 123–133.

42. Wiltschko, W., Weindler, P., Wiltschko, R., 1998. Interaction of magnetic and celestial cues in the migratory orientation of passerines. J. Avian Biol. 29 (4), 606–617.

43. Weindler, P., Liepa, V., Wiltschko, W., 1998. The direction of celestial rotation affects the development of migratory orientation in Pied Flycatchers, *Ficedula hypoleuca*. Ethology 104, 905–915.
    Weindler, P., Bohme, F., Liepa, V., Wiltschko, W., 1998. The role of daytime cues in the development of magnetic orientation in a night-migrating bird. Behav. Ecol. Sociobiol. 42, 289–294.

44. Mouritsen, H., 1998. Redstarts, *Phoenicurus phoenicurus*, can orient in a true zero magnetic field. Anim. Behav. 55, 1311–1324

45. Aborn, D., 2004. Activity budgets of summer tanagers during spring migratory stopover. Wilson Bull. 116, 64–68.

46. Moore, F.R., 1987. Sunset and the orientation behaviour of migrating birds. Biol. Rev. 62, 65–86.

47. Holberton, R.L., 1993. An endogenous basis for differential migration in the dark-eyed junco. Condor 95, 580–587.

48. Berthold, P., 2001. Bird Migration: A General Survey, second ed. Oxford University Press, 272p.
    Berthold, P., Gwinner, E., Sonnenschein, E. (Eds.), 2003. Avian Migration. Springer, Berlin.

49. Greenberg, R., Marra, P.P. (Eds.), 2005. Birds of Two Worlds: The Ecology and Evolution of Migration. Johns Hopkins University Press, Baltimore.

50. Pulido, F., 2007. The genetics and evolution of avian migration. Bioscience 57, 165–174.

51. Ramenofsky, M., Wingfield, J.C., 2007. Regulation of migration. Bioscience 57, 135–143.

52. Lehikoinen, E., Sparks, T.H., Zalakevicius, M., 2004. Birds and climate change. Adv. Ecol. Res. 35, 1–31.

53. Sparks, T.H., Bairlein, F., Bojarinova, J.G., Huppop, O., Lehikoinen, E.A., Rainio, K., et al., 2005. Examining the total arrival distribution of migratory birds. Glob. Change Biol. 11, 22–30.

54. Emlen, S.T., 1970. Celestial rotation: Its importance in the development of migratory orientation. Science 170 (3963), 1198–1201.

55. Berman, C.H., Quinn, T.P., 1991. Behavioural thermoregulation and homing by spring Chinook salmon, *Oncorhynchus tshawytscha* (Walbaum), in the Yakima River. J. Fish Biol. 39. 301–312.

56. Thorrold, S.R., Latkoczy, C., Swart, P.K., Jones, C.M., 2001. Natal homing in a marine fish metapopulation. Science 291, 297–299.

57. Scholz, AT., Horrall, R.M., Cooper, J.C., Hasler, A.D., 1976. Imprinting to chemical cues—basis for home stream selection in salmon. Science 192 (4245), 1247–1249.

58. Dittman, A.H., Quinn, T.P., Nevitt, G.A., 1996. Timing of imprinting to natural and artificial odors by coho salmon (*Oncorhynchus kisutch*). Can. J. Fish. Aquat. Sci. 53 (2), 434–442.

59. Roff, D.A., Fairbairn, D.J., 2007. The evolution and genetics of migration in insects. Bioscience 57, 155–164.

60. Lohmann, K.J., Hester, J.T., Lohmann, C.M.F., 1999. Long-distance navigation in sea turtles. Ethol. Ecol. Evol. 11 (1), 1–23.
    Lohmann, K.J., Lohmann, C.M.F., 1998. Migratory guidance mechanisms in marine turtles. J. Avian Biol. 29 (4), 585–596.

61. Papi, F., Luschi, P., Akesson, S., Capogrossi, S., Hays, G.C., 2000. Open-sea migration of magnetically disturbed sea turtles. J. Exp. Biol. 203 (22), 3435–3443.

62. Goff, M., Salmon, M., Lohmann, K.J., 1998. Hatchling sea turtles use surface waves to establish a magnetic compass direction. Anim. Behav. 55, 69–77.

63. Waser, P.M., 1985. Does competition drive dispersal? Ecology 66, 1170–1175.

64. Lena, J.-P., Clobert, J., de Fraipont, M., Lecomte, J., Guyot, G., 1998. The relative influence of density and kinship on dispersal in the common lizard. Behav. Ecol. 9, 500–507.

65. Dobson, F.S., 1982. Competition for mates and predominantly juvenile male dispersal in mammals. Anim. Behav. 30, 1183–1192.
    Dobson, F.S., 1985. Multiple causes of dispersal. Am. Nat. 126, 855–858.

66. Bowen, B., Koford, R. R., Vehrencamp, S.L., 1989. Dispersal in the communally breeding groove-billed Ani (*Crotophaga sulcirostris*). Condor 91, 52–64.

67. Clobert, J., Danchin, E., Dhondt, A.A., Nichols, J.D. (Eds.), 2001. Dispersal. Oxford University Press, Oxford, 480 p.

68. Holekamp, K., 1984. Natal dispersal in Belding's ground squirrels (*Spermophilus beldingi*). Behav. Ecol. Sociobiol. 16, 21–30.

69. Bollinger, E.K., Harper, S.J., Barrett, G.W., 1993. Inbreeding avoidance increases dispersal movements of the meadow vole. Ecology 74, 1153–1156.

70. Greenwood, P.J., 1980. Mating systems, philopatry and dispersal in birds and mammals. Anim. Behav. 28, 1140–1162.

71. Clarke, A.L., Saether, B.E., Roskaft, E., 1997. Sex biases in avian dispersal: a reappraisal. OIKOS 79 (3), 429–438.

72. Williams, D.A., Rabenold, K.N., 2005. Male-biased dispersal, female philopatry, and routes to fitness in a social corvid. J. Anim. Ecol. 74, 150–159.

73. Hammond, R.L., Handley, L.J.L., Winney, B.J., Bruford, M.W., Perrin, N., 2006. Genetic evidence for female-biased dispersal and gene flow in a polygynous primate. Proc. R. Soc. Lond., B, Biol. Sci. 273 (1585), 479–484.

74. Lucas, J.R., Waser, P.M., Creel, S.R., 1994. Death and disappearance: estimating mortality risks associated with philopatry and dispersal. Behav. Ecol. 5, 135–141.

75. Morelli, T.L., King, S.J., Pochron, S.T., Wright, P.C., 2009. The rules of disengagement: takeovers, infanticide, and dispersal in a rainforest lemur. Propithecus edwardsi. Behavior 146, 499–523.

76. Sherman, P.W., 1981. Reproductive competition and infanticide in Belding's ground squirrels and other animals. In: Alexander, R. D., Tinkle, D. (Eds.), Natural Selection and Social Behavior. Chiron Press, New York.

77. Zeil, J., 1998. Homing in fiddler crabs (*Uca lactea annulipes* and *Uca vomeris*: Ocypodidae). J. Comp. Physiol. A–Sens. Neural Behav. Physiol. 183 (3), 367–377.

78. Froy, O. Gotter, A.L., Casselman, A.L., Reppert, S.M., 2003. Illuminating the circadian clock in monarch butterfly migration. Science 300, 1303–1305.

Foraging

253

LEARNING OBJECTIVES

Studying this chapter should provide you with the knowledge to:

- See how dietary requirements, natural selection for dietary specialization, and the evolution of foraging behavior interact.

- Understand the diversity of dietary strategies employed by animals.

- Be able to use optimality theory to test and refine hypotheses about important resources—the currencies of animal lives. In the study of foraging, this is frequently energy.

- Understand the limits of optimality theory; animals are not expected to optimize every behavior. These constraints are the same as those that make adaptation and natural selection works in progress rather than finished portraits.

- Apply the marginal value theorem as an economic concept that makes predictions about where to forage.
- Realize that risk sensitivity, along with whether an animal is risk-averse or risk-prone, can influence foraging behavior.

9.1 INTRODUCTION

It is a truism that all animals have to eat. After all, the act of eating is part of what distinguishes animals from other living things. Animals do not photosynthesize; many other types of organisms capture energy from the sun and use solar energy to turn basic elements such as carbon, oxygen, hydrogen, and nitrogen into organic molecules. Animals then exploit these molecules for their own nutritional needs. Foraging is collecting food, and the fact that most potential food items resist being eaten makes this a dynamic and complex topic.

Foraging involves dietary choice, and this is the first topic of discussion in this chapter. Diet depends on what can be consumed, on the nutritional needs of the forager, and ultimately, on specialized adaptations for collecting and digesting specific types of food. Obtaining energy is usually the primary goal of foraging, but essential nutrients must also be consumed, and animals sometimes make surprising dietary selections to gain needed nutrients. Selection to avoid competition may drive animals to choose foods that are relatively undigestible, and this may result in behavioral adaptations related to digestion, such as passing digestive symbionts among animals or gardening fungi.

Foraging behavior defines an animal's trophic niche, such as carnivory, within its ecological community. Foraging specializations involve behavioral adaptations that coordinate with morphology and physiology. Big cats, such as lions and tigers, have stealthy predatory behavior and strong pursuit abilities, adaptations that are accompanied by cryptic coloration, sharp claws, slicing teeth, short digestive tracts, and a digestive physiology evolved to handle a diet high in protein and fats. Herbivores are no less well equipped with behavior and physiology appropriate to their diets.

Once dietary choice is established, food must be located. The first step is search, and the second step is handling the food. For herbivores, food handling is often as simple as consumption, but plant material can be difficult to digest and sometimes extra behavioral steps prepare the food for digestion. For carnivores, handling requires first capturing and then subduing and killing the prey. Complex games of "cat and mouse" have evolved as selection favors prey that are hard to locate, catch, and subdue. At the same time, selection favors countermeasures in predators. These evolutionary arms races are central to understanding foraging; over evolutionary time, adaptations that make foragers more efficient are met with counteradaptations of food species.

Food is not usually evenly distributed in the environment; it occurs in patches, and animals must make decisions about when to leave patches as they become depleted. A patch may be a small meadow in which a certain type of plant is found, the space under a rock in a stream, or a flock of birds. Optimal foraging theory, mathematical models that describe how an animal gains the maximum possible yield from its foraging efforts, yields important tools for understanding discovery and exploitation of food. For a sit-and-wait predator such as a snake or an orb-weaving spider, a patch is where prey are likely to walk or fly. Successful foragers must be able to find patches, exploit food in that patch, and then assess when the patch is depleted and move on to more profitable locations. From digestive physiology to changing dietary needs over a lifetime, and from the work of natural selection in predator–prey interactions to the information we glean from fossil teeth, foraging is an area in which all of Tinbergen's questions can be clearly identified, if not always answered.

CASE STUDY

A Cheetah's Prey Choice

How do animals choose what to eat? For most animals, and for predators like cheetahs in particular, foraging is not just one decision, but a series of discrete decisions; in this case study, those decisions involve whether to hunt and what to hunt. Discrete choice models, borrowed from economists, give us a way to ask how animals make these decisions. Cooper et al. did this with cheetah foraging in the Serengeti.[1] This technique calculates the probability of the choice to hunt, and given that choice, the probability of choosing a particular prey species in the context of prey on offer, environmental conditions, and predator conditions. Cheetahs hunt diurnally (during the day) and in the open, so observational data can be collected with some accuracy. Cooper and coworkers tested the following hypotheses about conditions that should positively or negatively influence the decision to hunt:

FIGURE 9.1
Two male cheetahs feast on a wildebeest. *Photo: Sarah Durant, ZSL.*

1. The presence of preferred prey (Thomson's gazelle in particular)—a positive influence
2. The presence of competitors (e.g., lions and hyenas)—a negative influence
3. Energy needs as reflected in the presence of cubs or hunger level (indicated by belly size)—a positive influence
4. Cover availability—a positive influence
5. Time of day (preferred times are 700–800 hours and 1600–1800 hours)

Additional predictions emerged from the facts that preferred prey are seasonal and male cheetahs hunt larger species than females do (see Figure 9.1).

Cooper and colleagues found broad agreement with their hypotheses, but there were some interesting nuances. For instance, cheetahs were reluctant to hunt in the presence of lions, but hyenas had no effect. Thomson's gazelles motivated hunting, but other prey such as wildebeest and Grant's gazelles did not increase the probability of hunting. The presence of cubs increased the probability of hunting, but hunger level did not. Moreover, although hyenas did not affect the decision to hunt, they did affect decisions about what prey to take, with Grant's gazelles increasing in preference. These and other results show us both the complexity of decision making by predators and the extent to which models can help us appreciate those decisions.

255

9.2 DIET CHOICE AND FOOD SELECTION

This section provides a brief overview of animal nutrition. Foraging behavior is the bridge between an animal's physiological needs and the potential foods in the environment that will meet those needs. A basic conclusion from this section is that a diverse diet is more likely to meet an animal's full nutritional needs. Given the advantages of a diverse diet, why do animals specialize? In some cases, natural selection favors dietary specialization because focused adaptations to collect specific food types increase foraging efficiency.

What Is an Adequate Diet?

Foraging behavior must fulfill two goals. The first is to gain enough energy to support growth, development, and reproduction. The second is to gain the right nutrients. The obvious energy-rich foods are carbohydrates and fats. Protein can also be used as an energy source and has the added benefit of being composed of nitrogen-containing amino acids. Thus, the bulk of most animal diets consists of carbohydrates, fat, and protein, and most of this chapter is about behavior that leads an animal to the energy it needs. In addition, for growth and development, an animal's diet must be much broader than just energy-rich sources; essential macro- and micronutrients are also targeted in foraging. For animals with carbohydrate-rich or fat-rich diets—including animals that eat foliage, fruits, seeds, and nuts—special strategies may be needed to gain enough dietary nitrogen. Sources of minerals such as sodium, potassium, and phosphorus are also important. For animals with protein-rich diets—carnivores and animals that eat decaying flesh—nutrients such as vitamin A may be in short supply. Behavioral strategies play important roles for animals as they seek to fulfill their dietary needs.

Essential Nutrients

All animals have specific organic molecules or minerals that must be in their diets for survival. Some such nutrients, particularly minerals such as iron, sodium, magnesium, and copper, are essential for all animal life. Nearly all animals require specific amino acids in their diets, but the list of essential amino acids differs among types of animals. Vitamins, such as ß-carotene (vitamin A), are also essential to many animals. Essential dietary components can be surprising; insects need cholesterol in their diet because they lack the ability to synthesize this compound. From a behavioral point of view, essential dietary elements are important because sometimes animals go to great lengths, and perhaps considerable risk, to gain needed nutrients. Elemental nutrient deficiency may not be all bad; for instance, phosphorus-limited *Daphnia* were also poorer hosts for a pathogenic bacterium.[2]

The sodium-limited world of herbivores has selected for some odd behaviors. Sodium is particularly valuable for herbivores because plant tissues have relatively little sodium, and an entirely plant-based diet is usually poor in sodium. Because of this, human perspiration attracts sweat bees. These small black or green bees lick the sweat, thereby gaining sodium and other minerals. Insects such as butterflies often gather at the edges of puddles where evaporation has concentrated the minerals.[3] This behavior is called "puddling". Salt licks, exposed patches of crystallized salt, attract mammalian herbivores such as cattle and deer. Nutritional constraints are revisited in Section 9.10, where salt again takes a starring role.

Digestion of Cellulose

Cellulose may be the most abundant organic macromolecule on the planet. It is also one of the least digestible. Like starch, cellulose is a string of glucose molecules, but unlike starch, the bonds between the cellulose glucoses are strong; they require considerable energy to create and equal energy to break down. (These are termed *ß bonds*.) Few animals have the enzyme needed to break these strong bonds, so gaining access to the energy tied up in cellulose is challenging. Animals that eat cellulose commonly solve this problem by having microbes in the gut that do have the enzyme. For this strategy to work, young animals must be "infected" with the symbiotic microbes; the obvious route is consumption of feces from infected animals. This is taken to an extreme in termites, in which anus-to-mouth feeding is a key element of their social structure (see Figure 9.2). Another solution is employed by leafcutting ants, which collect leaves and feed them to a fungus garden. The fungus digests the cellulose, and the ants then consume the fungus. (See Chapter 14 for a more complete discussion of fungus gardening ants.)

FIGURE 9.2
Termites practice anus-to-mouth feeding as a way to share essential intestinal symbionts.
Photo: USDA/ARS Formosan termite image gallery.

9.3 HOW ANIMALS GET FOOD

When animals are feeding, the first step is to find food through search behavior, covered in Chapter 8. The focus in this section is to expand on those fundamental ideas about search and to introduce the basic types of feeding strategies. Search strategies vary among species, but within a species, a given strategy may be invariable. These species-specific search behaviors may be genetically determined or influenced, but individuals can frequently become better foragers by incorporating experience into their species-specific strategy. Also, animals often take advantage of whatever foraging opportunities present themselves, so sometimes an animal takes unusual (for its species) prey or eats an unusual plant. The world of eating is so broad that feeding strategies defy easy categorization. Indeed, Manuel Molles suggested a broader term in place of typical categories of food acquisition (e.g., herbivory, predation);[4] that term—*exploitation*—refers to the habit of an organism living at another organism's expense.

> **KEY TERM** Exploitation refers to an animal living at another organism's expense. Almost all animals are exploiters in this sense.

Trophic Strategies and Styles of Foraging

An examination of diet first leads to a discussion of the ways in which animals capture energy from other organisms. The diversity of feeding styles among animals is almost intimidating. Foraging animals can be classified by their trophic roles:

- *Herbivores* (animals that eat plants)
- *Carnivores* (animals that eat other animals)
- *Saprophages* (animals that scavenge or eat nonliving organic matter)

These categories do not capture the full range because each of these can be broken down further if desired: *frugivores* eat fruit, *insectivores* eat insects, and so on. Another way of looking at this diversity places the focus on the effect that foraging has on the prey: there are predators—animals that kill and eat their prey—and there are parasites—animals that eat tissues but usually do not kill their hosts. Sometimes animals that eat seeds are called *seed predators*, expanding the usual meaning of the word. Often, though, seeds are not destroyed when consumed; they are dispersed when the animal defecates.[5] *Hematophagous animals*, such as mosquitoes, are an interesting variant; these species drink the blood of other animals. (Hematophagous animals are discussed in more detail later in this chapter.) If the unwilling (or unwitting) blood donor is large enough, these organisms are simply parasitic, but small animals may be exsanguinated and thus killed.

> **KEY TERM** A hematophagous animal consumes blood.

Mobility is a key issue in trophic behavior. Thus, many filter-feeders sit in one place to feed. They strain organic particles and small prey out of the water. They are very good at competing for space, and almost un-animal-like in their disinclination to move. Predators that chase prey are quite the opposite. They frequently attack very large prey, especially if they hunt cooperatively, and they run like the deer, which they must do, if they want to catch one. Predators that ambush prey are usually very good at hiding, and so on (see Figure 9.3).

Styles of Herbivory

Browsers glean leaves, bark, and green stems from plants, whereas grazers clip vegetation at or near ground level. Deer, such as the white-tailed deer in the Rocky Mountains, are browsers, a distinct advantage when grasses and other ground-level vegetation are covered by deep, wet snow. The disadvantage of browsing is that height may make vegetation inaccessible, and oftentimes browsers must glean parts of the vegetation that are low in nutrients, chemically defended, or both.

FIGURE 9.3
The ways that animals forage are diverse beyond imagining, as these examples illustrate. (A) Nectar-feeding, (B) tissue parasitism, (C) active hunting, (D) carrion feeding, (E, F) sit-and-wait prey capture. *Photo credits: (A) Michael Breed, (B) Jeff Mitton, (C) Ken Keefover-Ring, (D, E) Ben Pless, (F) Thomas Barnes, U.S. Fish and Wildlife Service.*

Grazers, such as sheep and cattle, can feed on the more nutrient-rich meristematic regions of the grasses when those grasses are available. The differentiation between browsers and grazers is one way in which ungulates (hoofed mammals) can partition food resources.[6] Resource partitioning, in response to competition for food, ultimately allows species to exploit different diets and to co-exist without competing.

FIGURE 9.4
These species represent a contrast in primate feeding styles. Howler monkeys (A) are thought to eat primarily leaves, a relatively poor diet, while white-faced monkeys (B) eat fruits, a richer but more difficult to discover foodsource. *Photo: Paul Bolstad, University of Minnesota, Bugwood.org.*

This type of differentiation also occurs in the behavioral and digestive adaptations of monkeys in Central America. The howler monkey mostly eats leaves (folivory) but will take fruits if available. Leaves are nutritionally difficult because most of the sugars are tied up in cellulose, which has already been discussed as a problematic nutrient. Folivores tend to be slower and less active; they must also have a longer intestinal tract to maximize extraction of energy from their food. White-faced monkeys live in many of the same habitats but prefer fruits (frugivory), which are rich in more easily digested carbohydrates. Frugivores are more active and have shorter intestinal tracts (see Figure 9.4).

Overcoming Plant Defenses

Plants have an impressive array of defenses to deter herbivores. Plant defenses include tough, thick, indigestible leaves, along with spines, thorns, stinging hairs (see Figure 9.5), and poisons (secondary compounds) produced in the plant tissues. Not surprisingly, herbivores evolve countermeasures to these defenses. Tough leaves may be countered by stronger jaws, muscular tongues, and large grinding molars. Spines and thorns can be overcome by thick skin. Chemical defenses may lead to specialized mechanisms for detoxification or to sequestration (storage) of the toxins in specialized body compartments. Sequestered plant toxins can actually become defenses for the animals that consume them, such as the cardiac glycosides produced by milkweeds and sequestered by monarch butterflies. Different types of foragers (e.g., carnivores, grazers, etc.) may have different tolerances for bitter (potentially toxic) compounds.[7]

Behaviorally, one of the primary methods of overcoming plant defenses is dietary selectivity. This involves finding the species of plant that an animal is adapted to eat; again, the example of milkweeds and monarchs comes to mind. Adult monarchs seek out milkweed plants when they lay their eggs; this means the developing caterpillars will feed on plants that fit their metabolic adaptations. Selectivity may also involve choosing susceptible plants from the population of possible targets; some individual plants may be less well defended than

FIGURE 9.5
The spines on this locust tree are examples of the defenses that plants employ to resist the attentions of herbivores. *Photo: Michael Breed.*

FIGURE 9.6

Adult tiger beetles (Family: Cicindelidae) visually track their insect prey, making short runs which are punctuated by stops that allow them to recalculate prey position and accommodate for prey movement. They quickly zero in on prey location using this process. Tiger beetle larvae (called *doodlebugs*) are also predators, but they use a sit-and-wait strategy, living within burrows in the ground. The larva blocks the burrow with the top of its head and the back of its thorax. When a prey animal comes close, it flips its head backward, exposing powerful jaws that grab and subdue the prey. As with the adult, visual tracking of the prey is an important component of being a successful predator. *Photo: Jeff Mitton.*

FIGURE 9.7

Tigers (*Panthera tigris*) are an interesting example of active predators because they quite willingly include humans in their menu. They seem to prefer medium to large prey; a study in India showed that half of the diet of tigers in a forest reserve was composed of chital (a spotted Indian deer species), sambar (another common species of deer in India), and wild pigs. Excellent camouflage and stealthy movements combine with powerful jaws and claws to make this a highly feared predator. As humans encroach on tiger habitats, the threat of tigers to human life creates a particularly difficult conservation challenge. Conservation plans that include multiple forest uses—usually low-level logging or harvesting of other forest products by humans—may be thwarted by tiger predation on human users of the forest. This creates pressure on conservation officials to remove the tigers from the forest. Modeling tiger/human movement patterns and probabilities of encounters may provide a tool for minimizing these problems. This tiger is maintained in an animal rescue facility. *Photo: Michael Breed.*

260

FIGURE 9.8

Not all spiders build webs for catching their prey. Hunting spiders use a variety of strategies, including active search and ambush, to obtain their food. This spider, a common resident of neotropical wet forests, hunts at night, seeking prey in the rich environment of the forest leaf litter. *Photo: Jeff Mitton.*

others of the same species. Pine bark beetles weaken host plants by attacking *en masse*; a large number of beetles burrowing into a single tree overwhelms the tree's ability to defend itself.

Some herbivores manipulate plants in a way that interferes with the delivery of defensive chemicals. Thus, the way a squash beetle "trenches" the leaves it eats does not trigger the host plant's defenses. This behavior allows leaves to remain palatable.[8,9]

Active Hunters

Catching prey is not a simple matter, as potential prey species are under constant pressure to evolve mechanisms to avoid becoming prey items (see Figures 9.6 through 9.8). In order to survive, predators must evolve in response to adaptations of their prey. One consequence of this co-evolutionary race is that predators are often endowed with excellent sensory capabilities.

Some carnivores eat, if they can, other carnivores in their home range. When a carnivore consumes another carnivore, it both obtains food and eliminates a competitor. One important implication of these observations is that links in food webs may often include carnivore–carnivore consumption, as well as consumption of herbivores by carnivores. For instance, carnivorous house pets, domestic dogs and cats, often become victims of wild carnivores, such as coyotes and mountain lions, in suburban areas; these pets are particularly vulnerable because they have lost much of their self-defense ability in the domestication process. Cheetahs in Africa are the target

BRINGING ANIMAL BEHAVIOR HOME: LIVING WITH PREDATORS

Many human companion animals are predators; this obviously includes dogs and cats, but extends to other commonly kept animals such as ferrets, frogs, snakes, and some species of fish. Some companion animals that are largely herbivorous, such as iguanas, eat meat when presented with the opportunity to do so. In isolated instances, people maintain larger mammalian predators such as wolves and lions in their households. Animal shows, zoos, and aquaria manage large predators like orca (killer) whales,[10] tigers, and lions and incorporate interactions between trainers and the predators into their shows. What are the risks to humans from co-existence with these carnivores?

One unfortunate possibility is for a person to be bitten by someone's companion animal. Bites may result from territoriality, defensive behavior, anxiety, or predatory instinct. For the human victim of a bite, the animal's motivation matters less than the injury, but understanding the circumstances that lead to bites can prevent injuries. In a sense, *umwelt* is as meaningful in this context as in the rest of nature; our perception of the events surrounding our attempt to pet a dog, for instance, may not match up with those of the dog.

Thousands of years of artificial selection have moderated carnivorous instincts in dogs and cats to present acceptable behavioral traits: dogs and cats are domesticated. Anxious, fearful, and defensive dogs often can be recognized by their posture—teeth bared, hackles raised, tail lowered, ears back. But sometimes a dog's reactions are so quick that they cannot be predicted, and all dogs do not express fear or defensiveness in the same ways. For this reason, never greet a dog, particularly a dog that is not familiar with you, with your face. Too many people rapidly move their face toward a dog's muzzle; if the dog is fearful or interprets this movement as a threat, the resulting bite to the human can be tragic, leading to plastic surgery and permanent facial scars. The dog often suffers a tragedy as well, either being euthanized as a "vicious animal" or being condemned to spend the rest of its life in a cage. It is best to let an unfamiliar dog take the lead in greeting and to let the dog sniff you where it wants.

Ferrets, frogs, and small snakes are too outsized by humans to pose threats beyond bites that potentially could become infected, and small carnivorous fish are also benign except for the occasional piranha bite. Iguanas have strong enough jaws to seriously injure a person, including completely severing a finger. Common sense in handling can help to avoid bites from these animals. With small- to medium-sized snakes and lizards, making sure the handler's hands do not carry a meat scent helps. In other words, do not eat a ham sandwich and then put your hand into a snake's or lizard's living quarters.

Large constrictors, such as boas, pythons, and anacondas are dangerous predators, and large captive snakes have caused human deaths from crushing and suffocation. Many cities forbid keeping snakes above a certain size, such as 5 feet in length, because of the hazards. Crocodiles and alligators also outgrow their interest as pets and become dangerous housemates. Unauthorized releases of these animals when they outgrow their homes extends the danger to the general public. It is hard to imagine the circumstances under which bringing a young, wild predator into the household would be a good idea; the outcome is usually bad for animal and human alike.

of predation by lions and hyenas; this contributes to the precarious status of cheetah populations. Finally, wolves were absent from the Yellowstone ecosystem for over half a century. During their absence, coyotes expanded their dietary range to include larger animals, such as deer, which formerly were preyed upon by wolves. The reintroduction of wolves into Yellowstone in the 1990s brought these two carnivore species back together. While the two species co-exist in many habitats, wolves prey upon coyotes, given the opportunity. After the wolf reintroduction, coyote diet preferences shifted back to smaller prey, probably as a result of competitive pressure from the wolves.

Sit and Wait Predation

OF SPECIAL INTEREST: SIT AND WAIT PREDATION IN SPIDERS

Some spiders build elaborate orb webs for catching prey. Like a fisherman choosing a spot to cast a fishing line, orb-weaving spiders must build their webs where prey (usually insects) are likely to fly. Spiders assess prey capture rates and relocate webs if they build in a spot with a low yield. For example, in the spider *Zygiella x-notata*, prey capture stimulates the construction of smaller webs, presumably because once the location of prey is discovered, the same yield can be obtained with a smaller capture area.[11]

Several hypotheses have been proposed to explain the purpose of structures, called stabilimentums, such as the one illustrated in Figure 9.9 in webs. In some cases, they give birds a visual cue of the web's presence, protecting the web from damage when birds fly through it.[12] However, such defense can be costly, both in construction materials and in making prey capture less likely. Webs with stabilimenta captured 30% fewer prey than webs that lacked stabilimenta, even though the same species of spider constructed them both.[13]

FIGURE 9.9

This spider has built a more densely webbed, opaque center, called a stabilimentum, in its web. *Photo: Michael Breed.*

In other spiders, the stabilimentum serves a defensive function, making the spider appear larger to potential predators.[14] Another hypothesis is that the stabilimentum actually attracts prey, but there is less experimental support for this idea. Schoener and Spiller noted that spiders with disk-shaped stabilimenta, like the one pictured in Figure 9.9, run to the other side when approached by a predator. Depending on the spider species, stabilimenta may serve to protect the web, to defend the spider itself, or serve both purposes.[15]

FIGURE 9.10
Chimpanzees are omnivorous; these are sharing a meal of meat. *Photo: John Mitani.*

Omnivory

Omnivory, the consumption of both animal and plant tissues, is an excellent strategy for obtaining the correct mix of nutrients, but it is a difficult lifestyle in terms of morphological and behavioral adaptations. An omnivore is often a "jack of all trades, but master of none." Some primates, including humans, successfully occupy omnivorous niches, largely because they can use their cooperative social skills in predation (see following sections and Chapter 13; Figure 9.10). Some ants employ another successful omnivorous strategy, in which they prey on insects and other small-bodied animals, but also collect nectar or **honeydew**, which are carbohydrate-rich foods. When compared with trophic styles that allow more specialization, omnivory is relatively rare among animals.

Mimicry and Luring

Mimicry will be discussed in more detail in Chapter 10, which focuses on antipredator behavior, because many animals use a variety of mimetic devices to help them avoid discovery by predators. In a fascinating symmetry, predators, especially predators that engage in ambush techniques, often employ mimicry as a way to disguise the threat they present. While many ambush predators simply hide and then pounce, some actually look like completely different objects, or use lures to decrease the distance between themselves and their prey.

Male burrowing owls line their burrows with shredded mammalian feces, which they also distribute around the opening of the burrow (see Figure 9.11). This habit has inspired numerous hypotheses and investigations about why they might do this. Smith and Conway tested four of these hypotheses:[16]

1. Were males trying to attract mates?
2. Were males trying to signal that the burrow was occupied?
3. Were males trying to camouflage the burrow, protecting it from predators?
4. Were the males trying to attract prey arthropods?

The first three hypotheses were not supported. The manure scattering behavior often began after pair formation and almost a month after the burrow was occupied. Predation did not differ between nests that were festooned with manure and those that were not. However, nests with manure attracted 69% more arthropods than those without manure, and included many arthropods reported from burrowing owl diets.

Meanwhile, 1600–2300 meters below the surface of the sea, a siphonophore (a colonial hydrozoan) may use bioluminescence to attract prey.[17] The behavior and sensory physiology of deep sea organisms is difficult to investigate, but the fact that this siphonophore eats fish (most siphonophores eat crustaceans) is probably relevant. The bioluminescence is emitted from side branches of tentacles.

As noted earlier, predators that sit and wait for prey are usually pretty good at hiding. In fact, in the case of orb-weaving spiny spiders (*Gasteracantha fornicate*), bright colors actually attract prey to the vicinity of the web! Hauber investigated the role of color when he dyed some of the spiders black, covering their conspicuous yellow and black dorsum.[18] Spiders that were dyed black caught fewer prey.

> **KEY TERM** Honeydew is secreted by sap-feeding insects, such as aphids. Aphids suck liquid from plant phloem; they ingest more carbohydrate than needed. The excess carbohydrate is then secreted. Some ants tend nectar-secreting aphids, protecting them from predators while they collect the nectar.

FIGURE 9.11
Burrowing owls attract arthropods to their burrows by lining the burrows with feces. urrowing owl. *Photo: Ben Pless.*

263

Cooperative Hunting

When two or more animals forage together, they are foraging socially. Social behavior is covered in more detail in Chapters 13 and 14, but it has special relevance to foraging; some behavioral biologists see foraging as a major selective force favoring social groups, whether they are loosely knit colonial nesters that share (or parasitize) information about food sources, predators that can take prey far larger than themselves by cooperative effort, or tightly regulated eusocial insects.[19]

Animals may cooperate to trap elusive prey, or to take down larger prey. While social bonds derived from kinship or mating may enhance cooperation, unrelated animals can also benefit from cooperation. An extreme example of unrelated animals cooperating to overwhelm a food item is the case of pine bark beetles, which alone cannot overcome a tree's defenses but can do so in a mass attack. More familiar examples of cooperative hunting involve felids and canids (wild species of cats and dogs, respectively). Once the prey item is secured, competition for food among the hunters can be severe and may be regulated by dominance relationships in the group.

Uninvited Guests during Foraging

Uninvited guests can be members of the same or different species. When one animal observes another eating, it is attracted and attempts to share the food. This behavior can lead to outright competition for food and to the loss of the food item by its original discoverer. For the discoverer, the problem is that adequate defense of the food takes time away from eating, even to the point of not eating at all. The animal must choose a balance between defense and consumption. For the guest, the advantages are numerous; it may get to eat without searching and may avoid the risks involved in capturing and killing prey.

9.4 WILLING FOOD
Predator Saturation

Sometimes animals are victims of their parents' reproductive strategies. From the point of view of the parent, it may be a better strategy to produce many defenseless offspring, or to fail to defend them, than to produce a few better-defended offspring. Selection may lead a species to produce so many seeds or offspring that consumers cannot possibly eat them all. This is reinforced by making the vulnerable stage available for short periods of time, separated by long periods of unavailability. This strategy is called *predator saturation*, because during the bursts of reproduction, predators cannot possibly eat all the offspring. The strategy might equally be termed *predator starvation*, because specialized predators may go long periods of time with nothing to eat.

> **KEY TERM** Predator saturation (also called *predator swamping*) is a reproductive strategy in which synchronized reproduction produces so many offspring that predators cannot consume all of them.

264

FIGURE 9.12
Mass emergence of cicadas. Many insects emerge simultaneously from their larval or pupal states, thus swamping predators. *Photo: John Cooley.*

Examples of predator saturation include both plant and animal prey. For instance, fruit- and seed-producing trees such as bamboos in southeast Asia "mast," producing large crops of fruits and/or seeds at several-year intervals. The adult stage of many mayfly species is present for only a few days in the spring; a very large number of individuals emerge. The mayflies are tasty to a variety of birds, but the birds cannot eat them all, leaving enough to reproduce and lay eggs (see Figure 9.12). Likewise, cicada species are hidden underground through most of their life cycle. Species that emerge as adults once every few years (a seven year cycle is common) saturate their potential predators.

Saprophagy

Animals that feed on dead or decaying material do not have to overcome their prey's resistance, but likely encounter substantial competition for the food. Saprophages feed on carrion and on more decayed plant and animal material. Competition among animals such as crows, vultures, eagles, coyotes, and wolves (in North America), or vultures, hyenas, wild dogs, and jackals (in Africa) is obvious and is usually resolved with aggressive interactions around the carcass. Carrion flies, burying beetles, and ants enter the competition; they are less obvious but equally important. Finally, microbes play important but often underappreciated roles in the competitive landscape, secreting chemicals that make dead material poisonous, noxious, or unpalatable. Microbes wage chemical warfare among themselves and with animals that seek to consume dead materials.

> **KEY TERM** In scramble competition, the animal that reaches the resource is the successful competitor; it is an ecological version of "first come, first served."

The behavioral result for saprophagous animals is *scramble competition*; often the first animal to find carrion "wins" or at least gains a disproportionate share of the carrion before being displaced by larger or more aggressive competitors. This puts a premium on the discovery of carrion, hence the stereotyped image of vultures circling a dying animal.

9.5 MANIPULATION OF PREY

After an animal's discovery of food, there are often behavioral steps before food can be actually consumed. These actions are termed *handling*. This discussion focuses on carnivores handling prey, but the same principles apply to bees handling flowers or leafcutter ants handling the plants they collect. The important aspects of handling are rapid physical control of the prey, avoidance of injury, and efficient use of time.

> **KEY TERM** Handling refers to all of the activities involved in converting a resource into usable nutrition; these activities can be everything from subduing prey to digesting food.

Two examples illustrate how prey can be rapidly controlled (see Figure 9.13). Mountain lions (also known as puma) are common in western North America and, while they can take a variety of prey, they specialize on white-tailed deer. The lions are well armed, with large canine teeth, powerful claws, and strong body musculature. The deer have surprisingly effective countermeasures; given the opportunity, their hooves are sharp and their antlers can be intimidating defenses. Mountain lions quickly subdue deer by severing the cervical spinal cord with their canines; the teeth have sensitive nerves that allow the lions to find the lethal spot.

Pit vipers, such as rattlesnakes, copperheads, and cottonmouth water moccasins, inject venom deep into their victim's tissues with long fangs through which the venom flows. The interval between the snakebite and the immobilization of the prey can range from a few seconds to a few minutes; during this time the snake could be bitten or stomped by the prey. To avoid this, the snake recoils and waits. The challenge for the snake, then, is to find the wounded prey, which may have traveled some distance. Heat and odors coming from the prey assist in this search.[20] In this case the snake sacrifices some time and even risks losing the prey to be able to protect itself from injury. This illustrates the often-underestimated threat of prey defensive behavior (see Chapter 10).

265

FIGURE 9.13
These examples hint at the wide range of food-handling challenges that confront animals, whether they eat plants or other animals. Upper left, gull cracking a crab; lower left, bird preying on an egg; right, squirrel harvesting seeds. *Photos: Ben Pless (gull) and Jeff Mitton.*

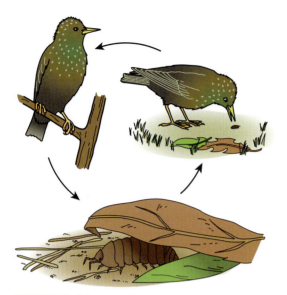

FIGURE 9.14

This is a typical acanthocephalan life cycle, with a vertebrate final host and an arthropod intermediate host.

266

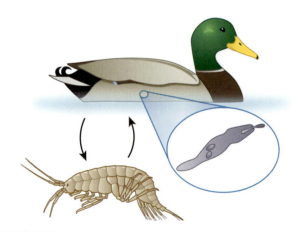

FIGURE 9.15

An acanthocephalan of ducks, showing the proboscis from which the worm gets its name (*acantho* = thorny, *cephalan* = head).

9.6 PARASITIC LIFE CYCLES

Much like herbivores and carnivores, parasites exploit other organisms (hosts) to gain nutrients. Many parasites have life cycles that involve at least two hosts: a final host, in which the parasite reaches adulthood and reproduces, and an intermediate host, in which the immature parasite develops. In such life cycles, the intermediate host must often be eaten by the final host for the parasite to complete the life cycle. In the case of intestinal worms, for instance, worm eggs pass out with final host feces. When they are consumed by an intermediate host, the egg hatches and the immature parasite invades that host and develops to an infective stage. When the intermediate host, along with the parasite that it contains, is eaten by the final host, the parasite takes up residence in the final host intestine and the life cycle continues.[21]

Animals, including animals that might become intermediate hosts, have been under strong selection to avoid being eaten, and this would seem to interfere with a parasite's goal of getting into the intestine of the final host. A large number of parasites have solved this problem by altering the behavior of their intermediate hosts. For instance, in the case of acanthocephalans ("thorny-headed worms" that use arthropod intermediate hosts; see Figure 9.14), every acanthocephalan investigated has been shown to modify the behavior of its arthropod host. These parasites modify the movements of the host in the environment, making the host more vulnerable to predation. This ensures the passage of the parasite to the next host in the life cycle. Other parasites (e.g., nematodes, tapeworms, trematodes, and a variety of protists) do likewise. In some cases, there are obvious physical reasons for the modification; it makes sense that sheep with cestodes in their brains might stagger around and fall prey to wild canids (the final host), and it makes sense that cestodes in the lungs of moose might make the moose winded during a chase. But many parasites do not produce such obvious morphological damage, yet they profoundly alter their hosts' responses to environmental cues. In so doing, they change habitat choice in ways that can be critical to predator–prey encounters. For instance, terrestrial isopods that are host to an acanthocephalan of songbirds spend more time on light surfaces and in areas of lower humidity than do uninfected isopods. They ignore overhanging shelter and tend to be more active. All of this may contribute to the fact that they suffer disproportionally high levels of predation (compared to uninfected isopods) in the field and in laboratory tests. Amphipods infected with an acanthocephalan of dabbling ducks have an equally hard time: they move into the lighted zone of the water column. When the water is disturbed, they swim to the surface (uninfected ones dive), and skim around in circles, attaching to floating debris and even to ducks (Figure 9.15).

OF SPECIAL INTEREST: BLOOD SUCKERS

Hematophagous (blood-sucking) insects are prime candidates to transmit blood-borne parasites such as malaria and sleeping sickness. They are living vectors of disease. (A vector, in biological terms, is something that transmits disease.) Biting flies (mosquitoes, horseflies, blackflies, and the like) are particularly well known for this, and they are also good candidates for manipulation by their protistan parasites. In general, they acquire the parasites when they ingest them while imbibing blood from a vertebrate host. The parasites alter their insect host's behavior in a variety of ways, but the end result of most of their machinations is a reduction in the host's ability to feed. This may seem counterintuitive; after all, wouldn't the parasite "want" the host to feed and thus be transmitted to the next host? That is correct, but when it interferes with feeding, the parasite does not inhibit it altogether. Instead, it simply causes feeding to be difficult, resulting in increased probing by the fly, and in the best case (for the parasite), probing multiple hosts in an attempt to get an adequate blood meal.

How can this happen? Most blood-feeding animals have a concoction of salivary chemicals that would make a pharmacist jealous. Blood-feeding is a dangerous way to make a living; for instance, host defense is thought to be the primary source of mortality in adult female mosquitoes. (Think of all the mosquitoes you have tried to slap and squash.) To the extent that the blood-feeder can probe, feed, and leave—preferably without discovery—it can live to reproduce and feed another day. Thus, the chemicals of the salivary glands contain anesthetics, blood thinners, and all manner of compounds that facilitate the feeding process. As a result, there are abundant opportunities for parasites to alter blood-feeding, and there are many examples of vector-borne parasites doing so in a variety of ways. For example, plague can interfere with flea feeding by blocking the anterior intestine (see Figure 9.16); some protists transmitted by tsetse flies interfere with the receptors that tell the fly about how much blood it has consumed.

One of the most sophisticated forms of interference is seen in *Plasmodium gallinaceum*, a malarial parasite. Most mosquitoes have an enzyme, an apyrase, that inhibits platelet aggregation and thus inhibits blood clotting. As you might imagine, this makes for much easier blood acquisition for the mosquito; if blood clots, the general outcome must be very much like trying to suck an old-fashioned malted milkshake, made with real ice cream, through a straw—a lot of work, as the delicious ice cream "clots"

FIGURE 9.16
A blood-sucking arthropod. *Photo: Jeff Mitton.*

slow the progress of milkshake up the straw. *Plasmodium gallinaceum* interferes with the production of this enzyme by damaging the region of the salivary glands that produces it. The rest of the salivary gland is left untouched, all the better to transmit the infective malarial propagules. But the enzyme that would make feeding fast and efficient—the apyrase—is not present in the infected mosquito's saliva. As a result, blood does not pool under the skin where the mosquito is biting, and feeding is difficult and time consuming.[22]

Are all of these behavioral alterations (and there are more than can possibly be described here) simply a result of "pathology"? That is unlikely; behaviors that involve hyperactivity or that cause an animal to alter its responses to environmental stimuli are not usually termed *pathology*; the host is not too sick to move or run away. Instead, in some cases, parasites may be actively manipulating the behavior of their hosts by secreting or interfering with neurologically active compounds. Moreover, the distinction between pathology and active manipulation may be less important than it might seem. Pathology itself is subject to natural selection, and the evolution of virulence of a pathogen depends in large part on what that virulence does to the pathogen's fitness.[23] Thus, any pathology that is induced by a parasite may be favored by natural selection on the parasite if it improves parasite transmission. It is not necessarily a side effect of infection, somehow invisible to selection.

This discussion gets into fascinating but dimly lit territory. The evolutionary origins of parasite-induced behavioral alterations are not well understood; neither are the mechanisms that cause them. Some may benefit the parasite; others, the host (see Chapter 10). In the case of parasites with intermediate hosts, some behaviors that benefit the host (e.g., crawling into exposed areas to bask and induce fever) may be difficult to discern from behaviors that benefit the parasite (e.g., crawling into exposed areas and encountering more final hosts). The one clear thing is that an animal with a parasite is unlikely to behave like an animal without that parasite. Beyond that, much remains to be learned.[24,25]

9.7 FORAGING AND OPTIMALITY THEORY

How does food distribution affect food discovery? In 1966, two independent papers published in *The American Naturalist* set the stage for optimal foraging theory (OFT). Written by R. H. MacArthur and E. R. Pianka and by J. M. Emlen, they asked how animals made decisions about food preferences and places to seek food.[26,27] In general, the expectation of optimal foraging theory is that by making adaptive foraging decisions, an animal will seek to maximize its inclusive fitness. (*Optimal* comes from a Latin word meaning "best.") Before continuing, however, a longer visit with optimality theory is in order. This theory can be applied to any one of a number of behaviors, but it is covered here because historically it has been most closely associated with foraging.[28]

> **KEY TERM** In optimal foraging theory, the currency is the measurement being optimized. In most animals, time and energy are the most important currencies.

The application of optimality theory to foraging or any other behavior does not mean that all scientists expect animals to optimize fitness outcomes. There are many *constraints* that can prevent an animal from achieving an optimal state. These are typically the constraints that prevent natural selection from producing perfectly adapted organisms, but they bear special consideration in a discussion of optimality because they limit what can be expected as optimal behavior. For instance, because of heterogeneity in time and space, what is optimal in one location or at one time may not be optimal under different circumstances. The environment may change too rapidly, evolution may proceed too slowly, or genetic variation may be insufficient. There may be conflicting selective forces that exclude optimal adaptations. The animal itself may simply be unable to gather the information that is required to make optimal decisions. In other words, the use of optimality theory to predict what animals should do if natural selection optimized their behavior is not the same as testing for natural selection, nor is it the same as expecting a perfect outcome.[29]

> **KEY TERM** A constraint is a limitation imposed by existing behavioral, morphological, or physiological adaptations. For example, an animal that is unable to learn how to efficiently open a certain kind of seed cannot be a completely optimal forager on those seeds.

Instead, optimality theory provides a way to test hypotheses about what is important in an animal's behavioral decisions. This involves identifying the *currency* used to investigate behavior. For instance, if animals are assumed to maximize energy intake (and this assumption is a common approach when investigating foraging), then energy acquisition is also assumed to be crucial to the animal's fitness and its foraging decisions should reflect that.

What if these predictions are not met by the data? If the experiment is well designed and the resulting data are nonetheless inconsistent with predictions about energy acquisition based on optimality theory, then a return to the assumptions on which the predictions were based is in order. How can they be changed or refined to better reflect reality? Of course, this general approach is true of most hypothesis testing, but the added advantage of bringing optimality theory to one's investigation is that it provides a fairly clear road map for generating predictions. There are usually a limited number of ways in which an animal can reach an optimal outcome…and an infinite number of ways to achieve suboptimal ones.

Optimality theory—and modeling in general—has benefits beyond those of generating predictions and testing assumptions. It demands clear thinking about the ideas involved in investigating a behavior. For instance, an animal may be expected to feed its young high-energy prey that are easy to digest, but as those prey are depleted, at what point should that animal switch to a different prey? How long should the forager keep looking in one area for a diminishing number of food items? Under what conditions should it begin to look elsewhere? (Please note that for these purposes, prey can be any food item, not only an animal.) These and other refinements are core concepts in the optimal foraging literature and introduce more precision into thinking and investigation than a vague notion about yummy caterpillars.

Optimal foraging theory concerns itself primarily with where an animal eats (patch choice) and what it eats when it gets there (prey choice). These appear to be separate considerations, and they frequently are. However, keep in mind that in some cases (e.g., a relative large prey that takes a long time to eat), *prey item* and *patch* can become the same thing. For practical purposes, prey and patch will be considered separately here. Within these two seemingly simple decisions lies a wealth of discovery about how animals sustain themselves.

OF SPECIAL INTEREST: HOLLING'S DISC EQUATION

Studies of both prey choice and patch choice emerge from a surprisingly simple experiment conducted by C. S. Holling.[30] In this experiment, he asked how predators respond to changes in prey density. After all, such changes are inevitable if the predator is successful at finding and consuming prey.

To investigate this question, Holling tacked 4 cm (diameter) sandpaper discs to a 3 × 3 ft table (see Figure 9.17). A blindfolded assistant (the "predator") was given one minute to locate the discs by tapping her finger across the table; as each disc was encountered, it was removed and the search continued. The initial density of discs varied from 4 to 256, and the experiment was replicated 8 times for each density.

Holling found that as the food supply increases, an animal increases the rate at which it feeds, but this eventually levels off. The relationship between the amount of food and the rate at which it is consumed is called a *functional response*. The most common type of functional response—and one that was demonstrated by Holling's "predator"— results in a reduction in the rate of feeding as the discs are collected. Ecologists have investigated perhaps thousands of exploiter–prey relationships, and most fall into this category.

FIGURE 9.17
The *functional response* is the relationship between the amount of food and the consumption rate of that food. This curve shows the rate at which Dr. Holling's "predatory" (and blindfolded) assistant located and removed sandpaper discs spread across a table. When more discs are available, the rate at which they are located increases; eventually, this levels off. *Adapted from Holling, 1959. The Canadian Entomologist 91, 386.*

9.8 OPTIMAL PATCH CHOICE

Different types of predators have different challenges facing them as they find prey, but for predators that seek habitat-specific prey in a heterogeneous ("patchy") environment, their first decision must be where to look.[31,32] This decision requires that they know quite a bit about their environment so that they can decide which patch might be most profitable. As a result of this, scientists interested in optimal foraging have investigated how animals sample their environment when given a variety of patches. They have done this in the laboratory under fairly artificial conditions. For example, Smith and Sweatment used trained birds foraging in grids with different densities,[33] but even under those conditions, birds are able to comprehend which patches are better for foraging and when to switch patches. *How* animals discover and process this information in the field is not well understood, but learning about the environment is a big part of the business of being an animal. Animals that can do so have many advantages. They can avoid predators and avoid abiotic extremes by knowing about hiding places, and they can certainly find food more efficiently by knowing where and how much of it there is.

Given the fact that animals have some information about their environments and about the relative foraging value of different patches, an animal might be expected to choose a particularly rich patch and stay there, eating to its heart's content. In many instances, however, this will not work out for very long; as the animal consumes prey, the value of the patch—that is, the number of prey it contains—will diminish. At some point, the patch that was so rich and attractive will become less so. Assuming that an animal knows about the richness and location of surrounding patches, what influences an animal to leave the patch it has been exploiting and enter another one? Eric Charnov proposed an answer to that question in the marginal value theorem, a solution that bears closer examination.[34]

The Marginal Value Theorem

Imagine a berry-picking expedition, complete with a bush that is laden with delicious berries. After a spree of rapid collection (and eating), things begin to slow down—the remaining berries become harder and harder to find. All such berry collectors and eaters have faced this quandary: when to leave and find another bush and when to persist at the current one.

Making this decision requires the same information that other animals probably employ when they forage in patches. Foraging in patches is not that unusual. Many food items, from termites in nests to fruit on trees, are patchily distributed. To decide whether or not to move on to the next patch, an animal needs to know (at minimum) how far away the next patch is. The further away the next patch is, the longer the forager will stay in the current patch. (It also helps to know how rich the next patch is, but for now, assume all patches are equivalent. In Charnov's model, the comparison is the average amount of food in the other patches.)

In essence, an animal should move to the next patch when the net gain from remaining in the current patch decreases so that it equals the net average gain available from the environment.

FIGURE 9.18
Bumblebee foraging in a patch of flowers. Bumblebees use experience gained over a few hours of foraging to inform decisions about patch choice and time spent in patches. *Photo: Michael Breed.*

This rule is modified somewhat when the cost of traveling to and exploiting the next patch are considered. The berry harvester should stay at one bush until it can do better elsewhere, even when considering travel, at another. To do this, animals must be able to learn the average value of patches in the environment or to be equipped with this information as a result of natural selection. This requirement is usually assumed in foraging studies but has been tested in only a few species, such as bumblebees (see Figure 9.18).[35]

This decision point is illustrated in Figure 9.19. The decision about the berries may seem obvious, but the beauty of Charnov's approach is that his graphical model allows for both precision and a large variety of conditions. For instance, if travel time is decreased, the slope of the tangent is increased and optimal time in the patch

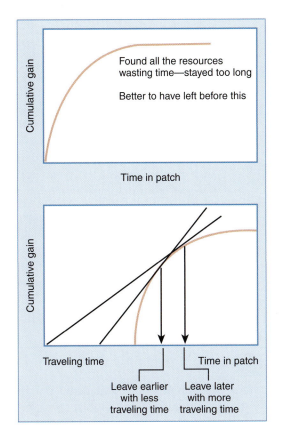

FIGURE 9.19
When to leave a patch? The solid curved line represents the cumulative energy gain. The greater the distance to the next patch, the greater the transit time. Drawing a line from the origin so that it is tangential to the gain curve reveals the point in the gain curve at which the animal should shift to the next patch. Note that the greater the transit time, the more time the animal is expected to remain in the initial patch.

271

is reduced. In the berry-picking example, if the next bush is a few steps away and equally rich, the current bush will not be attractive for very long. If travel time is increased, then a slower energy gain will be tolerated before moving on; the tangent will intersect the energy gain curve after it begins to flatten. Animals tend to linger a bit longer in a patch than theory would predict, perhaps because of the likely variability in travel time to discover the next patch.

A surprising variety of animals have been tested and found to conform to the expectations of the marginal value theorem. Of course, there is no reason to expect animals to have perfect knowledge of their environments. There may be patches that they have not sampled and, as the next chapter reveals, prey are under strong selection not to be found. Nonetheless, the marginal value theorem provides a framework for exploring how animals make decisions about where to forage.

Some environments will present greater challenges in obtaining needed levels of energy. Forage quality and availability may also change with season. American Bison search for high-energy patches of sedge in the summer,[36] but dietary preferences change seasonally as food availability and snow cover impact movement and foraging. Across the broader landscape, choice of meadow by bison herds is less influenced by the presence of particular plant species and more influenced by abiotic ecological factors, travel distance among meadows, and probably the presence of other bison herds (see Figure 9.20).

FIGURE 9.20
The large-scale foraging movements of bison are influenced by such things as abiotic factors and travel time. *Photo: Phyllis Cooper, US Fish and Wildlife Service.*

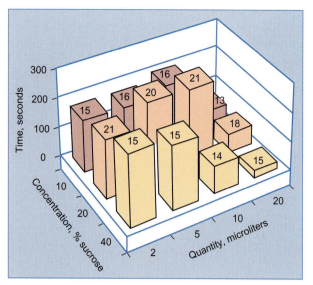

FIGURE 9.21
Giving-up times for ants foraging on sugar solutions. Ants linger the longest when presented with high concentration rewards of intermediate amounts; these factors predict that the patch is worth exploiting. At the highest quantity (20 microliters) the ants have collected all they can carry and quickly return to their colony with the food. *Adapted from Breed et al., 1996. J. Insect Behav. 9, 659–672.*

Optimal foraging theory can also be used to estimate resource abundance at a landscape level. Assuming the marginal value theorem—that is, that animals leave patches when the yield declines below the average yield of the environment—the route to this calculation is simple. The food abundance in a patch, as measured when animals depart the patch, should represent a slight underestimate of the food availability in the landscape. (The underestimate is caused by the effects of travel time and conservative decisions by foraging animals.) This gives ecologists a tool for assessing landscape-level resource levels that otherwise might be difficult to measure.

Another approach to optimal foraging is to assess how an animal decides to give up on a patch and leave in search of the next patch. In some cases this is time dependent, so the appropriate measure is giving-up time. In other cases, density, and not time, is the relevant measure; the animal assesses the density of food items in the patch, so giving-up density is measured. In a few cases, animals assess changes in the food composition of the patch, so there is a giving-up composition (see Figure 9.21).[37]

Central Place Foraging

Central place foraging addresses the fact that many organisms, for all or part of their life cycles, leave a central place to forage and return to that place with provisions. Central places can be as diverse as bee hives, wolf dens, squirrel caches, and bird nests. This topic is an extension of the marginal value theorem, in that the animal must know not only when to leave a patch, but how much to carry back to the central place. Travel distance is the significant variable. The further the animal has to carry the load, the longer it will remain in the patch and the more it will carry.

DISCUSSION POINT: CENTRAL PLACE FORAGING

You can think of central place foraging in terms of your own shopping. It costs more energy and time to shop at distant stores than to shop nearby. If you have a corner grocery, you may go there almost every day for something; keeping an organized grocery list is unnecessary. But if only certain stores carry the dog food that your best friend clearly prefers, and if those stores are further away, you will go to them infrequently and stock up on everything you think you might need for a while. When you do that, you are conforming to the predictions of central place foraging. Can you think of other instances (not related to food) in which the principles of central place foraging might apply? How do you make decisions in those instances?

Risk Sensitivity

What if patches are variable, and what if animals can comprehend that? There is good evidence that both are true. Charnov's marginal value theorem dealt with the average amount of food in patches, but average values can obscure some information. Imagine a choice of two games at an arcade. Either game is included in admission to the arcade, but only one can be played. In one, the prize is guaranteed to be a $10 gift card. (Admittedly, this is sort of a boring game.) In the other game, there is an equal chance of winning a $20 gift card, or absolutely nothing. If each game is played many times, the average prize in either place would be the same: $10. But because only one game is allowed, the player is faced with a decision that causes acute awareness of the variability in these two patches. What is better—to take the $10 and run, or to hope for the best with the highly variable version of the game?

Researchers have asked animals to make similar decisions. For instance, a tray that always has 5 seeds in it has an average of 5 seeds. So does a tray that alternates seed number so that it has 0 seeds for one trial and 10 the next. If a little bird like a junco (see Figure 9.22) is exposed to both sets of trays and given an opportunity to learn that the two trays differ, it can distinguish between the two types of trays. An animal that can make this distinction is called *risk-sensitive*. The junco can then be asked to choose between the trays. A junco (or any other animal) that chooses the tray that always has the same amount of food in it is called *risk-averse*. If the junco prefers the variable tray, it is considered *risk-prone*. Note that in this context, risk is related to uncertainty, not necessarily danger.

Thomas Caraco and coworkers showed that juncos can tell the difference between the trays, but they also discovered that the junco's response varied depending on its nutritional state.[39] Hungry, cold juncos were more risk-prone than well-fed juncos. This response would be expected as a survival strategy. If the stable tray offered enough food to get by on, then the juncos chose that tray. When it did not, under adverse conditions, they were willing to gamble. The interaction of risk perception and foraging has been investigated in a wide range of organisms, from humans to bees.[40,41]

The consideration of risk sensitivity leads to a discussion of ***utility functions***. This is another term that sounds more complicated than it is. A utility function is simply a graph of a resource's value (utility) to an animal plotted against its abundance (see Figure 9.23). A linear utility function, say, a steadily sloping line, indicates that the value of the

FIGURE 9.22
Dark-eyed juncos are risk-sensitive. Whether they are risk-prone or risk-averse depends on their nutritional state. *Photo: Lee Karney, U.S. Fish and Wildlife Service.*

> **KEY TERM** A utility function is a graph that plots the value of a resource against its abundance.

273

FIGURE 9.23
These two curves plot the value of a resource to an animal (its utility) against units of resource. In the case of concave-down functions (left), the more units of resource the organism consumes, the less valuable those units are. This is a common occurrence when foraging on a batch of morsels. After a certain point, eating another one is not all that beneficial. In the process of such foraging, risking the current morsel (and losing) costs more than gaining the next one. Animals faced with this situation should be risk-averse. The curve on the right shows the opposite situation: units of a resource increase in value (again, utility) as they are acquired, a situation that often occurs when some threshold must be attained before the resource can be useful. In this case, an animal is likely to risk losing a unit of resource in order to gain more, because the potential gain outweighs the possible loss. Such animals are risk-prone. *Adapted from Stephens, D., Krebs, J., 1986. Foraging Theory. Princeton University Press, Princeton. p. 131.*

resource in question (liters of water, number of seeds, units of protein, etc.) does not change with its amount. This is not the most common case. It is more common to see curved utility functions. When the curve is "concave down," the utility of a resource for an animal diminishes as more is consumed. Ingested food is a good example of this; under most conditions, it is less useful as more is consumed. On the other hand, a "concave-up" utility function indicates that each unit of resource is more valuable than the last. This can happen when an animal needs a certain number of resource units to reach a threshold; as that threshold is approached, each unit is more valuable than the previous one. In the case of the juncos, the well-fed birds that chose the stable tray had a concave-down utility function; each seed was less valuable than the previous seed. On the other hand, the cold, hungry juncos had a concave-up utility function because each additional seed was more valuable.

Discussions of risk and utility functions reveal that foraging decisions can be nuanced. They may change over a lifetime or as a result of shifting environmental conditions. Much like human decision makers, animals may balance long-term and short-term gain.[42]

9.9 OPTIMAL PREY CHOICE

Should animals be less selective about food choice and save time and energy in their search for food? In most cases, that expectation would be wrong. Animals do not eat food in proportion to its availability in the environment. Some foods may be abundant but less nutritious or require more effort to eat and digest, and there is clear evidence that animals have definite preferences in the food that they eat. How can optimal foraging theory be used to address prey choice?

To begin with, whether an animal forages for seeds or antelope, there are some basic elements in a foraging event. Some of these elements are the amount of time it takes to find the food item, how frequently the item might be encountered, the energy in the food item, and how long it takes to eat the item. The list would be very long, of course, but the goal of using models based on theory is to identify the most influential elements.

To understand such models, begin with the currency—again, using energy (E). Of course, prey do not give up their energy willingly; they require some chasing and subduing—that is called handling time (H). In addition, there are search times (S) associated with prey. Beginning with the first two variables, the profit (P) that a predator experiences upon eating the prey can be described as

$$P = E/H \tag{9.1}$$

Now assume that there are several types of prey. Each prey will have its own energy value, its own handling time, and its own search time, so to make these equations informative in all cases, we must identify these different types of prey for comparison. Subscripts are handy for that. Thus, prey type i has a profitability of P_i, an energy value of E_i, a handling time of H_i, and a search time of S_i. When we compare two prey types (designated 1 and 2), a notation that $P_1 > P_2$ simply means that prey type 1 is more profitable than prey type 2.

Assume that the predator knows these values; again, given the knowledge that animals have about their world, that is not an unreasonable assumption, even if how they comprehend such information is not precisely understood.

Suppose that a predator has indeed encountered prey type 1. What should it do? Given the inequality of the two prey items, the answer is easy: it should eat the item. Of the two items on offer, this behavior is the more profitable. So the *first prediction* of the model is not terribly difficult to embrace: the most profitable prey, when encountered, should always be eaten.

What happens, however, if the predator encounters prey type 2? Answering this question is more difficult. Prey type 2 is within the predator's grasp and requires no searching, but there still may be handling time. In contrast, Prey type 1 will require a search, so

$$P_1 = E_1/(S_1 + H_1) \tag{9.2}$$

This yields a *second prediction*, that prey type 2 should be eaten when its profitability exceeds that of prey type 1, something that can occur if H_1 is large enough. In that case,

$$E_2/H_2 > E_1/(S_1 + H_1) \tag{9.3}$$

Clearly, the search time for prey type 1 is highly influential in this decision. If prey type 1 is easy to find, then S_1 will be very small, and profitability of prey type 1 will remain high. If that profitability is high enough, then it should remain the preferred prey, and prey type 2 should be ignored, no matter how often it is encountered. The reason is that given the small value of S_1, the profitability of prey type 1 still exceeds the profitability of prey type 2, no matter which one was encountered. This yields a *third prediction*: The foraging choice is solely based on the encounter rate with the most profitable prey; the abundance of the less-profitable prey is irrelevant.

The concept of optimal foraging is not without its critics, and it is not clear if all examples of optimal foraging in the literature are good representations of what might happen in a natural setting. (In fact, it is pretty clear they are not.) Lest these predictions seem too simple or unrealistic, remember the benefits of using models. If organisms ignore the predictions generated by theory, then there is an opportunity to revisit and refine the assumptions of the model being tested, not to mention the experimental design. Of course, one goal is that with increased understanding, models approximate reality, but even if they do not, they offer useful insights into problem-solving techniques used with models.

Of course, many things can impinge on an optimal diet; a few of those constraints were discussed in the introduction to optimality. Three deserve special mention: nutritional constraints are discussed in the following section, whereas predator risk and self-medication are explored in Chapter 10. Search image formation can also affect the discovery of prey items.

9.10 NUTRITIONAL CONSTRAINTS

The classic study of nutritional constraints in foraging was done not with a highly specialized insect or rodent devoted to a limited range of dietary items, but with a moose! Vertebrates have important dietary requirements for sodium, and moose are no exception. In addition, moose face a daunting trade-off: Aquatic plants contain the most sodium, but have little energy, whereas terrestrial plants offer the greatest energy, but are sodium-poor. In addition, aquatic plants are not available in the winter when the lakes are frozen. Therefore, a moose must acquire all of its winter sodium during summer foraging bouts. All of this adds up to a critical foraging decision for a moose.

Gary Belovsky used a linear programming model to investigate this decision.[43] He included variables such as the moose's daily energy requirement, the energy in the plants, the sodium needed, and time for digestion. His model predicted that 18% of a moose's summer foraging schedule should be devoted to aquatic plants, despite the fact that they are a poor source of energy—and this was borne out by observations of moose. Clearly, energy is not the sole concern of a foraging moose and a foraging model that did not take sodium into account would have been off the mark.

OF SPECIAL INTEREST: FOSSIL FORAGING

Foraging models can be used to explore the diets of long-extinct organisms. Sauropod dinosaurs hold the record for being the largest herbivores (up to 70 tons), yet they lived on plant taxa such as gymnosperms and ferns whose modern relatives are thought to have little nutritional value. By using analytical techniques that have been developed to investigate modern herbivore nutrition, Hummel and colleagues discovered that (assuming fermentation) at least some of the plants descended from those that coexisted with sauropods could have yielded enough energy to support the giant animals.[44] While this study certainly didn't address optimal foraging in dinosaurs, it set forth interesting questions: *Araucaria* is a large tree that had global distribution during the Jurassic Period. Hummel and collaborators discovered that modern relatives ferment slowly, but when they do, they yield significant energy. Other plants such as *Ginkgo* offer more protein. In contrast, yet other plants available to the sauropods (e.g., cycads, podocarp conifers, tree ferns) would not have sustained the large animals. Given the nutritional information that we can gather from the descendants of plants that were available to sauropods, how do you think that those dinosaurs made foraging decisions? What might the currency have been?

Using a different approach, Liu and colleagues have used diet indicators to determine that an ancestor of elephants (*Mesotherium*) spent at least some of its time in the water.[45] They analyzed the chemical composition of *Mesotherium* tooth enamel, which revealed the aquatic source of at least some of the diet. (The skeleton, built to carry the animal's 500–700 pound body on land, argued against a fully aquatic existence.)

SUMMARY

Foraging is a fundamental behavior that defines animals. Animals do not photosynthesize; we capture energy from other organisms. There are many ways to forage, but many of them can be placed into the framework of optimality theory. This theory is not limited to foraging behavior—almost any decision can be optimized—but it has perhaps had the biggest impact there. When we consider optimality and animal behavior, or optimality and any product of evolution, it is important to understand that many constraints can prevent optimal outcomes in nature. Nonetheless, models are useful for what they teach about the assumptions on which they are based and for how they demand clarity of thinking.

In the study of foraging, optimality models have classically been applied to two large areas: decisions about where to forage (patch choice) and decisions about what to eat (prey choice). Both of these areas are based on several assumptions, including the fact that animals are likely to have sampled their environment and be familiar with the choices on offer.

The marginal value theorem predicts that moving from patch to patch should be a function of both the average richness of the patches in the area and the distance to the nearest patch that equals the current one in richness. A related concept, central place foraging, is concerned with the fact that many animals must make foraging decisions about how to provision core areas that can be some distance from foraging locations. Perception and response to risk can influence an animal's choice of where to forage. Theory about prey choice yields some simple, testable predictions that may be influenced by constraints such as nutritional demands.

Parasites can manipulate hosts in ways that take advantage of host foraging behavior. One common manipulation involves changing intermediate host responses to environmental stimuli in ways that make them more conspicuous to predators (final hosts). Vectors of blood-borne parasites also exhibit altered patterns of blood-feeding that may well favor parasite transmission.

The ways that animals get food are so diverse as to defy description or categorization. From fungal gardens to central place foraging (and hoarding), from antelope grazing to cheetah decisions, this diversity is both a wonderful showcase for how evolution encourages animals to convert nutrients into offspring and also a schoolhouse, teaching the importance of clear assumptions and careful experiments.

STUDY QUESTIONS

1. Using the marginal value theorem, we have examined the influence of travel time on the decision to leave a patch and go to another one. What is the influence of patch quality? How would you investigate this aspect of patch choice and time spent in patch? Can you apply the marginal value theorem to your own foraging behavior?

2. We have discussed some of the more widely acknowledged constraints on optimizing prey choice. Choose an animal in which you are particularly interested. Read about that animal and see if you can come up with a list of constraints that are likely to influence that animal's foraging. Your list may be very different from our general one!

3. What is your preferred method of categorizing the ways in which animals get food? Why do you like this method? How does it help you think about foraging? Are some categorization methods better at stimulating thoughts and hypotheses than others?

4. Foraging can, at times, trump phylogeny. By that, we mean that certain foraging tactics can promote convergent evolution, with animals using specific resources converging on traits that help them exploit those resources. Consider the following groups: blood-drinkers, wood-eaters, intestinal parasites, filter-feeding invertebrates. Do some research on any one of them and discover what foraging traits they have in common!

Further Reading

Abramson, C.I., Singleton, J.B., Wildon, M.K., Wanderley, P.A., Ramalho, F.S., Michaluk, L.M., 2006. The effect of an organic pesticide on mortality and learning in Africanized honey bees (*Apis mellifera L.*) in Brasil. Am. J. Environ. Sci. 2, 33–40.

Ahearn, S.C., Smith, J.L.D., Joshi, A.R., Ding, J., 2001. TIGMOD: an individual-based spatially explicit model for simulating tiger/human interaction in multiple use forests. Ecol. Modell. 140 (1–2), 81–97.

Biswas, S., Sankar, K., 2002. Prey abundance and food habit of tigers (*Panthera tigris tigris*) in Pench National Park, Madhya Pradesh, India. J. Zool. 256, 411–420.

Borrell, B.J., 2007. Scaling of nectar foraging in orchid bees. Am. Nat. 169, 569–580.

Catania, K.C., 2006. Olfaction: underwater 'sniffing' by semi-aquatic mammals. Nature 444, 1024–1025.

Clark, R.J., Harland, D.P., Jackson, R.R., 2000. Speculative hunting by an araneophagic salticid spider. Behaviour 137, 1601–1612.

Darmaillacq, A.-S., Chichery, R., Dickel, L., 2006. Food imprinting, new evidence from cuttlefish *Sepia officinalis*. Biol. Lett. 2, 345–347.

Dees, N.D., Hofmann, M., Bahar, S., 2010. Physical constraints and the evolution of different foraging strategies in aquatic space. Anim. Behav. 79, 603–611.

Filippi, L., Hironaka, M., Nomakuchi, S, 2005. Kleptoparasitism and the effect of nest location in a subsocial shield bug *Parastrachia japonensis* (Hemiptera: Parastrachiidae). Ann. Entomol. Soc. Am. 98, 134–142.

Gilbert, C., 1997. Visual control of cursorial prey pursuit by tiger beetles (Cicindelidae). J. Comp. Physiol. A. 181 (3), 217–230.

Gillette, R., Huang, R.-C., Hatcher, N, Moroz, L.L., 2000. Cost-benefit analysis potential in feeding behavior of a predatory snail by integration of hunger, taste and pain. PNAS 97, 3585–3590.

Giraldeau, L.-A., Caraco, T., 2000. Social Foraging Theory. Princeton University Press.

Jacobs, L.F., Spencer, W.D., 1994. Natural space-use patterns and hippocampal size in Kangaroo rats. Brain Behav. Evol. 44, 125–132.

Kevan, P.G., Greco, C.F., 2001. Contrasting patch choice behaviour by immature ambush predators, a spider (*Misumena vatia*) and an insect (*Phymata americana*). Ecol. Entomol. 26 (2), 148–153.

Molles, M., 2008. Ecology: Concepts and Applications, fourth ed. WCB/McGraw-Hill.

Mull, J.F., MacMahon, J.A., 1997. Spatial variation in rates of seed removal by harvester Ants (etc). Am. Midl. Nat. 138, 1–13.

Nelson, X.J., Garnett, D.T., Evans, C.S., 2010. Receiver psychology and the design of the deceptive caudal luring signal of the death adder. Anim. Behav. 79, 555–561.

O'Brien, T.G., Kinnaird, M.F., Wibisono, H.T., 2003. Crouching tigers, hidden prey: sumatran tiger and prey populations in a tropical forest landscape. Anim. Conserv. 6, 131–139.

Okamura, J.Y., Toh, Y., 2001. Responses of medulla neurons to illumination and movement stimuli in the tiger beetle larva. J. Comp. Physiol. A 187 (9), 713–725.

Pasquet, A., Ridwan, A., Leborgne, R., 1994. Presence of potential prey affects web-building in an orb-weaving spider *Zygiella x notata*. Anim. Behav. 47 (2), 477–480.

Pyke, G.H., 2010. Optimal Foraging Theory: Introduction, In: Michael D. Breed and Janice Moore (Eds.), Encyclopedia of Animal Behavior, Academic Press, Oxford, Vol. 3, pp. 601–603.

Raubenheimer, D., 2010. Foraging Modes, In: Michael D. Breed and Janice Moore (Eds.), Encyclopedia of Animal Behavior, Academic Press, Oxford, Vol. 3, pp. 749–758.

Samelius, G., Alisauskas, R.T., Hobson, K.A., Larivi`ere, S., 2007. Prolonging the Arctic pulse: long-term exploitation of cached eggs by Arctic foxes when lemmings are scarce. J. Anim. Ecol. 76, 873–880.

Schmalhofer, V.R., 2001. Tritrophic interactions in a pollination system: impacts of species composition and size of flower patches on the hunting success of a flower-dwelling spider. Oecologia 129 (2), 292–303.

Stephens, D., Krebs, J., 1986. Foraging Theory. Princeton University Press, Princeton.

Stephens, D.W., Brown, J.S., Ydenberg, R.C. (Eds.), 2007. Foraging: Behavior and Ecology. University of Chicago Press, Chicago.

Tanaka, K., 1989. Energetic cost of web construction and its effect on web relocation in the web-building spider *Agelena limbata*. Oecologia 81 (4), 459–464.

Watwood, S.L., Miller, P.J.O., Johnson, M., Madsen, P.T., Tyack, P.L., 2006. Deep-diving foraging behaviour of sperm whales (*Physeter macrocephalus*). J. Anim. Ecol. 75, 814–825.

Wirsing, A.J., Cameron, K.E., Heithaus, M.R., 2010. Spatial responses to predators vary with prey escape mode. Anim. Behav. 79, 531–537.

Notes

1. Cooper, A.B., Pettorelli, N., Durant, S.M., 2007. Large carnivore menus: factors affecting hunting decisions by cheetahs in the Serengeti. Anim. Behav. 73, 651–659.

2. Frost, P.C., Ebert, D., Smith, V.H., 2008. Responses of a bacterial pathogen to phosphorus limitation of its aquatic invertebrate host. Ecology 89, 313–318.

3. Adler, P., Pearson, D.L., 1982. Why do male butterflies visit mud puddles? Can. J. Zool. 60, 322–325.

4. Molles, M., 2008. Ecology: Concepts and Applications, fourth ed. WCB/McGraw-Hill.

5. Sork, V.L., 1983. Mammalian seed dispersal of pignut hickory during three fruiting seasons. Ecology 64, 1049–1056.

6. Post, D.M., Armbrust, T.S., Horne, E.A., Goheen, J.R., 2001. Sexual segregation results in differences in content and quality of bison (*Bos bison*) diets. J. Mammal. 82, 407–413.

7. Glendinning, J.I., 1994. Is the bitter rejection response always adaptive? Physiol. Behav. 56, 1217–1227.

8. Tallamy, D.W., 1985. Squash beetle feeding behavior: an adaptation against induced cucurbit defenses. Ecology 6, 1574–1579.

9. Dussourd, D.E., 2009. Do canal-cutting behaviours facilitate host-range expansion by insect herbivores? Biol. J. Linn. Soc. 96, 715–731.

10. Goley, P.D., Straley, J.M., 1994. Attack on gray whales (*Eschrichtius robustus*) in Monterey Bay, California, by killer whales (*Orcinus orca*) previously identified in Glacier Bay, Alaska Can. J. Zool. 72, 1528–1530.

11. Venner, S., Pasquet, A., Leborgne, R., 2000. Web-building behaviour in the orb weaving spider *Zygiella x-notata*: influence of experience. Anim. Behav. 59, 603–611.

12. Eisner, T., Nowicki, S., 1983. Spider web protection through visual advertisement—role of the stabilimentum. Science 219 (4581), 185–187.

13. Blackledge, T.A., Wenzel, J.W., 1999. Do stabilimenta in orb webs attract prey or defend spiders? Behav. Ecol. 10 (4), 372–376.

14. Schoener T.W., Spiller, D.A., 1992. Stabilimenta characteristics of the spider *Argiope-argentata* on small islands—support of the predator-defense hypothesis. Behav. Ecol. Sociobiol. 31 (5), 309–318.

15. Eberhard, W.G., 1990. Function and phylogeny of spider webs. Annu. Rev. Ecol. Syst. 21, 341–372.

16. Smith, M., Conway, C.J., 2007. Use of mammal manure by nesting burrowing owls: a test of four functional hypotheses. Anim. Behav. 78, 65–73.

17. Haddock, S.H.D., Dunn, C.W., Pugh, P.R., Schnitzler, C.E., 2005. Bioluminescent and red-fluorescent lures in a deep-sea siphophore. Science 309, 263.

18, Hauber, M.E., 2002. Conspicuous colouration attracts prey to a stationary predator. Ecol. Entomol. 27, 686–691.

19. Kurland, J.A., Beckerman, S.J., 1985. Optimal foraging and hominid evolution: Labor and reciprocity. Am. Anthropol. 87, 73–93.

20. Dole, J.W., Rose, B.B., Tachiki, K.H., 1981. Western toads (*Bufo boreas*) learn odor of prey insects. Herpetologica 37, 63–68.

21. Papaj, D.R., Vet, L.E.M., 1990. Odor learning and foraging success in the parasitoid, *Leptopilina heterotoma*. J. Chem. Ecol. 16, 3137–3150.

22. Rossignol, P.A., Ribeiro, J.M.C., Spielman, A., 1984. Increased intradermal probing time in sporozoite-infected mosquitoes. Am. J. Trop. Med. Hyg. 33, 17–20.
Rossignol, P.A., Ribeiro, J.M.C., Spielman, A., 1986. Increased biting rate and reduced fertility in sporozoite-infected mosquitoes. Am. J. Trop. Med. Hyg. 35, 277–279.

23. Anderson, R.M., May, R.M., 1992. Infectious Diseases of Humans: Dynamics and Control. Oxford University Press, Oxford.

24. Moore, J., 2002. Parasites and the Behavior of Animals. Oxford Series in Ecology and Evolutionary Biology. Oxford University Press, New York.

25. Camp, J.W., Huizinga, H.W., 1979. Altered color, behavior and predation susceptibility of the isopod *Asellus intermedius* infected with *Acanthocephalus dirus*. J. Parasitol. 65, 667–669.

26. MacArthur, R.H., Pianka, E.R., 1966. On the optimal use of a patchy environment. Am. Nat. 100, 603–609.

27. Emlen, J.M., 1966. The role of time and energy in food preference. Am. Nat. 100, 611–617.

28. Schmidt, K.A., 1999. Foraging theory as a conceptual framework for studying nest predation. Oikos 85, 151–160.

29. Cooper Jr., W.E., 1995. Foragine mode, prey chemical discrimination and phylogeny in lizards. Anim. Behav. 50, 973–985.

30. Holling, C.S., 1959. Some characteristics of simple types of predation and parasitism. Can. Entomol. 91, 385–398.

31. Waddington, K., 2004. Flight patterns of foraging bees relative to the density of artificial flowers and distribution of nectar. Oecologia 44, 199–204.

32. Raguso, R., Wills, M., 2002. Synergy between visual and olfactory cues in nectar feeding by naïve hawkmoths, *Manduca sexta*. Anim. Behav. 64, 684–695.

33. Smith, J.N.M., Sweatman, H.P., 1974. Food searching behaviour of titmice in patchy environments. Ecology 55, 1216–1232.

34. Charnov, E.L., 1976. Optimal foraging: the marginal value theorem. Theor. Popul. Biol. 9, 129–136.

35. Biernaskie, J.M., Walker, S.C., Gegear, R.J., 2009. Bumblebees learn to forage like Bayesians. Am. Nat. 174, 413–423.

36. Fortin, D., Fryxell, J.M., O'Brodovich, L., Frandsen, D., 2003. Foraging ecology of bison at the landscape and plant community levels: the applicability of energy maximization principles. Oecologia 134, 219–227.

37. Barrette, M., Boivin, G., Brodeur, J., 2010. Travel time affects optimal diets in depleting patches. Behav. Ecol. Sociobiol. 64, 593–598.

38. Breed, M.D., Bowden, R.M., Garry, M.F., Weicker, A., 1996. Search time variation in response to differences in nectar quantity and quality in the giant tropical ant, *Paraponera clavata*. J. Insect Behav. 9, 659–672.

39. Caraco, T., Martindale, S., Whittam, T.S., 1980. An empirical demonstration of risk-sensitive foraging preferences. Anim. Behav. 28, 820–830.

40. Ludvico, L.R., Bennett, I.M., Beckerman, S., 1991. Risk-sensitive foraging behavior among the Bari. Hum. Ecol. 19, 509–516.

41. Perez, S.M., Waddington, K.D., 1996. Carpenter bee (*Xylocopa micans*) risk indifference and a review of nectarivore risk-sensitivity studies. Am. Zool. 36, 435–446.

42. Stephens, D., Krebs, J., 1986. Foraging Theory. Princeton University Press, Princeton.
Stephens, D.W., Brown, J.S., Ydenberg, R.C. (Eds), 2007. Foraging: Behavior and Ecology. University of Chicago Press, Chicago.

43. Belovsky, G.E., 1978. Diet optimization in a generalist herbivore: the moose. Theor. Popul. Biol. 14, 105–134.

44. Hummel, J., Gee, C.T., Sudekum, K.-H., Sander, P.M., Nogge, G., Clauss, M., In vitro digestibility of ferm and gymnosperm foliage: implications for sauropod feeding ecology and diet selection. Proc. R. Soc. 275, 1015–1021.

45. Liu, A.G.S.C., Seiffert, E.R., Simons, E.L., 2008. Stable isotope evidence for an amphibious phase in early proboscidean evolution. PNAS 105, 5786–9105.

279

Self-Defense

281

LEARNING OBJECTIVES

Studying this chapter should provide you with the knowledge to:

- Recognize the three elements of predator avoidance: avoiding detection, evading capture, and fighting back.

- Understand how animals hide from predators, including background matching, countershading, brokenness and disruptive coloration.

- Understand a prey animal's options if it fails to escape a predator's notice and then runs away: confusing predators, engaging in flash behavior, taking evasive action, or advertising prowess at fleeing.

- Describe how prey fight back, and the models of self-defense that incorporate the costs that predators incur at these times.

- Explain other ways to escape predator notice, including diversion and the types of mimicry.

- Understand the importance of vigilant behavior and the role of group formation in vigilance.

- Describe the parallels between the avoidance and deterrence of predation and the ways that animals cope with parasites and pathogens.

10.1 INTRODUCTION

Self-defense is about not being eaten; put another way, it is about not becoming lunch. There are three key elements of self-defense. The first is avoiding detection, which typically involves either hiding or camouflage. The second is evading capture; if an animal is detected, then it must take evasive actions to avoid capture. The third is fighting back if an animal is in the grasp of a predator.

Self-defense is a deadly serious game, similar to "Hide and Seek" between predators and prey, that plays itself out over evolutionary time. The current strategies that potential prey use to evade being eaten reflect countless generations of evolution. The evolutionary game results in adaptation, counteradaptation, counter-counteradaptation, and so on, leading to ever more sophisticated defenses and counterdefenses. Co-evolution between the hunter and the hunted is the powerful force underlying self-defense.

Much of self-defense is about an animal's structure (morphology) and its coloration,[1] and a major portion of this chapter is necessarily devoted to defenses that are largely structural or visual (e.g., color), with behavior playing a smaller, but nonetheless essential, role. Often behavior makes a morphological adaptation effective. Looking like a dead leaf works for a moth only if it locates itself among dead leaves and is motionless, and looking like a stick is protective for a walking stick only if it blends in among actual sticks (see Figure 10.1).

The section at the end of this chapter addresses how animals avoid and respond behaviorally to diseases and parasites. There are some similarities between antipredator and antiparasite behavior that connect these two topics. After all, parasites and predators are on a continuum defined by exploiting another animal. Predators do so directly and with lethal results; parasites, more often, do so by consuming host resources or host tissues. Once parasites are included in the crowd of animals that exploit other organisms, it is a small logical step to consider the behavioral effects of being parasitized or infected by a pathogen. Responses to parasitism and disease complete the picture of how animals defend themselves. Each of these elements of self-defense will be examined in greater detail in the pages that follow. All of the strategies are subject to a number of constraints that influence all of behavior. Those constraints have appeared in this text before—the realities of trade-offs and the limits of evolution. Large animals may resist some predators, but they have a hard time being cryptic, and spines and toxins are not equally available, evolutionarily speaking, to every kind of animal.

Finally, it is useful to recall that self-defense—especially predator avoidance—can have profound effects on the distribution of prey species. Most animals base habitat selection and activity patterns at least in part on the presence, absence, and habits of their predators.[2]

FIGURE 10.1
Dead leaf moths and walking sticks are almost impossible to see if they choose the right backgrounds. *Photos: Left and right, Jeff Mitton; center, Michael Breed.*

Hide and Seek

The game of Hide and Seek can teach us more than we might imagine about how animals escape predation. Children all over the world play this game in some form or other. In general, one or more seekers try to find one or more hidden individuals who, when discovered, race for "home" or some other designated safe place in an effort to avoid being tagged by the hunter. While the consequences of discovery in a game of Hide and Seek are a good deal less dire than when a predator finds its prey, the game contains at least two elements of antipredatory behavior: avoiding detection and running away. A third element—predator deterrence, or fighting back—is considered bad form on most playgrounds.

A child crouching in the bushes can certainly experience an adrenaline rush as the "hunter" approaches, but—thankfully—this game may be as close as most of us ever get to being hunted. Indeed, humans have probably never appreciated the fine points of predator avoidance as thoroughly as do field mice, many caterpillars, or a variety of zooplankton (to name a few dietary staples of the natural world). After all, we are relatively large animals, and size limits the kinds of predators that can successfully attack us. Moreover, stretching back far into prehistory, humans and their ancestors have lived in social groups, which are effective predator detection/deterrent units, as we will learn in Chapter 13. Finally, our ancestors minimized the threat of predation many generations ago, modifying the surrounding environment to discourage such activity. In short, once again we are bumping into the concept of *umwelt*, which should never be far from our consciousness as behavioral biologists. When it comes to the threat of being pounced upon, shredded, eaten, devoured, and/or consumed, most humans don't have a clue, relatively speaking. This makes what we will learn about how other animals cope with predation that much more amazing.

10.2 CRYPTIC BEHAVIOR: CAMOUFLAGE

Camouflage: An Introduction

The first step in avoiding the predicament of becoming someone else's lunch or dinner is fairly simple: do not be discovered by a predator. *Crypsis*—being hard to see—is a successful strategy for many animals. Being cryptic includes both hiding and *camouflage* strategies; whatever makes an animal hard to find contributes to crypsis. Predators take advantage of camouflage as well; sit-and-wait predators such as mountain lions and rattlesnakes blend into their backgrounds so that unsuspecting prey can wander dangerously close without detecting the predator.

> **KEY TERM** Crypsis is behavior, color, or shape that makes an animal difficult for predators to detect.

283

> **KEY TERM** Camouflage is color or shape that helps to hide an animal from visual predators when it is on its normal substrate.

Many small animals and some large ones (herding animals being one notable exception) have nests, caves, and crevices in which they hide. Although there are certainly predators that enter burrows in search of prey, the use of hiding places is on average a good predator-avoidance strategy; it does impose a cost, however, in that most animals cannot forage or seek mates while hiding. Darkness provides shelter from predators that rely on vision, and for nocturnal animals, foraging and other activities often peak during dark nights and wane on moonlit nights, all the better to

> **KEY TERM** Mimicry occurs when one species evolves to look, sound, smell, or act like another species.

avoid detection, especially by visual predators. Many animals "hide" in plain sight by using camouflage to become more or less invisible or by taking on the appearance of other animals that are less interesting to the predator. (This is *mimicry*, from the word *mime*.)

The word *camouflage* is an unexceptional one—common in everyday use. It is a military term, and refers to disguising something so that it is hidden from an enemy. Its military origins are hardly surprising; they are consistent with its appearance in a chapter about predator–prey interactions. Surprisingly, the first record of the word in the English language

FIGURE 10.2
A male and female Red-Rumped Tanager in Costa Rica. Note the strikingly different visibility of the camouflaged female and the male, dramatically apparent against the green background. Sexual dimorphism in camouflage is quite common in birds.
Photo: Michael Breed.

is in 1917, when the editors of the *Daily Mail* felt the need to define the term for readers. Digging into its French origins reveals that during World War I (1915), the French military actually hired artists to help camouflage equipment and people.[3]

Is it curious that armies had to wait until World War I to discover camouflage? Was this some tragic oversight? Paintings of warfare before the twentieth century are awash with colorful uniforms and brilliant battle standards; there are no muted greens and browns. This great divide in military attire may have more biological significance than is initially apparent. World War I was an unusual war in that it was largely stationary; countless people fought and died in trenches. (Fatality estimates are in the vicinity of 9 million, but no one knows the precise number.) Prior to that time, the flashy appearance of uniforms points to another great biological function of color: communication (see Chapter 7). The brilliant battle standards and the brightly colored jackets identified the troops to friend and foe at a time when there were no telephones of any sort, much less wireless communication. In the trenches, on the other hand, invisibility was more important than identification. The analogy between human uses of camouflage and crypsis in animals is clear and direct. Humans have improved their own camouflage strategies by copying what animals do.

One irony to consider in this discussion of camouflage is that natural selection must balance ability to hide from predators with the ability to attract mates. This often results in sexual dimorphism in camouflage; one sex in a species (usually the females) is cryptic, whereas the other sex (usually the males) is showy (see Figure 10.2). The evolutionary reasons for this await further illumination in Chapter 12, but for now, consider that dimorphism means that males and females differ dramatically not only in appearance, but also in risk from predation.

> **KEY TERM** Countershading is a gradation of color from the dorsal (top side) to ventral (underside) surface of an animal.

> **KEY TERM** Background matching occurs when an animal has evolved to have the same color as its normal substrate.

> **KEY TERM** Brokenness is the use of pattern to break larger blocks of body color into smaller areas.

> **KEY TERM** Disruptive coloration is patterning of an animal, such as zebra stripes, which disrupts its larger outlines.

The following sections analyze the different ways in which natural selection has favored the use of physical appearance to limit detection. These adaptations are largely a matter of color and pattern—and again, although color and pattern may seem to be primarily morphological properties, camouflage works only when facilitated by behavior. Throughout these sections, think about the way behavior interacts with morphology to provide efficient protection from predators.

Types of Camouflage

Camouflage can act in several ways, including *countershading*, *background matching*, *brokenness*, and *disruptive coloration*. Countershading is the gradation from dark color on the top of an animal to a light-colored underside. Animals use background matching when their color resembles that of their normal substrate. Brokenness helps animals to blend into their background by using different shades of color to break large blocks of color into smaller areas. Disruptive coloration also breaks up the larger pattern of an animal's shape, making it more difficult to see; a zebra's stripes are the classic example of disruptive coloration (see Figure 10.3).

Countershading

In the late nineteenth century, the artist Abbott Thayer noticed that many animals show a gradation in color, top to bottom, from dark to light.[4] This sort of color differential is called *countershading*. Because light comes from above, shadow is cast on the ventral surface of an animal. A light ventral surface mitigates this and makes the overall outline of the animal more difficult to see. This coloration works not only for animals such as pronghorn antelope, but also for aquatic animals and flying animals, which may be observed from below. In these cases, countershading effects may overlap with those of background matching (see Figure 10.4). Again, because color can serve several biological functions, it is possible that countershaded animals are also using their color patterns for thermoregulation or signaling.

Countershading is a simple form of camouflage that allows an animal to blend into the darker backgrounds (soil, rock, dried leaves) when viewed from above and into the lighter backgrounds presented by sky or vegetation when viewed from below. Fish that swim upside down, such as some catfish, have reverse countershading. Flatfish, such as flounder and halibut, which have both eyes on one side of the body, have dark shading on their "up" side and light on their "bottom," thus following the general rule for countershading.

Background Matching

If an animal's habitat is generally green, as is true for animals that occupy living leaves, or if the habitat is generally brown, as is the case for animals that crawl on bare soil, rock, or sand (see Figure 10.5), then the behavioral challenge for the animal that exploits background matching is to end up on the appropriate background. A bright green caterpillar may disappear against a healthy leaf but be as conspicuous as bright green can be if it crawls onto a branch.

An animal can simply look like its background; this is called *general background matching*. Sometimes animals that match backgrounds change color with the seasons; polar and high-altitude animals seem particularly adept at this, perhaps because winter in these locations is

285

FIGURE 10.3
These stripes on a zebra (top), and an okapi (bottom)—so noticeable at close range—make them difficult to see at a distance. *Photos: Pen-Yuan Hsing (top) and Michael Breed (bottom).*

FIGURE 10.4
Countershading can be seen in these antelope and penguins. *Photos: left, Jeff Mitton; right, Michael Breed.*

FIGURE 10.5
For this animal, choosing the appropriate background color is critical to its successful crypsis. *Photo: Jeff Mitton.*

FIGURE 10.6
Tobacco hornworms are almost invisible against their host plants, as anyone can attest who has tried to find the thing that is taking big chunks out of the tomato plants! *Photo: R.J. Reynolds Tobacco Company Slide Set, R.J. Reynolds Tobacco Company, Bugwood.org.*

predictably white. Of course, background matching means that the animal must choose a background that matches itself, for in most cases, that choice is going to be faster and less complicated than changing appearance to match a variety of backgrounds. (Some intriguing exceptions to this will be discussed later in this section.) Indeed, many moths and other insects, including the famous peppered moth (*Biston betularia*) of freshman biology fame, not only choose their background, but choose specific locations that enhance their ability to blend into their surroundings (see Figure 10.6). Although crypsis and camouflage in general are elements of appearance, the success of such appearance depends on appropriate behavior. A wriggly, twitchy caterpillar that matched its background would probably fare worse than a less-rambunctious one.

Brokenness and Disruptive Coloration

Vision—this applies to nearly all visual systems—emphasizes edges. Long straight or slightly curved edges, such as the outlines of an animal, are particularly noticeable to the eye and the brain. One of the keys to being hidden is to break up the edges; this makes it more difficult for the visual system to discover the outline of an animal (see Figure 10.7).

Disruptive coloration that breaks up an animal's outline is another form of camouflage (see Figure 10.8). Unlike an animal that looks like its background (background matching), a disruptively colored animal exhibits bold patterns that serve to hide it in a different way, by making its outline difficult to perceive.[5,6] Disruptive coloration differs from a random assortment of contrasting patterns because it works best when the disruptive patches are at the margins of the animal (as it might be viewed by a predator) and not in the middle of the body.

DISCUSSION POINT: CAMOUFLAGE

Find images of military camouflage from various time periods (WWI, WWII, etc.) on the Web. How has the concept of effective camouflage changed over time? Can you critique contemporary camouflage strategies that are digitally generated? How might these be better than older patterns, and how might they be improved? What might an animal behaviorist bring to a discussion about effective camouflage?

FIGURE 10.7
Crypsis and warning coloration in deadly snakes. (left) The fer-de-lance is a deadly poisonous snake that is common in Central and South America. The diagonal slashes across this snake's body break up the edges and, particularly when the snake is coiled, make it cryptic against a background of dead leaves. (Center and right) These eyelash vipers occur in two strikingly different color morphs in Central America. In the middle picture, the snake's green and brown tones, combined with brokenness that disrupts the animal's edges, are effective in making this snake cryptic. The morph in the right picture is highly visible, probably to warn away birds that might attempt to prey on it. Notice how the edges of this morph (type) clearly stand out against the background. *Photos: Michael Breed.*

Changing Colors and Patterns

Recall that cryptic animals often choose backgrounds that match their own appearances. However, it is possible to change appearance to match a changing background and achieve the same undetectable result. As mentioned earlier, many residents of high mountains and the far north, such as snowshoe hares (see Figure 10.9), do this by seasonal color change, turning white in the winter. However, changing appearance can be rapid as well. For this, consider cephalopods, those intelligent mollusks that learn to unscrew jars for food and to escape from their aquaria. They can also match the pattern, color, intensity, and even texture of their backgrounds and, in so doing, become functionally invisible.

FIGURE 10.8
Zebra may be the champions of disruptive coloration. In the distance, individuals are difficult to see and to tell apart. *Photo: Mike Huffman.*

As if that were not enough, octopus can do all of this quickly, within milliseconds![7] How? *Octopus vulgaris* is a typical cephalopod in that it has tiny chromatophores. A chromatophore is a cell in an animal's surface that contains pigment and that has contractile fibers that can expand the cell, thus increasing that pigment across the surface. These contractile fibers are innervated by neurons that proceed directly from the octopus brain to the chromatophore. There may be hundreds of chromatophores in a square millimeter of octopus dermis; any given chromatophore contains one of several pigments. Thus, if a dark patch occurs, that is the result of the contraction of fibers around the margins of chromatophores that contain dark pigments. The pigment cell expands, and the dark color spreads. Thus, the octopus can match the background precisely and quickly. . .and if that were not enough, an octopus can do all of this in three dimensions! The octopus skin papillae can change shape or disappear entirely, all the better to match background texture as well as color.[8]

287

Keep in mind that while the cephalopod is matching its background, it is aware of potential predators and aware of potential prey. It may be moving slowly, foraging, as it shifts its appearance slightly but continuously. How does it know how to match its background? What neurophysiology surrounds this ability to perceive the environment in a way that allows an animal to almost instantly disappear, for all practical purposes? Amazingly, Hanlon and coworkers discovered that all the great variety of cephalopod camouflage patterns can be sorted into three categories: uniform patterns (very little variation), mottle patterns (light-dark contrast on a small scale), and disruptive patterns (larger-scale contrast).[9]

FIGURE 10.9
Seasonal color change is common in high elevation and high latitude animals such as the snowshoe hare and arctic hare. In summer, the brown, mottled fur is effective camouflage in the subalpine woods. In the winter, the white fur blends well with the snow. *Photos: left, Michael Breed; right, US Fish and Wildlife Service.*

They raised an intriguing question: If the Masters of Camouflage use only three basic camouflage patterns, are predators that easy to fool? Would Professor Tinbergen be pleased with the number of answers we can bring to his questions?

Finally, cephalopods can be used to introduce the next strategy for hiding: mimicry. Mimicry differs from background matching; although in a stretch, an animal engaged in background matching might be said to mimic the background, mimicry usually involves looking like some other organism. In background matching, the animal often escapes notice. A mimic, on the other hand, may be noticed, but it is perceived to be something it is not.

OF SPECIAL INTEREST: RELYING ON THE PROTECTION OF OTHERS

By taking advantage of another species' tendency toward vigorous antipredator behavior, less-combative species can improve their own survival. This seems to be the case with Azure-winged magpies (*Cyanopica cyana*), which synchronize their reproduction with that of Japanese lesser sparrowhawks (*Accipiter gularis*). They do this only if they are in the vicinity of the sparrowhawks; magpies nesting far from sparrowhawks do not breed synchronously. Choughs (*Pyrrhocorax pyrrhocorax*) also select nesting sites near those of lesser kestrels (*Falco naumanni*) in Spain; in a study, the kestrels noticed a stuffed predator (an eagle owl, *Bubo bubon*) more quickly than the choughs did and responded more aggressively.[10,11]

Tests for the Effectiveness of Camouflage

Given the problems just described, scientists have turned to artificial prey in an attempt to ask questions about the function of camouflage. One such test was conducted by Schaefer and Stobbe, who used artificial moths to determine that in a heterogeneous environment, disruptive coloration was a more effective method of concealment than background matching.[12] Indeed, because disruptive coloration obscures an animal's outline, it is effective on many backgrounds, whereas background matching requires a specific background to work well. In the case of disruptive coloration, they found that the same color patterns, when distributed on the interior of the "moth" wings rather than the margins of the wings, did not conceal the animal as well.

There are surprisingly few tests of the efficacy of camouflage. Remember that what seems almost invisible to a person might be obvious to a predator, so what a scientist imagines as camouflage may be less effective than it seems. Another problem with testing hypotheses about camouflage is the fact that color itself can have several functions, ranging from concealment to thermoregulation, intraspecific communication to interspecific warning. It is difficult to devise experiments that sort out the relative importance of these influences. Another drawback is the difficulty of experimental comparison; if an animal is conveniently polymorphic, the morphs can be compared under different conditions, but if such variety is absent from natural populations, then scientists are faced with comparing the naturally camouflaged animal to a conspecific that they have somehow manipulated. In addition, not all color is equal; that is, a given hue may have a wide range of properties (e.g., reflectance) that affect its visibility. Again, humans may not be aware of all of this variation.

Distraction and Dazzle

Before leaving camouflage, let us consider how an animal can discourage predators by making itself more, rather than less, obvious. This is called *startle behavior*. Good examples are nonvenomous snakes, such as bull snakes, which rattle their tails in dead leaves when disturbed. Bull snakes also do a good job of coiling and gaping their mouths; these behaviors attract attention, but because the bull snakes use behavior to disguise themselves as venomous snakes, they gain protection against predators (see Figure 10.10). Another way

KEY TERM Startle behavior is the sudden production of movement, sound, or another stimulus that distracts a predator.

288

to startle a predator is to have a feature that is hidden but surprising when revealed. The eyespots on the wings of many kinds of moths may fit into this category (see Figure 10.11). A momentary pause on the part of the startled predator may provide just enough time for the prey to escape. Squeaks, screams, and buzzes produced by captured animals may also have startle value.

Aposematism

Chemical deterrents can be used before or after a predator's attack. In the latter case, poisons, along with repellants, are effective. They can give favor to the evolution of aposematic coloration, in which a brightly colored, easy-to-remember appearance, when combined with the disastrous experience of tangling with such a noxious prey, causes predators to avoid an animal with that appearance in the future. *Aposematism* (Greek, *apo* = away, *sematic* = sign) is the use of warning coloration to inform potential predators that an animal is poisonous, venomous, or otherwise dangerous. Oftentimes orange or red patterns may be warnings (as in coral snakes), but do not assume that red is always a warning (see Chapter 7). Examples of warning colors and patterns are shown in Figure 10.12.

FIGURE 10.10
Bull snakes can be mistaken for rattlesnakes. Their behavior encourages such mistakes; they coil and behave as if they will strike . . . but unlike rattlesnakes, have no venom.

Although many foul-tasting prey are aposematic, some animals use chemical defense as a last resort and do not advertise it. This seems to be true in the well-camouflaged silkmoth caterpillars. When discovered and attacked, they emit a series of clicks using their mandibles, accompanied by regurgitation of deterrent liquid. This phenomenon was first discovered by neuroethologist Jayne Yack,[13] who brought caterpillars home with her when no one was on campus to care for them. In truth, it was first discovered by Yack's cat, who was, in turn, discovered (by Yack) gagging in the presence of a caterpillar and its regurgitant. This led to a series of experiments that confirmed the hypothesis that both acoustic aposematic signals and chemical deterrents are used by these caterpillars when camouflage fails. (The cat, which gagged for years thereafter any time it saw a similar caterpillar, also seemed to confirm the hypothesis.)

FIGURE 10.11
The eyespots of this *polyphemus* moth are hidden until the wings are spread; this is thought to startle predators. *Photo: Lacy L. Hyche, Auburn University, Bugwood.org.*

10.3 VIGILANCE AND ALARM

When a predator is nearby, it helps tremendously to know where the predator is and, if possible, what its intentions are. Behavioral biologists call this *vigilance.*[14,15,16] Much of what

FIGURE 10.12
Three examples of potentially aposematic coloration. Right: The poison frog is mildly toxic and certainly stands out against the monotonous greens and browns of the rainforest. Center: The seed-feeding bug (family *Coreidae*) has red and yellow patterning that may warn that it contains toxins. Left: A toxic tetrio sphinx caterpillar. This caterpillar may also mimic the color pattern of a coral snake! *Photos: left and center, Michael Breed; right, Jeff Mitton.*

> **KEY TERM** Vigilance is the awareness that an animal has of its surroundings and the potential presence of predators.

FIGURE 10.13
Prairie dogs chatter, whistle, and sound the alarm when predators are in the neighborhood. Prairie dog coteries are matrilineal, and animals that give alarm calls tend to be animals that are related to others in the area. *Photo: Jeff Mitton.*

is known about vigilance is related to social behavior; for instance, vigilance is usually more effective in groups than for solitary animals. Groups have a hard time hiding, but they are good at detecting predators. Nonetheless, seeing and observing a predator is only the beginning of how animals cope with predation. Clearly, it is advantageous for the (vigilant) prey if the predator is not equally aware of the prey. A vigilant animal is attentive to predators or parasites. Being vigilant takes effort away from other important tasks, such as feeding, but it is a critical part of the time budget of many animals (see Figure 10.13). Because vigilance plays a key role in self-protection, it is a central topic that crops up in several chapters. Here, vigilance is placed into the array of self-defense behaviors that an animal might employ. Chapter 13, on social behavior, explores how vigilance is shared among members of social groups. Chapter 15, on conservation, discusses how measuring vigilance helps to assess the effects of human disturbances on animal populations.

To further consider vigilance, imagine camping out—alone—in a wilderness area, one of the few remaining parcels that is home to wolves, bears, and other carnivorous megafauna. It is dark; the fire (if it is even allowed) is not going to last the night; and the tent is a wonderfully lightweight, high-tech marvel that is great for rain but no match for teeth and claws. And by the way, the wilderness area abuts a maximum security facility, and there is an escaped convict on the loose. Now imagine camping with friends in a circle of 10 such tents. Does it feel safer to be camped alone or as part of the circle?

BRINGING ANIMAL BEHAVIOR HOME: VIGILANCE AND HOUSEHOLD ANIMALS

What are the results of vigilance in household animals? Dogs and cats are both vigilant about intrusions into their space, but the results differ dramatically between the species. Both dogs and cats can distinguish familiar animals and humans from unfamiliar intruders. Watchful cats often freeze or flee when a stranger enters a room. Within the long list of animals kept in people's houses, only one, the dog, shows a strong desire to communicate alerts about intrusions by strangers to other members of the household. This is hypothesized to have been one of the benefits for early humans as they began to form associations with dogs, and continues to this day.

Despite the security that a vigilant dog can offer, barking at intruders often becomes an issue for dog owners. While vigilance and barking are a normal part of dog behavior, dogs that bark can annoy neighbors. Stranger intrusion is only one possible stimulus for vigilance barking, but this can be a particularly difficult issue in managing dog behavior. Barking can be discouraged by distraction, followed by praise and rewards when the dog does not bark; owners can "stage" situations in which friends walk their dogs by the owners' house and the dog can be trained to associate calmer reactions with reward. Punishments are not very effective in this context.

Some trainers recommend clicker training as a way to deal with barking behavior. They encourage training the dog to bark on command, that is, putting the bark on cue. If the command to bark is then rarely or never given, the dog's impulse to bark may be extinguished. "No bark" may also be taught once the bark can be cued; the dog must understand what the behavior of interest is before it can learn commands aimed at ceasing the behavior.

The urge to be among more tents reflects one of the benefits of group living—enhanced vigilance, also known as the "many eyes" effect. Assuming that the escaped convict is not among the 10 friends inhabiting that group of tents, being with those friends is a much safer situation than being alone. Humans are not the only organisms to have figured this out. A 1970s classic study with wood pigeons and a trained goshawk showed that birds in larger flocks respond more quickly to the appearance of the predator than ones in smaller flocks; that is, they take off when the goshawk is farther away.[17] Indeed, there is good evidence that in many group-foraging birds (e.g., geese), the amount of time that each organism is vigilant (head up and watching versus head down and foraging) is inversely proportional to the size of the flock. Animals in groups can share vigilance and can profit if another animal becomes alarmed, or sends out an alarm signal. Antipredator systems that rely on vigilance and alarm are greatly enhanced if animals are near to each other.

Obviously, alarm is an important result of vigilance. When animals in groups perceive danger, they frequently behave in ways that alert other animals to the danger.[18] Here it is important to understand the difference between cues picked up from other animals and actual alarm signals, which have evolved for the purpose of conveying alarm. Sometimes the alert is simply a matter of being tuned in to neighboring animals; in a feeding flock of geese, one or more birds will be vigilant and the rest feeding. If a goose becomes alarmed, its head pops up, then the goose next to it looks up, and so on, so the heads-up alarm passes as a wave through the flock.[19,20] Indeed, the trade-off between feeding and vigilance is one of the major constraints of antipredatory behavior.[21,22,23] In other cases, alarm is signaled—using ways of communicating alarm that have evolved specifically for that purpose (see Chapter 7 on communication). Alarm signals can be chemical, visual, or auditory.

Signaling alarm can be costly; when an animal makes itself more visible, makes a sound, or emits an odor, it may attract the predator to itself. Because alarm signaling is often costly, alarm signals are expected to have evolved only when there is an evolutionary benefit to the signaler. Generally this benefit comes from alerting closely related animals—siblings or offspring (see Chapter 13 on social behavior). In other words, overt alarm signals should occur in family groups. Often animals can "eavesdrop" on alarm signals, gaining valuable information that may not have evolved for their benefit. Eavesdropping can happen between family groups or even across species lines.

DISCUSSION POINT: HUMAN VIGILANCE

Even humans display vigilance behavior in ways that might be surprising. Of course, we are alert in dark alleys and strange, forbidding places, but there is evidence that we are routinely vigilant in ways that perhaps elude consciousness. For instance, when eating in a group setting (e.g., cafeteria), individual humans looked up (vigilance scanning) less often and for shorter time periods as the size of their group(s) increased. People did not alternate scans or show any other evidence of coordinated scanning, which is consistent with our understanding of scanning behavior in animals that have fluid societies.[24] How might scanning be compared between groups of friends and groups of strangers? Should friendship make a difference in shared scanning?

10.4 MIMICRY AND DIVERSION

Unpleasant experiences and even noxious tastes have given rise to one of the most fascinating areas of mimicry because animals experience lower predation risk when they look like bad-tasting, dangerous, or poisonous animals. This mimicry seems to work whether or not the mimic itself tastes bad. Henry Walter Bates, a contemporary of Darwin's, first described the mimicry that bears his name—*Batesian mimicry*. In this type of mimicry, a palatable prey (the mimic) is favored by natural selection if it resembles an

> **KEY TERM** Batesian mimicry is a behavior in which a benign food item (prey) looks like or behaves like a distasteful or poisonous species.

unpalatable species (the model); it will benefit if predators learn that an animal that looks like the model is not worth eating. For this to occur, the model needs to be relatively abundant. At the other extreme, *Muellerian mimicry*, named after Fritz Mueller, occurs when two unpalatable species converge on an appearance; both may benefit by increasing the number of "teachers" that are informing predator-"students" about these bad-tasting prey. In sum, mimicry covers a broad range of evolutionary possibilities (see Figure 10.14).

> **KEY TERM** Muellerian mimicry is a behavior in which noxious animals converge on the same appearance or behavior.

Batesian mimicry, in which a benign food item looks like or behaves like a distasteful or poisonous species, and Muellerian mimicry, in which noxious animals converge on the same appearance or behavior, are important self-defenses; examples range throughout the animal world. One hypothesis for the brightly colored but nontoxic king snake is that it may mimic the highly poisonous coral snake. Both have red, yellow, and black bands in differing arrangements: "Red and yellow, kill a fellow; red and black, venom lack." Many butterflies are Batesian mimics of other butterflies, whereas yellow or orange stripes on the abdomen denote Muellerian mimicry among stinging insects.

Probably every student has experienced a wasp or bee sting and learned, through that experience, to associate yellow/orange and black banding in a flying insect with the pain of the sting. Birds and nonhuman mammals make the same association; this results in protection for stinging species with similar appearances (Muellerian mimicry) and for nonstinging species that look like a bee or wasp (Batesian mimicry). Color and shape tell part of the story, but for a species to be a really convincing mimic of a stinging insect, behavior comes into play. Flies that look like bees add to their deceptive story by being around flowers. Flies give themselves away to knowledgeable humans (particularly entomologists) by having a different pitch to their buzz and a propensity to hover rather than to dart when they fly. The extent to which a good morphological mimic is also a good behavioral mimic often determines its success.

FIGURE 10.14
Stinging insects and their mimics. (A) This is a yellowjacket, a wasp with a very painful sting. (B) A honeybee. (C, D) Flies that mimic stinging insects. *Photos: (A) Whitney Cranshaw, Colorado State University, Bugwood.org; (B) Deng Xiaobao; (C) Susan Ellis, Bugwood.org; (D) Johnny N. Dell, Bugwood.org.*

DISCUSSION POINT: BATESIAN OR MUELLERIAN MIMICS?

Experimentally, it can be difficult to discover whether a mimic is Batesian or Muellerian.[25] For many years, scientists thought that Viceroy butterflies were Batesian mimics of Monarch butterflies (see Figure 10.15). More recently, studies have shown that Viceroys can be quite unpalatable and that some populations of Monarchs can be tasty to birds. There are actually many instances all along this continuum between palatable/unpalatable and unpalatable/unpalatable pairings in which mimics may be unpalatable, but not as unpalatable as models. In fact, palatability can vary even within a population, so the story in at least some cases is not as neat as that presented here. Nevertheless, mimicry does happen, and in most cases, the thing that is mimicked is the model's aposematic advertisement of unpalatability. Do your own literature search to explore this issue further, and discuss with your classmates the interaction between Batesian and Muellerian mimicry.

FIGURE 10.15
A monarch (left) and a Viceroy butterfly (right), long thought to be a tasty mimic of the noxious Monarch butterfly. Recent research has blurred this distinction. *Photos: left, Jeff Mitton; right, Thomas G. Barnes/U.S. Fish and Wildlife Service.*

FIGURE 10.16
In addition to *Hemeroplanes*, other caterpillars such as this spicebush swallowtail (*Papilio sp*) have benefitted from looking like snakes. *Photo: Ronald F. Billings, Texas Forest Service, Bugwood.org.*

Sometimes animals defend themselves by imitating dangerous things. A number of caterpillars have developed an alarming resemblance to snake heads. Perhaps the most realistic example of this is the *Hemeroplanes* caterpillar (also known as a "viper worm"). When startled, it inflates its head and appears remarkably snake-like (Figure 10.16). The mimic octopus (see earlier description) comes to mind again.[26] When disturbed or in the presence of predators, this octopus is known to mimic other animals, such as sea snakes; tellingly, the animals it mimics are often dangerous ones.

A rather dramatic octopus was recently discovered in a decidedly undramatic habitat. Indeed, the ordinary and even uniform appearance of the background (silt and sandy littoral areas near the mouths of rivers in Indonesia) may account for both the reason that the octopus remained undiscovered until 2001 and also the adaptive influences that cause the octopus's mimicry strategy to be so unusual. This mimic octopus (*Thaumoctopus mimicus*) mimics a variety of poisonous vertebrates, ranging from poisonous fish to sea snakes, possibly because camouflage against such a uniform background might not be highly successful. (Be ready to consider the phenomenon of mimicking dangerous animals later, in the context of deterrence.) In addition, however, when fleeing, this octopus can mimic a flounder, but then match its background (camouflage) when it stops. Clearly, although predator avoidance strategies can be categorized for ease of organized discussion, to the animal doing the predator avoidance, they blend together with remarkable effectiveness.

OF SPECIAL INTEREST: MASS OVERWINTERING BY SNAKES

Garter snakes (*Thamnophis sirtalis parietalis*) overwinter in dens. When they first emerge, they are cold and sluggish. In addition, emerging females are almost immediately set upon by hordes (>100) of males in mating balls. Emerging males may mimic females, producing a pheromone that is typical of a female. It was first thought that this behavior was a reproductive tactic that allowed the female-mimicking males (called *she-males*) increased access to females. Further research with larger sample sizes and temperature-sensing devices revealed that this was a male strategy that could have two benefits: protection from predators in the mating ball and faster posthibernation warm-up. The reproductive males seeking females have a body temperature of over 25°C; the ground is more than 15° cooler. Females that are the subject of male attention increase their body temperatures from 4° to 20° in 30 minutes. Males in a similar situation receive similar advantages. The she-male behavior is transient and disappears as a she-male warms up.[27]

Animals also mimic a variety of substances. Some caterpillars look deceptively like bird droppings, others like twigs; leafhoppers may look more like leaves than like insects, and a variety of treehoppers mimic thorns. Perhaps even more deceptive, some animals disguise parts of their bodies so that they resemble other parts.[28] This behavior is called *diversion*. For instance, some Lepidoptera (e.g., *Thecla togarna*) have wing markings that look like heads (complete with long "antennae") on the posterior parts of their wings. There is some debate about how redirecting attack benefits the prey, but the consensus is that the predator's attack is indeed redirected by such mimicry. In lizards, tail-flicking may serve a similar purpose. In this case, the tail does not mimic anything in particular, but because it moves more than the rest of the animal, it draws the predator's attention away from more vulnerable areas. If the tail is seized by the predator, the lizard can autotomize (break off) the tail and escape.[29,30] In at least one species of lizard, *Acanthodactylus beershebensis*, this diversion is correlated with ontogenetic changes in foraging behavior.[31] Young (<3 weeks old) lizards are active predators, moving about more than older lizards and frequenting more open microhabitats. Because of this, young lizards probably expose themselves to predators more than older lizards do, which tend toward sit-and-wait foraging behavior. The young lizards also have blue tails, which they wiggle and wave more often than adults do; the blue color of the tails fades with age, along with the wiggling and waving behavior. Thus, conspicuous tail color, conspicuous tail movement, and conspicuous foraging habits are all common in young lizards, and not common in older lizards. Two other lizards in the same family do not exhibit such changes, nor do they shift foraging behavior with age. Finally, even scent mimicry is possible: It is thought that ground squirrels engage in defensive scent mimicry when they chew rattlesnake skins that have been shed and then lick their fur.[32]

FIGURE 10.17
A killdeer feigning injury.
Photo: Tracy Thomas.

Animals may even mimic a wounded and vulnerable version of themselves. Killdeers are well known for their "broken wing" displays. In this display, a parent bird, upon spotting a predator, will move away from its nest, dragging an "injured" wing. It will do this, continuing to move just out of the predator's reach, until it is a safe distance from the nest; then it will fly away. The predator, in the meantime, has been lured away from the bird's offspring (see Figure 10.17).

Finally, some animals go so far in self-mimicking as to mimic their dead selves. This behavior is called *thanatosis*, from a

Greek word meaning "a putting to death," and is common not only among insects, but also vertebrates. Virginia opossums (*Didelphis marsupialis*) are perhaps most famous for this behavior and have given rise to the expression "playing possum," in reference to not only death-feigning, but almost any kind of duplicity, especially that involving health.[33] The adaptive benefits of the behavior have been rigorously tested in flour beetles. Researchers selected two strains of *Tribolium* beetles for long and short death-feigning episodes, and showed that after 10 generations of selection, which resulted in a clear thanatosis difference between the two lines, those selected for thanatosis were far more likely to survive the attentions of a predatory spider.[34]

10.5 EVASION

Not all prey are cryptic, and not all crypsis succeeds. In this event, when a prey animal attracts the unwelcome attention of a predator, two major strategies remain: fleeing and fighting. These strategies are not mutually exclusive; fleeing offers fewer choices than fighting, which can assume many forms, ranging from outright combat to chemical defenses (e.g., venom and repellants), morphological defenses (spines, armor), and simply very bad taste.

In the case of fleeing, the choices are usually fairly straightforward, although mimicry can once again play a role, usually by confusing a predator. Cephalopods (again!) are experts at this confusion strategy, using not only chromatophores but also ink sacs. These ink sacs are a modification of part of the lower intestine; they empty near the siphon, so when the mollusc jet-propels itself to safety, the ink (if released) squirts out with the mantle cavity water. The large dark cloud created by the ink may have several effects on the predator. First, recall that the cephalopod can change colors rapidly; if the cephalopod turns a very light color as it flees (after the ink release), the predator may be more attracted to the dark image where the cephalopod was (and what the cephalopod looked like a moment ago) than to the light-colored apparition receding in the distance. Even if there is no color change, the dark mass of ink may be confused with the cephalopod, by now rapidly decamping. Second, the ink contains chemicals that can hamper the olfactory abilities of the predator or be otherwise unpleasant. Some workers have suggested that the ink also acts as an alarm pheromone, but this is difficult to distinguish from other antipredatory properties of the ink; if it is chemically unpleasant, other cephalopods, as well as predators, may react negatively to it.

"Flash" behavior also mixes predator confusion with fleeing. In flash behavior, an animal flees and then freezes. One common example of this is the flight of grasshoppers. When fleeing a disturbance, a grasshopper shows different colors (e.g., colored second wings), which then disappear when the insect alights and remains motionless.

Animals that run away without freezing often tend to do so in an erratic, zig-zag fashion, introducing an element of unpredictability into their escape. This makes it difficult for the predator to know where the prey is and also makes the prey difficult to chase with great speed, even if it can be followed. Noctuid moths are some of the champions of this behavior. They have membranes on each side of the thorax that act as "ears" that are connected to the A1 and A2 neurons. The A1 neuron alone is stimulated when the predator, in this case a bat, is far away and the sound is not loud; under these conditions, the moth simply flies away, a feat that is made possible by a comparison of input from right and left neurons. When the right and left A1 neurons are responding equally, then the moth is moving directly away from the bat. If the sound from the bat is loud, however, both A1 and A2 neurons fire; in this case, the moth wingbeats are desynchronized, and the moth flies erratically and falls through the air in an unpredictable pattern, perhaps the ultimate in evasive action.[35]

Finally, many long-legged animals simply run very fast. Deer and their relatives are excellent examples of this strategy. Although unadorned running away is an obvious escape strategy, its importance should not be underestimated. The intense selection on predators for speed and stealth (e.g., cheetah, leopard) is testimony to the value of a fast getaway.

Stotting is an intriguing behavior in which the prey is thought to actually advertise its readiness to take on a race with a predator.[36] Ungulates such as antelope and deer perform this behavior by jumping, stiff-legged, with all four feet leaving the ground simultaneously. Of course, like many behaviors, it might also function as a warning display or a diversion tactic if young are in the area, but observations of the effect of Thomson's gazelle stotting on cheetahs showed that cheetahs were less likely to pursue a gazelle that had stotted, and if they did pursue it, they were more likely to fail in the attempt to catch it.

Stotting therefore combines potential fleeing behavior with a deterrent; most predators do not voluntarily waste time or energy in high-speed chases. They may also be deterred if the prey seems dangerous.

10.6 PREDATOR DETERRENCE AND FIGHTING BACK
Physical Deterrence

Spines, horns, teeth, large size, and a variety of other weaponry can keep predators at bay. Many animals use hair, feathers, or posture to appear larger than they actually are; they do this in intraspecific conflict (see Chapter 11 on mating) and when confronted by other threats, including those from predators. Although large animals such as porcupines and elk are usually associated with this defensive equipment, small crustaceans can also be armored. One of the functions of spines on cladocerans may be to thwart predators; in some experimental situations, cladocerans grow longer spines in the presence of predators.[37,38] Spiny rotifers are also less likely to be eaten (by other rotifers) than less well-protected conspecifics.

Predators do well to be cautious of armored or aggressive prey. If predators are injured in the process of acquiring a meal, such wounds may, at best, hamper future foraging and, at worst,

296

OF SPECIAL INTEREST: SPINY EXOSKELETONS

As we have emphasized, few benefits are without their costs, and this is certainly true of the arthropod exoskeleton. Although its tough construction is thought to have played a major role in making arthropods the most successful phylum on the planet, the fact that it is hard and largely inflexible requires that it be molted periodically if the animal is to grow. The molt is a hazardous time for arthropods; emergence from the old exoskeleton may be fraught, and even if that proceeds smoothly, their new body armor is soft and easily punctured until it hardens. For this reason, many arthropods behave secretively and hide during the time surrounding their molt.

Spiny lobsters (*Panulirus argus*) can use their exoskeletons for defense even when they are newly molted (see Figure 10.18). When alarmed by a predator, they make a loud rasp that is thought to startle the predator. Unlike many arthropods that make sounds by rubbing hard ridges of the exoskeleton together (see Chapter 7), spiny lobsters use structures that do not need to be hard in order to produce sound.

FIGURE 10.18

Despite their name, spiny lobsters defend themselves with their hardened exoskeleton, but when it is soft and newly molted they make a noise that repels predators, rubbing an extension of the antenna across a "file" like area near their eyes.

They rub an extension of the antenna across microscopic shingles on a "file" located below the eye, producing sound in much the way a violin does, although perhaps with less-pleasant results for the predator.[39]

cause death if infected. Since the discovery of penicillin, humans have lost much of their awareness of the dangers of infection, yet it was always a lurking danger prior to the discovery of antibiotics, and may be again if we continue to ignore the impact of widespread antibiotic use on the evolution of drug resistance in pathogens. Death from infection is not an uncommon occurrence in a world without antibiotics, and that is the predator's world.

Individual animals are known to react aggressively toward predators; a variety of birds and small mammals are noticeably unfriendly to predatory snakes. A special case of physical deterrence involves group defense; this is often called *mobbing*. Smaller birds, for example, may mob a crow or hawk attempting to steal eggs or hatchlings from a nest.[40] In turn, crows sometimes use mobbing against larger birds, such as eagles (see Figure 10.19). Each of the individual prey may be relatively defenseless on their own, but as a group they can put up an effective defense, dive-bombing and harassing the predator. Some behavioral biologists see mobbing as a type of aggression (see Chapter 12).

Chemical Deterrence

There are two fundamental kinds of chemical deterrence: those that repel a predator before it attacks and those that create unpleasant experiences for the predator after an attack. Many organisms secrete, emit, squirt, or exude what collectively amounts to a pharmacopoeia of noxious repellants. The skunk is perhaps the most dramatic of the group. Although biologists have dutifully described four genera and several species of skunk, the names of the family (Mephitidae) and type genus (*Mephitis*) translate from Latin as "noxious vapor" and leave no doubt about the skunk's chemical deterrence strategy. When disturbed, it stomps its feet, hisses, growls, and displays its vivid striped pelage, all prior to an accurate 2–5 meter squirt of an obnoxious mixture of sulfides and mercaptans. Recent observations of responses of carnivores to models of skunks and gray foxes showed that mammalian carnivores approached skunk-like models more cautiously than they approached gray fox models (both pattern and shape were important to the carnivores), and that the degree of caution is positively related to the abundance of skunks in the area. Thus, skunk appearance qualifies as aposematism, that is, a conspicuous warning from a distance that this animal might be best left alone.[41] Aposematism is due more consideration shortly, in a discussion of negative postattack experiences.

Animals other than skunks use chemicals to fend off attackers. Beetles, in general, need good defenses because, given their elytra, they are frequently slower to take flight than some other insects. For example, carabid beetles have paired abdominal glands containing chemicals that are mixed upon release to impressive effect. The bombardier beetle (e.g., *Stenaptinus insignus*) stores hydroquinones and hydrogen peroxide in one chamber and enzymes including peroxidases in the other. When these come together, a rather intense chemical reaction takes place, with oxygen released from the hydrogen peroxide and reacting with the hydroquinones to produce both great heat ($100\,°C$) and noxious chemicals (benzoquinones) that can be aimed from the abdominal tip with surprising accuracy (see Figure 10.20). Attacking a bombardier beetle cannot be a good experience, and eating one (should an animal somehow persist through this deterrence) must be downright miserable. Bombardier beetles are high on the list of Animals We Do Not Wish to Consume.[42]

FIGURE 10.19
Mobbing is a common occurrence among birds. Smaller birds form groups and harass larger birds, including birds of prey, making it difficult for the larger bird to hunt or remain unnoticed.

297

FIGURE 10.20
A bombardier beetle spraying noxious, heated, defensive secretions. This behavior effectively deters ants and other predators. *Photo: Copyright (1999) National Academy of Sciences, U.S.A.*

OF SPECIAL INTEREST: CHEMICAL DEFENSES

A wide assortment of poisonous or distasteful animals get their noxious chemicals from their diets. For instance, when feeding on milkweed, monarch butterfly caterpillars sequester the cardiac glycosides and other poisons that the host plant contains, thus discouraging many (but not all!) predators. A variety of nudibranchs have special evaginations (external pouches called *cerata*) of the dorsal and lateral body wall that contain the undischarged stinging cells from the corals on which they graze. These cerata—along with other noxious substances—discourage and distract predators and provide a good deal of protection for these mollusks that have (over evolutionary time) lost their shells. Lovely poisonous frogs get their toxic compounds from their arthropod diets.

The use of another animal's chemical defense can be related to other aspects of antipredator behavior. The Asian snake *Rhabdophis tigrinus* has special nuchal glands that contain bufadienolides (cardiotonic compounds) in the skin of the dorsal neck area. (*Nuchal* indicates the nape of the neck.) When confronted by a predator, the snake actually directs this area of the neck toward the attacker. Oddly, the nuchal glands do not possess secretory cells, and few of their cells contain organelles associated with secretory activity (e.g., Golgi apparatus and ER); they are, however, heavily vascularized. This histological evidence led researchers to hypothesize that the glands store chemicals acquired from an external source.

By comparing snakes from an island that had abundant toads with snakes from a toad-free island, Deborah Hutchinson and colleagues showed that the nuchal glands of the former group were rich in bufadienolides, whereas the latter group had none. However, the snakes from the toadless island had not lost their ability to sequester these poisonous compounds. They, too, could accumulate bufadienolides when fed toads. In addition, female Asian snakes with a good supply of such compounds can pass them on to their offspring. Finally, while snakes from the toad-rich island used their nuchal glands as part of a defensive posture, snakes from the toad-free island were more likely to flee from a predator.[43]

FIGURE 10.21
Predators such as this rattlesnake can use their predatory equipment as defensive equipment as well. In the case of rattlesnakes, some defensive bites may not deliver venom. *Photo: Jeff Mitton.*

Venom, Stings, and Defensive Bites

Predators often become prey. As discussed in Chapter 9 on foraging, predators that are higher in the food web benefit from consuming lower predators because this eliminates some of potential competition. Ecologists refer to middle-sized predators as *mesopredators*; coyotes, for example, consume domestic dogs, foxes, and smaller predators but are themselves vulnerable to predation by wolves. Top predators, such as mountain lions in North America, are typically not potential prey because of their size and armament. Much of the weaponry used by predators can be turned to defensive purposes. A classic example is the venom of pit vipers, which evolved to allow the snakes to immobilize prey, but which is also a potent deterrent against potential predators (see Figure 10.21). Sometimes defensive bites by vipers are "dry," meaning that no venom is transferred; the dry bite has the effect of inducing fear in the potential predator while at the same time preserving venom for later use in predation.

> **KEY TERM** A mesopredator is a middle-sized predator that occupies an intermediate position within a food web. Mesopredators are potential prey of top predators, as well as being predators.

10.7 PATHOGEN AVOIDANCE/DETERRENCE AND SICKNESS BEHAVIOR

Pathogen Avoidance

Although many behavioral biologists like to think about lions and tigers more than tapeworms and ticks, parasites, by definition, have negative effects on fitness. The ways that animals alter behavior so as to avoid infection are just now becoming apparent to behavioral biologists. Humans tend to think of host defense as something that happens after an animal acquires a parasite, when the immune system is activated. That certainly occurs, and it has behavioral components that will be addressed shortly. However, even as hiding successfully from a predator may be less risky than fighting against one, so, too, the use of behavior to avoid infection altogether can be at least as valuable as fighting off the infection.

Ectoparasites provide some of the best examples of parasite avoidance. Although humans are inclined to view ectoparasites as more of an annoyance than a serious threat, this (again) is not representative of the vast majority of animals that have no shelter from, say, biting flies, and no medicines with which to combat the diseases that they transmit. To most animals, ectoparasites can be a substantial threat and are worth avoiding. Indeed, fly swatting by howler monkeys (*Alouatta palliata*) proves this point nicely: these monkeys engaged in over 1,500 slaps or other avoidance maneuvers during a 12-hour period, using 24% of their metabolic budget in the process (less basal metabolism).[44] For smaller animals, blood loss itself can be significant. Wood rats can use bay leaves—nibbling them so as to release the volatile chemicals they contain—to "fumigate" their nests, probably reducing the number of flea larvae in the nest.[45]

> **KEY TERM** An ectoparasite lives on the outside of its host; a flea is an example of an ectoparasite. Endoparasites, such as tapeworms, live within their host.

Parasite avoidance could have some serious implications for behavioral ecology. It is clear, for instance, that organisms may change habitat preference in an attempt to avoid parasites. Like many animals, juvenile stickleback fish (*Gasterosteus spp.*) frequented areas near the bottom and near vegetation in a tank that was devoid of the branchiuran parasite *Argulus canadensis*. In the presence of the parasite, however, the fish spent more time near the surface and in open areas, where the parasite was less abundant.[46] Clearly, the presence or absence of the parasite would influence a study of stickleback habitat preference.

Social behavior is affected by parasites, much as it is affected by predators. Both the dilution effect and the selfish herd principle (see Chapter 13) can be seen in animal responses to biting flies. During fly season, many animals "bunch," that is, form larger and tighter herds. In experimental tests of bunching, repellant-sprayed Holstein heifers did not form groups as often as control (water-sprayed) ones did.[47] Group formation may lower the likelihood of any one individual being bitten.

Even camouflage, so prominent in predator avoidance, may function in parasite avoidance as well. In 1981, Jeffrey Waage hypothesized that the stripes of zebras may contribute to protection from tsetse flies, which have distinct preferences for some patterns over others and, in general, prefer large, dark objects to lighter ones.[48] (The boldest zebra patterns co-occur with tsetse.)

Avoidance of microscopic parasite propagules is more difficult to study; if nothing else, time and extent of exposure are frequently unknown. Nonetheless, parasite avoidance has been hypothesized to account for xenophobia; when strangers are kept away from the group, so are their pathogens. Hygienic behavior is also widespread, and many animals are fastidious about where they deposit urine or feces. This may be in part related to the function of each in identification and territory marking, but in addition, some species will not forage where they have defecated. Social insects have developed special alarm behaviors to alert nestmates to contagion.[49]

Sexual selection can be influenced by parasites. There are several reasons why choice of a parasite-free mate is beneficial: first, the absence of parasites may indicate "good genes," that is, a genetically based resistance to parasites. Second, a parasite-free mate will not transmit parasites to mates or offspring. Parasite-free mates may also be better at parental care, if such exists.

OF SPECIAL INTEREST: PARASITES AND BEHAVIOR

As we have noted (see Chapter 9 on foraging), parasites are adept at manipulating host behavior, and they often do so in ways that enhance parasite transmission. Perhaps the most surprising manipulation is that of reduced antipredator behavior. This has been seen in a wide range of parasite and host taxa, ranging from isopods parasitized by acanthocephalans (they are more likely to be around a fish predator) to fish with cestodes (they recover more quickly from fright response). However, in no system has this phenomenon been more thoroughly investigated than in rodents with *Toxoplasma gondii*. This protistan parasite can infect a variety of hosts but undergoes a sexual phase in felids. We have known since 1980 that mice with *Toxoplasma* were less fearful, but recent studies on rats have revealed an even more amazing scenario: rats infected with *Toxoplasma* are not only unafraid of cat urine, but are actually attracted to it. In addition, this is not because of a general reduction in fear; the infected rats continued to display more general fear responses (e.g., edge preference, shock avoidance) and spatial learning was unaffected. Although the parasites were encysted in a variety of areas of the brain, they were most concentrated in the amygdala (see Chapter 2).[50] What have you learned in earlier chapters about the amygdala that makes this fact particularly intriguing?

Sickness and Fever

What happens if, despite an animal's best efforts, avoidance fails, and it becomes parasitized? Behavior is also important in this event, and being "sick" is often just a part of getting well. Humans who are sick are frequently sleepy, depressed, inactive. Food may not be very attractive. Pets behave in a similar manner; they groom themselves less, sleep a lot, and are not interested in the food bowl. These behaviors, hereafter called *sickness behavior*, are frequently viewed as undesirable, and when viewed as symptoms of disease, they are subject to "treatment," with the goal of "feeling better." Veterinary medicine professor Ben Hart realized that, far from being signals of physiological systems gone awry,

> **KEY TERM** Sickness behavior is a set of behavioral adaptations that cause an animal to conserve energy and protect itself while it is fighting infection or parasitism.

these behaviors are adaptive.[51] They can be viewed as a syndrome, much of which functions to conserve energy so that the fever response can be supported. Fever helps the body defend against pathogens by both inhibiting their reproduction and stimulating the immune system.

Elements of the immune system, such as fixed tissue macrophages and lymphocytes, release cytokines when they are exposed to foreign substances such as bacteria. These cytokines raise the thermoregulatory setpoint of the organism, causing the generation of fever. The cost of raising body temperature varies with the animal and its environment; in humans, a temperature increase of 1°C can cost a 13% increase in metabolism. Remaining inactive can furnish the extra energy needed for such an increase. Moreover, by not foraging, the animal can sequester its existing iron and keep new iron sources from pathogens that are inhibited by iron deficits. By not grooming, the animal can reduce the substantial water loss, as well as evaporative cooling, that is associated with grooming. Of course, in the absence of grooming, ectoparasites will inevitably increase, but they can be managed by resumed grooming after the threat of illness has passed.

Of course, not every type of animal can generate fever metabolically like an endotherm can. For ectotherms, there is *behavioral fever*. Many ectotherms move to hot places when faced with an infection. They may bask, crawl up high on vegetation, or become hyperactive. In one particularly dramatic example, a fungus (*Entomophthora muscae*) that kills houseflies (*Musca domestica*) 5 days after infection can itself be killed if the fly can get to a hot enough environment during the first 3 days of infection. Infected flies do bask and move toward sources of heat; this is called behavioral fever.[52] The way that the animal raises its temperature differs from that seen in endotherms, but the end result is the same and, in the case of the fly and the fungus, is a matter of life and death.

Self-Medication

Animals have a rich arsenal of pathogen deterrents, but one of the more intriguing ones is self-medication. Some of the best evidence for self-medication comes from two far-flung organisms: chimpanzees and wooly bear caterpillars. Chimpanzees are known to eat a variety of plants that have medicinal properties, but perhaps the most interesting example is whole-leaf swallowing. Whole-leaf swallowing is unlike other instances of chimpanzee self-medication, such as bitter-pith chewing, during which the juice of medicinal plants is consumed. In whole-leaf swallowing, the entire leaf of certain plants that have "hairy" (trichomes) leaves is folded and swallowed whole. Michael Huffman examined the dung produced by a whole-leaf swallowing chimpanzee and discovered that living pathogenic roundworms (*Oesophagostomum stephanostomum*) had passed out with the leaves, often entangled in their rough surfaces. It may be that animals use a variety of mechanical "medications" to clean out intestinal passages.[53]

In the case of wooly bear caterpillars, *Platyprepia virginalis*, unparasitized caterpillars thrived on the host plant lupine (*Lupinus arboreus*). In the field, however, these caterpillars forage on hemlock (*Conium maculatum*) if they are parasitized by the tachinid fly *Thelairas americana*.[54] Hosts that were allowed to exert this preference and eat hemlock could survive the tachinid infection; moreover, the fly itself did better than did those caterpillars forced to eat lupine. The lethal effects of tachinids are legendary, but the reason may be that when collected in the field as caterpillars and then reared in the lab, the host organisms are not given a choice of food, but are given their assumed host plant (lupine). In other words, they are not allowed to display the behavior that could inform scientists about self-medication and about the fact that the parasitized caterpillars need different nutrients.[55]

The issue of medicinal plants presents something of a puzzle to conservation biologists. If such plants are important to animal species that are the targets of conservation efforts, then clearly, these plants need to be present in conserved habitats…if only they could be identified. Scientists have just begun to investigate what may be the rich natural pharmacy that surrounds animals challenged by pathogens.

Host Behavior Controlled by Parasites

In some fascinating cases, an animal behaves at the command of the infectious organism or parasite that has attacked it (see Figure 10.22). This was discussed in greater detail in Chapter 9 on foraging. In the case of a directly transmitted parasite, its host may move into closer contact with other animals of the same species (encouraging the passage of infection) or move to a place from which the disease or parasite is best dispersed into the environment. Ants are intriguing in that they respond to many (but not all!) parasites by climbing up on plants.

FIGURE 10.22
This tropical ant is infected by a *Cordyceps* fungus. The fungus has commandeered the ant, stimulating it to walk up a twig and then to fasten its jaws firmly on the twig. The fungal fruiting bodies have then grown out of the ant—the behavior of the ant, under the influence of the fungus, has perfectly placed the fruiting bodies for dispersal of spores in the wind. *Photo: Erich G. Vallery, USDA Forest Service-SRS-4552, Bugwood.org.*

301

UGA0014287

Moreover, many viruses cause caterpillar hosts to crawl to high places—so much so that this behavior is called *summit disease* or *tree-topping*. When parasites disperse by shedding propagules such as spores from the host, such behavior can be seen to aid transmission. However, one also wonders if it is an unsuccessful attempt at defense, a fever response perhaps too late, too little, or a response captured by the parasite?

In diseases that are transmitted via sexual contact, the animal in some cases may be driven to have unusually frequent sexual contacts.[56,57] Sexually transmitted diseases (STDs) often sterilize the host, but there is little evidence that potential mates are able to avoid copulation with carriers of STDs. One reasonable hypothesis is that STDs are under strong selection NOT to interfere with traits that attract mates.

These are only samples of a rich literature that addresses how animals cope with parasites and pathogens. What is clear is that behavior is a fundamental part of the host-parasite interaction; hosts use behavior to avoid and deter parasites just as they use it to avoid and deter predators.[58]

SUMMARY

This journey through self-defense shows that evolution has used many different strategies and tactics to provide animals with tools for avoiding predation. One important theme has been that behavior interacts with nonbehavioral defenses. For example, hiding is behavior, but being well hidden requires the adaptations of color and shape involved in camouflage. Being a convincing mimic may start with color and shape, but for mimicry to work, the mimic needs to behave like its model.

Avoidance of predators has three elements: avoiding detection, evading capture, and fighting back. Crypsis (camouflage) can employ background matching, countershading, brokenness, and disruptive coloration. These different patterns of color and form typically have the same result: they make the animal difficult to see.

If a potential prey fails to escape notice, it may run away from predators. In so doing, it may confuse predators, engage in flash behavior (sudden stops) or evasive action, or produce displays such as stotting that advertise prowess at fleeing.

Mimicry and diversion are other ways to escape notice. Mimicry differs from camouflage in that the mimic usually appears to be another living organism. Diversion may direct the predator toward a different (and more expendable) prey body area, or it may direct predators away from nest locations and young. Batesian mimics are palatable or defenseless prey that mimic unpalatable or dangerous animals. Muellerian mimics are noxious, unpalatable prey that mimic other noxious, unpalatable prey, thus increasing the negative experience a predator is likely to have with these organisms.

Vigilance behavior is best studied in groups. An alarm can be raised on any one of a number of communication channels. The existence of a group can minimize the risk of a given individual through the dilution effect or via the structure of the selfish herd.

Many of the behaviors that are expressed in the avoidance and deterrence of predation have parallels in animals' relationships with parasites and pathogens. If parasitized or infected with a pathogen, behavioral responses of the host can play an important role in survival.

STUDY QUESTIONS

1. Camouflage: If we come to these phenomena thinking about predators, we can begin to see considerable adaptive significance in the appearance of many animals, but once again, we must weigh alternate hypotheses before concluding that an animal's appearance

is a result of natural selection favoring invisibility to predators. In many cases, we do not know the adaptive significance of an animal's color. For instance, in the case of diurnal animals and for animals that live in open areas, color may be important in thermoregulation. Color also serves important roles in communication. Tim Caro has written a thought-provoking review that proposes and, when possible, tests, several hypotheses for mammalian coloration.[59] Can you develop a group of hypotheses for color in other groups?

2. Tests of camouflage: Kettlewell, in his famous experiments on the peppered moth, assumed that birds would see what we see when they foraged for moths.[60] They seem to do just that. Can you think of a different predator that might not fit such anthropocentric assumptions? How would you devise a predation test of camouflage for a nonvisual predator, or for a predator that sees wavelengths for which we are blind? Think of forms of camouflage that involve sensory information other than vision.

3. In an entry on mimicry in the *Encyclopedia of Evolution*, John Turner says, "Mimicry is the parasitic or mutualistic exploitation of a communication channel."[61] What do you think he means by that?

4. Should sickness behavior invariably accompany an infection? Can you think of social, seasonal, or reproductive circumstances that might mitigate the behavior? Develop one such hypothesis and describe how you might test it. Would some species be more informative than others in this regard? Can you think of human diseases in which it would be helpful to mitigate the sickness behavior syndrome?

Further Reading

Hart, B.L., 2010. Beyond Fever: Comparative Perspectives on Sickness Behavior, In: Michael D. Breed and Janice Moore (Eds.), Encyclopedia of Animal Behavior, Academic Press, Oxford, Vol. 2, pp. 205–210.

Rowe, C., 2010. Defense Against Predation, In: Michael D. Breed and Janice Moore (Eds.), Encyclopedia of Animal Behavior, Academic Press, Oxford, Vol. 3, pp. 106–111.

Stoner, C.J., Caro, T.M., Graham, C.M., 2003. Ecological and behavioral correlates of coloration in artiodactyls: systematic analyses of conventional hypotheses. Behav. Ecol. 14, 823–840.

Turner, J.R.G., 2002. in Mimicry. The Encyclopedia of Evolution, Vol. 2, Oxford University Press, Oxford.

Wickler, W., 1968. Mimicry in Plants and Animals. McGraw-Hill, New York.

Notes

1. Caro, T., 2005. The adaptive significance of coloration in mammals. BioScience 55, 125–136.

2. Lima, S.L., 1998. Nonlethal effects in the ecology of predator–prey interactions. BioScience 48, 25–34.

3. Rankin, N., 2009. A Genius for Deception: How Cunning Helped the British Win Two World Wars. Oxford University Press, Oxford. p. 24.

4. Thayer, G.H., 1909. Concealing Coloration in the Animal Kingdom. Macmillan, New York.

5. Schaefer, H.M., Stobbe, N., 2006. Disruptive coloration provides camouflage independent of background matching. Proc. R. Soc. B 273, 2427–2432.

6. Dimitrova, M., Stobbe, N., Schaefer, H.M., Merilaita, S., 2009. Concealed by conspicuousness: distractive prey markings and backgrounds. Proc. R. Soc. Lond., Biol. Sci. 276, 1905–1910.

7. Hanlon, R., 2007. Cephalopod dynamic camouflage. Curr. Biol. 17, R400–R404.

8. Finn, J.K., Tregenza, T., Norman, M.D., 2009. Defensive tool use in a coconut-carrying octopus. Curr. Biol. 19, 1069–1070.

9. Hanlon, R., 2007. Cephalopod dynamic camouflage. Curr. Biol. 17, R400–R404.

10. Ueta, M., 2001. Azure-winged magpies avoid nest predation by breeding synchronously with Japanese lesser sparrowhawks. Anim. Behav. 61, 1007–1012.

11. Blanco, G., Tella, J.L., 1997. Protective association and breeding advantages of choughs nesting in lesser kestrel colonies. Anim. Behav. 54, 335–342.

12. Schaefer, H.M., Stobbe, N., 2006. Disruptive coloration provides camouflage independent of background matching. Proc. R. Soc. B 273, 2427–2432.

13. Brown, S.G., Boettner, G.H., Yack, J.E., 2007. Clicking caterpillars: acoustic aposematism in *Antheraea polyphemus* and other Bombycoidea. J. Exp. Biol. 210.

14. Haff, T.M., Magrath, R.D., 2010. Vulnerable but not helpless: nestlings are fine-tuned to cues of approaching danger. Anim. Behav. 79, 487–496.

15. Makowska, J.J., Kramer, D. L., 2007. Vigilance during food handling in grey squirrels, *Sciurus carolinensis*. Anim. Behav. 74, 153–158.

16. Périquet, S., Valeix, M., Loveridge, A.J., Madzikanda, H., Macdonald, D.W., Fritz, H., 2010. Individual vigilance of African herbivores while drinking: the role of immediate predation risk and context. Anim. Behav. 79, 665–671.

17. Kenward, R.E., 1978. Hawks and doves: factors affecting success and selection in goshawk attacks on wood pigeons. J. Anim. Ecol. 47, 449–460.

18. Childress, M.J., Lung, M.A., 2003. Predation risk, gender and group size effect: does elk vigilance depend upon the behaviour of conspecifics? Anim. Behav. 66, 389–398.

19. Blumstein, D.T., Anthony, L.L., Harcourt, R., Ross, G., 2003. Testing a key assumption of wildlife buffer zones: is flight initiation distance a species-specific trait? Biol. Conserv. 110, 97–100.

20. Abbott, K.R., 2006. Bumblebees avoid flowers containing evidence of past predation events. Can. J. Zool. 84, 1240–1247.

21. Clark, J., 1983. Moonlight's influence on predator/prey interactions between short-eared owls (*Asio flammeus*) and deermice (*Peromyscus maniculatus*). Behav. Ecol. Sociobiol. 13, 205–209.

22. Maerz, J.C., Panebianco, N.L., Madison, D.M., 2001. Effects of predator chemical cues and behavioral biorhythms on foraging activity of terrestrial salamanders. J. Chem. Ecol. 27, 1333–1344.

23. Arenz, C., Leger, D., 2000. Antipredator vigilance of juvenile and adult thirteen-lined ground squirrels and the role of nutritional need. Anim. Behav. 59, 535–541.

24. Wawra, M., 1988. Vigilance patterns in humans. Behaviour. 107, 61–71. Stable URL: http://www.jstor.org/stable/4534719.

25. Ritland, D.B., Brower, L.P., 1991. The viceroy butterfly is not a Batesian mimic. Nature 350, 497–498.

    Ritland, D.B., 1991. Revising a classic butterfly mimicry scenario: demonstration of Mullerian mimicry between Florida viceroys (*Limenitis archippus floridensis*) and queens (*Danaus gilippus berenice*). Evolution 45, 918–934.

26. Norman, M.D., Finn, J., Tregenza, T., 2001. Dynamic mimicry in an Indo-Malayan octopus. Proc. R. Soc. B 268 (1478), 1755–1758.

27. Shine, R., Phillips, B., Waye, H., LeMater, M., Mason, R.T., 2001. Benefits of female mimicry in snakes. Nature 414, 267.

28. Rassmussen, A.R., Elmberg, J., 2009. Head for my tail—venomous sea snakes avoid becoming prey. Mar. Ecol. 30, 385–390.

29. Punzo, F., 1997. Autotomy and avoidance behavior in response to a predator in the wolf spider, *Schizocosa avida* (Araneae, Lycosidae). J. Arachnol. 25, 202–205.

30. Formancowicz, D.R., 1990. The antipredator efficacy of spider leg autotomy. Anim. Behav. 40, 400–401.

31. Hawlena, D., Boochnik, R., Abramsky, Z., Bouskila, A., 2006. Blue tail and striped body: why do lizards change their infant costume when growing up? Behav. Ecol. 17, 889–896.

32. Clucas, B., Rowe, M.P., Owings, D.H., Arrowood, P.A., 2008. Snake scent application in ground squirrels , *Spermophilus spp.*: a novel form of antipredator behaviour? Anim. Behav. 75, 299–307.

33. Gerald, G.W., 2008. Feign versus flight: influences of temperature, body size and locomotor abilities on death feigning in neonate snakes. Anim. Behav. 75, 647–654.

34. Miyatake, T., Katayama, K., Takeda, Y., Nakashima, A., Sugita, A., Mizumoto, M., 2004. Is death-feigning adaptive? Heritable variation in fitness difference of death-feigning behaviour. Proc. R. Soc. B 271, 2203–2296.

35. Fullard, J.H., Dawson, J.W., Jacobs, D.S., 2003. Auditory encoding during the last moment of a moth's life. J. Exp. Biol. 206, 281–294.

36. Caro, T., 1986. The functions of stotting in Thomson's gazelles: some tests of the predictions. Anim. Behav. 34, 649–662.

37. Relyea, R.A., 2001. Morphological and behavioral plasticity of larval anurans in response to different predators. Ecology 82, 523–540.

38. Dewitt, T.J., Sih, A., Hucko, J.A., 1999. Trait compensation and cospecialization in a freshwater snail: size, shape and antipredator behaviour. Anim. Behav. 58, 397–407.

39. Patek, S.N., 2001. Spiny lobsters stick and slip to make sound. Nature 411, 152–153.

40. Shedd, D.H., 1982. Seasonal variation and function of mobbing and related antipredator behaviors of the American Robin (*Turdus migratorius*). Auk 99, 342–346.

41. Hunter, J., 2009. Familiarity breeds contempt: effects of striped skunk color, shape and abundance on wild carnivore behavior. Behav. Ecol. 20, 1315–1322.

42. Eisner, T., Aneshansley, D. J., 1999. Spray aiming in the bombardier beetle: photographic evidence. Proc. Natl. Acad. Sci. 96, 9705–9709.

43. Hutchinson, D.A., Mori, A., Savitsky, A.H., Burghardt, G.M., Wu, X., Meinwald, J., et al., 2007. Dietary sequestration of defensive steroids in nuchal glands of the Asian snake *Rhabdophis tigrinus*. PNAS 104, 2265–2270.

44. Dudley, R., Milton, K., 1990. Parasite deterrence and the energetic costs of slapping in howler monkeys, *Alouatta palliata*. J. Mammal. 71, 463–465.

45. Hemmes, R.B., Alvarado, A., Hart, B.L., 2002. Use of California bay foliage by wood rats for possible fumigation of nest-borne ectoparasites. Behav. Ecol. 13, 381–385.

46. Poulin, R., Fitzgerald, G.J., 1989. Risk and parasitism and microhabitat selection in juvenile sticklebacks. Can. J. Zool. 67, 14–18.

47. Schmidtmann, E.T., Valla, M.E., 1982. Face-fly pest intensity, fly-avoidance behavior (bunching) and grazing time in Holstein heifers. Appl. Anim. Ethol. 8, 429–438.

48. Waage, J.K., 1981. How the zebra got its stripes—biting flies as selective agents in the evolution of zebra coloration. J. Entomol. Soc. South Afr. 44, 351–358.

49. Rosengaus, R.B., Jordan, C., Lefebvre, M.L., Traniello, J.F.A., 2000. Pathogen alarm behavior in a termite: a new form of communication in social insects. Naturwissenschaften 86, 544–548.

50. Vyas, A., Kim, S.-K., Giacomini, N., Boothroyd, J.C., Sapolsky, R.M., 2007. Behavioral changes induced by Toxoplasma infection of rodents are highly specific to aversion of cat odors. PNAS 104, 6442–6447.

51. Hart, B.L., 1988. Biological basis of the behavior of sick animals. Neurosci. Biobehav. Rev. 12, 123–137.

52. Watson, D.W., Mullens, B.A., Petersen, J.J., 1993. Behavioral fever response of *Musca domestica* (Diptera: Muscidae) to infection by *Entomophthora muscae* (Zygomycetes: Entomophthorales). J. Invertebr. Pathol. 61, 10–16.

53. Huffman, M.A., 1997. Current evidence for self-medication in primates: a multidisciplinary approach. Yearb. Phys. Anthropol. 40, 171–200.

54. Karban, R., English-Loeb, G., 1997. Tachinid parasitoids affect host plant choice by caterpillars to increase caterpillar survival. Ecology 78, 603–611.

55. De Moraes, C.M., Mescher, M.C., 2004. Biochemical crypsis in the avoidance of enemies in an insect herbivore. PNAS 101, 8993–8997.

56. Knell, R.J., Webberley, K.M., 2004 Sexually transmitted diseases of insects: distribution, evolution, ecology and host behaviour. Biol. Rev. 79, 557–581.

57. Abbot, P., Dill, L.M., 2001. Sexually transmitted parasites and sexual selection in the milkweed leaf beetle, *Labidomera clivicollis*. Oikos 92, 91–100.

58. Moore, J., 2002. Parasites and the Behavior of Animals. Oxford University Press, Oxford, 338 pp.

59. Caro, T., 2005. The adaptive significance of coloration in mammals. BioScience 55, 125–136.

60. Cook, L.M., 2003. The rise and fall of the carbonaria form of the peppered moth. Quarterly Review of Biology 78, 399–417.

61. Turner, J.R.G., 2002. Mimicry. In: Pagel, M. (Ed. in chief), The Encyclopedia of Evolution, Vol. 2, Oxford University Press.

305

Mating Systems

LEARNING OBJECTIVES

Studying this chapter should provide you with the knowledge to:

- Understand the importance of gametic investment in predicting behavior.
- Predict how and why animals choose mates.
- Classify mating systems by observing social interactions.
- Understand how gametic investment can help in predicting the choosy sex.
- Appreciate the effects of intrasexual competition for mates.
- Understand that animals that do not succeed in obtaining mates may opt for alternate strategies to gain reproductive success.

11.1 INTRODUCTION

A mating system is a pattern of male–female pairings. The term *mating system* captures the broad sweep of reproductive biology, including the nature of the *gametes*, location of mates, strategies for ensuring the correct choice of a mate, patterns of male–female pairings, and behavioral strategies for individuals to attain reproductive success and mate even if they are not chosen. In short, mating systems are an unavoidable consequence of sexual reproduction and offer a rich field of inquiry. Mating systems continue to be a major focus of study in animal behavior.

> **KEY TERM** A mating system describes how male and female interactions are built around choosing mates.

> **KEY TERM** A gamete is the haploid reproductive cell produced in meiosis. Two gametes must merge to produce a zygote, which develops into a new animal.

Understanding mating systems starts with the underlying asymmetry between females, which typically invest heavily in each of a relatively small number of gametes, and males, which typically invest little in each of a large number of gametes. This fundamental difference between the sexes means that males and females approach mating with differing and sometimes conflicting interests. This chapter begins with a hypothesis about how this disparate condition came to be, and continues with descriptions and examples of the different types of mating systems, such as *monogamy* (a male and a female), *polygyny* (one male, many females), and *polyandry* (one female, many males). It then delves deeply into the intriguing issue of how animals choose their mates (see Figure 11.1).

CASE STUDY

308

Mate Choice, Monogamy, and Human Sexual Behavior

Human behavior has not been a primary focus in this text, but everyone is curious about how humans fit into the mating behavior picture. Perhaps most obvious from a physical perspective is that sexual selection has affected both male and female human phenotypes. This suggests that both sexes actively choose their mates and that mate choice in humans has selected secondary sexual features in males and females. Scientific studies and personal experience support the idea that each person has a distinct view of what makes an "optimal" mate.

Individual perspectives on what constitutes "beauty" and "sexiness" vary and are strongly influenced by cultural background. Some scientists have argued that people judge more symmetrical faces ("even" or "balanced" features) to be attractive. Others note that symmetry seems to have little to do with actual mate choice. Symmetry may convey information about genetic variation and a person's health; the reasons for this are presented in this chapter. A counterargument is that asymmetries bring character and interest to a person's appearance. Mate choice revolves not only around physical appearance, but involves perceptions of future fecundity (ability to have children) and abilities as a parent, including possession of resources that will aid in rearing children.

The question of whether or not humans are naturally monogamous is fascinating and definitely serves as a flashpoint in moral, legal, and religious debates. The two ape species most closely related to humans, chimpanzees and bonobos, are strikingly promiscuous in their sexual behavior, with sex playing roles in bonding, social affiliations, and perhaps appeasement in both species. Copulation serves a variety of nonreproductive social functions in these apes. The validity of looking to chimpanzees and bonobos to understand human sexual behavior is an open question because the ecology of humans differs substantially from that of chimpanzees and bonobos, and ecology plays a major role in mating systems.

Across a broad range of human cultures, the most common mating systems are monogamy and polygyny; the latter occurs when a male has multiple female mates (sometimes this is called a *harem mating system*). However, from a lifetime perspective, both males and females are likely to have multiple mates, and serial monogamy—pair bonds that are nonpermanent—is a common mating system in humans. Extrapair copulations—sexual encounters outside a pair bond—are certainly frequent in humans. Extrapair copulations may allow a person to produce genetically diverse offspring while avoiding the costs of parental care.

Human mating patterns resist analysis because biological influences are difficult to separate from social and cultural influences. Morals, laws, and religion often seek to modify or restrict what might be natural impulses. From evolutionary and sociocultural perspectives, control over the reproduction of others is an extreme power. In this chapter, the same biological dynamics that are so much a part of human life play themselves out across the animal world; in many ways animal mating systems are much easier to understand than human mating systems because insofar as we know, biology alone is the driving force.

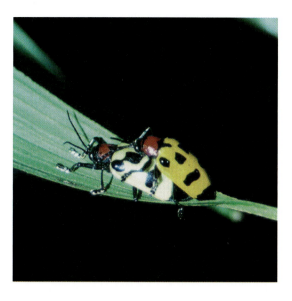

FIGURE 11.1
This chapter is all about very complicated questions that come from a simple act—mating. Look at this pair of beetles. It seems as though it would be an easy picture to interpret; they are mating, and the male is on top of the female. Are they paired for life, or will each have other partners over time? Did the male choose the female? Did the female choose the male? Will the male behave in ways that limit whether the female can mate with other males? This chapter focuses on how and why biologists ask Tinbergen's questions about mating in animals. *Photo: Jeff Mitton.*

11.2 THE EVOLUTION OF SEX: WHY SOME ANIMALS ARE CALLED MALE AND OTHERS FEMALE

Biologists define *maleness* and *femaleness* by the relative size of the gametes that individuals produce. Species in which all gametes are the same size are termed *isogamous* and lack identifiable males and females. Isogamy is common in algae and protists, but virtually all animal species are *anisogamous*, producing small motile gametes, or sperm, and large gametes, or eggs. By convention, organisms producing large, nutrient-rich gametes are termed *female*, and organisms producing small, motile gametes are termed *male*. The isogamous nature of many unicellular organisms leads to the inference that isogamy may be the ancestral state for gametes. One look at an ostrich egg or a brooding octopus reveals that much has changed since that time. What caused this change? What caused the evolution of male and female?

As is the case in all evolution, variation played a significant role. Gametes that were slightly smaller than average were more nimble and faster than other gametes; they could travel quickly to other gametes and produce diploid zygotes. In the world of motile gametes, if everything else was equal, small would often win the race to fertilization. True, the small gametes had reduced nutrient content for the zygote, but that could be balanced by fertilizing larger, nutrient-rich gametes. Meanwhile, large, nutrient-rich gametes also could influence the development of the zygote and could exert control over the fate of the offspring. Intermediate-sized gametes were neither nutrient rich, nor were they particularly speedy. Thus, natural selection on gametes favored the extremes—small and large. This sort of selection, called *disruptive selection*, is thought to be the selective force that resulted in the sexes (see Figure 11.2).

Some animals are capable of producing both eggs and sperm; these are *hermaphrodites*. Hermaphroditism can be further subdivided into animals such as some sea slugs that are simultaneously male and female (simultaneous hermaphrodites) and others that are first one sex and then the other (sequential hermaphrodites). For example, clownfish are

> **KEY TERM** By definition, a male produces small, motile gametes called *sperm*. Males produce large numbers of these gametes, with small investment in each.

> **KEY TERM** By definition, a female produces small, motile gametes called *eggs*. Females produce small numbers of these gametes, with large investment in each.

> **KEY TERM** Isogamous species produce gametes that are all the same size. There is no differentiation of male and female. Anisogamous species produce gametes of two distinct sizes; these species have males and females.

309

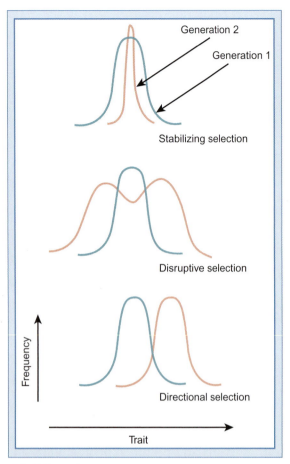

males early in adult development and become females as they grow larger; this is called *protandrous*, or male-first, *hermaphroditism*. In contrast, wrasses are females when smaller and become males as they grow larger; this is called *protogynous*, or female-first, *hermaphroditism*.

> **KEY TERM** A hermaphrodite is an animal that has both male and female reproductive organs and produces both eggs and sperm.

> **KEY TERM** Protandrous hermaphrodites are male early in their lives and females later. Protogynous hermaphrodites are female early in their lives and males later.

Why does sex exist? Superficially, it may seem obvious that organisms should reproduce sexually, but sex carries a high evolutionary cost. Recall freshman biology lessons about *diploid* reproductive cells becoming *haploid* through a special cell division process called *meiosis*; in other words, each time meiosis occurs, half the genome is "lost." This is called the *cost of meiosis*. As a result, an *asexually reproducing* animal is able to pass along all of its genes to each of its offspring, but animals participating in *sexual reproduction* pass only half of their genes to each offspring. In addition, there is a cost associated with producing male and female offspring (both are necessary in sexually reproducing species), instead of putting all reproductive effort into asexually reproducing female offspring. Given this reasoning, why does evolution so often fail to favor organisms that transmit 100% of their genes to each offspring—that is, organisms that asexually reproduce? What forms of counterbalancing selection could favor sexual reproduction?

310

FIGURE 11.2

This diagram shows the effects of disruptive, stabilizing, and directional selection. The x axis represents a measure of the trait—in this case, gamete size. The blue curve shows the distribution of the trait before selection; the red curve shows the distribution of the trait after selection has had an effect. Under disruptive selection, animals producing either large or small gametes are favored by selection, while animals producing middle-sized gametes are selected against. Stabilizing selection reduces the variation in a trait by favoring a narrow range of phenotypes. Directional selection shifts the phenotype of the population in one direction.

> **KEY TERM** The cost of meiosis is the loss of half an animal's genetic material when it produces haploid gametes.

> **KEY TERM** Asexual reproduction is the production of offspring that are genetically identical to a single parent.

> **KEY TERM** Sexual reproduction is the production of offspring by the combination of gametes from two parents.

In other words, why be sexual? The major argument for evolution and maintenance of sex is that the selective advantages from *genetic recombination* and *genetic diversity* among the offspring outweigh the costly nature of sex. In addition, deleterious genes do not accumulate in a sexual lineage. Indeed, this genetic diversification is so beneficial that it occurs in many organisms that do not have "male" and "female" forms, such as mating strains of protists. In a strict sense, then, sex refers to the mixing of genetic material, either within an organism (e.g., recombination) or between organisms. It is this mixing that proves to be highly advantageous.

Sexual reproduction is a fact of life for birds and mammals, which, with one or two rare exceptions, reproduce sexually; this yields the hypothesis that sexual reproduction is probably the ancestral state in vertebrates. There are nonetheless numerous examples of asexual reproduction in fish and a few examples in amphibia and reptiles. Asexual reproduction is likely to evolve in environments that change little from generation to generation, because genetic recombination is not an advantage in predictably constant environments. Sometimes asexual reproduction crops up in invasive or pioneering species, in which single individuals may migrate to a suitable habitat; if no potential mates are present, then asexual reproduction is obviously the only way to produce offspring, and selection will favor it. (Indeed, unless it arrived with fertilized gametes, a solitary invader that was incapable of asexual reproduction would leave few traces and no descendants). At least hypothetically, it is possible that due to physiological and/or neurobiological constraints, birds and mammals may have lost the ability to revert to asexual reproduction, or at least greatly diminished that probability.

> **KEY TERM** Genetic diversity results from different offspring having different genetic combinations. Each egg from a female is genetically unique, and each sperm from a male is genetically unique. Thus, when a male and female mate, if they have multiple offspring, their offspring will differ genetically; in other words, their offspring are genetically diverse.

> **KEY TERM** Basal taxa are types of organisms that arose earlier in evolution. We use this term to avoid using the misleading words *primitive* or *lower* when describing animals.

Among invertebrates asexual reproduction is more common, and in some invertebrate taxa, there appears to be considerably more flexibility, in terms of evolutionary change between sexual and asexual patterns of reproduction. Given the ubiquity of sexual reproduction in animals and the fact that the more *basal animal taxa* reproduce sexually, the hypothesis that sexual reproduction is basal for animals as a whole is reasonable. Sea anemones produce dispersing offspring sexually, but when one of those offspring arrives on a rock and develops into an anemone, it can also reproduce by *budding*, producing clones of itself. This alternation of sexual and asexual reproduction allows a species to take advantage of maintaining genetic combinations in stable environments through asexual reproduction, while using sexual reproduction for recombination and offspring diversity when colonizing new habitats (see Figure 11.3). Similarly, in parts of their life cycles, many aphid species reproduce asexually (see Figure 11.4). In these organisms, sex seems to be *facultative*; that is, it may or may not occur, depending on environmental stability. Stable environments favor asexuality.

311

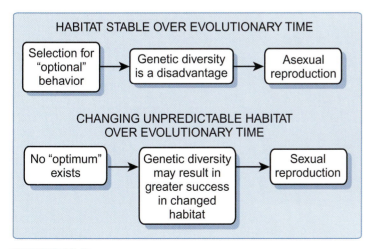

FIGURE 11.3
The effects of habitat stability on the evolution of reproduction. Stable habitats may favor asexual reproduction, while unstable (or unpredictable) habitats may favor sexual reproduction.

FIGURE 11.4
Aphids such as these may reproduce asexually, resulting in clones of the insect. *Photo: Michael Breed.*

Having established that sexual reproduction is common, and probably basal, in animals, it should be clear that the evolution of mating systems probably began before the evolutionary origin of animals, among their *protistan* predecessors in which proto-sperm swam in the ocean searching for proto-eggs to fertilize. This means that the use of chemical cues to find mates and competition among sperm are ancient behavioral features.

> **KEY TERM** An egg is a gamete that is large and relatively immobile.

> **KEY TERM** A sperm is a small and relatively mobile gamete.

Much of the theory that underlies the understanding of mating systems derives from the fact an *egg* is more costly to produce than a *sperm*. This means that females produce a few individually expensive gametes, while males can produce a huge number

OF SPECIAL INTEREST: THE RED QUEEN AND THE EVOLUTION OF SEX

In Lewis Carroll's *Through the Looking Glass*, Alice finds herself in the company of the Red Queen (see Figure 11.5). The charming story of Alice climbing a hill with the Red Queen is accompanied by the Red Queen's confusing explanation of how hills can become valleys can become hills. In the same conversation Alice learns that you have to run faster and faster just to keep up with the competition:

> Alice didn't dare to argue the point, but went on: "And I thought I'd try and find my way to the top of that hill — "

> "When you say 'hill,'" the Queen interrupted, "I could show you hills, in comparison with which you'd call that a valley."

FIGURE 11.5
The Red Queen lectures Alice about the difficulties of running as fast as you can, only to realize that you are not really moving forward relative to your surroundings.

FIGURE 11.6
Behavioral evolutionary arms race.

"Well, in our country," said Alice, still panting a little, "you'd generally get to somewhere else—if you ran very fast for a long time, as we've been doing."

"A slow sort of country!" said the Queen. "Now here, you see, it takes all the running you can do, to keep in the same place. If you want to get somewhere else, you must run at least twice as fast as that!"

Leigh van Valen introduced the Red Queen to evolutionary biology, pointing out that a race which becomes faster and faster without a sign of a winner is a perfect analogy for co-evolutionary arms races (see Figure 11.6).[1] In such co-evolutionary races, for example, a prey species evolves a defense, the predator evolves to defeat the defense, the predator evolves another defense, and on and on…. Moving fast seems to never do more than get you back to the starting point, but stopping during an evolutionary arms race leads to extinction.

Other evolutionary biologists saw the Red Queen's tale as relevant to the evolution of sex, explaining that the value of sexual reproduction fluctuates over time; genetic recombination gained from sexual reproduction produces new defenses against diseases and parasites. When disease and parasite pressure is low, the cost of meiosis and the benefit of proven genetic combinations favor asexual reproduction, but when diseases and parasites surge, sexual reproduction is favored. Thus, the race goes up and down hill, and the fast-moving species are again condemned to run in the same place.

of relatively cheap gametes. Theoretically, then, eggs are valuable and worth protecting, while sperm are much less valuable. Another way of looking at this is to contemplate the numerical disparity between eggs and sperm: because fewer eggs are made than sperm, many, perhaps most, eggs should be fertilized, while many sperm will never fertilize an egg. The result of this observation is the evolutionary argument that, in general, females should protect their investment in eggs by caring for them, while males should be less likely to care for either their sperm or for their offspring. This is the evolutionary argument for the root of differences between the sexes. Laying aside sociocultural arguments about sex and gender roles in humans (recall the discussion of nature and nurture in Chapter 3), this characterization is a good starting point in exploring the evolution of mating systems across animal taxa. In particular, these concepts lead to the prediction that in the absence of other influences, females should be very choosy about who fertilizes their gametes, whereas males should be relatively indiscriminate in their mating. This is an important point that will form the core for a large part of the discussion in this chapter. Indeed, when these predictions are not met, there are frequently other forces at work, such as circumstances that limit male options for mating; with limited mating options, the number of potential matings for the males is more or less equal to the number of potential matings for females, and male choosiness about mates should equal that of females. Males that invest more than females do in offspring are often more choosy than females (see Figure 11.7).

FIGURE 11.7
Female mormon crickets eat spermatophores such as this one. Spermatophores are sperm containers that males of many arthropod species produce; in the case of the mormon cricket, the spermatophore is nutrient-rich and left with the female after mating. Because the mormon cricket spermatophore is costly to make, a male carefully chooses his mate. *Photos: Darryl Gwynne (left) and Michael Breed (right).*

313

Evolutionary theory developed by R. A. Fisher[2] and expanded by R. L. Trivers[3] postulates that in a population the investment in male and female gametes should be equal. In essence, this theory concerns sex ratios in populations and applies not only to the parental investment of caring for offspring in species that do so, but also to the parental investment in sperm and eggs alone for species that provide no additional care. At the population level, equal time and energy should be invested in male and female offspring, but the number of male and female offspring may differ; the amount of investment in males and females at the population level is what matters. Because this equality is at the population level, rather than the individual level, individuals may produce male- or female-biased broods, as long as across the population, the ratio is equal. This should not be confused with population-level selection. As Fisher explained, if the ratio of males to females deviates very much from 1:1, the rarer sex becomes more valuable, and selection on individual parents favors increased production of that sex.

Equal investment in males and females is based on the assumption of equal genetic relationship of the parent to its male and female offspring. Nature does not always honor this assumption. For instance, there are at least three major exceptions to this condition of equal genetic relationship:

1. The chapters on sociality (Chapters 13 and 14) explore interesting examples in ants, bees, and wasps in which females are more related to their sisters than to their daughters. This disparity in genetic relationship leads to possible asymmetries in investment in males and females.
2. Another exceptional condition comes from local mate competition, when brothers may end up competing with each other to mate. In this case, theory predicts that the population sex ratio should be strongly female biased, so that only enough males are produced to ensure that every female mates.
3. A final exception is local resource competition, in which females remain near their birth site, and males disperse. The females end up competing with sisters for resources, while males probably do not compete with their brothers; this results in male-biased sex ratios.

314

11.3 SEXUAL SELECTION

Darwin argued for two kinds of selection: natural and sexual. In modern biology, *sexual selection* is divided into *intersexual selection*—the effects that mating preferences of one sex have on characteristics of the other sex—and *intrasexual selection*—the effects of competition within a sex for mates. Intersexual selection can result in the emphasis of features that attract mates, such as vocalizations or the peacock's tail. Intrasexual selection can yield features such as horns or antlers that are used in intrasexual combat. However, sometimes intersexual and intrasexual selection can both act on a trait, so the antlers of red deer serve in both combat (intrasexual) and female choice among males (intersexual; see Figure 11.8). For that matter, some traits used in combat can also be favored by natural selection if they enhance predator defense, for instance. Either intersexual and intrasexual selection can result in sexual dimorphism—different "morphs"—or appearances—for males and females.[4,5]

> **KEY TERM** Sexual selection results from differential reproduction of males or females within a species. Male deer, for example, may have large antlers as a result of sexual selection.

> **KEY TERM** Intersexual selection results when males choose female mates based on specific traits or vice versa.

> **KEY TERM** Intrasexual selection results from mating-related competition within a sex. Horns, antlers, and other weaponry may result from intrasexual selection.

Natural selection can also limit the power of sexual selection,[6] especially in the case of "runaway" sexual selection. *Runaway sexual selection* refers to

FIGURE 11.8
Right, a male and female pair of Canada Geese, which differ little in appearance. Left, male and female ducks, showing stronger sexual dimorphism, which may be the result of either intrasexual selection (competition among males) or intersexual selection (female mate choice). *Photos: Michael Breed.*

FIGURE 11.9
Examples of runaway sexual selection. Left, a male earwig, with large forceps used in mating; right, a bull elk with large antlers. *Photos: Jeff Mitton.*

315

circumstances in which intersexual preference is so consistent that selection on the trait produces an extreme form (see Figure 11.9). If females persist in choosing males with large tail feathers, for instance, then the males with the largest tail feathers will leave the most offspring. These offspring will then have very large tail feathers (if male) or prefer males with very large tail feathers (if female). As can easily be seen, this could eventually result in gargantuan tail feathers that, past a point, would be detrimental were there any predators in the vicinity. Natural selection curbs the excesses of runaway selection, although not before some spectacular traits appear.[7,8,9]

KEY TERM Runaway sexual selection occurs when one sex, chooses mates based on a characteristic of the other sex, resulting in an extreme expression of that characteristic. Over generations this causes that trait to become amplified, as in the peacock's tail.

DISCUSSION POINT: SEXUAL SELECTION

How do you think intersexual and intrasexual selection have affected phenotypes in a range of animals with which you are familiar? If you compare humans to common domestic animals, such as dogs and cats, are humans more or less sexually dimorphic than those species? (Hint: Can you determine the sex of a dog without looking at its genitalia?) Can a dog determine the sex of another dog without sniffing and looking? Are any human, dog, or cat traits the result of sexual selection? What might prevent runaway sexual selection from taking over a trait used in mate choice?

The Handicap Principle

If the sexually selected trait is actually a handicap for the animal carrying it, then it exacts a price in energy or vulnerability to predation. In this case, the sexually selected trait, such as huge antlers, becomes an honest advertisement of the male's ability to carry the handicap. Zahavi argued in the mid 1970s that sexually selected traits were handicaps for the animal carrying them and that, as such, sexually selected traits reflect good genes even if the trait itself has been selected to the point of having no residual additive genetic variance on which selection might work.[10]

If one sex consistently chooses mates with extravagant decoration (like a peacock's tail), the costs of producing the decoration can become very high. The ornament may be metabolically costly to produce and difficult to carry. Ornaments may also put an animal at risk for predation by making it more conspicuous or making it move awkwardly. The ability of an animal to pay these costs may then be an honest measure of that animal's metabolic and physical capabilities. It is "honest" because the animal cannot pretend to pay the costs; they are real. This is the "handicap principle" developed by Zahavi and explored later in the chapter. For the purposes here, the important point to understand is that runaway sexual selection can ultimately be limited by natural selection acting on the costs of signal production and that natural selection defines the cost of a handicap.

OF SPECIAL INTEREST: MATE CHOICE IN PEACOCKS

In a classic series of experiments, Petrie explored the function of the peacock's tail in mate choice.[11] Peafowl are highly sexually dimorphic; the peahen is relatively drab, whereas the peacock carries a large, beautifully colored tail that is decorated with an eyespot pattern (see Figure 11.10). Petrie and coworkers discovered that removing eyespots from a male's train made him less attractive to females, and adding eyespots made him more attractive. Recently, Petrie's results have been criticized by Takahashi and his colleagues as not fully explaining female choice in peafowl.[12] Petrie and her colleagues responded by noting that Takahashi had not actually provided measures of the density of eyespots in the males' trains in his study.[13] On balance, the evidence seems to favor Petrie's interpretation. Nonetheless, Takahashi's point is interesting: might females have abandoned (over evolutionary time) their response to the males' trains because all of the males have them? A trait that all of the animals in a population express is not informative for choosing a mate or, indeed, for making any sort of decision. In addition to using the male's train in choosing mates, peahens also consider the male's vocal and physical displays, as well as his location. Mate choice is complex, and these studies show that oversimplifying the analysis by focusing on single features or behaviors of animals during mate choice may be misleading.[14] That said, the peacock tail is emblematic of sexual selection; it would be the odd biologist who would not respond to it with ardent curiosity.[15]

FIGURE 11.10
A peacock. *Photo: Michael Breed.*

The Effects of Sexual Selection on the Heritability of Traits

Strong *directional selection* usually exhausts *additive genetic variance* for a trait in three to five generations. (In this context, this means traits governed by *polygenic inheritance*, or *quantitative trait loci;* see Chapter 3 on genetics.) After that, there can be no further response to selection because the remaining phenotypic variation is from either environmental or nonadditive genetic variation. In theory, sexual selection on a trait such as antler size

> **KEY TERM** Directional selection causes one form of a trait in a population, over generations, to be favored (see Figure 11.2).

should rapidly eliminate the additive genetic variance for the trait. In other words, the trait will be genetically fixed. In practice, many traits that seem to be under strong sexual selection still have considerable heritability.[16]

There are a number of possible explanations for why selection does not eliminate all of the additive genetic variance for traits involved in mate choice. They include the following:

1. Sexual selection is strong only under extreme environmental conditions in which survivorship is low, thus maintaining variation during periods of relaxed selection.
2. Interactions (e.g., linkage effects, viability effects) with other traits limit sexual selection before the additive variation is exhausted.
3. Mate choice relies on many factors, rather than one trait; when selection is acting on multiple traits, variation will remain for each of the traits.
4. Counterbalancing selection maintains additive genetic variance by limiting the elaborateness of a signal.[17,18] As with the study of mate choice in peafowl, it is hard to overemphasize the complexity of mate choice and the need to consider multiple factors involved in any mate choice decision.

Courtship: The First Step in Sexual Behavior

In most animals courtship precedes mating. Courtship is an exchange of information between a male and a female, and sexual selection often acts on signals used in courtship. Typically, courtship includes signals that allow the male and female to ensure that they are mating with an animal of the same species.[19] It may also include information that would prevent mating with a closely related animal (inbreeding). Finally, courtship gives animals the opportunity to present information about their quality as a potential mate.[20] The issue of mate quality is complex, and there is no simple answer to the question: Which potential mate is best?[21] Through the rest of this chapter, the assessment of mate quality comes up repeatedly, offering opportunities to form a synthesis of how animals assess the quality of potential mates.

317

Honesty and Dishonesty in Mating Systems

The mating arena promotes considerable *dishonesty*. In many species, males run a high risk of never having the opportunity to mate, and females suffer from the risk of choosing a low-quality mate. This situation leads to strong selection for two types of dishonesty. The first is misrepresentation of quality in mate choice. This is the animal version of human male misrepresentation of monetary and social status. In species in which males participate in parental care, males may also misrepresent whether they already have a mate and are spending time caring for young from that mating. Mated males would be expected to conceal this fact, as their attractiveness to additional mates depends in part on their potential to contribute to offspring care.

> **KEY TERM** Dishonesty is a false representation of an animal's genotype or of resources it holds.

> **KEY TERM** Infanticide occurs when an adult animal kills a young animal.

The second type of dishonesty occurs in females. In contrast to male dishonesty, females should represent themselves as mating only with one male. Having chosen a mate, however, a female might be somewhat uncertain that her choice was correct, or unpredictable conditions might make it impossible for her to know which male is appropriately matched to the environment their offspring will encounter. Such uncertainties about her choice may select for surreptitious matings with other males. Females hide these other matings to protect their offspring from *infanticide*; males of some species often kill young if they are uncertain about paternity. (See Chapter 12 on nesting and parental care for more information on this topic.) These two forms of dishonesty combine to make the study of mating systems in animals full of surprises.[22]

In some cases, of course, male and female roles are more alike, or the stereotype of philandering males and choosy females may be reversed. The more similar male and females roles are, the more similar male and female deceptive practices will be (see Figure 11.11). If male and females roles are reversed, then patterns of deception and cheating may also be reversed.

FIGURE 11.11
In fiddler crabs, the enlarged claw can be an honest signal of male fighting ability. However, if a male loses a claw, it may regrow a large, but weak claw that is a dishonest signal of fighting ability. *Photo: Allen Bridgman, South Carolina Department of Natural Resources, Bugwood.org.*

BRINGING ANIMAL BEHAVIOR HOME: MATING SYSTEMS IN DOGS AND CATS

Dog and cat mating systems contrast in interesting ways. Domestic cats are not much different from their closest relatives, the European wildcat. Domestic cats are essentially solitary animals. Tom cats are not involved in parental care; they associate with a female only during her estrous period, when they compete ferociously for her. Sometimes cats that have been released on farms or in urban areas (these are called *feral cats*) congregate around food, and then adult females may show more mutual tolerance. Copulation stimulates ovulation in cats, favoring tom cats that kill kittens in order to father a new litter. Females become receptive again shortly after copulation, and litters of mixed paternity are relatively frequent in cats.

Life for dogs is more pack oriented, and wolves, the closest wild species to the domestic dog, have a mating system much like dogs. A pair-bonded male and female typically head a wolf pack, and their offspring form the core of the pack. Male wolves participate in parental care, at least to the extent that the pack shares food and that the male wolf is active in defending the pack. Perhaps because humans often have only one dog as a member of their household, the mating system of domestic dogs is not so strongly based on pair bonds. Of course, domestic dogs are often neutered, so they do not participate in mating. Estrous females attract a number of males, and, as in cats, a female may mate with more than one male while she is in season and mixed litters are common. The dog's penis expands during copulation, tying the male and female together, so that an attempt to forcibly separate copulating dogs can injure the animals. This may be a form of mate guarding, helping to ensure that some of the female's eggs are fertilized from that copulation.

11.4 VARIANCE IN MATING SUCCESS

Classic experiments on fruit flies by Angus Bateman extended our knowledge well beyond what might be applied to these useful little insects. Bateman learned that variation among females in mating success is likely to be low, while male variance is high. The reason is that one mating is likely to be sufficient to fertilize the eggs of a female, while male reproductive success increases each time the male mates (see Figure 11.12). As a result of this difference, competition for mates should be higher among males than females, and in general, sexually selected traits, such as antlers in deer or plumage in birds, should be more elaborate in males than females.

Are a female's best interests really served by mating only once? Is mating actually so cheap for males that they can mate repeatedly? As will become apparent in this chapter, there are ample exceptions to these "rules." For instance, natural selection can favor females mating with a number of males, and mating can be costly enough to limit variance in male reproductive success. Females may mate with more than one male to ensure fertility (the first male may be infertile), to increase genetic variation among their offspring, or to seek a better genetic match. The effort of locating and courting a mate can be high for males; the energy expenditure involved in achieving a mating may be so high that efforts to find subsequent mates are limited.

11.5 MATE CHOICE

Most of the rest of this chapter deals with how mates are chosen. This section introduces the basic principles of mate choice. To give a concise preview, animals can choose mates based on

1. What the potential mate possesses, such as a territory.
2. The genetic quality of the potential mate; this is used to forecast the genetic qualities of the pair's offspring. The genetic quality of a mate can be viewed fairly broadly to include factors such as the ability of the potential mate to avoid parasites.
3. A tangible nutritional offering by the potential mate; these are called nuptial gifts.[23]

Some scientists argue that all of these factors basically reduce to genes: for instance, an animal cannot hold a territory if it does not have the genes to make it strong enough to do so.

Female Mate Choice

For most species, females exert more choice concerning their mates than do males. The theoretical basis of female choosiness was explained at the beginning of this chapter and is simple: females invest more in their gametes (eggs), make fewer gametes, and are more likely to be present when the young emerge than are males. The difference between female and male gametic investment and the relative rarity of female gametes, combined with the likelihood that in many species the female provides parental care, favor females that carefully select their mates.

What effect does female mate choice have on males? Any trait that is preferred by females will be strongly favored. This type of selection, called *sexual selection*, explains the elaborate ornamentation of males of many species, such as the peacock's tail (see Section 11.3). With some exceptions,[24] females should make their choices among males based on the likelihood that mating with the chosen male will enhance the survival of the female's offspring. This could be because the male possesses resources, such as food within a territory, or because the male has "good genes," which will then be passed on to their offspring. "Good genes" models are complicated because any trait chosen by females will be under strong selection, and the usefulness of the trait for assessment will be lost if selection results in all males being the same. Also, separating honest signals representing good genes from deceitful signals may be difficult for females. Finally, some female animals are known to copy other female choices;[25,26] males do this as well.[27]

In contrast to mate choice that reflects genetic quality, another prediction about mate choice states that females should prefer to mate with males that will father sons that will be attractive to females in the next generation. In other words, nothing matters except male attractiveness—not gene quality, not potential as a parent, and not resources held by the male. This is known as the *bad boy* phenomenon or the *sexy son* principle. It does not take too much imagination to apply this line of reasoning to a broad range of animals.

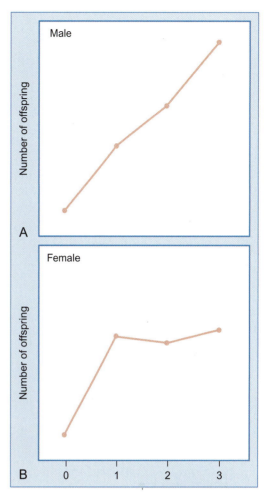

FIGURE 11.12
These figures illustrate the traditional idea of sexual differences in mating success as a function of number of mates. Initially, biologists thought that while a male's reproductive success increased with the number of females he mated with (A), a female's success would not (B). Subsequent research has shown that there are many exceptions to this "rule."

319

KEY TERM The sexy son model argues that females should prefer to mate with males whose sons are likely to mate. In other words, sexy fathers produce sexy sons. This could lead the female to ignore traits that suggest the father would be a good parent, and so on.

FIGURE 11.13
Female baboons are famous for their ornamentation; male primates are known to choose females on the basis of qualities that indicate high fertility.

Male Mate Choice

Male mate choice is less well documented than female mate choice, but plays an important role in some mating systems. Male mate choice is likely to appear when the male has substantial involvement in parental care or when females vary greatly in their qualities as mates (see the account of Mormon crickets, Figure 11.7). If males of a species are choosy about their mates, females should develop secondary sexual characteristics and ornamentation, much as males do in other species (see Figure 11.13). Close observation of hyenas revealed that high-ranking males associated most closely with high-ranking females and exhibited mate choice, favoring fecund females.[28]

11.6 MATING SYSTEMS: HOW MANY MALES, HOW MANY FEMALES?

Early animal behaviorists observed that some species lived in male–female pairs, whereas others lived in groups with a single, dominant male and several females. Orians and Emlen and Oring systematized ways of looking at how animals come together to mate.[29,30] This section elaborates on these mating systems.

For several decades, animal behaviorists and ethologists thought the key to understanding mating systems lay in documenting the number of males and females in a group of mating animals. For example, wild horses live in herds with one stallion and many mares; informally, this might be called a *harem*; more formally, this is a called a *polygynous* (many females) *mating system*. Mating systems tend to be a species-level characteristic, although alternative mating strategies may be expressed by individuals within a species.[31]

KEY TERM A social mating system is measured by the time spent by animals in pairs or groups.

KEY TERM A genetic mating system is measured by the actual genetic outcome of matings. For example, a mating system may be socially monogamous (male and female pairs spending most of their time together and cooperating), but genetic analysis may reveal that the male and/or the female had several mating partners.

KEY TERM An extrapair copulation (EPC) occurs when a member of a monogamous pair copulates with an animal outside the pair bond.

Contemporary biology has shown that there are dramatic differences between observations of *social mating systems* and *genetic mating systems*. Until the 1980s, animal behaviorists assumed that social bonding translated directly into mating and a genetic relationship with juveniles associated with the bond. As is often the case in science, the advent of sophisticated genetic techniques and the widespread use of these new tools, particularly genetic techniques for testing paternity, reveal that this assumption is often far from the truth.

In many species the social father may not be the biological father of the young, and surreptitious matings, usually called *extrapair copulations* (EPCs), are unexpectedly frequent.[32,33,34,35] Because extrapair copulations are often difficult to observe directly, investigations of mating systems prior to DNA testing generally credited animals with a higher level of fidelity to their mate than they deserved.

Birds are great examples of the potential for disparity between social and genetic mating systems. Most bird species are socially monogamous; a male and female spend much of their time together, they may collaborate in nest construction, and both birds often contribute to the care of the nestlings. Genetic testing reveals, however, that in many species of birds, the social father is not the genetic father of the nestlings (see Figure 11.14).

FIGURE 11.14
Plovers (left) are typically socially and genetically monogamous, with fewer than 10% of the chicks deriving from extrapair copulations. In wild turkey (right), on the other hand, nearly half the nests contain chicks resulting from extrapair copulations, suggesting a lack of genetic monogamy in this species. *Photos: Ben Pless (left) and Paul Bolstad, University of Minnesota, Bugwood.org (right).*

Nonetheless, despite the revolution in the understanding of mating systems led by DNA testing, the "how many males, how many females" approach to classifying mating systems is still valuable. Bonds associated with mating are at the core of animal social behavior, and the interplay between social and biological parentage is a key aspect of contemporary studies of mating systems. The following section outlines the different possibilities for male–female bonds and discusses the evolutionary dynamics that lead to each type of mating system.

Monogamy

Monogamy (Gr. *mono*, once; *monogamy*, marry once)—the pairing of a single male with a single female—is common in birds but rare in most other animals (see Figure 11.15). Monogamy usually requires that animals be able to recognize their mate as an individual, a cognitive task that is not particularly easy. In theory, a monogamous pairing ensures that both of the mates will contribute to care of their offspring and to mutual defense. Monogamy may evolve when the cost of acquiring mates is very high, when females have the ability to restrict male behavior, or when offspring survival requires more intensive care than can be provided by a single animal.

FIGURE 11.15
A pair of wood ducks, *Aix sponsa*, resting. The bright coloration of the male suggests that females of this species use this characteristic to choose their mates. The female's drab coloration makes her less conspicuous when she incubates eggs. In wood ducks, removal of the male from the pair during breeding season does not affect the ability of the widowed female to incubate the eggs, nor does it affect the survival of the eggs to hatching. However, the continued presence of the male makes production of a second brood more likely. This supports the hypothesis that the male may remain in the monogamous pairing because his chances of future reproduction are higher than if he leaves.[36] *Photo: Michael Breed.*

The major cost of true monogamy is that the entire *reproductive investment* of an individual depends on the fitness of its mate. This puts a high premium on choosing an appropriate mate. Choosing an inappropriate mate could have catastrophic fitness costs for an animal. Thus, monogamy is usually more apparent, or social, than real, and monogamous animals commonly use two methods to *hedge their bets* on mate choice. First, they may actually engage in *serial monogamy*, bonding with a mate for one mating season, but choosing a different mate in a subsequent season (see Figure 11.16). Second, many seemingly monogamous pairings are often subject to infidelities, or extrapair copulations. While over 90% of bird species appear to be monogamous, genetic studies show that in most populations at least a few offspring in each generation result from matings with partners other than a pair member. In some species, extrapair copulations produce over half of the offspring.

> **KEY TERM** Reproductive investment is the amount of time, energy, and nutrition a parent invests in its offspring.

> **KEY TERM** Bet hedging is when some investment is reserved and placed elsewhere. Extrapair copulations can be a form of bet hedging.

> **KEY TERM** Serial monogamy occurs when an animal pair bonds for some period of time but then switches to a new mate.

> **KEY TERM** Mate guarding is when, after copulation, one member of the pair prevents the other from seeking additional mates, or prevents potential additional mates from approaching.

Extrapair copulation (EPC) and extrapair paternity (EPP) are best studied in birds, where they are surprisingly common in species that were presumed to be monogamous or polygynous (see below). From a biologist's perspective, EPCs can be observable behavior, while EPPs can generally only be measured by doing genetic testing. Ninety-five percent of birds are presumed monogamous, but across species the mean rate of EPP is 13% (range = 0% to 76%). Extrapair copulations are not restricted to monogamous species; they also occur in polygynous species when females mate with males other than the dominant male in the group (see Figure 11.17).[37]

Given the frequency of EPCs, it should not be surprising that partners in a socially monogamous pairing might have evolved to attempt to enforce genetic monogamy. Among the possibilities are *mate guarding* (see Figure 11.18), *copulatory plugs*, prolonged

322

FIGURE 11.16
Beavers probably engage in serial monogamy, changing mates during their lifetime. *Photo: Jeff Mitton.*

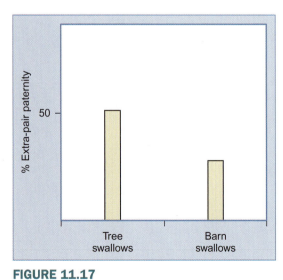

FIGURE 11.17
Extrapair paternity is higher in tree swallows than in barn swallows.[38]

FIGURE 11.18
Prolonged copulation in arthropods is often a form of mate guarding. *Photos: Upper left, tiger beetles; upper right, robber flies; lower left, water striders (Jeff Mitton). Lower right, millipedes (Michael Breed).*

copulation, aggressive behavior, and withholding parental care (or worse—infanticide) when paternity is uncertain.[39,40,41,42]

323

Polygyny

Polygyny (Gr. *poly*, many; *gyny*, female) is the association of one male with many females. The most common mating system in mammals, it is found in a few birds, insects, and other animals. The study of polygynous mating systems focuses on male–male competition prior to mating, which is sometimes called *precopulatory competition*. If a female mates with more than one male within a short time span, the male–male competition extends to competition among sperm. Along with copulatory plugs and postcopulatory mate guarding, this is *postcopulatory competition*. Sperm competition is an important aspect of mating systems, discussed in Section 11.9 later in this chapter.

There are several types of polygyny; the two most common are

1. Groups of females centered around resources, or a male that holds those resources.
2. Membership in a *harem*, or a defended group of females. While the pattern of one male and many females is similar between these two types, they differ in very important ways.

> **KEY TERM** A copulatory plug is a secretion or object left by the male in the female's reproductive tract with the goal of preventing further matings by the female.

> **KEY TERM** A harem is a group of females associated with a single male.

> **KEY TERM** Resource defense polygyny occurs when male territoriality centers on a resource needed for breeding success. If multiple females join the male on the territory, then it is a polygynous mating system.

Resource Defense Polygyny

Resource defense polygyny occurs when male territoriality centers on a resource needed for breeding success. In resource defense polygyny, male resource holding power (RHP) is

FIGURE 11.19
Red-winged blackbirds are well-known examples of resource defense polygyny. The male defends a territory and the quality of that territory determines the size of his harem. *Photo: Michael Breed.*

critical for male reproductive success. This means that males must be competitively equipped to dominate resources, and that females may sample and make choices among potential mates, using either direct measurements of male quality (size, plumage color, and so on) or measurements of territory quality. Because male movements are restricted by the actions of males on adjacent territories, any limitations on female movements (and choices) are imposed by competition among females for males. This competition can be intense, especially on better territories (see Figure 11.19). A female may find that her potential reproduction is greater if she chooses a male on a high-quality territory, even if that male already has a mate, than it would be on a lesser territory, even if no other female is present on that territory. The *polygyny threshold model*[43,44,45] makes specific predictions about when females should choose to be a secondary mate in one territory rather than be the sole female in another. Note that in this form of polygyny, females exercise choice about males and resources.

OF SPECIAL INTEREST: MATING SYSTEM AND TERRITORIALITY IN THE YELLOWHEADED BLACKBIRD

Yellowheads (Figure 11.20), like their relatives, redwinged blackbirds, have a polygynous mating system in which a male territory holder may have more than one female nesting within his territorial boundaries.

Why might the male blackbird be so brightly colored? Here are three hypotheses:

- The color may be the result of runaway sexual selection of a feature that attracts females.
- The color may signal the male's good health to females. Unhealthy males may, due to poor nutrition, parasites, or disease, be unable to produce the bright colors. If the color is costly to produce, it may be a handicap.
- The color may warn other males of the presence of a territorial male.
- If the color attracts predators, it would also be a handicap.

How would you test these hypotheses?

FIGURE 11.20
The yellowheaded blackbird, *Xanthocepalus xanthocephalus*, perched near a marsh. *Photo: Jeff Mitton.*

Because male–male competition is central to resource defense polygyny, it is important to consider how competition among males occurs. Numerous studies have paired males and tested their fighting prowess.[46,47] Not surprisingly, generally larger males have an advantage in contests in neutral arenas; experience, or age, also can play an important role, as can the ability to search for widely distributed females, a form of scramble competition.[48,49,50] When contests are staged within a male's territory, though, the resident male often has a substantial edge in winning the encounters. Thus, the outcome of the resource defense polygyny may not be predictable in terms of variables that might influence non-territorial competition. This resident male effect can outweigh the effects of size, age, and experience (see Figure 11.21). Because of this, the "best" male by measurable standards in a given area may not possess the richest territory. Overall, though, this mating system tends to favor larger, stronger, older males, and variance in mating success is skewed to these males.

Harems, or Female Defense Polygyny

If females are clustered around a resource, then males may compete to monopolize that group of females. This type of polygyny puts the emphasis on male–male competition and minimizes the role of mate choice, either by males or females (see Figure 11.22). In other cases, a band, or herd, of females is not associated with any resource but may have aggregated for other reasons (e.g., defense against predators). Regardless of the reason for the aggregation, the group itself becomes the focal point of competition among males. Because a female's movement is not limited by a male's hold on valuable resources and her evaluation of those resources or that male, direct male dominance of females themselves may come into play (see Figure 11.23).

Competition among females within a polygynous group is also potentially important and may be reflected by "pecking order" among the females. For instance, in domestic cows, priority of entrance through a gate or stable door is a reflection of dominance in the herd; while it is hard to imagine that getting into the stable first carries survival value, this is a reflection of other dynamics that may affect fertility, nutrition, and ultimately survival of a cow's calf. The best-known dominance hierarchies among females, such as those in flocks of domestic hens, are linear, meaning that each individual's rank can easily be determined by observation. In addition to hens, dominance hierarchies are known in various primates, in herds of ungulates, and in cockroaches. Because this sort of dominance is not as easily associated with the often conspicuous traits of competing males, this cause of variance in female mating success has probably not been sufficiently assessed. We will discuss types of dominance relationships in Chapter 13.

Leks

A *lek* is an aggregation of males, each seeking to attract a mate, and each displaying fervently to do

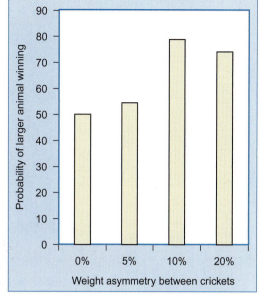

FIGURE 11.21
In these crickets, the greater the weight asymmetry between competing males, the greater the probability that the larger male would win the contest.

325

> **KEY TERM** A lek is an aggregation of males, each seeking to attract a mate, that is not associated with a key resource.

FIGURE 11.22
A zebra harem. *Photo: Mike Huffman.*

FIGURE 11.23
Harem polygyny in wild horses. *Photo: Allen Bridgman, South Carolina Department of Natural Resources, Bugwood.org.*

FIGURE 11.24
A lekking male sage grouse. Males aggregate and females choose among the males. *Photo: Dave Menke, U.S. Fish and Wildlife Service.*

so. Leks are not associated with resources, such as food, water, or shelter. Conventional wisdom holds that males sacrifice their needs for food, water, and shelter by being near their competitors because the lek attracts more females than isolated males do.[51,52] Leks may be aggregations of males around a particularly attractive male (hotshots), or aggregations of males in a location in the landscape that females are likely to travel near (hotspots; see Figure 11.24).

326

OF SPECIAL INTEREST: BACHELOR HERDS

What do surplus males do in a polygynous mating system? If male reproductive success has high variance, then younger, weaker, and less-experienced males may be left without either mates or territories. Oftentimes these males herd together, forming bachelor groups. A bachelor herd is a group of adolescent or adult males that do not have harems. They travel together and gain benefits from mutual protection and food discovery. Such groups can also coalesce outside the mating season in species that form polygynous groups for only parts of the year. Membership in bachelor herds confers advantages in shared vigilance and group defense against predators and allows group members to take advantage of each other's capabilities in finding food. Sparring—male–male combat with less danger of injury—can set the stage for more intense competition during mating season (see Figure 11.25).

FIGURE 11.25
A bachelor herd of walruses. *Photo: U.S. Fish and Wildlife Service.*

FIGURE 11.26
Honeybees are polyandrous; queens mate with ten to twenty males. *Photo: Jessica Lawrence, Eurofins Agroscience Services, Bugwood.org.*

Polyandry

Polyandry (Gr. *poly*, many; *andry*, male) is the reverse of polygyny—one female, many males (see Figure 11.26). Because female choice of mates seems to drive many mating systems, polyandry, at least superficially, seems to be rare. There are a few cases of resource defense polyandry, in which the female is territorial and males coexist on her territory. This happens in American jacanas and spotted sandpipers. In the case of cooperative polyandry, seen in the Galapagos hawk, for instance, females have several mates and cooperate with them to raise a brood.

Polygynandry and Promiscuity

Polygynandry occurs when females mate with several males and the males care for the broods of several females. This stands in contrast to *promiscuity*, in which there are no pair bonds, and males and females in a population seem to mate randomly and with multiple partners. In general, the theory of mating systems suggests that promiscuity should be rare or absent in animals, as it is usually to the advantage of one or both of the sexes to choose their mates. Promiscuity might occur in species that live in completely unpredictable environments, that is, environments in which it is impossible to judge which mate might carry genes that would be good in the next generation. Another factor favoring promiscuity would be the inability of any animal to hold a territory and monopolize access to critical resources. In other words, promiscuity is expected when choice becomes unlikely or impossible.

> **KEY TERM** Polygynandry occurs when females mate with several males and the males care for the broods of several females.

Mere observation of seemingly random matings does not establish that a mating system is, in fact, genetically promiscuous. (Recall the distinction between the social appearances of a mating system and the actual genetic outcomes of that system.) Copulation can serve social functions other than reproduction; they include bonding, dominance, or even concealment of actual paternity of a female's offspring.[53,54] Genetic studies (usually using DNA fingerprinting or microsatellite marker techniques) establish true paternity of juveniles. If the norm for males in a population is to sire offspring with more than one female, and each of the males is more or less equally successful in fathering young, then the mating system is truly promiscuous. If there is substantial variation among males in their success in siring young, then even if mating frequencies suggest promiscuity, the system is probably actually polygynous or polygynandrous.[55]

> **KEY TERM** Promiscuity occurs when there are no pair bonds, and males and females in a population seem to mate randomly.

The snowshoe hare, *Lepus americanus*, while best known for its dramatic population cycles in Canada, offers a good example of a possibly promiscuous mating system. Snowshoe hare populations rise and fall on a 10-year cycle. While once thought to represent a classical predator–prey cycle, it is now recognized that hare population cycles are more complex, regulated not only by predators but also by food supply.[56] Snowshoe hares show no sign of being territorial, or of having any sort of complex social system, even at the peak of their population cycle. Going along with this lack of social complexity is a promiscuous mating system. Behavioral observations suggest that female and male snowshoe hares mate with many partners, and genetic analyses confirm that in many litters more than one father is represented. Some males, though, fathered more offspring by being the only father of some litters. Burton interprets these results to indicate that the mating system is not fully promiscuous, in that paternity is not entirely random.[57] Some sort of competition may occur among males. (Perhaps some males are more attractive to females and gain more matings.) The number of fathers represented in the offspring, and the frequency of litters representing more than one father, does make this mating system as close to true promiscuity as has been observed in any mammal.

DISCUSSION POINT: MONOGAMY

Now that we have completed our discussion of the types of mating systems, can you form some hypotheses about why social monogamy is the most common mating system among birds and polygyny is the most common mating system among mammals? Can you point to some basic difference between the biologies of birds and mammals that might account for this? Does sexual dimorphism seem more common in birds or mammals? When sexual dimorphism is present, what form does it typically take in birds? What about mammals? Returning to the original question about social monogamy versus polygyny, do you think this difference in social mating systems causes differences in sexual dimorphism between birds and mammals?

11.7 HORMONES AND SEXUAL BEHAVIOR

Chapter 2 reviewed the roles that hormones play in behavior and promised more detail about reproductive hormones in this chapter. Of all aspects of behavior, reproduction may be the most subject to hormonal influences. This section applies to mammals and birds.

327

To begin with, luteinizing hormone (LH) and follicle-stimulating hormone (FSH) are both neuropeptides produced in the hypothalamus and act on the ovaries and testicles. Behaviorally, LH is linked to sexual receptivity. A surge in LH occurs before ovulation, so it makes perfect sense that LH would prime female sexual behavior. In female rats, LH releases lordosis, a posture exhibited by an ovulating female in response to contact by a male. The male, in turn, mounts the female. FSH is less associated with behavioral responses. The gonads produce steroid hormones in response to LH and FSH from the pituitary—primarily testosterone by males and progesterone and estrogen by females, although all three hormones are produced by both sexes. Endocrinologically speaking, males are fairly uninteresting as testosterone levels increase with sexual maturity and then remain constant or cycle seasonally depending on the species. In females, estrogen provides feedback to the pituitary, so that LH and FSH production increases with maturity of the egg or eggs. This culminates with the LH peak, ovulation, and a return to low levels of estrogen. If the female becomes pregnant, progesterone levels increase; behaviorally and physiologically, progesterone is the hormone of pregnancy in mammals and of brooding in birds. This fits into a larger context of overall hormonal regulation, of course. For instance, stress hormones can interact with pair bonding in rodents such as prairie voles.[58]

Pituitary hormones that ultimately affect mating and parental behavior include luteinizing hormone (LH) and follicle-stimulating hormone (FSH), which activate the ovaries and testes of vertebrates, stimulating the production of estrogens and testosterone. Oxytocin and vasopressin, also both produced by the pituitary, play key roles in priming animals for pair bonding. Behaviorally, the vertebrate hormones, the estrogens and testosterone, have major impacts. These hormones affect the expression of mating behavior, parental behavior, and dominance and aggression.

The development of female sexual behavior in mammals is more complex than that of males. In most mammals, female sexual receptivity is linked to ovulation, so that females are primed and physiologically able to copulate when the probability of fertilization is high. In vertebrate development, the production of estrogen by the ovaries guides appropriate development of adult sexual morphology. Ovulation is stimulated by progesterone, a steroid secreted by the pituitary, and in many mammals progesterone is a prime trigger for female sexual receptivity. Once the female is pregnant, progesterone takes on an inhibitory role for sexual behavior, and after birth, hormones associated with milk production and suckling (particularly oxytocin and prolactin) further inhibit sexual behavior. In nonmammalian vertebrates, these hormones serve other functions.

DISCUSSION POINT: HORMONES AND GOVERNMENT

Until the last few years of the twentieth century, virtually no medical research was performed on women. In fact, women were not allowed in most clinical trials until after 1990. Researchers defended this practice by citing the observation that hormone levels in men changed more slowly than those in women; men were therefore thought to be less-complicated research subjects. In 1990, the Government Accountability Office (GAO) issued a highly critical report, resulting in the formation of the Office of Research on Women's Health within NIH and a much more balanced approach to medical research.

Like LH and FSH, vasopressin and oxytocin are neuropeptides released into the circulation from the pituitary gland. Behaviorally, oxytocin has become known as the hormone of pair bonding and maternal behavior. Many of the adjectives associated with oxytocin (*love, trust,* and so on) are laden with human emotion and culture, but a well-known study of voles showed that one species, the prairie vole, has vasopressin and oxytocin receptors in the brain in locations that might associate monogamy (having a single mate) with dopamine receptors, a source of positive neural reinforcement. In the montane vole, which has multiple mates, this association with dopamine receptors and centers of positive neural reinforcement is lacking. Vasopressin may facilitate recognition of social partners, a key element of pair bonding.[59]

11.8 HORMONES, TERRITORIALITY, AND AGGRESSION

The role of testosterone in regulating aggression and a general suite of male physical behavioral phenotypes is well known. Early awareness of this relationship is reflected in the agricultural practice of castrating male horses, cattle, swine, and chickens. The result is an animal that is both more manageable and, in the end, more edible. (In the absence of testosterone, the muscles are less well developed, more tender, and less sinewy.) Experience taught early domesticators of animals that castration was most effective if performed prior to the animal's attaining sexual maturity, and that once dominance and territoriality had been asserted in a male, castration was relatively ineffective in moderating these behaviors. This empirical knowledge of the testes as the seat of maleness led to the use of testes and testicular extracts to manipulate male traits (including treatment of impotence in humans), long before the chemical nature of the testicular factor was identified. The chemical structure of testosterone was discovered in the 1930s, and in 1939 the chemists Adolf Butenandt and Leopold Ruzicka received the Nobel Prize for synthesizing testosterone.

A long history of animal studies using testosterone injections or implants clearly links sexual behavior, aggression, and territoriality with this hormone (see Figure 11.27). Note, though, that in some cases exposure to testosterone during maturation is sufficient to trigger irreversible adult behavioral patterns. While largely regarded as a male hormone, testosterone is produced in vertebrate ovaries as a derivative of estrogen and is important in modulating female sexual responses, as well as impacting some sexually selected traits in females.

Given current societal concerns about testosterone use as a performance-enhancing drug in humans, it is appropriate to ask what might be gleaned from knowledge about supplementation of natural testosterone in animal studies. It is clear from a broad range of animal studies that testosterone levels are associated with territoriality and aggression. Thus, "'roid rage" has a real biological basis. In particular, rats treated with androgens respond much more aggressively to a provocation, such as a tail pinch, than do untreated controls. This effect in rats persists after the termination of the drug treatment. Human behavior, of course, is very much modulated by cultural effects, so care must be taken in extending conclusions from studies of nonhuman animals to humans.[60]

11.9 SPERM COMPETITION

Sperm competition is a form of male–male competition that warrants its own section in a textbook because it is indeed so basic and important.[61] Many marine animals release their gametes into the ocean (this is called *spawning*), but this does not mean there is an absence of competition in mating, nor does it mean that the resulting zygotes are the results of random meetings of eggs and sperm. As with many topics in science, the absence of much evidence for competition among small objects in the vast sea does not mean competition does not occur. Spawning is most common in aquatic or semiaquatic environments, where sperm can easily swim to

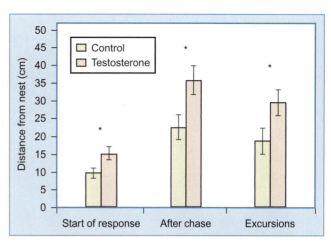

FIGURE 11.27
Typical responses of vertebrate males to testosterone implants. In this case, the study species is a fish, the rock-pool blenny. The pink bars represent individuals receiving supplemental testosterone. "Start of response" is the distance of the intruder from the nest when the territorial response starts, "after chase" is the distance the intruder is chased, and "excursions" compares the distance that treated and untreated males patrol from their nests. *Adapted from Ros, et al., 2004, Hormones and Behavior, 46:491–497.*

329

KEY TERM Sperm competition occurs among sperm after they are released from the male. The sperm from a single male may compete to reach an egg first; if the female has mated with more than one male, then sperm competition occurs among the males.

FIGURE 11.28
Large testicular size relative to body size may be an indication of an evolutionary history that includes sperm competition. Why might this male be displaying/exposing his scrotum? *Photo: Michael Breed.*

FIGURE 11.29
The bluegill sunfish and nest. *Photo: Eric Engbretson, US Fish and Wildlife Service.*

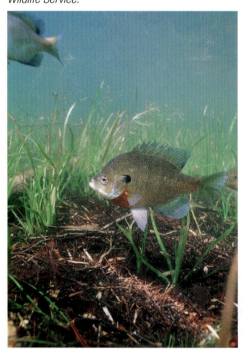

eggs. In this system, animals that release the largest amounts of sperm with strong swimming abilities are most likely to sire young. Sperm competition was probably the first type of sexual competition to evolve, and as anisogamy evolved, sperm competition was central to the evolution of mating systems.

Sperm competition, coupled with external fertilization, is the simplest and, by far, the most common type of fertilization across the broad spectrum of the animal world. In many species, male and female never need meet; the female deposits her eggs, and later a male finds and fertilizes them. Even under these uncomplicated conditions, two processes make mating nonrandom. First, there is variation in the ability of males and females to find appropriate locations to deposit their gametes. Second, there is variation in sperm swimming speed and energy resources that allow it to reach the egg first.

Sperm competition has been studied in many species of fish with external fertilization. Selection strongly favors all males that produce sperm with maximum speed and energy resources (ATP), but in many species there are still large enough differences among sperm from different males to affect fertilization probabilities. The amount of sperm, or sperm volume; the distance of sperm release from the eggs; and the timing of sperm release are also critical factors in sperm competition. The more sperm a male is able to produce, the higher the probability that one of his sperm will win the competition. Males that are able to get closer to eggs are generally more likely to be successful, and a few seconds' difference in sperm release can make all the difference in whether a male's sperm reaches the eggs first.

Generally, sperm volume is correlated with testicular size (see Figure 11.28); larger testicles are necessary for production of larger amounts of sperm. A basic prediction in the biology of mating systems is that males of species with more intense sperm competition will have larger testes. The evolutionary responsiveness of testicular size to sperm competition is an important example of the effects of sexual selection on morphology and anatomy.

In many cases external fertilization is combined with courtship; even though the eggs and sperm are released into the environment, the male and female first have the opportunity to assess the quality of their potential mate before committing their gametes. Bluegill sunfish (*Lepomis macrochirus*), a common fish of lakes in North America, exemplify how sperm competition can play a role in a more complex mating system that involves sophisticated behavioral interactions. Bluegill males can adopt a variety of mating strategies (see Figure 11.29). Some, termed "parental" males, build nests in the bottom of the lake and court females, who spawn in the nest. Parental males then guard the eggs. "Sneakers" dash in and release sperm near the nests of parental males. And male "satellites" are able to come close to eggs because they mimic females. Sperm of all three types of males had similar speed and swimming ability, but even though sneaker males release their sperm farther from the eggs and later than parental males, they have an advantage over parental males.[62] This suggests that sneaker males produce higher-quality sperm. Satellites also have an advantage over parentals, but this could be explained by the fact that they can get closer to the eggs when they release their sperm.

Internal fertilization facilitates mating in terrestrial environments and, compared to external fertilization, allows for more control over gametes by both males and females; most terrestrial animals have evolved some form of internal fertilization. With internal fertilization, males do not need to respond so much to the effects of dilution and distance from the eggs on their gametes, and both sexes can exert a greater degree of choosiness over their mates. However, sperm competition remains an important mechanism in animals with internal fertilization. Sperm from a male will compete among themselves for fertilization opportunities, but any time a female mates more than once within a short time period, sperm competition among the sperm from the different males is likely.[63] In some cases, males even remove the sperm deposited by previous males.[64]

11.10 GOOD GENES MODELS FOR CHOOSING A MATE

As pointed out previously, most of mate choice seems to reflect the genetic quality of the potential mate. This is seen in considerations of the handicap principle and the idea that costly signals may be honest indications of a mate's quality. This section covers three other ways that animals might assess the genetic qualities of their mates. These models assume that mate choice is based on indicators of the mate's *good genes*. The possibility of mate choice based on good genes can be extended by considering characteristics that correlate with the viability of the resulting offspring. Ironically, strong mate choice based on a trait should render the trait noninformative; there will be no significant heritability for the trait if genetic variation associated

> **KEY TERM** Good genes models hypothesize that mate choice relies on determining the potential of mates to pass strong genetic properties on to their offspring.

with that trait is eliminated. This irony was noted on page 11, along with the fact that despite this expectation, variation in sexually selected traits was often persistent. Table 11.1 shows causes of variation among animals in mating preferences. Note that all of these mechanisms preserve additive genetic variation.[65]

331

Inbreeding and Outbreeding

Laboratory experiments often show that inbreeding, that is, mating between closely related animals, results in lowered reproductive success. Inbreeding depression, evidenced by lowered survival rates in offspring of inbred matings, occurs across a wide range of taxa. Inbreeding exposes deleterious recessive genes by increasing the likelihood that both of an animal's copies of an allele will be the same. In the ants, bees, and wasps, all members of the insect order Hymenoptera, inbreeding results in lethal genetic combinations. These observations suggest that animals should have mechanisms for avoiding mating with relatives. Inbreeding avoidance can be as simple as sex-biased dispersal: if males disperse from their birth area and females remain, then brother–sister matings will be unlikely. Learning the identities of littermates and parents early in development can also provide essential information for avoiding inbreeding. Sensory cues that allow genetic matching provide another way to facilitate inbreeding avoidance. If, for example, related animals smell

TABLE 11.1 Causes of Variation among Animals in Mating Preferences

Advantage

Genetic compatibility
Genetic complementarity
Heterozygosity
Mating facilitation
Variation among optimal offspring phenotypes
Phenotypic compatibility
Rare offspring advantage

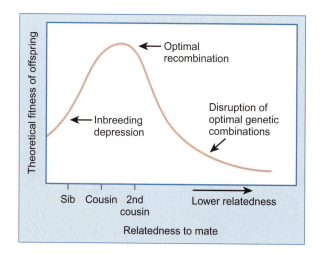

FIGURE 11.30

Patrick Bateson proposed that there was a level of outbreeding that would maximize fitness. He envisioned it as a compromise between the inbreeding depression caused by insufficient outbreeding and the disruption of optimal combinations of genes resulting from high levels of outbreeding. *Source: Bateson, P. 1983. Optimal outbreeding in mate choice. P. Bateson, ed. Cambridge University Press, Cambridge. pp 257–276.*

alike, an animal may be able to avoid inbreeding by not mating with animals that smell similar to itself, or to its littermates. This type of social recognition is phenotype matching, which is discussed in more detail in Chapter 13. Females can also ensure against infertility resulting from inbred matings by mating with multiple males.

Of course, if an animal is performing well in a habitat, then selection would favor having genetically similar offspring. Sex has the potential for creating less-than-optimal genetic recombinations in this circumstance. Is inbreeding all bad under these conditions? The solution to this problem lies in optimal outbreeding, an idea developed by Patrick Bateson. By mating with animals that are related enough to share gene combinations, but not so related that inbreeding depression occurs, favorable genetic combinations can be maintained across generations (see Figure 11.30).

If females exert choice among mates, does this really have an evolutionary effect on males? Moller and Alatalo presented a meta-analysis in which they tested the correlation between male traits and offspring survival.[66] The mean correlation coefficient was 0.122, which is significant, but explains only 1.5% of the variation in offspring survival. While small, the effect may be substantial over extended evolutionary timescales. (An aside: The authors also found a significant effect of publication year on the results of the original studies; this could be due to biases in the studies resulting from paradigm shifts in the field!)

Why not choose the same mate every time? If good genes are good, then why vary mating choice? The answer is simple and is the explanation at the heart of all biological diversity: Environments vary, and as the environment changes, so does the ideal genetic combination.

The Major Histocompatibility Complex

The major histocompatibility complex loci (MHC) play an interesting role in vertebrate mating systems.[67,68] Their primary function, of course, is one of mediating a vertebrate's immune response to diseases and parasites. MHC loci are hypervariable, meaning that a wide array of different alleles is possible at each locus. Many mammals detect MHC genotype by smell, giving them cues that can be used in behavioral decisions, including mate choice. MHC heterozygotes express twice the number of alleles as homozygotes and are hypothesized to have a broader range of resistance to disease. Because variability in MHC is a presumed advantage, heritability should be high. Honest signaling may be desirable or at least unavoidable. These are the ultimate good genes.

FIGURE 11.31
The evolution of hyperdiversity at MHC loci can result from gene duplications, then mutations at the duplicated loci. *Adapted from Potts, W.K., & Wakeland, E.K., 1993, Evolution of MHC genetic diversity: a tale of incest, pestilence, and sexual preference, Trends in Genetics, Vol. 9, pp. 408–412.*

Hypothetically, pathogens evolve to evade common MHC genotypes, giving rare MHC genotypes a selective advantage. Thus, another prediction for female mate choice is that females should mate with males that are immunologically dissimilar. This can actually happen, as odors may correlate with immune system diversity.[69] Such choices would give their offspring high immune diversity and, in theory, make their offspring more disease resistant (see Figure 11.31).

Fluctuating Asymmetry

Fluctuating asymmetry (FA) is the deviation from perfect bilateral symmetry caused by environmental stresses, developmental instability, and genetic problems during development.[70,71,72,73,74] It is thought that the more nearly symmetrical an organism is, the better it has been able to handle developmental stress and the more developmental stability it has, because environmental effects on development produce asymmetrical body form in animals. Females that favor symmetry may therefore be using an honest signal of "good genes." The relationship of this type of asymmetry to mating behavior is a controversial area of research in animal behavior. Fluctuating asymmetry, then, may be another honest signal of good genes.

> **KEY TERM** Fluctuating asymmetry is the difference(s) within an organism from right to left. For example, an animal's right legs may be shorter than the left. The legs on each side are identical, genetically, so the length difference must be due to environment. Scientists have argued that more symmetrical animals have shown that they have better genetic combinations for their environment and, as a result, aren't so challenged in their development by the environment. Asymmetrical animals, following this argument, have genes that do not allow them to cope so well with the environment, resulting in developmental variation.

333

DISCUSSION POINT: SYMMETRY AND ATTRACTIVENESS

In humans, symmetrical men are more attractive to women, particularly during high fertility phases of the ovulatory cycle. Scents from symmetrical men are also more attractive to women. Conversely, men show no preferences among women based on either symmetry or scent. This is interpreted as support for good-genes preferences, with the scents being an "honest signal of phenotypic and genotypic quality of the human male." In men, there is a high correlation between symmetry and their lifetime number of sex partners. "One possible good-genes explanation is that symmetric men may have particularly rare major histocompatibility (MHC) genotypes making them attractive to many fertile women."[75] Find some photos of politicians, actors, and celebrities. Are these individuals symmetrical in their facial features? Do you think facial symmetry contributes to the kind of public success and attention that politicians, actors, and celebrities receive?

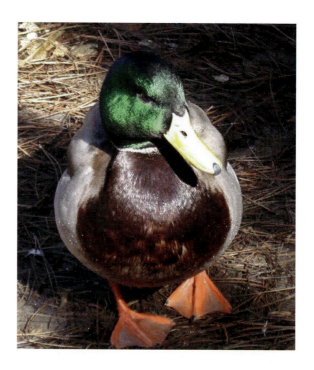

FIGURE 11.32

This male mallard duck (*Anas platyrhynchos*) is an excellent example of how males of many bird species use bright colors to attract mates. Note the contrast of the vibrant green head, rich brown bib, and reddish-orange feet. The ability of males to produce these colors may be affected by their nutritional state and by whether or not they are infected by parasites. Males that are effective foragers with good nutrition or that have well-functioning immune systems may be more brightly colored than malnourished or infected males. Females who choose brightly colored males may be keying in on features that signal "good genes" in their prospective mate. Poorly fed or infected males, in theory, are unable to produce the bright colors of healthy plumage; coloration, then, becomes an honest signal that the male is carrying genes which make him successful in his environment. *Photo: Michael Breed.*

Parasites, Plumage, and Choosing a Good Mate

Another reason that animals may be unable to falsify signals of their fitness is that they cannot falsely alter their ability to cope with parasites.[76,77] Plumage color in some birds may be brightest in individuals with lower parasite loads and dullest in highly parasitized birds (see Figure 11.32). If "good genes" are responsible for lower parasite loads, then bright plumage color reflects good genes.[78]

11.11 FORCED COPULATIONS

Forced copulation is a common reproductive strategy in ducks and geese but is much rarer in other types of animals.[79,80] The threat of forced copulations may be a driving force in selecting for monogamy, at least in ducks and geese.[81] In this model, when the female chooses to pair bond with a male, she gains the male's services in protecting her from forced copulations by other males. Females can also combat forced copulations by evolving methods of expelling or not using the resulting sperm.[82] The evolutionary dynamics surrounding forced copulations are complex; Clutton-Brock and Parker review the causes, costs, and modes of evading forced copulation.[83,84]

> **KEY TERM** Forced copulation occurs when one animal subverts another's mating choice and uses physical means to force copulation on that animal.

SUMMARY

Mating is a simple act with the most profound possible outcome: the production of the next generation. Mating is the point at which individual animals make a key evolutionary choice: how to combine their genes with the genes of another animal to give their offspring the greatest possible genetic advantage. Gametic investment determines sex roles; this is fundamental to defining maleness and femaleness. Many animal species produce two types of gametes: small, mobile sperm and large, nutrient-rich eggs. Males are defined as the producers of sperm; females are the producers of eggs. Mate choice and parental investment follows from these differences in investment in gametes.

Biologists have classified mating systems using observations of social interactions. Social monogamy is the pairing of a single male with a single female. Social polygyny is the pairing of one male—generally either a territory holder or a dominant male with a harem—and many females. Social polyandry, which is more rare, is one female associated with a number of males. Polygynandry and promiscuity result from more complex patterns involving multiple mates for both males and females. Animals may attempt to enforce monogamy; for example, mate guarding by males can limit mating opportunities for females.

Genetic characterization of mating systems leads to a far different picture. In many socially monogamous mating systems, organisms participate in extrapair copulations (EPC) that result in extrapair paternity (EPP). Sometimes the majority of the offspring of a socially monogamous pair are genetically fathered by other males. Females may have multiple mates during their lifetime to increase the genetic diversity of their offspring or to hedge their bets against poor mate choices. Mating with more than one male may also ensure against male infertility. Males may mate with more than one female to produce more offspring than they might have with one mate.

Female gametic investment generates the prediction that females are often choosier than males when selecting mates. Females may prefer males that have "good genes," males who hold valuable resources, or males who are likely to produce sons that will have high chances of mating ("sexy sons"). The stereotype that males will be less choosy is sometimes not true; if a male's investment in his offspring is high or if he is unlikely to be able to obtain more mates in the future, then he, too, should be choosy.

Intermale competition plays a key role in obtaining mates for males of many species. Sexual selection resulting from this competition can result in large body size, enhanced musculature, and evolution of armament such as horns or antlers. Sperm competition among males extends intermale competition to the postcopulation interval. Interfemale competition can be as important as competition among males, and can result in selection for competitive abilities in females. Finally, forced copulations can subvert processes of choice. While they give males who would otherwise not mate opportunities to gain reproductive success, they are vigorously opposed by females and their mates.

STUDY QUESTIONS

1. Find a nonbiologist and try to explain to him or her why sex evolved and what the evolutionary function of sex is. Doing so will force you to organize your thoughts on this conceptually difficult question.
2. What are some attributes of females that increase the likelihood that they will be the caregiver for the offspring?
3. What is the role of dishonesty in courtship? Can you form a list of circumstances in which courtship signals should be honest? You may wish to refer to the discussion of honest signaling in Chapter 7 on communication as you answer this question.

4. What are the benefits to a female from social monogamy? As you think about this, focus on monogamy as a strategy the female might use to involve a male in parental care of their offspring. What are the disadvantages, from a female's point of view, of genetic monogamy? Can you explain some examples of how females may overcome these disadvantages?

5. What circumstances might favor the evolution of polyandry?

6. Investigators have argued that symmetry is a measure of "good genes" in an animal. A more detailed explanation is that symmetry reflects developmental stability and perhaps greater allelic diversity in the animal. If this is true, then animals should prefer to have symmetrical mates. How does this line of reasoning interact with thinking about environmental variability? Would you expect symmetry to be more important for mate choice in stable or unpredictable environments? Can you design an experiment to test your prediction?

7. Greek mythology states that the "eyes" on a peacock's tail are derived from Argus, whom Hermes killed after boring him to sleep with dull stories and insipid songs. Hera then cut the eyes off the corpse of her valued friend and scattered them on the tail of the bird now called the peacock. Can you give an explanation for the existence of these "eyes" that is more consistent with modern scientific thought?

Further Reading

Ahnesjo, I., 2010. Mate Choice in Males and Females, In: Michael D. Breed and Janice Moore (Eds.), Encyclopedia of Animal Behavior, Academic Press, Oxford, Vol. 2, pp. 394–398.

Clutton-Brock, T.H., Parker, G.A., 1995. Sexual coercion in animal societies. Anim. Behav. 49 (5), 1345–1365.

Dixson, A.F., 2009. Sexual Selection and the Origins of Human Mating Systems. Oxford University Press, Oxford, 232 pp.

Emlen, S.T., Oring, L.W., 1977. Ecology, sexual selection and the evolution of mating systems. Science 197, 215–233.

Grether, G.F., 2010. Sexual Selection and Speciation, In: Michael D. Breed and Janice Moore (Eds.), Encyclopedia of Animal Behavior, Academic Press, Oxford, Vol. 3, pp. 177–183.

Orians, G.H., 1969. On the evolution of mating systems in birds and mammals. Am. Nat. 103, 589–603.

Shuster, S.M., Wade, M.J., 2003. Mating Systems and Strategies. Princeton University Press, Princeton, 520 pp.

Thornhill, R., Alcock, J., 1983. The Evolution of Insect Mating Systems. Harvard University Press, Cambridge, Mass. 564 pp.

Zahavi, A., Zahavi, A., 1997. The Handicap Principle: A Missing Piece of Darwin's Puzzle. Oxford University Press, Oxford, 304 pp.

Notes

1. Van Valen, L., 1973. A new evolutionary law. Evolutionary Theory 1, 1–30.

2. Fisher, R., 1930. The Genetical Theory of Natural Selection. New York: Dover, 1957.

3. Trivers, R.L., Willard, D., 1973. Natural selection of parental ability to vary the sex ratio of offspring. Science 179, 90–92.

4. Lande, R., 1976. Models of speciation by sexual selection on polygenic traits. PNAS (USA) 78, 3721–3725.

5. Grether, G., 1996. Intrasexual competition alone favors a sexually dimorphic ornament in the rubyspot damselfly Hetaerina americana. Evolution 50, 1949–1957.

6. Hogg, J.T., Forbes, S.H., 1997. Mating in bighorn sheep: frequent male reproduction via a high-risk "unconventional" tactic. Behav. Ecol. Sociobiol. 41, 33–48.

7. Basolo, A.L., 1990. Female preference predates the evolution of the sword in swordtail fish. Science 250, 808–810.

8. Rosenthal, G.G., Evans, C.S., 1998. Female preference for swords in Xiphophorus helleri reflects a bias for large apparent size. PNAS 95, 4431–4436.

9. Coleman, S.W., Patricelli, G.L., Coyle, B., Sianil, J., Borgia, G., 2007. Female preferences drive the evolution of mimetic accuracy in male sexual displays. Biol. Lett. 3, 463–466.

10. Zahavi, A., 1975. Mate selection—selection for a handicap. J. Theor. Biol. 53 (1), 205–214.
   Zahavi, A., 1977. Cost of honesty (further remarks on handicap principle). J. Theor. Biol. 67 (3), 603–605.

11. Petrie, M., Halliday, T., 1994. Experimental and natural changes in the peacocks (Pave cristatus) train can affect mating success. Behav. Ecol. Sociobiol. 35 (3), 213–217.
   Petrie, M., Halliday, T., Sanders, C., 1991. Peahens prefer peacocks with elaborate trains. Anim. Behav. 41, 323–331.

Petrie, M., Krupa, A., Burke, T., 1999. Peacocks lek with relatives even in the absence of social and environmental cues. Nature 401 (6749), 155–157.

12 Takahashi, M., Arita, H., Hiraiwa-Hasegawa, M., Hasegawa, T., 2008. Peahens do not prefer peacocks with more elaborate trains. Anim. Behav. 75, 1209–1219.

13. Loyau, A., Petrie, M., Saint Jalme, M., Sorci, G., 2008. Do peahens not prefer peacocks with more elaborate trains? Anim. Behav. 76, e5–e9.

14. Widemo, F., Saether, S.A., 1999. Beauty is in the eye of the beholder: causes and consequences of variation in mating preferences. Trends Ecol. Evol. 14, 26–31.

15. Manning, J.T., Hartley, M.A., 1991. Symmetry and ornamentation are correlated in the peacocks train. Anim. Behav. 42, 1020–1021.

16. Pomiankowski, A., Moller, A.P., 1995. A resolution of the lek paradox. Proc. Roy. Soc. Lond. B 260, 21–29.

17. Zuk, M., Kolloru, G.R., 1998. Exploitation of sexual signals by predators and parasitoids. Quart. Rev. Biol. 73, 415–438.

18. Zuk, M., Johnson, K., Thornhill, R., Ligon, J.D., 1990. Parasites and male ornaments in freeranging and captive red jungle fowl. Behaviour 114, 232–248.
Zuk, M., Kolloru, G.R., 1998. Exploitation of sexual signals by predators and parasitoids. Quart. Rev. Biol. 73, 415–438.

19. Hall, M.L., 2000. The function of duetting in magpielarks: conflict, cooperation, or commitment? Anim. Behav. 60, 667–677.Hall, M.L., Magrath, R.D., 2000. Duetting and mateguarding in Australian magpielarks (*Grallina cyanoleuca*). Behav. Ecol. Sociobiol. 47 (3), 180–187.

20. McGraw, K.J., Mackillop, E.A., Dale, J., Hauber, M.E., 2002. Different colors reveal different information: how nutritional stress affects the expression of melanin- and structurally based ornamental plumage. J. Exp. Biol. 205, 3747–3755.

21. Jennions, M.D., Petrie, M., 1997. Variation in mate choice and mating preferences: a review of causes and consequences. Biol. Rev. 72, 283–327.

22. Gabor, C.R., Halliday, T.R., 1997. Sequential mate choice by multiply mating smooth newts: females become more choosy. Behav. Ecol. 8, 162–166.

23. Sakuluk, S.K., 1984. Male crickets feed females to ensure complete sperm transfer. Science 223, 609–610.

24. Searcy, W.A., 1992. Song repertoire and mate choice in birds. Am. Zool. 32, 71–80.

25. Pruett-Jones, S., 1992. Independent versus nonindependent mate choice: do females copy each other? Am. Nat. 140, 1000–1009.

26. Alonzo, S., 2008. Female mate choice copying affects sexual selection in wild populations of the ocellated wrasse. Anim. Behav. 75, 1715–1723.

27. Schlupp, I., Ryan, M.J., 1997. Male sailfin mollies (*Poecilia latipinna*) copy the mate choice of other males. Behav. Ecol. 8, 104–107.

28. Szykman, M., Engh, A.L., Van Horn, R.C., Funk, S.M., Scribner, K.T., Holekamp, K.E., 2001. Association patterns among male and female spotted hyenas (*Crocuta crocuta*) reflect male mate choice. J. Behav. Ecol. Sociobiol. 50, 231–238.

29. Orians, G.H., 1969. On the evolution of mating systems in birds and mammals. Am. Nat. 103, 589–603.

30. Emlen, S.T., Oring, L.W., 1977. Ecology, sexual selection and the evolution of mating systems. Science 197, 215–233.

31. Morris, M.R., Rios-Cardenas, O., Brewer, J., 2010. Variation in mating preference within a wild population influences the mating success of alternative mating strategies. Anim. Behav. 79, 673–678.

32. Petrie, M., Doums, D., Moller, A.P., 1998. The degree of extrapair paternity increases with genetic variability. PNAS (USA) 16, 9390–9395. Could be characterized as cryptic polyandry.

33. Schwagmeyer, P.L., Ketterson, E.D., 1999. Breeding synchrony and EPF rates: the key to a can of worms. Trends Ecol. Evol. 14, 47–48.

34. Stutchbury, B.J.M., 1998. Breeding synchrony best explains variation in extrapair mating system among avian species. Behav. Ecol. Sociobiol. 43, 221–222.

35. Westneat, D.F., Sherman, P.W., 1997. Density and extrapair fertilizations in birds: a comparative analysis. Behav. Ecol. Sociobiol. 41, 205–215.

36. Manlove, C.A., Hepp, G.R., 1998 Effects of mate removal on incubation behavior and reproductive success of female wood ducks. Condor 100 (4), 688–693.

37. Webster, M.S., Pruett-Jones, S., Westneat, D.F., Arnold, S.J., 1995. Measuring the effects of pairing success, extra-pair copulations and mate quality on the opportunity for sexual selection. Evolution 49, 1147–1157.

38. Kempenaers, B., Congden, B., Boag, P., Robertson, R.J., 1999. Extrapair paternity and egg hatchability in tree swallows: evidence for the genetic compatibility hypothesis? Behav. Ecol. 10, 304–311.

39. Sparkes, T.C., Keogh, D.P., Pary, R.A., 1996. Energetic costs of mate guarding behavior in male stream-dwelling isopods. Oecologia 106, 166–171.

40. Emlen, D., 1997. Alternative reproductive tactics and male-dimorphism in the horned beetle *Onthophagus acuminatus* (Coleoptera: Scarabaeidae). Behav. Ecol. Sociobiol. 41, 335–341.

41. Alonzo, S.H., Warner, R.R., 2000. Allocation to mate guarding or increased sperm production in a Mediterranean wrasse. Am. Nat. 56 (3), 266–275.

42. Zamudio, K.R., Sinervo, E., 2000. Polygyny, mateguarding, and posthumous fertilization as alternative male mating strategies. Proc. Nat. Acad. Sci. USA. 97 (26), 14427–14432.

43. Orians, G.H., 1969. On the evolution of mating systems in birds and mammals. Am. Nat. 103, 589–603.

44. Verner, J., Willson, M.F., 1966. The influence of habitats on mating systems of North American passerine birds. Ecology 47, 143–147.

45. Verner, J., 1964. Evolution of polygamy in the long-billed marsh wren. Evolution 18, 252–261.

46. Bronstein, P.M., 1985. Predictors of dominance in male *Betta splendens*. J. Comp. Psych. 99, 47–55.

47. Petrulis, A., Weidner, M., Johnston, R.E., 2004. Recognition of competitors by male golden hamsters. Physiol. Behav. 81, 629–638.

48. Kovach, A.I., Powell, R.A., 2003. Effects of body size on male mating tactics and paternity in black bears. Ursus americanus. Can. J. Zool. 81, 1257–1268.

49. Duhrkopf, R.E., Hartberg, W.K., 1992. Differences in male mating response and female flight sounds in *Aedes aegypti* and *Ae. Albopictus* (Diptera: Culicidae). J. Med. Entomol. 29, 796–801.

50. Duvall, D., Schuett, G.W., 1997. Straight-line movement and competitive mate searching in prairie rattlesnakes. Crotalus viridis viridi, Anim. Behav. 54, 329–334.

51. Lank, D.B., Smith, C.M., Hanotte, O., Burke, T., Cooke, F., 1995. Genetic polymorphism for alternative mating behavior in lekking male ruff *Philomachus pugnax*. Nature 378 (6552), 59–62.

52. Aspbury, A.S., Gibson, R.M., 2004. Long-range visibility of greater sage grouse leks: a GIS-based analysis. Anim. Behav. 67, 1127–1132.

53. Drnevich, J.M., Papke, R.S., Rauser, C.L., Rutowski, R.L., 2001. Material benefits from multiple mating in female mealworm beetles (*Tenebrio molitor L.*). J. Insect. Behav. 14, 215–230.

54. Eakley, A.L., Houde, A.E., 2004. Possible role of female discrimination against 'redundant' males in the evolution of colour pattern polymorphism in guppies. Proc. Roy. Soc. Biol. Sci. 271, S299–S301.

55. Moscovice, L.R., Di Fiore, A., Crockford, C., Kitchen, D.M., Wittig, R., Seyfarth, R.M., et al., 2010. Hedging their bets? Male and female chacma baboons form friendships based on likelihood of paternity. Anim. Behav. 79, 1007–1015

56. Krebs, C.J., Boutin, S., Boonstra, R., Sinclair, A.R.E., Smith, J.N.M., Dale, M.R.T., et al., 1995. Impact of food and predation on the snowshoe hare cycle. Science 269 (5227), 1112–1115.

57. Burton, C., 2002. Microsatellite analysis of multiple paternity and male reproductive success in the promiscuous snowshoe hare. Can. J. Zool. 80 (11), 1948–1956.

58. DeVries, A.C., DeVries, M.B., Taymans, S.E., Carter, C.S., 1996. The effects of stress on social preferences are sexually dimorphic in prairie voles. PNAS 93, 11980–11984.

59. Ross, H.E., Cole, C.D., Smith, Y., Neumann, I.D., Landraf, R., Murphy, A.Z., et al., 2009. Characterization of the oxytocin system regulating affiliative behavior in female prairie voles. Neuroscience. 162, 892–903.

60. McGinnis, M.Y., Lumia, A.R., Breuer, M.E., Possidente, B., 2002. Physical provocation potentiates aggression in male rats receiving anabolic androgenic steroids. Horm Behav. 41, 101–110.

61. Fisher, H.S., Hoekstra, H.E., 2010. Competition drives cooperation among closely related sperm of deer mice. Nature 463, 801–803.

62. Stoltz, J.A., Neff, B.D., 2006. Sperm competition in a fish with external fertilization: the contribution of sperm number, speed and length. J. Evol. Biol. 19 (6), 1873–1881.

63. Corley, L.S., Cotton, S., McConnell, E., Chapman, T., Fowler, K., Pomiankowski, A., 2006. Highly variable sperm precedence in the stalk-eyed fly, *Teleopsis dalmanni*. BMC Evol. Biol. 6, 53.

64. Waage, J.K., 1979. Dual function of the damselfly penis: sperm removal and transfer. Science 203, 916–918.

65. Widemo, F., Saether, S.A., 1999. Beauty is in the eye of the beholder: causes and consequences of variation in mating preferences. Trends Ecol. Evol. 14, 26–31.

66. Moller, A.P., Alatalo, R.V., 1999. Good genes effects in sexual selection. Proc. Roy. Soc. Lond. B 266, 85–91.

67. Penn, D.J., Potts, W.K., 1999. The evolution of mating preferences and the major histocompatibility complex genes. Am. Nat. 153, 145–164.

68. Yamazaki, K., Beauchamp, G.K., Singer, A., Bard, J., Boyse, E.A., 1999. Odortypes: Their origin and composition. PNAS (USA) 96, 1522–1525.

69. Jordan, W.C., Burford, M.W., 1998. New perspectives on mate choice and the MHC. Heredity 81, 239–245.

70. Blanckenhorn, W.U., Reusch, T., Muhlhauser, C., 1998. Fluctuating asymmetry, body size, and sexual selection in the dung fly *Sepsis cynipsea*—testing the good genes assumptions and predictions. J. Evol. Biol. 11, 735–753.

71. Ligon, D.J., Kimball, R., MerolaZwartjes, M., 1998. Mate choice by female red junglefowl: the issues of multiple ornaments and fluctuating asymmetry. Anim. Behav. 55, 4150.

72. Manning, J.T., Hartley, M.A., 1991. Symmetry and ornamentation are correlated in the peacocks train. Anim. Behav. 42, 1020–1021.

73. Morris, M.R., 1998. Female preference for trait symmetry in addition to trait size in swordtail fish. Proc. Roy. Soc. Lond. B 265, 907–911.

74. Oakes, E.J., Barnard, P., 1994. Fluctuating asymmetry and mate choice in paradise whydahs, *Vidua paradisaea*: an experimental manipulation. Anim. Behav. 48, 937–943.

75. Thornhill, R., Gangestad, S.W., 1999. The scent of symmetry: A human sex pheromone that signals fitness? Evol. Hum. Behav. 20, 175–201.

76. Hamilton, W.D., Zuk, M., 1982. Heritable true fitness and bright birds a role for parasites. Science 218 (4570), 384–387.

77. Zuk, M., Johnson, K., Thornhill, R., Ligon, J.D., 1990. Parasites and male ornaments in freeranging and captive red jungle fowl. Behaviour 114, 232–248.
Zuk, M., Kolloru, G.R., 1998. Exploitation of sexual signals by predators and parasitoids. Quart. Rev. Biol. 73, 415–438.

78. Buchholz, R., 1995. Female choice, parasite load and male ornamentation in wild turkeys. Anim. Behav. 50, 929–943.

79. Mckinney, F., Derrickson, S.R., Mineau, P., 1983. Forced copulation in waterfowl. Behaviour 86, 250–294.

80. Seymour, N.R., 1990. Forced copulation in sympatric American black ducks and mallards in Nova Scotia. Can. J. Zool. 68 (8), 1691–1696.

81. Gowaty, P.A., Buschhaus, N., 1998. Ultimate causation of aggressive and forced copulation in birds: Female resistance, the CODE hypothesis, and social monogamy. Am. Zool. 38 (1), 207–225.

82. Dunn, P.O., Afton, A.D., Gloutney, M.L., Alisauskas, R.T., 1999. Forced copulation results in few extrapair fertilizations in Ross's and lesser snow geese. Anim. Behav. 57, 1071–1081.

83. Clutton-Brock, T.H., Parker, G.A., 1995. Sexual coercion in animal societies. Anim. Behav. 49 (5), 1345–1365.

84. Shields, W.M., Shields, L.M., 1983. Forcible rape: an evolutionary perspective. Ethol. Sociobiol. 4, 115–136.

339

Nesting, Parenting, and Territoriality

LEARNING OBJECTIVES

Studying this chapter should provide you with the knowledge to:

• Appreciate nest site selection and function, and the complexities of animal architecture.

• Understand how parents control their investment in offspring and how these choices involve balances between current and future reproductive opportunities.

• Know the origins of parent–offspring conflict over investment and the relationship of infanticide to paternal investment.

- Understand that conflict over resources extends to interactions among siblings that in extreme cases can lead to siblicide.
- See that animal aggression has many causes and emerges from conflicts over resources, territories, or in the contexts of mating and predation.

12.1 INTRODUCTION

The main consequence of mating and reproduction is the production of young. This chapter covers topics that surround where to live and how to care for young. For students of animal behavior, one of the most fascinating aspects of this material comes from the study of parent–offspring conflict and sibling–sibling (sib–sib) conflict, things that are well known to many people from personal experience. *Parental care* can be as simple as choosing a safe location for laying eggs, or as complicated as caring for young for months or years while they mature.

> **KEY TERM** Parental care is behavior directed to feeding, protecting, and in some cases teaching offspring.

CASE STUDY

Meerkat Instruction

Teaching survival skills to offspring can be a considerable investment. The meerkat is one of the few nonhuman animals in which the teaching of young has been demonstrated.[1] While it doubtlessly occurs in many species, it is difficult to demonstrate in the field, because it is difficult to distinguish from observational learning. (In observational learning, there is no attempt to teach; instead the learner simply observes.) Teaching young how to handle scorpions, a prey item that makes up approximately 5% of meerkat diets, is especially useful since some of the scorpion species consumed by meerkats can be lethal. Thornton and McAuliffe[2] set out to test the hypothesis that meerkats engaged in opportunistic teaching of young meerkats about preying upon scorpions. Much of this study hinged on the fact that young meerkat calls change with age. By using age-specific playbacks of meerkat calls as controls, Thornton and McAuliffe showed that helper meerkats responded differently to meerkat young of different ages, and the younger the meerkat, the more help it received

FIGURE 12.1
Adult meerkat with young. *Photo: Katherine McAuliffe.*

in preying upon scorpions. This help ranged from removal of stingers to nudging and retrieving escaped prey (see Figure 12.1).

Some important topics relating to parenting were covered in earlier chapters. The endocrine basis of parental care and parent–offspring bonding was discussed in Chapter 2 (in the section on hormones). Chapter 5 on learning included the way offspring learn food preferences from parents, and Chapter 8 on navigation reviewed parental roles in orientation within the habitat and transmission of information about migratory routes in some species. Refer to these discussions when reading this chapter because much of what is presented here is a natural outgrowth of that material.

Often, parental care includes construction of a *den* or *nest*—a good starting place for this chapter. In many

> **KEY TERM** A den or nest is a location that has been modified by the animal so that it is appropriate for rearing young. The modification can be minimal, as in clearing stray material from the area, or can be extensive, as exemplified by beaver construction of lodges and dams.

ways, this is "land use," from an animal perspective. The architecture of these structures is often fascinating and sometimes puts human construction attempts to shame. For generation after generation, without external blueprints, many animals construct nests that typify their species. Animal construction is a key element in keeping young safe from harm and in the right microclimate for development.

Continuing from nesting and architecture to the question of parental investment in offspring, notice how parents must make choices about time and energy use, and whether to invest in current offspring or to save energy for investment in future offspring. This naturally leads to conflict between parents and their offspring over these investments. The two parties may have strikingly different viewpoints about allocation of resources within a family; understanding this tension is part of the appeal of studying parent–offspring conflict. The same dynamics drive competition among sibs for parental investments, a competition that in extreme cases can become lethal. Infanticide is also covered in this chapter; it most frequently involves males killing offspring, especially if the males are uncertain about paternity.

Discussion of siblicide and infanticide naturally leads to a broader consideration of aggression and then to territoriality, the final two topics of this chapter. In a way, the journey from nesting, at the outset of this chapter, to territoriality, at the end, describes a circle that begins with a family in its nest and ends with that family's interactions with neighbors or with the family's offspring staking out its own claim to the future. The next chapter (13 on social behavior) expands this focus to extended families and larger social structures.

12.2 NESTS AND NESTING
Habitat Choice

Habitat choice in nesting is largely an issue of finding the appropriate microclimate for growth of the young and right level of protection from predators and parasites.[3] Behavioral strategies for finding appropriate habits are covered in Chapter 8 on navigation. If appropriate nesting sites are scarce, then aggressive competition with other animals may be the first step in nesting.

Consider animals that nest in tree holes. Cavities in hollow trees, used as nests by many species, are classic examples of competition for nesting sites. Cavity-nesting birds, such as flickers, may compete among themselves and with other species for good nest sites. Honeybees nest in hollow cavities, and some scientists hypothesize that their introduction into North America contributed to the decline and extinction of the Carolina Parakeet due to competition for nest sites. Honeybee swarms land on a branch and send out scouts to search for good potential nest sites. Nest sites are largely assessed by their size, and scouts dancing on the surface of the swarm effectively organize a "vote" to determine the best nest site.[4] In a habitat lacking adequate nesting sites, honeybees may end up nesting in marginal locations, such as hanging their comb from a tree branch; these marginal nests do not survive the winter.

Nest Construction and Animal Architecture

The simplest shelter is obtained by occupying a protected place, such as a cave, hollow tree, or crevice between rocks. The beauty of such accommodations is that they come labor free; no time or energy is invested in making modifications to create a more habitable space. This section highlights some species with interesting and remarkable architectural and construction abilities (see Figure 12.2).

Bat Roosts

Bats exemplify the simplest type of nesting. Most bat species live in sheltered spots, such as hollow trees or caves, and they do not modify these homes (see Figure 12.3). The drawback to such an arrangement is that the only control of *microclimate* comes from

KEY TERM Microclimate is the specific set of climatic conditions at a defined (micro) location. Conditions could include factors such as temperature, humidity, or light level.

FIGURE 12.2
Nests and nest construction in birds. (A) A cormorant swimming with nesting material *(Photo: Michael Breed).* (B) Cormorant nest *(Photo: Michael Breed).* (C) Black sandpiper nest *(Photo: U.S. Fish and Wildlife Service).* (D) Rusty blackbird on nest *(Photo: Robin Corcoran, U.S. Fish and Wildlife Service).* (E) Bald eagle in nest *(Photo: Donna Dewhurst, U.S. Fish and Wildlife Service).*

FIGURE 12.3
Big brown bats in a roosting site. *Photos: Rick Adams.*

choosing a suitable spot and perhaps benefiting from heat and moisture generated by companions. Tent bats, on the other hand, choose large leaves for their nest; they modify this nest by cutting the leaves so that they fold over and protect the bats.

Beaver Lodges

Beavers are among the most famous animal engineers. Beaver construction usually includes a dam and a lodge, but if trees are not around the stream or pond, the beavers simply dig a burrow in the bank; this construction flexibility gives them a broad range of habitat. Prior to European colonization of North America, burrowing beavers were common in prairie ecosystems. The lodge and the dam are built with branches and logs from trees, usually

FIGURE 12.4
A diagrammatic depiction of a beaver lodge (top), beaver swimming in a pond (left), and a beaver lodge (right). *Photos: Jeff Mitton.*

soft-wooded deciduous trees such as poplars and aspen. Mud may be used to fill cracks and cement the construction together. The beavers strip the bark from their construction materials and feed on the inner bark, as well as on nearby aquatic and terrestrial plants. The submerged entry to the lodge provides protection against predators (see Figure 12.4). Beavers are remarkable for their persistence in constructing dams across large spans of streams and rivers, and for their ability to maintain their dams even during flood conditions.

345

Prairie Dog Burrows

Prairie dogs re-engineer vast expanses of prairie; their towns can extend up to 100 square miles. Above ground, they construct a mound of excavated soil, which provides a vantage point for observation of the surroundings (see Figure 12.5). The mound also prevents rainwater from running into the burrow and flooding their living quarters. The area around the mound is cleared of grass, improving the view of approaching potential predators. Below ground, the tunnel leads to a chamber in which the young prairie dogs are reared. The tunnel and chamber may be lined with grass, which acts as both insulation and food store. The prairie dog's primary natural enemies are rattlesnakes, coyotes, and large birds of prey, such as eagles and hawks. The tunnel provides a defensible escape route that is effective against some of these predators, such as coyotes and birds. The young are vulnerable to snakes, and the adults work to prevent snakes from entering the chamber via the burrow.

FIGURE 12.5
A prairie dog at its burrow entrance. *Photo: Michael Breed.*

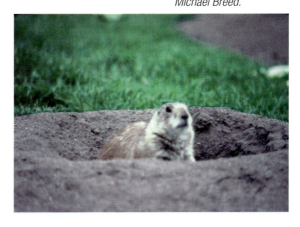

Magpie Nests

The Black-billed Magpie is native to the American Northwest. Like many birds, it builds its nest of sticks. The nest appears to be a fairly sloppy construction, but quite sturdy and holds up well against breezes that toss around the high branches where it is located. A major threat to magpies is predation of their eggs by crows, and magpies have responded by evolving to build

A

B

FIGURE 12.6
In habitats where crows are absent, magpies build open nests of sticks (A). When crows, which eat magpie eggs and nestlings, are members of the community, magpies build covered nests (B).

stick roofs over their nest; unlike the traditional "cup" style of bird nest, magpie nests, at least in areas where crows occur, have specially designed architecture to thwart their foes (see Figure 12.6).

Paper Wasp Nests

Paper wasps, as well as yellowjackets and hornets, collect plant materials that they chew and glue together with saliva to form a nest with cells.[5] An egg is laid in each cell, and the growing larvae are fed by adults on the nest. Architecturally, small variations in building pattern can result in big differences in nest shape. In Figure 12.7, each new cell is begun near the base of the adjacent cell; this results in a flat comb. In some other species, the new cells are begun near the top of an older cell, resulting in a long, string-like nest that resembles a dead stick.

Termite Mounds

Most termites construct their mounds using a combination of feces, saliva, mud, and bits of chewed-up plant (see Figure 12.8). The amazing thing about termites is that although each animal is only a few millimeters long, thousands of termites working collectively can build monumental nests that tower over a human. Such a large nest, occupied by so many animals, presents major architectural problems. It must support itself against collapse while providing enough porosity to allow air exchange: carbon dioxide and water from the metabolism of the termites will accumulate in the nest, while oxygen will be depleted. The termites' architectural solution is to build passageways specifically for airflow and to rely on convective movement of air (hot air rising, cool air sinking) to maintain an ideal

FIGURE 12.7
A paperwasp nest, Polistes sp. The nest is constructed of chewed plant material and hangs by a thin connection, or pedicel, from the substrate. The queen lays eggs in the cells within the comb, and the workers then feed the larvae. *Photo: Michael Breed.*

FIGURE 12.8
Airflow, temperature, and gas exchange in a termite mound. Oxygen comes into the nest with fresh air. As the termites metabolize, they use the oxygen and release carbon dioxide, water, and heat, which all leave the nest. The air circulation within the nest ensures that water and carbon dioxide do not build up, while adequate oxygen to support life is maintained. *Adapted from Turner, J.S., The Extended Organism. The Physiology of Animal-Built Structures. 2000, Cambridge: Harvard University Press. 234 pp.*

microclimate within the nest. Why build such a big mound? The mound protects the termites from voracious predators, such as aardvarks.

12.3 PARENTAL INVESTMENT

There were hints about *parental investment* in the preceding chapter, with the discussion of relative investment of males and females in offspring. Robert Trivers brought the idea of parental investment to the forefront of evolutionary thinking about reproduction in a 1972 paper that defined parental investment as the investment that a parent makes in an offspring that reduces the parent's future fitness.[6] It would be wrong to think of this as some species-specific number because the impact of such behavior on future fitness depends, in part, on a variety of factors (e.g., food abundance) that can change without regard to the parent–offspring interaction. Nonetheless, Trivers's concept provides a useful framework for understanding the allocation—and limits—of parental investment.

> **KEY TERM** Parental investment is a measure of how much time, energy, and other resources are given to a particularly young animal.

This investment can take many forms. Some can be as seemingly routine as provisioning gametes; the provisions can come from unexpected sources, such as nutrients associated with sperm transfer, or even cannibalistic mating. Indeed, when provisioning gametes is included, it is difficult to think of many species that do not invest in offspring. Parental investment does not stop with gamete provisioning, however. Once fertilized, eggs can be guarded and protected, either in the nest or inside the mother, and once the offspring are hatched/born, parents can continue to provision and/or guard them (see Figure 12.9). Guarding of nests and territories where young are reared overlaps with mating behavior as well, because nest and territory quality may be used in mate assessment (see Chapter 11 on mating).

347

OF SPECIAL INTEREST: FOSSIL PARENTAL BEHAVIOR

Parental care is not a new behavior on the evolutionary scene, nor is all evidence of behavior gone when an animal dies. We can infer behavior from a variety of fossil structures (e.g., dentition is a clue to food, bone density to running speed, foot structure to habitat, footprints to social behavior); likewise, we can infer parental behavior from the existence of nests and careful egg placement. In fact, a recently discovered ostracod from the Silurian Period shows that ostracod parental care was probably practiced in much the same way 425 million years ago as it is today. An ostracod is a small crustacean that is shaped like a clam, with two valves forming its carapace, which almost entirely encloses its body.[7] In this group, as in some other crustaceans, a space within the carapace called the marsupium is used to brood eggs. This 425 million-year-old fossil had eggs, and possibly juveniles, in its marsupium, thus indicating that, much like today, Silurian ostracods directed parental care at their young.

OF SPECIAL INTEREST: DOLPHINS CARRYING YOUNG—THE COST OF PARENTAL CARE

Carrying young can be an energetically expensive proposition, as evidenced by the number of times you may have witnessed human parents "trade" that responsibility. It turns out to be no different for aquatic mammals. Cetacean mothers engage in "echelon swimming" with young, during which the calf swims very close to the mother's mid-lateral flank. This has considerable hydrodynamic benefits for the calf, which can then keep up with its mother, and it probably minimizes the risk of separation. However, it is not without cost for the mother, who swims at about 76% of the speed of her mean maximum speed when unencumbered, and whose distance per swimming stroke is reduced by 13%.[8,9]

FIGURE 12.9
Parents and offspring.
(A) Owl *(Photo: Jeff
Mitton)*. (B) Juvenile
giraffe *(Photo: Michael
Breed)*. (C) Adult giraffe
(Photo: Michael Breed).
(D) Blue-footed boobie
and chick *(Photo: Randy
Moore)*. (E) Mallard and
her ducklings *(Photo:
Michael Breed)*. (F) Plover
and chick *(Photo: Ben
Pless)*.

While the nature of parental investment is more variable than is easily summarized, that variability is often the result of a combination of phylogenetic and ecological factors, and as such has significance for *life history traits*. Minimum investment per offspring, especially minimum nutritional investment, is most often seen in animals with high reproductive rates, whereas organisms that invest heavily in offspring have fewer of them. This is a logical association; if resources are limited, then it is impossible to invest a lot in hordes of offspring.

> **KEY TERM** Life history traits are a concept from ecology used to describe a species' or population's reproductive strategies. They include age at first reproduction, number of offspring in a clutch, number of clutches in an animal's lifetime, and so on. This concept is explored in more detail in many ecology textbooks.

This "choice" (often made over evolutionary time) is part of a "life history strategy," that is, characteristics that determine the course of a life. Such characteristics include, but are not limited to, a variety of parameters that determine maturation, reproduction, and longevity. Life history theory places these investment extremes on a continuum. Low-investment, highly fecund organisms are said to be r-selected (the *r* comes from the term for reproductive ratio in population growth equations) and high-investment, less-fecund organisms are said to be K-selected (the *K* refers to carrying capacity in those equations).

Generally speaking, r-selected species with their low investment and high fecundity seem particularly well adapted to unpredictable environments that allow fast invasion, maturity, and reproduction. In contrast, K-selected species do well in more stable, highly competitive environments, where high investment in each offspring is necessary for that offspring to get a foothold in the habitat.

Clutch size—the number of young in a brood or litter—is another important life history trait. The number of young requiring care at any one time clearly has behavioral implications, but clutch size is determined by evolutionary responses to long-term food availability and life history strategy, so it is more a question of evolution and ecology than of behavior. Behavior becomes morbidly interesting when clutch size exceeds the ability of parents to feed the offspring, a situation addressed in the section on sibling conflict later in this chapter.

The preceding paragraphs paint with the broadest brush imaginable, but the continuum does provide a context for beginning to predict shifts in investment. Such shifts can occur even within a lifetime.[10] In species that experience senescence, older organisms may invest more in offspring because the probability of future reproduction is reduced, along with the negative impact of current investment on that future reproduction. Thus, reproductive behavior may change within a lifetime or may change across habitats. Life history theory is one way to understand those shifts and make predictions about when they might occur and what they might look like.

12.4 PATTERNS OF PARENTAL CARE

Most behavioral research seems to have focused not so much on the existence of parental investment or the type of investment, but rather who invests and why. To begin, review the discussion in the preceding chapter about the evolution of sex. For a simple case of uniparental care, the basic rules of anisogamy seem to apply: the sex that invests the most in the offspring and has the most to lose should provide the most care. Thus, eggs are more expensive to produce than sperm, so the initial working hypothesis is that females should provide parental care. Things are not always simple, however, and a variety of extenuating circumstances can cause modification of that hypothesis.

For instance, even if females do invest, they should take steps to assure the identity of the target of their investment; any investment is a mistake in evolutionary terms if the female misidentifies her young and pours maternal investment into an unrelated animal. As might be expected, there are recognition cues that help prevent this misidentification, and they are most highly developed in social animals for which the need to identify individual young is most pressing.[11,12] For instance, Tinbergen's classic work shows that in colonially nesting gulls, the ability to identify offspring develops simultaneously with the offspring's ability to leave the nest and move around. In contrast, because kittiwakes nest on cliff ledges, their young are under strong selection not to wander too far, and are unlikely to be confused with other young from other ledges.[13] As a result, recognition of the young by their parents even at the species level, has not evolved in kittiwakes.[14]

For many birds, as well as for bottlenose dolphins,[15] auditory cues are involved in recognition and may develop into "signature" vocalizations that are used into adulthood for individual recognition[16] (see Chapter 13). As might be expected, many terrestrial mammals seem more likely to use olfactory cues (see "Herding Behavior, Young Recognition, and the Old West"); these cues from young can affect maternal behavior by influencing hormonal state and even structures in the olfactory region of the brain. For instance, odors emitted from young animals (e.g., rodents) promote maternal behavior in females that have recently given birth and are lactating, but may be aversive to females that do not have young (see Chapter 2).[17]

BRINGING ANIMAL BEHAVIOR HOME: HERDING BEHAVIOR, YOUNG RECOGNITION, AND THE OLD WEST

In herds of ungulates (hooved mammals), switching of young can easily occur, and it is not surprising that mothers can easily recognize their offspring. For cattle, there is considerable lore supporting the idea that cows use the odors of their calves to make this identification.

For instance, calves often had difficulty keeping up with a cattle drive. One solution was to give them a ride in a wagon and to reunite them with their mothers at the end of the day. This passage from a book by J. Frank Dobie describes how nineteenth century cowboys handled this situation to ensure that mothers would take back their calves:

> Charlie Goodnight said to me that when he began taking a "calf wagon" along, to pick up calves born on the trail, he had much difficulty in getting mother cows to accept their young after they were unloaded from the wagon in the evening. A half-dozen or twenty calves jostling together during the day would get their scents mixed, thus making each a stranger to its mother. Goodnight overcame the difficulty by putting over each calf a loose sack, so marked that it was used by the same calf day after day, being removed at evening; thus the scent proper to a calf would be retained for recognition by its mother.[18]

In some instances calves became orphaned. Here are two passages, from very different ranching situations, a century apart, which describe virtually the same procedure used to induce a cow to adopt an orphan calf. First, we have J. Frank Dobie reporting on 19th century cattle drives:

> If a cow lost her calf and at the same time there was a calf that had lost its mother, the bloodless hide of the dead calf fastened loosely over the orphan would influence the cow to adopt it.[19]

In her account of life on a modern ranch, Judy Blunt describes a nearly identical method used to ensure the adoption of orphan calves: "He would live or die by morning.... If he died we'd peel his hide and make a jacket for one of the orphan calves in the barn, convince the cow it was hers."[20]

The similarities between these passages is striking, and emphasizes the importance of mother-calf recognition in the delivery of parental care.

Returning to the question of who should provide parental care, it only takes a short exposure to nature shows on television to realize that the "rule" of maternal care was meant to be broken in the natural world. There are obvious examples of male parental care; although female parental care is more common, male parental care is not rare. Why this deviation from the anisogamy-based prediction? For one thing, in many organisms, sperm are not the tiny, inexpensive propagules that work well for theory. Instead, they are produced in energetically costly packets (*spermatophores*). In other organisms, males make serious investments in territory holding and resource guarding. Once investment extends beyond the gamete, understanding which parent invests more can be a challenge.[21]

> **KEY TERM** A spermatophore is a nutritive glandular product passed from male to female during mating.

In addition, certainty of paternity may influence the extent to which males are willing to provide care. In animals with internal fertilization, the female is more certain of her parentage than the male, whereas in externally fertilized organisms, males may be certain, especially if the fertilized eggs are maintained on the male's territory.[22,23] In addition, proximity to offspring can obviously offer opportunities for care, and males that guard territories may coincidentally guard the eggs within those territories that they have fertilized.

Finally, many organisms exhibit biparental care. How does this evolve? An ecological hypothesis seems to explain most cases. The reasoning begins with female uniparental care, which is the default case for parental care in general, based on the greater investment that females usually have in young. In a world where provisioning and guarding are not limiting reproductive success—that is, where females can rear offspring alone as successfully as they can with male assistance—the male is free to leave the female and her eggs or young and seek other reproductive opportunities for increasing his fitness.

Not all environments are so benign, and in some species, there are steep requirements for producing young that are likely to survive and reproduce themselves. These requirements can be environmentally imposed (scant resources and/or high predation pressure) and exacerbated by life history traits (young that are helpless for long periods of time or that mature at rates requiring intense provisioning). In these conditions, a male that leaves is likely to have no fitness; the female cannot do the task alone. Such conditions favor the evolution of monogamy and subsequent biparental care.[24,25,26,27]

For example, a comparison of paternal care in two congeneric species of hamsters, the Campbell's Russian dwarf hamster (*Phodopus campbelli*) and Siberian (*P. sungorus*), revealed that the male Djungarian hamsters not only care for their young, but actually assist in the birth, whereas Siberian hamsters attend young much less often and do not assist at birth.[28] This is hypothesized to reflect the greater environmental challenges experienced by the Siberian hamsters, including thermoregulatory ones.

Thus, evolutionary history as well as ecological factors influence both the degree of parental care and who performs it. Female mammals, for instance, gestate and then feed young. This makes male uniparental care almost impossible, although biparental care might be beneficial in some cases, often depending on ecological conditions. In addition, life history traits such as maturity at birth can strongly influence the opportunity for males to contribute significantly to offspring care. Altricial organisms are relatively immature at birth and require a great amount of parental care; in such animals, biparental care might increase fitness by increasing the survival probability of the offspring. Contrast this with precocial animals that are able to move and forage shortly after birth. These young need to be guarded from predators and they may need some time to learn survival skills, but they do not need to be given nutrients. Compared to altricial young, they require much less investment, so that biparental care may be less critical for precocial than for altricial young.

Parental Care When Fertilization Is External

In many aquatic species, the female lays her eggs prior to fertilization. The male then deposits his semen in the water, and both mother and father may have left by the time gametes have formed. In fish, some species simply scatter their eggs in moving water. Other species deposit their eggs on a surface. This opens the door for rudimentary parental care, as the female can then guard her eggs and young hatchlings (very small fish are called *fry*). Still other species of fish, such as Bettas, build nests of bubbles in which spawning takes place. A few species of fish keep their fry in their mouths to protect them; these are *mouth-brooders*, usually found within the family Cichlidae (see Figure 12.10). Male participation in parental care is common in fish,[29] with males often constructing nests, guarding eggs or fry, or even participating in mouth brooding.

Amphibia—salamanders, frogs, and toads—also have external fertilization (see Figure 12.11). Again, egg guarding is a possibility. In one well-known case, the midwife toad, the male takes over parental care, carrying the fertilized eggs on his body to protect them from predation. In salamanders, species living in streams lay larger eggs and are likely to guard their eggs and embryos, whereas pond-living species' eggs are smaller and remain unguarded.[30]

351

FIGURE 12.10
Spawning sockeye salmon. *Photo: U.S. Fish and Wildlife Service.*

Uniparental Caregiving

In terrestrial animals, internal fertilization is more common. In many, perhaps most, terrestrial animal species, the male and female do not maintain contact after mating. This leaves the female responsible for caring for the eggs. In many cases, this simply means finding an appropriate microclimate for egg development and then laying the fertilized eggs (see Figure 12.12). For plant-feeding insects such as moths and butterflies, this typically involves ovipositing on a host plant; the emerging caterpillars then have a ready-made food source at hand. Cockroaches illustrate the range of possibilities for parental care in terrestrial animals. Cockroaches lay their eggs in leathery masses called *oothecae*; the female cockroach may take some care in finding a hidden spot for her eggs, where she glues the egg mass in place. In some cockroach species, the ootheca is carried externally by the female until the eggs hatch. In still others, the ootheca is retracted into the body and protected in a brood pouch. In one cockroach genus, nutrients are transferred directly from the female to the embryos in the pouch, much like a placental mammal.

These same patterns play out in vertebrates with internal fertilization. Female reptiles may lay their eggs in a protected place; sea turtles do this when they bury their eggs on beaches (see Figure 12.13). Snakes, such as the milk snake, may guard their eggs, or some species give live birth, such as the fer-de-lance.

FIGURE 12.11
A frog egg mass. *Photo: Michael Breed.*

FIGURE 12.12
A cluster of insect eggs laid on a leaf. *Photo: Michael Breed.*

FIGURE 12.13
A pair of Canada geese with their young. *Photo: Michael Breed.*

DISCUSSION POINT: EXCEPTIONAL PARENTAL CARE

Can you, with a little web-based research, come up with examples of interesting and unexpected patterns of parental care? You might look specifically for interspecific adoptions. You'll find lots of cute interspecific adoption videos on YouTube, news sites, or other video archiving sites. As you watch the videos, try to form evolutionary hypotheses for the extension of parental care from adults to young of different species.

FIGURE 12.14
Male parental care in seahorses. The male protects the young in a brood pouch. *Photo: Rebekah D. Wallace, Bugwood.org.*

Examples of male-only caregiving are rare enough to attract attention when discovered. In addition to the midwife toad, other well-known examples include seahorses, pipefish, and sea spiders.

Biparental Caregiving

Biparental caregiving is common in birds and is sometimes seen in reptiles and mammals. For the evolution of male parental behavior, the balance is fairly simple; the fitness gained by added survival of young due to the male's care is weighed against the fitness that might be gained from other potential matings. If male behavior can add to the survival of his young and he has little chance of mating with another female, then the balance favors male support of his offspring. In some cases, the prospect of the young being killed by another male (infanticide; see later section) is enough to keep a father present. In some cases, the father's only role in care may be to ward off other males from the family group; this is a limited but crucial duty (see Figure 12.14).

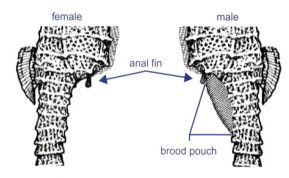

353

Intra- and Interspecific Adoptions

An odd and puzzling phenomenon is **adoption** of other females' young, within a species or across species lines. Adoptions within species, *intraspecific adoptions*, sometimes come about as a result of shared nests and communal brood care (see Chapter 13). *Interspecific adoptions* also occur: news stories report a cat that adopts and nurses puppies, a foal reared by a pig, and so on (see Figure 12.15). There is no evolutionary advantage associated with adopting young from another species. Rather, these incidents are a testament to the power of maternal drive and to the impact of hormones in priming females, at certain times, to place their maternal response above other behavioral priorities.

> **KEY TERM** Adoption involves giving parental care to a genetically unrelated animal.

FIGURE 12.15
An interspecific adoption. In this illustration, piglets have been adopted by a cat. Stories of cross-species adoptions illustrate the strength of parental responses in mammals.

12.5 HORMONES AND PARENTAL BEHAVIOR

While maternal behavior in vertebrates is strongly associated with prolactin, a peptide hormone produced in the anterior pituitary, in male vertebrates parental care is associated with declining levels of testosterone. In female birds, prolactin stimulates nesting behavior and operates, via positive feedback, to promote the development of the brood patch. The brood patch is an area from which the female plucks the feathers so that she has direct skin contact with the eggs or hatchlings, improving her ability to warm the brood. In management of domestic poultry, such as chickens, females displaying nesting behavior and a brood patch due to high levels of prolactin are referred to as *broody*. If a farmer's object is to have the female produce more eggs, behavioral interventions (preventing hens from spending time in their nest, for example) are used to attempt to prevent broodiness and thus keep the birds laying. Drugs that block dopamine may also prevent broodiness. In birds, prolactin also inhibits the ovaries from developing more eggs and promotes foraging behavior. Prolactin-induced weight gain may be an important preparation for migration.

In mammals, prolactin has many of the same effects. During pregnancy, prolactin levels increase, stimulating glandular development of the breasts. Prolactin primes maternal behavior and modulates stress that is associated with pregnancy and maternal care. Repeated pregnancies reinforce better maternal care, partly because each pregnancy induces more prolactin receptors in the brain; the larger number of receptors then enable stronger maternal responses. The inhibitory effect of prolactin on ovulation makes it unlikely that a nursing mammal will become pregnant; nursing can be an effective natural contraceptive. At least in humans, prolactin is also associated with sexual arousal and pleasure, due to its inhibition of dopamine secretion. The precise behavioral effects of prolactin and testosterone on parental behavior vary among species, but the general pattern is that prolactin enhances maternal behavior while testosterone inhibits or interferes with paternal behavior. Finally, oxytocin also causes muscular contractions in the uterus associated with birth and milk-letdown in milk production, and is thought to be important in maternal bonding with young animals.

12.6 PARENTING AND CONFLICTS OF INTEREST

Family life, from an animal behavior perspective, might be seen as rooted in this assessment of investment (cost) versus fitness outcomes (benefit) that can often lead to conflict if the participants differ in these parameters. Not only is mating behavior and parenting heavily influenced by cost and benefit, but parent–offspring relationships and sibling relationships are thought to reflect such influences.

For behavioral biologists, one central question in parental care focuses on timing: when should it cease? Offspring and parents should differ in their assessment of the costs and benefits associated with the cessation of parental care; this is reflected in nature every time a recently fledged bird squawks loudly as it hops energetically after its (presumed) parent, which in turn studiously ignores the young bird's existence. From a fitness perspective, it is to the young bird's advantage to extract larger amounts of resources from the parent bird than benefits the parent bird. By restricting parental care once the young can meet most of its survival requirements, the parent can allocate those resources to its own future fitness. Trivers[31] addressed this issue, and it can be easily illustrated in graphical form (see Figure 12.16). This is most easily seen in offspring that are nutritionally provisioned by parents. In general, as the offspring gets older, it depends less on parental resources to be able to survive. At the same time, it requires more resources in general because it is getting larger and more active. At some point, there will be conflict between parent and offspring because the increasing demands of the offspring are not matched by increasing fitness benefits to the parent.

FIGURE 12.16
The upper line in this graph represents the benefit to a young animal from parental care. From the parent's point of view, the middle line expresses the cost of the care. The parent should stop giving care when the greatest benefit per unit cost has been achieved. From the young animal's point of view, the cost to the parent of delivering this care is represented by the lower line; the young animal values itself twice as much as a sib, so the optimum investment from the young animal's point of view is higher. The gap between the parent's and the young animal's assessment of the optimum investment is the region in which theory predicts parent-offspring conflict. *Adapted from Trivers, R.L., 1974. Parent–offspring conflict. Am. Zool. 14, 249–264.*

354

The manipulative potential of offspring demand for resources was convincingly demonstrated by Matthew Bell's study of banded mongoose food solicitation.[32] Banded mongoose breed communally, and each pup associates with an "escort" that cares for it. Pups begged more vigorously when resources had been withheld than when they were well provisioned, but they also begged more vigorously when cared for by escorts that were more likely to provide. Thus, female pups, which receive better care than males, begged more vigorously than males did given the same degree of food deprivation, and even more tellingly, pups with generous escorts begged more vigorously than pups with more stingy escorts. This raises the possibility that demand for resources among these offspring (and probably others) reflects both need and an assessment of the likelihood that the need will be met.

Similar reasoning can be applied to the distribution of resources among siblings by parents. Parents are equally related to all of their offspring, sharing 50% of their genetic material, and thus are expected to treat them equally when they all have equal expectations of survival. On the other hand, each offspring may be related, on average, to its siblings by 50% (some more, some less, given the whims of independent assortment of chromosomes), but it is related to itself by 100%. Thus, siblings of most species are not terribly excited by the prospect of sharing resources. In contrast, parents benefit greatly when siblings do cooperate.

There are times when it is in the parents' interest for some offspring to die, or at least become nonreproductive. Nonreproduction (e.g., sterile castes or helpers) is covered in Chapter 13, and is a hallmark of eusociality. Outright infanticide, however, differs in that it usually occurs more opportunistically and results from either male–male competition (see Chapter 11 on mating) or offspring provisioning strategies.[33] The latter case is an extreme version of sibling rivalry, discussed later in this chapter.

12.7 BEGGING AND WEANING CONFLICT

Young birds and mammals ask for food, sometimes almost continuously. It would seem to an observer that the food requests could exceed the actual need for food and likely exceed the parent's ability to deliver food. This section reviews the function of *begging* and the time when the parent determines it no longer needs to feed the young—*weaning*.

> **KEY TERM** Begging is signaling to induce feeding.

> **KEY TERM** Weaning is the process of removing parental nutritional support from a young animal. The young must replace parental support by developing their own foraging abilities.

Functions of Begging

From the chick's point of view, begging is a way of attracting food (see Figure 12.17). Begging may reflect the actual needs of young for food (honest signaling), or it could reflect the

FIGURE 12.17
Gaped-mouth begging in a cormorant (A) and barn swallows (B). *Photos: Michael Breed (A) and Joanna Hubbard (B).*

young's point of view in parent–offspring conflict (dishonest signaling?). In most birds and mammals, there are multiple young in a litter or brood, so begging could be important in scramble competition among the offspring for food. In a recent study of redstarts, an honest signaling model was supported.[34] Hungrier nestlings moved to the front of the nest, which is the preferred location for parents to feed offspring, and there was no evidence of scramble competition (see Chapter 11). Studies of other species sometimes suggest scramble competition; begging is probably sensitive to evolutionary effects of food availability in habitat, with scramble competition likely in species that have an evolutionary history of food shortages.

Begging: Parental Point of View

From the point of view of the parents, begging may help to ensure that they are feeding their offspring enough. Weakened or sick young that are unable to beg may not receive food, reducing "waste" in feeding young that are fated to die. A recent theoretical study argued that responding to begging reduces two risks: The first is wasting time with too-rapid returns to the nest; if nestlings are not begging intensively, then more time can be spent foraging. The second is overlooking nestlings by not responding to their begging.[35] A study in house sparrows supports the argument that parents should be responsive to begging.

Remember that if parents respond to begging, food will not necessarily be wasted. This is similar to filling the gas tank in a car when the tank is half full (as compared to waiting until it is empty); this does not waste gasoline. However, time may be wasted if filling is done too often, so an argument can be made that begging should be an honest signal of condition. Begging may also carry a cost if the noise and commotion attract predators, so there may be some counter-selection against continuous begging.

Begging by Cuckoo-Type Parasites

Sometimes brood parasites manage to get one of their eggs into the nest of a host. Once the parasite hatches, its begging is almost always more intense than nestlings of the host. The parasite begs whenever food is available (when the host parent arrives), instead of begging as a signal of need, and thus deprives the host's nestlings of food. In cowbirds, the nestling begs more intensely if host nestlings are large, suggesting an effect of competition on begging.[36]

Weaning Conflict

For birds and mammals, there comes a point in the development of a young animal at which it might feed itself, but it is likely to pursue being fed by its parents. This conflict derives from the young animal seeking a guaranteed supply of food while the parent may benefit from diverting food to other, less completely developed, offspring or to its own metabolism (to build fat resources for overwintering or to have another set of young; see Figure 12.18).

12.8 SIBLING CONFLICT

Returning to first principles, a parent shares the same proportion of its genes with each of its offspring (50%). As a result, parents might be expected to treat all offspring equally. That would be true if genetics were everything. However, the parent may discern that some offspring are more likely to survive and reproduce than others. In this case, the parent may create a situation in which siblings are in conflict for resources. Siblings may not all have an equal chance for many reasons, ranging from developmental problems to resource limitations that preclude investment in all offspring.

FIGURE 12.18
Illustration of approach and rejection in parent-offspring interactions. Above, young animal solicits nursing. Below, parent pushes young away. Interactions following this pattern are called weaning conflict, and may occur when the parent's and offspring's interests do not coincide.

356

In many mammals, sibling conflict plays itself out in competition for access to the mother's nipples and for position among the nursing young. Some teats do, indeed, produce less milk than others (thus, the saying "sucking hind teat"); smaller, less-well-developed young will lose out in this competition. An animal that is the runt of the litter has little chance of surviving the competition.

But why would there be more offspring than the parents can support? Environments are unpredictable, and from the parental point of view, it is probably best to produce the number of newborns or hatchlings that could be supported in an extremely good year. The "extra" offspring also provide some level of insurance against the loss of one of the young to predation or disease. For birds, this is sometimes called the *insurance egg hypothesis*. If the insurance is not needed and food availability is constrained, then from the parent's point of view, it would be better for a few strong members of the litter to survive than to have them all survive as malnourished weaklings. In other words, it might be more profitable to invest in fewer offspring and raise them to fledging age than invest in the entire clutch only to see complete failure.

> **KEY TERM** The insurance egg hypothesis states that a female may have more offspring than can be supported, as insurance against mortality within the group of offspring.

Parents generally do not kill "surplus" offspring. (Infanticide can occur, but for different reasons that are discussed later.) Instead, the issue is resolved either by death stemming from malnourishment, or in extreme cases by larger, older, members of the brood killing their sibs (siblicide). Siblicide has been particularly well studied in Laughing Kookaburras, boobies,[37] and egrets.[38] In these bird species, insurance eggs are laid. At least in boobies, high testosterone in the chicks is associated with fight behavior. Only under very favorable environmental conditions do all the chicks survive; usually, the last emerging chick is pecked to death.

357

OF SPECIAL INTEREST: ARE BABY BIRDS CUTE?

Baby American Coots look like bizarre Halloween party favors and, at least in terms of plumage and bare heads, do not resemble their more distinguished-looking parents (see Figure 12.19). Bruce Lyon and co-workers[39] manipulated this spectacular plumage and showed that it did indeed influence parental provisioning. Chicks with bright plumage received more food, grew faster, and had higher survival rates than chicks that had the orange tips of their feathers removed. The bright plumage may trigger stronger parental responses.

FIGURE 12.19
American Coot. *Photo: U.S. Fish and Wildlife Service.*

Why do parents not intervene? There is no parental benefit to intervention because the parents' fitness is best served by having fewer well-nourished young survive. Because of this, the idea that parents benefit if siblings cooperate (see earlier) needs some qualification. Sometimes that is not true; those times tend to occur when rivalry is taken to the extreme and siblings compete to the death for limited resources. Under those conditions, parents essentially allow sibling conflict to determine which offspring are most likely (and probably best equipped) to survive to reproductive age.

12.9 INFANTICIDE

Infanticide by Males

Infanticide, killing of young animals, is common among mammals and has been observed in some birds. Members of the cat family (Felidae), including the domestic cat, are notorious for incidents of infanticide by roving males. Infanticide by males is also a feature of some primate societies. Infanticide by a male is almost always the result of paternity uncertainty—the male cannot be sure the young are his—or certainty that the young are not his.

One interesting example comes from Osorio-Beristain and Drummond, who found that male blue-footed boobies that were experimentally held so they could not return to their nest for several hours were likely to expel the eggs from their nest and remate with the female.[40] This is probably a response to opportunities the females had for extra-pair copulations while the male was gone. In primates, there are two mechanisms that at least hypothetically should prevent infanticide. The first is *concealed ovulation*, coupled with promiscuous copulation that may prevent infanticide because males do not want to risk killing young that are potentially theirs (chimpanzees). Pair-bonding also creates an impression of paternity certainty but keeps group size small, as in gorillas, because the larger the group, the more difficult it is for a male to track the females' activities. As a result, risk of infanticide may be a constraint that limits social group size. This may be true in many animals, not just primates.

> **KEY TERM** Concealed ovulation occurs when a female is physiologically able to mate at times when she is not ovulating. Humans and a few other primates have concealed ovulation.

When Might Females Commit Infanticide?

Infanticide by females is predicted when resources for their offspring are limited; the female may then kill other females' infants. In Columbian ground squirrels, females kill other females' infants to increase the chances of their own infants maturing and entering the adult population.

The Bruce Effect

In at least some rodents, the presence of a new male causes a female to physiologically abort pregnancy; this is termed the *Bruce effect*. In mice, the Bruce effect is caused by odors from the new male, which allow the female to recognize him as new (different). Replacement of a male with a genetically identical (from the same clone) male does not cause the Bruce effect. The odors are associated with major histocompatibility complex (MHC) differences, which are discussed in Chapters 11 on mating and 13 on social behavior. The Bruce effect makes the female available more quickly to mate with the new male, and the aborted young likely would have been victims of infanticide if the female carried them to birth.

> **KEY TERM** The Bruce effect is physiological abortion induced by the presence of a new male.

358

12.10 AGGRESSION AND TERRITORIALITY

The behavior of young egrets points to a topic that has fascinated (and repelled) countless students and fans of animal behavior. Aggression is one of those subjects in animal behavior that might be inserted in any one of a number of chapters. It is considered in this chapter simply because the majority of aggressive behaviors are related to reproductive and parenting concerns, broadly defined. Aggression occurs for a variety of reasons and is therefore frustratingly difficult to define and categorize. E. O. Wilson, in a 1973 book titled *Sociobiology* that inspired a new, vibrant field within animal behavior, attempted both definition and categorization; his approach is used here, with modification.[41] Aggression is considered in this chapter simply because the majority of aggressive behaviors are related to reproductive and parenting concerns. Other ethologists and popular writers have also published books on the behavioral biology of animal and human aggression, stimulated by a desire to understand and explain why humans fight and sometimes kill one another. The titles of these books illustrate the depth of thought and concern about human aggression (examples are *On Aggression* by Konrad Lorenz; *The Territorial Imperative* by Robert Ardrey; and *Aggression and War: Their Biological and Social Bases*, edited by Groebel and Hinde). Humans have struggled to understand aggression among animals and to use that understanding to illuminate human conflict.

DISCUSSION POINT: IS AGGRESSION INEVITABLE?

Perhaps the deepest question we can ask is whether aggression and war are inevitable for human societies. Is conflict a necessary outcome of the evolutionary origins of humanity, or can we use cognition and reasoning to allow human societies to overcome propensities for conflict? In other words, can moral and ethical sensibilities somehow separate humans from other animals? For that matter, do other animals compete in ways that have overtones that we fail to recognize as "ethical"?

What Is Aggression?

Wilson noted that among humans aggression occurs when one individual surrenders something it owns or might have owned as a result of a physical act or a threat. When applied to animal behavior, the damage is calculated in terms of fitness. Broadly considered, conciliatory behavior and surrender can be part of this mix, and the term *agonism* is frequently used in animal behavior circles to include those actions and to diffuse some of the unpleasant connotations that might accompany "aggression." *Agonism* comes from a Greek word that refers to contenders in games, an origin that is reflected in more commonly used words such as *protagonist* and *antagonist*. Agonism usually refers to intraspecific interactions. For purposes here, the categories used in the following sections (loosely based on E. O. Wilson's categorization in his book *Sociobiology*)[41] may illuminate this set of behaviors more than any definition. Above all, aggression—or agonism or conflict—is not one thing.

Territorial Conflict

Territorial conflict is related to the acquisition or maintenance of a *territory*. The distinguishing element of territoriality is agonism associated with a space or location.[42,43] Unlike any other space that an animal may occupy or use, a territory is defended. That defense is what distinguishes a territory from a *home range*. Male–male interactions are often either territorial or dominance, which is discussed in the following section; aggression among males can determine access to mates or can give males resources that are attractive to

> **KEY TERM** A territory is any defended space.

> **KEY TERM** A home range is the area in which an animal normally moves. A home range is not defended but may have a defended territory within it. Birds, for example, do not usually defend their feeding home ranges but are territorial around their nests.

mates[44] (see Chapter 11 for more on this behavior). Territorial conflict also includes defense of young; mothers, in particular, may be very aggressive when they are guarding young.

Dominance Conflict

We will consider the type of interaction we call dominance conflict in more detail in Chapter 13, which deals with social behavior. Dominance interactions include aggression, threats of aggression, and submission. Oftentimes, dominance occurs when the subordinate animal finds it advantageous to remain within a group, even if it has lower access to resources in the group. However, some forms of dominance do not occur in social groups. In these instances, dominance is exerted simply to withhold a resource from a competitor.

Sexual Conflict

Males can use force or threats to ensure that females remain with them or remain in their harems. For example, male bottlenose dolphins threaten, bite, tail-slap, body-slam, and otherwise "herd" females. They do this much more frequently to females that are not pregnant, which allows the inference that the motivation is indeed reproduction. Two or three males form alliances to herd the females and defend them from other males; they give chase when females attempt to escape.

Parent-Offspring Conflict

Parents may display mild forms of aggression toward young that are perhaps straying too far away or, at weaning time, remaining too close. In general, this is fitness-related behavior, either because it promotes the safety of the offspring (offspring fitness) or, at weaning time, assists the parent in shifting parental investment to future offspring (parental fitness). Parents also defend young against potential enemies with antipredator aggression (see below).

Predatory, Pain-, and Fear-Based Conflict

We do not consider interspecific predation as a purely aggressive act, although it is certainly one that results in a loss of fitness for one participant and a gain for the other. However, in the case of intraspecific predation, or cannibalism, especially cannibalism associated with territoriality, there is definite overlap with other forms of agonism. In addition, fear-related aggression is the unrecognized source of many human–animal conflicts, including conflict with domestic animals. This is yet another illustration of how the realities of nature can defy human categories.

Antipredatory Conflict

Antipredatory conflict is considered more fully in Chapter 10. It encompasses not only fighting back when pounced upon, but also (potential) prey-initiated attacks, such as mobbing. Animals at risk of being eaten have nothing to lose and use all of their available weaponry to fight a predator.

What ties these categories of aggression together? In many cases the armaments evolved for use in one type of aggression work equally well in other contexts. Venomous snakes and predatory wasps bite or sting to subdue their prey, but they can also use these weapons to defend themselves. Bears are well armed as carnivores but use their teeth and claws in defense of their young. In vertebrates, the hormonal underpinnings of aggression may be the same across these categories (see Chapter 2).

We stress that the widespread assumption that aggression in animals is monolithic in nature, a single behavior, is one of the bigger mistakes that can be made in animal management. There is no single cause for aggression, and because of that, there is no single solution for managing aggression in animals. An animal with offspring in the area may be much more aggressive than that same animal without offspring. A fearful animal may be

BRINGING ANIMAL BEHAVIOR HOME: RITUALIZATION AND DOG BEHAVIOR

One way in which evolution tones down animal conflict is *ritualization*. In studies of aggression, this term is used when animals seem to apply symbolic value to their behavior. A dominant dog bares its teeth, threatening in a way that seems to symbolize its ability to bite. A submissive dog rolls on its back and exposes its vulnerable throat (see Figure 12.20). Conflict among animals may be resolved by ritualized displays, without resort to physically damaging violence. It is important to avoid, though, the romantic fallacy that animals do not fight to injure or kill—they do, but ritualization has evolved because avoidance of fighting can benefit the fitness of both participants in a conflict. Human conflicts are similarly ritualized, with threat displays such as "spear-waving," which has evolved into "missile-waving," playing important roles in human territorial conflicts. Note the ritualization of aggressive signals in this verse: "The shield of his mighty men is red; his soldiers are clothed in scarlet. The chariots come with flashing metal on the day he musters them; the cypress spears are brandished" (Nahum 2:3, The Bible, English Standard Version).

FIGURE 12.20
A dominant dog and a submissive dog. *Source:* Darwin, C. 1890. *The Expression of the Emotions in Man and Animals*, London: J. Murray, pp. 52–53.

361

> **KEY TERM** In this context, ritualization means assigning symbolic value to signals of aggressive intent. Ritualized aggression may reduce the risk of injury and death in encounters.

extremely aggressive and give the appearance of dominance to the uninformed observer. A territory defender may be content, even meek, away from its territory. To be able to manage aggression in any species, one MUST identify the motivation; in the absence of that knowledge, failure is very likely, and the results for the animal, especially a domestic animal, may be not only catastrophic, but sadly unnecessary.

BRINGING ANIMAL BEHAVIOR HOME: SHOULD A DOG BE "TOP-DOG"?

Here is an example for you to ponder as you consider animal conflict: many people believe that dog aggression is a "dominance" issue and that the true solution to such problems is to show the dog who is boss. It's true that some aggressive dogs insist on their dominance, but many more are thought to be fearful. Let's assume that you accept the dominance hypothesis and do everything possible to convince the dog that you are the boss, or "alpha," as the television shows like to say. What results might that have for a fearful dog? Will its fear-based aggression diminish as it is increasingly dominated?

FIGURE 12.21
Animal distributions:
(A) an even distribution
that could reflect
territoriality; (B) a
clumped distribution;
(C) a random distribution.

Territoriality

A territory is a defended space, and that defense is what distinguishes it from any other use of space. Once again, the origins of the word *territory* inform its use in animal behavior. While the Latin *terra*, or earth, may have influenced this word, there is an intriguing, early link to a thirteenth century French word that means "a place from which people are warned off" (*OED*, 2nd ed.) and that is also related to *terror*. The earliest use of *territory* in English has to do with land that is under the control of some town or ruler. All of this points strongly toward some essentials about the concept of territory: It is not just any space; it is a defended one. Nesting almost always creates a valuable resource for an animal, one that requires territorial defense.

To test the hypothesis that a species is territorial, one needs to develop two lines of evidence. First, observations need to establish that the territory is defended. Second, the spatial distribution of the animals must be assessed (see Figure 12.21). Are the animals clumped so that they are grouped together in their environment, as in a flock or herd? Are they nonrandomly spaced so that they are evenly dispersed in the environment? The third alternative is that they are randomly spread through their environment. A truly random distribution is sometimes called an *ideal free distribution*.

KEY TERM An ideal free distribution occurs when animals are located in the environment without being influenced by other animals or other environmental factors.

An ideal free distribution means that the location of each animal is not influenced by either other animals or by environmental features. If a random numbers table is used to assign x and y coordinates to locations on a map, the result is an ideal free distribution. In contrast, throwing marbles into a box does not result in an ideal free distribution because the marble's resting spots are influenced by their collisions with other marbles and with the walls. Statistical tests can be employed to test if animal distributions are random or nonrandom.

Both of these types of evidence—territory defense and nonrandomness of distribution—must be met to make an argument for territoriality. Birds may settle on a telephone wire at equally spaced intervals, but they do not fight over locations along the wire. Thus, birds on a wire are not territorial. In many animals, though, "personal space" within a social group is important, and animals will threaten or attack if a group member comes too close. Fights may break out among baboons in a troop, but the fights do not determine possession of specific locations or result in defended spaces; thus, within the troop the baboons are not territorial. Analyses of territoriality also need to take into account spatial scale. At a landscape level, bird species such as magpies may be concentrated—clumped—in subhabitats, such as cottonwoods along streams. But if the spatial scale is changed to consider only the cottonwood habitat, then nests may be evenly spaced within the subhabitat, which would suggest territoriality, especially if aggression is observed.

Territoriality is a driving force in animal behavior and almost always needs to be considered in behavioral field studies. In this chapter, the focus has been on the relationships among nesting, parenting, aggression, and territoriality. In Chapter 15, on conservation, this topic will come to the front again because territorial behavior often determines how large an area is needed to preserve a species.

SUMMARY

Animals choose nest sites based on protection and microclimates. Manipulation of the nest site, through excavation and the addition of materials from the surrounding environment, feces, or glandular secretions, gives animals the capability for complex and elaborate architecture. Nest design incorporates protection from the elements, concealment and protection from predators, and manipulation of microclimate.

Parental investment choices involve balancing current and future reproductive opportunities, and allocating resources among young within a group of siblings. Parental decisions are made to maximize lifetime parental reproductive output, which sometimes results in less-than-optimal conditions from a young animal's point of view. Even infanticide can become a male's strategy for controlling his paternal investment. This conflict over resources includes interactions among siblings; the gain of resources by one sibling may result in diminished survivorship for others in the group. In extreme cases, this conflict can lead to siblicide.

Much of animal aggression emerges from these conflicts over investment and resources within families, from territorial conflicts between family groups, or in the context of mating and predation. Aggression among animals generally occurs as a result of conflicts over resources, and across the broad range of animals, conflicts are much more likely to be resolved by some form of aggression than by negotiation. However, animal combat may evolve to be ritualized, thereby reducing the risk of injury to participants. Territoriality is defense of a location, often a nest. Finishing this chapter with a mention of territoriality as it relates to nests completes the circle and returns to parenting behavior.

STUDY QUESTIONS

1. Thinking about animal nests, are there unifying themes that make nest construction strategies similar among all animals? How is microclimate within nests regulated? What provisions for escape are included in nest design?
2. What are the causes of conflict over parental investment between parents and their offspring?
3. Why is it that parents often do not interfere in attacks among their offspring, even when an attack leads to the death of one of the young? What circumstances might favor interference on the parents' part?
4. When would you expect weaning conflict to occur? Would you expect weaning conflict to be more intense between a young mother, who might have litters in following years, and her offspring; or an older mother, who is near the end of her lifespan, and her offspring? Why?
5. We discussed the possible causes of aggression and the potentially tragic consequences of misidentifying them. How might you go about discerning the causes of aggression in an animal? What animal behavior tools would you bring to bear on this important question?

Further Reading

Buntin, J.D., 2010. Parental Behavior and Hormones in Non-Mammalian Vertebrates, In: Michael D. Breed and Janice Moore (Eds.), Encyclopedia of Animal Behavior, Academic Press, Oxford, Vol. 2, pp. 664–671.

Clutton-Brock, T.H., 1991. The Evolution of Parental Care. Princeton University Press, Princeton, 368 pp.

Hansell, M.H., 2009. Built by Animals: The Natural History of Animal Architecture. Oxford University Press, Oxford, 280 pp.

Trivers, R.L., 1974. Parent–offspring conflict. Am. Zool. 14, 249–264.

Lorenz, K., 1969. On Aggression. New York: Bantam Books, 306 pp.

Ardrey, R., 1966. The Territorial Imperative. New York: Atheneum, 390 pp.

Groebel, J., Hinde, R.A., eds, 1989. Aggression and War: Their Biological and Social Bases. Cambridge: Cambridge University Press, 237 pp.

van Schaik, C.P., van Noordwijk, M.A., 2010. Infanticide, In: Michael D. Breed and Janice Moore (Eds.), Encyclopedia of Animal Behavior, Academic Press, Oxford, Vol. 2, pp. 138–143.

Wynne-Edwards, K.E., 2010. Parental Behavior and Hormones in Mammals, In: Michael D. Breed and Janice Moore (Eds.), Encyclopedia of Animal Behavior, Academic Press, Oxford, Vol. 2, pp. 657–663.

Notes

1. Storm, J.J., Lima, S.L., 2010. Mothers forewarn offspring about predators: a transgenerational maternal effect on behavior. Amer. Nat. 175, 382–390.

2. Thornton, A., McAuliffe, K., 2006. Teaching in wild meerkats. Science 313, 227–229. Tinbergen, N., 1953. The Herring Gull's World. Collins, London, 255 pp.

3. Farnsworth, G.I., Simmons, T.R., 1999. Factors affecting nesting success of wood thrushes in Great Smoky Mountains National Park. Auk. 116, 1075–1082.

4. Seeley, T.D., 2003. Consensus building during nest-site selection in honey bee swarms: the expiration of dissent. Behav. Ecol. Sociobiol. 53, 417–424.

5. Downing, H.A., Jeanne, R.L., 1990. The regulation of complex building behaviour in the paper wasp, *Polistes fuscatus* (Insecta, Hymenoptera, Vespidae). Anim. Behav. 39, 105–124.

6. Trivers, R.L., 1972. Parental investment and sexual selection. In: Campbell, B. (Ed.), Sexual Selection and the Descent of Man. pp. 139–179.
   Trivers, R.L., 1974. Parent–offspring conflict. Am. Zool. 14, 249–264.

7. Siveter, D.J., Siveter, D.J., Sutton, M.D., Briggs, D.E.G., 2007. Brood care in a Silurian ostracod. Proc. R. Soc. B 274, 465–469.

8. Noren, S., 2008. Infant carrying behaviour in dolphins: costly parental care in an aquatic environment. Funct. Ecol. 22, 284–288.

9. Connor, R.C., Smolker, R.A., Richards, A.F., 1992. Two levels of alliance formation among male bottlenose dolphins (*Tursiops sp.*). PNAS 89, 987–990.

10. Sarno, R., Franklin, W.L., 1999. Maternal expenditure in the polygynous and monomorphic guanaco: suckling behavior, reproductive effort, yearly variation, and influence on juvenile survival. Behav. Ecol. 10, 41–47.

11. Brown, C.R., 1984. Lay eggs in a neighbor's nest: benefit and cost of colonial nesting in swallows. Science 224, 518–519.

12. Lahti, D.C., Lahti, A.R., 2002. How precise is egg discrimination in weaverbirds? Anim. Behav. 63, 1135–1142.

13. Burger, J., 1974. Breeding adaptations of Franklin's gull (*Larus pipixcan*) to a marsh habitat. Anim. Behav. 22, 521–567.

14. Cullen, E., 1957. Adaptations in the kittiwake to cliff nesting. Ibis 99, 275–302.

15. Mann, J., Smuts, B., 1999. Behavioral development in wild bottlenose dolphin newborns (*Tursiops sp.*). Behaviour 136, 529–566.

16. Chaiken, M., 1992. Individual recognition of nestling distress screams by European Starlings (*Sturnus vulgaris*). Behavior 120, 139–150.

17. Le'vy, F., Keller, M., Poindron, P., 2004. Olfactory regulation of maternal behavior in mammals. Horm. Behav. 46, 284–302.

18. Dobie, J.F., 1961. The Longhorns. Castle Books, New Jersey, p. 178.

19. Dobie, J.F., 1961. The Longhorns. Castle Books, New Jersey, p. 178.

20. Blunt, J., 2002. Breaking Clean. Vintage Books, New York, p. 290.

21. Albo, M.J., Costa, F.G., 2010. Nuptial gift-giving behaviour and male mating effort in the neotropical spider *Paratrechalea ornata* (Trechaleidae). Anim. Behav. (In press).

22. Delehnty, D.J., Oring, L.W., 1993. Effect of clutch size on incubation persistence in male Wilson's Phalaropes (*Phalaropus tricolor*). Auk. 110, 521–528.

23. Emlen, S.T., Wrege, P.H., 2004. Division of labour in parental care behaviour of a sex-role-reversed shorebird, the wattled jacana. Anim. Behav. 68, 847–855.

24. Thomas, J.A., Birney, E.C., 1979. Parental care and mating system of the prairie vole, *Microtus ochrogaster*. Behav. Ecol. Sociobiol. 5, 171–186.

25. Fillater, T.S., Breitwsch, R., 1997. Nestling provisioning by the extremely dichromatic Northern Cardinal. Wilson Bull 109, 145–153.

26. Eggert, A.-K., Reinking, M., Muller, J.K., 1998. Parental care improves offspring survival and growth in burying beetles. Anim. Behav. 55, 97–107.

27. Spoon, T.R., Millam, J.R., Owings, D.H., 2006. The importance of mate behavioural compatibility in parenting and reproductive success by cockatiels, *Nymphicus hollandicus*. Anim. Behav. 71, 315–326.

28. Hume, J., Wynne-Edwards, K.E., 2006. Paternal responsiveness in biparental dwarf hamsters (*Phodopus campbelli*) does not require estradiol. Horm. Behav. 49, 538–544.

29. Blumer, L.S., 1979. Male parental care in the bony fishes. Q. Rev. Biol. 54, 149–161.

30. Nussbaum, R.A., 1987. Parental care and egg size in salamanders: an examination of the safe harbor hypothesis. Res. Popul. Ecol. 29, 27–44.

31. Trivers, R.L., 1972. Parental investment and sexual selection. In: Campbell, B. (Ed.), Sexual Selection and the Descent of Man. Aldine Transaction, Piscataway, New Jersey, pp. 139–179.
   Trivers, R.L., 1974. Parent–offspring conflict. Am. Zool. 14, 249–264.

32. Bell, M., 2008. Strategic adjustment of begging effort by banded mongoose pups. Proc. R. Soc. B 275, 1313–1319.

33. O'Connor, R.J., 1978. Brood reduction in birds: Selection for fratricide, infanticide and suicide? Anim. Behav. 26, 79–96.

34. Porkert, J., Spinka, M., 2006. Begging in Common Redstart Nestlings: scramble competition or signalling of need? Ethology 112 (4), 398–410.

35. Grodzinski, U., Lotem, A., 2007. The adaptive value of parental responsiveness to nestling begging. Proc. R. Soc. B 274, 2449–2456.

36. Rivers, J.W., 2007. Nestmate size, but not short-term need, influences the begging behavior of a generalist brood parasite. Behav. Ecol. 18, 222–230.

37. Ferree, E.D., Wikelski, M.C., Anderson, D.J., 2004. Hormonal correlates of siblicide in Nazca boobies: support for the Challenge Hypothesis. Horm. Behav. 46 (5), 655–662.

38. Mock, D.W., 1987. Siblicide, parent–offspring conflict, and unequal parental investment by egrets and herons. Behav. Ecol. Sociobiol. 20, 247–256.

39. Lyon, B., Eadie, J.M., Hamilton, L.D., 1994. Parental choice selects for ornamental plumage in American coot chicks. Nature 371, 240–243.

40. Osorio-Beristain, M., Drummond, H., 2001. Male boobies expel eggs when paternity is in doubt. Behav. Ecol. 12, 16–21.

41. Wilson, E.O., 1975. Sociobiology: The new synthesis. Cambridge, Mass: Belknap Press of Harvard University Press, 697 pp.

42. Hazlett, B.A., 1978. Shell exchanges in hermit crabs: aggression, negotiation or both? Anim. Behav. 26, 1278–1279.

43. Guerra, P.A., Pollack, G.S., 2010. Colonists and desperadoes: different fighting strategies in wing-dimorphic male Texas field crickets. Anim. Behav. (In Press).

44. Dubois, N.S., Kennedy, E.D., Getty, T., 2006. Surplus nest boxes and the potential for polygyny affect clutch size and offspring sex ratio in house wrens. Proc. R. Soc. B 273, 1751–1757.

Social Behavior, Cooperation, and Kinship

LEARNING OBJECTIVES

Studying this chapter should provide you with the knowledge to:

- Understand that animals live in groups for a variety of reasons, including shared protection from predators, cooperation in discovering and hunting food, and opportunities for taking advantage of public information.

Animal Behavior.

- Know that the simplest social groups are often families, which can be expanded by including multiple generations.
- See that kinship structure is often important in determining cooperative and competitive interactions.
- Understand the mutual benefits that accrue to social groups of non-kin.
- Distinguish among the evolutionary explanations for living in social groups and know that these are not mutually exclusive hypotheses.
- Apply the concept of extreme altruism, or eusociality, and understand the remarkable mechanisms for using communication to enhance cooperation and colony efficiency.
- See social recognition as a key to formation and maintenance of many social groups.

13.1 INTRODUCTION

Some of the most ecologically successful species on the planet are social. What causes animals to group together? How is group membership determined? What are the benefits and costs of group living? Being in a group can give an animal significant advantages over its solitary counterparts.[1]

Groups allow some predators to exploit prey that would be impossible for them to capture on their own (see Chapter 9). Group membership also provides protection (see Chapter 10). Social dynamics keep family groups together, and social factors may eventually lead families to break up and individuals to disperse to establish independent life (see Chapter 12). An expanded view of groups includes all of the ways in which animals can come together and explores the evolutionary models that explain group living.

Most evolutionary questions can best be analyzed using a cost-benefit approach; social behavior is no exception. The benefits of social living can be enormous; self-defense, foraging, and parental care may all be easier or more efficient in groups. To these benefits can be added potential efficiencies gained from dividing labor among members of the social group, so that individuals specialize in tasks. Group living also has significant costs: Groups may be more visible and therefore attract predators. Parasites and diseases may spread more easily within groups than among isolated animals, and inevitably, group members compete for food and perhaps for mates.

For any one animal, the decision to join a group must be the result of a situation in which the benefits outweigh the costs. Evolution, of course, can make this decision for an animal; selection favors behavior that has higher benefits than costs. In extreme cases, some animals within groups pay the very high cost of lifetime sterility; these animals spend their lives working for others and gain reproductive benefits only if their relatives reproduce. Species with sterile workers are termed *eusocial*.

KEY TERM Eusocial species live in colonies with overlapping generations in which the mother (queen) plays the reproductive role and the offspring are workers.

Social behavior is a big topic within animal behavior, so this coverage is divided into two chapters. This chapter focuses on the theories that aid in understanding the existence of social groups and then explores social recognition, a primary component of the glue that holds social groups together. Chapter 14 addresses how labor is divided within social groups, along with fascinating examples of behavior that is at once elaborate and highly social.

The discussion of theories begins with theories of selfishly driven cooperation, primarily Hamilton's *kin selection* model.[2] Kin selection plays a key role in conversations about *cooperation* and group living, and this theory deserves substantial emphasis. Group selection is an interesting set of ideas initially based on the thought that animals might behave "for the good of the species." We must emphasize that modern ideas about group selection differ substantially from earlier models. Equally important are models for *reciprocal altruism*—the trading of aid in the present for potential returned aid in the future.[3] This leads to more complex transactional and social contract models for cooperation. Models for *selfish teamwork* and delayed competition can be considered next; they can be powerful in explaining why animals with competing interests might cooperate in the short term. Finally, aid can simply be stolen.

> **KEY TERM** Kin selection occurs when an animal behaves in ways that add to the fitness of its close relatives.

> **KEY TERM** Cooperation is joint activity among animals to achieve a shared goal.

> **KEY TERM** In selfish teamwork, animals work together for some time so that the group as a whole benefits, but at a later date they compete for individual benefits.

CASE STUDY

Social Network Analysis

Social network analysis is an exciting new frontier in animal social behavior.[4,5,6] Social network analysis uses an analogy to communication networks. Each animal is considered a *node* in the network, and social relationships are diagrammed as links among nodes. A famous hypothesis related to network analysis is that all humans are separated by six degrees, or links (see Figure 13.1). (See also the discussion of this concept in the context of information theory in Chapter 7.)

> **KEY TERM** Social network analysis portrays social relationships as connections in a web and analyzes the effects of the network on communication and social behavior.

> **KEY TERM** A node in a social network is a single animal that is connected with other animals.

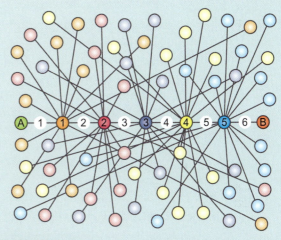

FIGURE 13.1

Six degrees of separation exist between animal A and animal B. Note that six comes from the number of links, not the number of individuals in the contact chain. For a message to reach animal B from animal A, it must pass through six links (five individuals).

369

Several measures of social structure come out of network analysis. More central animals in the group, in terms of social interactions, will have more than the average number of links with other animals; likewise, loss of a central animal is more disruptive to information flow and more damaging to network structure than the loss of a peripheral animal. Networks can be fairly open, with each animal having an equal probability of linkage to others, or they can be cliquish, with subgroups that are relatively isolated from the overall network.

Within a network, not all animals have the same amount of information. The locations of potential predators, food, and shelter may be better known to some individuals than others. Information can flow only between linked animals, so animals removed by several links from information centers are at much greater risk. Often the group leader is the animal with the most information.

This leads to the prediction that animals should compete to have close linkages with leaders and that leaders might manipulate information flow within a network to maintain their status.

In human social networks, people are very tuned to discovering which network members are better informed.[7] Slight social cues, such as location within the group, allow members to identify information centers. Groups in which leaders have been identified to group members prior to performing a task do better than groups in which leaders are not identified, provided that the leader has information that is useful in the task.

Network analysis has considerable potential for answering questions about how public information is passed around in animal groups, how social exclusivity impacts survivorship, and how disease is transmitted via social contacts.

The last section of the chapter explores the fascinating ability of animals to recognize group members and to discriminate group members from nongroup members. In vertebrates, evolution has led to social recognition that is intimately tied with immune recognition of self. Insect and crustacean social groups have equally unexpected and intriguing ways of knowing who belongs in their group and who does not. The combination of theory and social recognition then leads into the amazing array of social life inhabiting Chapter 14.

13.2 ALTRUISM OR SELFISH INTERESTS?

Are animals generous or do they act out of self-interest? Altruism is behavior performed without regard for self-interest. Biologists define a truly altruistic act as one that involves one animal giving aid to another with no opportunity for payback. Does this ever occur, or can apparently altruistic behavior ultimately be explained by self-interest? Animals often cooperate in foraging, defending themselves, and caring for young, but true altruism is entirely contrary to Darwinian thinking, which claims that everything an organism does is shaped by the drive to maximize an individual's fitness. Most cases of apparent altruism actually have roots in self-interest, so examples of truly altruistic aid-giving behavior are quite rare.[8,9] While altruism is clearly one possibility for explaining aid-giving behavior, the alternative hypotheses consider aid-giving and cooperation to be the results of self-interest.

In studies of cooperation and aid-giving behavior, the term *donor* signifies the animal that gives the aid. The *recipient* receives the aid and benefits. The donor usually incurs a cost for giving the aid; it gives up food, puts itself at risk for predation, or delays its own reproduction as a result of aiding another. This is true whether the donor is motivated by altruism or self-interest. The cost to the donor translates into a benefit for the recipient, usually measurable in food obtained, protection from predation, or increased chances of reproduction. But is there also a benefit to the donor? In true biological altruism, the donor pays the cost without any consideration of potential benefits it might later receive. In contrast, any evolutionary explanation of aid-giving behavior must include benefits to both the donor and the recipient.

370

DISCUSSION POINT: ALTRUISM AMONG SPECIES

Through a little searching on the Internet, you'll be able to find a few seemingly miraculous cases of true altruism between species. They include events such as dolphins saving drowning humans or dogs rescuing their human family members from fire or other catastrophes. It's interesting to think analytically about these events. Certainly, the potential for true altruism is there, but can you approach the event scientifically, forming alternative hypotheses that can be tested to explain such altruism? Are there hypotheses for this type of behavior that don't involve altruism? Might these events be more legend or wishful thinking than reality? When thinking about this, you might want to refer to the section on emotions in Chapter 6 on cognition.

FIGURE 13.2
Mutual grooming in monkeys. Note the parallel in behavior with mutual grooming in chimpanzees shown in a later chapter (see Figure 14.8). *Photo: Pen-Yuan Hsing.*

Sometimes the benefits to the donor are nearly immediate; mutual grooming among monkeys (see Figure 13.2) is an example of aid-giving with a rapid return. Benefits may also be returned to the donor at a later date. In most examples of cooperation among animals, the donor expects to receive benefits that at least balance the cost. These benefits may come directly to the donor during its lifetime, such as reciprocated protection from predators, or over a longer evolutionary span, due to increased propagation of the donor's genes in following generations.

This evolutionary trade-off can actually be quantified. For natural selection to favor cooperation and aid-giving, the costs of being a donor must always be counterbalanced by current or future benefits to the donor. The very simple *cost-benefit equation*, $B > C$, describes the circumstances under which animals should become donors. The benefit (B) is the increase in the donor's genetic contribution to the next generation, and the cost (C) is any loss in contribution to the next generation that the donor incurs. A modification of this equation formulated by William D. Hamilton takes into account the genetic relationship between the donor and the recipient (see the box below).

> **KEY TERM** A cost-benefit equation is a quantitative analysis of the balance between the cost of a behavior and the benefits gained from that behavior.

OF SPECIAL INTEREST: WILLIAM D. HAMILTON

Only a small handful of scientists contribute enough to a field to merit special mention in a textbook: William D. Hamilton is one of those select few in animal behavior (see Figure 13.3). Hamilton contributed several major ideas to animal behavior, including three central concepts in this chapter: kin selection, cooperation, and selfish herds. Hamilton made other important contributions, notably the Hamilton-Zuk hypothesis that parasite load affects coloration and consequently mate choice. He also improved our understanding of dispersal and sex ratios.[10]

Hamilton is credited with having more influence on evolutionary thought than any other individual biologist since Darwin. He made his points with elegant simplicity, and while some of those arguments are supported by mathematical models, the real genius of his work is that he developed principles that have survived nearly half a century of scrutiny. Hamilton, a humble and gentle-spirited person, received nearly every conceivable honor and award in his field during his lifetime. He died in 2000 of complications from malaria, an infection he contracted in Africa, where he was working on discovering the origin of HIV.

FIGURE 13.3
William D. Hamilton.

371

13.3 SCHOOLS, FLOCKS, HORDES, AND HERDS

Why live in a school, flock or herd?[11,12,13] Groups of a few, a dozen, tens, hundreds, or even thousands of animals are a common phenomenon, one that is easily observed across major taxa ranging from flatworms to Crustacea, from insects to birds and mammals. Basic family groups (female, perhaps a male, and their offspring) or slightly extended families (for example, related females cooperatively rearing brood) were covered in Chapter 12. Evolution, though, can draw together other kinds of family associations, such as schooling sibling tadpoles. People have crafted exotic descriptive names for animal groupings, but the terms *school*, *flock*, and *herd* capture the essential focus of this section.

OF SPECIAL INTEREST: ANIMAL GROUPS

We'll use the rather pedestrian word *group* to describe associations of three or more animals. The English language abounds with colorful words to describe groups of specific kinds of animals (see Figure 13.4). For example.

- Quail: *covey*
- Fish: *shoal*
- Geese: *gaggle*

- Crows: *murder*
- Lions: *pride*
- Dolphins: *pod*
- Cats: *clutter* or *nuisance*
- Dogs or wolves: *pack*

372

FIGURE 13.4
Groups of animals.

Social Groups and Public Information

Animals rarely act in complete privacy, and an observable behavior becomes *public information*, meaning that other animals can exploit what they see (or smell or hear; see Chapter 7). An animal may find it advantageous to conceal some of its discoveries, such as food, the location of a food cache, the identity of its mate, or the location of its nest. The advantages of privacy and concealment are lost in social groups; an animal in a group makes most of its life "public information." On the other side of the coin, membership in a group gives an animal the ability to observe and gain leverage from information revealed by other animals.[14] The costs and benefits of revealing and gaining public information are part of the interplay that leads to the evolution of social groups.

> **KEY TERM** Public information is any feature that other animals can perceive. A wide range of attributes, such as signals, health status, mating condition, and size can become public information.

Within such groups, alarm signals can alert members to threats. Alarm signals are public information;[15] once the signal is given, any animal that can perceive the signal can take advantage of it, including the potential predator. These signals also can attract costly and unwelcome attention to the signaler, and under the rules of evolution, benefits such as improved survival of close kin are expected to at least balance that cost. Alternatively, alarm signals may cause a visually distracting commotion in the group, drawing attention away from the signaler. Such signals often have evolved so that it is difficult to locate their source, publicizing the information without revealing too much about the source. For instance, many audible alarm signals are short and high-pitched calls, and it can be particularly difficult to find the source of such calls. Alarm signals serve as a great example of the balance between a need to create public information about the location of threat and the potential need to conceal the private information of the signaler's location.

Food Discovery

Following this line of reasoning about public information, there are two basic ways to find food: An animal can discover food for itself, or it can take advantage of food discovered by other animals. Taking advantage of public information within a group extends far beyond a situation in which more than one animal is attracted to the same food item; it includes group members' exploitation of the food discovery abilities of their group-mates.

A conspicuous example of food theft is the behavior of herring gulls on beaches. The first gull to discover a crab or other morsel is often not the gull that ends up enjoying the feast. Nearby gulls are always keenly interested in food discoveries by their companions, and a gull with a piece of food is a magnet for the attentions of its voracious flock-mates.

Similar behavior abounds in magpies, ravens, and crows, all members of the bird family Corvidae. Exploiting public information appears to draw crows or seagulls together (see Figure 13.5); they stay close to other crows or seagulls to watch for food discovery. After one animal has found food, the others try to steal it. A bird that wants to forage in "private" may have a very difficult time evading observation because completely driving off the other birds is a fairly hopeless task. Thus, the group may contain unwilling participants.[16]

This grouping behavior can also involve multiple species. In fall and winter, mixed-species flocks of birds are common; in North America, house sparrows, house finches, and juncos frequently occur together. Discovery of food by one bird in the flock inevitably leads to exploitation by the other members. Ultimately, the best foragers in the group may lose food; they cannot spend all their time and energy trying to drive away other animals, so groups formed around exploitation of other animals' foraging abilities are common.

Mutual Protection

In mutually protective groups, one possible advantage is shared vigilance. Perhaps all animals in the group benefit from having others in the group spend time watching for predators.[17]

373

FIGURE 13.5
A flock of seagulls competing for a food item, pelicans, which often forage in flocks, and a school of fish. These are examples of groups that might coalesce because individuals evolve to exploit public information. *Photos: Michael Breed (left and right), and Ben Pless (center).*

In groups of genetically related animals—a school, herd, or flock that is an extended family—mutual aid, including shared vigilance, can be explained by kin selection (see the section later in this chapter). Tadpoles often school in family groups, as do some species of trout and salmon. However, most schools, flocks, and herds consist of unrelated animals, requiring explanations other than kin selection—such as reciprocal altruism or *selfish herds* (see below)—for why the flocks come together.

> **KEY TERM** A selfish herd is one in which animals group together but are motivated by their own interests, rather than aid-giving.

In addition to vigilance, members of groups may benefit if predators are confused by the group. Chasing a single animal can be much simpler than trying to select a prey item from a rapidly shifting group in which the animals' paths interweave. The flurry of motion when a flock of birds takes wing may make it much more difficult for the predator to catch any one animal. Related to predator confusion is predator saturation, a topic already discussed in some detail in Chapter 9.

Finally, sometimes the lead, or reproductively dominant, members of the group also assume protective roles. For example, in the herds of wild horses that thrived in the American West in the eighteenth and nineteenth centuries, the dominant stallion in the herd was also the chief protector of the mares and foals. Stallions actually hunt and kill mountain lions with their sharp front hooves if they catch the lion's scent (see Figure 13.6). Bulls are similarly aggressive against interlopers in their herd's area. The balance between costs and benefits favors the dominant male that protects his reproductive investment in the herd. This type of protectiveness is similar to of the fierce defense that many animal mothers show for their young.

FIGURE 13.6
Mountain lions.
Photo: U.S. Fish and Wildlife Service.

Selfish Herds

An animal can join a group for its own selfish reasons, seeking to push the risk of being a victim of predation onto other group members. A herd formed in this manner is aptly named a *selfish herd*; the name is self-explanatory. Selfish herds are groups of animals in which group membership confers protection on each animal in the group, but within the herd the animals compete for the value gained from group membership.

William D. Hamilton formalized the concept of the selfish herd in 1971. The idea is charming in its simplicity. It is based on the fact that while for social animals, group membership in any form may be more beneficial than remaining solitary, but nonetheless good and bad locations exist within a group. A bad location is one in which exposure to predators is high and access to public information about predators is hard to come by. A good location has low exposure to predators and high access to public information.

Generally speaking the good locations are at the center of the group, in positions with considerable social connectivity (see the earlier section on social networks) and with a good buffer of groupmates between the animal and any potential predator. Selfishness comes into play because not all positions within the flock are equally protected, so animals selfishly compete for position within the group.

For instance, the trailing fish in a school may be more vulnerable to predation; if the predator focuses on the last fish in the group, the others will scatter forward and to the sides. In this scenario, the scattering fish do not interfere with the predator's visual fix on the last fish, making the chase easier for the predator. In a selfish herd, animals should constantly jockey for position. Their constantly shifting positions reflect competition among group members for these favored positions. The importance of position distinguishes a selfish herd from a group that depends primarily on the dilution effect for defense.

How Large Should a Group Be?

What is the right number of animals in a group?[18] Begin by thinking about a herd, flock, or school that has come together for shared vigilance, and the value of mass movements confusing predators. If each member of the group has an equal chance of being caught by a predator, then the more the merrier—10 members means a 10% chance of being eaten, 100 members translates into a 1% chance, and so on. This is called a dilution effect, and in contrast to a selfish herd, group members may depend on group size, not location in the group, for relative safety.

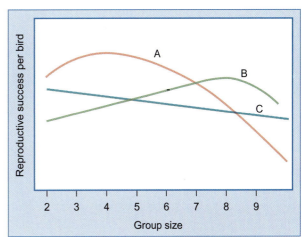

Following the logic of the dilution effect, there might be no upper limit: A school might then include all the fish in the ocean, a herd should include all the ungulates on a continent. What might establish upper limits? Three factors seem most obvious; first, food supply will be limiting at some point, so the amount of food per animal establishes an upper group size. Second, larger groups may attract more predators. Third, the benefits of shared vigilance probably plateau after the group has a certain number of animals. The balance between the value of being in the group and the potential costs can result in an optimal group size.[19,20,21]

Another way of looking at group size is to consider reproduction of each group member related to the number of animals in the group.[22] Theoretically, animals should be willing to participate in a group as long as their reproduction is maximized. Figure 13.7 shows models for how group size might relate to reproduction. If animals can assess their potential reproduction, then these models may give insight into whether a species may exhibit social behavior, and if it does, what group size might be.

Dominance Hierarchies in Social Groups

The best known *dominance hierarchies*, such as those in flocks of domestic hens, are linear, meaning that there is one individual (A) that dominates the others, there is a second individual (B) that dominates all others except A, there is a third individual that dominates all the others except A and B, and so on, so that if they were to be ordered by their ranks, relative to others in the group, they would form a straight line. A linear hierarchy is one in which each animal is dominated by higher-ranking group members and, in turn, dominates lower-ranking members. Other types of dominance hierarchies are known in various primates, in herds of ungulates, and in cockroaches.[23] Whether the group is composed of related or unrelated animals, dominance hierarchies can structure social interactions in groups.

FIGURE 13.7
Group size as it relates to reproduction. In A, reproduction per individual peaks in small groups, while in B larger groups are advantageous. In C grouping is disadvantageous at all group sizes. Ecological factors, such as effort needed to collect food for young animals or increased effectiveness of group defense by larger groups affect which of these models apply in a given species. Adapted from *Koenig, W.D., 1981. Reproductive success, group size, and the evolution of cooperative breeding in the acorn woodpecker. Am. Nat. 117, 421–443.*

375

KEY TERM A dominance hierarchy describes situations in which animals are physically or chemically dominant over other animals in their social group.

High-ranking animals in dominance hierarchies may have preferred access to food, mating, grooming services of other group members, and perhaps to protected positions within the group (although when horses are on the move, mature mares often lead the way, and the stallion brings up the rear, a risky position from which he can defend the herd). Clearly, the benefits of group membership are not the same for all the animals when there is a dominance hierarchy (see Figure 13.8); lower-ranking animals pay a much higher cost for group membership than the alpha animal does. A couple of dynamics help to keep groups together in these circumstances. First, the costs of not being a member (loss of access to food, higher predation risk) may be much higher than the costs associated with membership. Second,

Linear hierarchy

A (α – animal)
↓ Dominates
B (β – animal)
↓ Dominates
C
↓ etc.

Non-linear hierarchy

A
↓ ↓ ↓   Dominates
B C D   Dominates

A
↓ Dominates
B
↓ Dominates
C

FIGURE 13.8
Linear and non-linear dominance hierarchies.

KEY TERM A bachelor herd is a social group of reproductively mature males. This can occur either during the nonbreeding season or can consist of males that do not have harems.

low-ranking animals may play a waiting strategy, buying a chance at rising to a higher rank later by paying the costs of low rank for the moment. In groups with multiple males, low-ranking animals, particularly males, may also have sneaker or satellite opportunities for mating (see Chapter 11 on mating). In addition, an animal's dominance with respect to one kind of resource may not mean it dominates all resources; the dominant animal at a watering hole may not be the dominant animal where some other resource is involved.

Generally, dominance hierarchies play out differently for males and females. Alpha, or dominant, males are much less likely to tolerate the presence of potentially reproductive males in the group than the presence of females. Holding the dominant position is very taxing, physically, for the male, and monopolizing reproduction in the group is the key benefit for the alpha male. Many females, on the other hand, can mate with one male, so they are more tolerant of each others' presence. This is not to say that dominance hierarchies do not exist among females; they do, and they play important roles in access to food and timing of reproduction in many species. One result of this difference between males and females in mammals is the formation of *bachelor herds*: outside the mating season, groups of males affiliate and gain benefits of group living, such as protection from predators. Bachelor herds break down with the onset of the mating season and intensified competition among males.

BRINGING ANIMAL BEHAVIOR HOME: DOGS AS MEMBERS OF HUMAN PACKS

Wolves, the wild ancestors of domestic dogs, live in packs led by an adult male and female; most of the pack members are the offspring of that pair. Much has been made about how a dog in a human household views itself as a pack member. Because early studies on wolves emphasized dominance relationships within packs, it seemed reasonable to interpret a dog's responses to the humans within its social group as a reflection of its dominance status in the group.

To a certain extent, the idea that dogs act like members of a human pack is true; dogs can be very territorial in protecting a house and the members of the household. Dogs respond to leadership within a human household and are more likely to follow direction from an older, deep-voiced male human than from a female or from younger group members. Unfortunately, many humans interpret this to mean that they should somehow establish their dominance over household dogs, and do so by physical means, such as hitting, kicking, and forced roll-overs, so-called "alpha rolls." These approaches may have disastrous consequences.

Early studies of behavior in wolf packs focused on captive packs that were composed of unrelated individuals who, because they were captive, were unable to disperse even if they had so desired. In these packs, agonistic behavior occurred more frequently than would be expected in a hierarchical social system. These studies formed the basis for much of the "wolf pack" view of domestic dogs. In the wild, however, members of wolf packs exhibit little aggression, although when it does exist, it can be lethal. In these packs, agonistic behavior does not conform to hierarchical expectations, but reflects a variety of attributes, including age, weight, and the nature of the contested resource.

One might then ask if wolves are good models for domestic dogs. Do dogs, perhaps by virtue of domestication, engage in social interactions that differ significantly from wolf social interactions? Studies of both feral dogs and neutered domestic dogs—animals that approximate the average family pet more than a wolf does—show that dogs differ substantially from wolves in their social behavior. Concepts like hierarchies and suppressed reproduction do not seem to apply to domestic dog interactions, and dogs do not seem to strive at every turn for dominance.

Instead, "dominance" in dogs may shift depending on motivation (how much do I value this contested resource?) and context (where have I seen this black dog before and did it mean trouble?). The Resource Holding Power (RHP) model (see Chapter 11) seems to offer more promise for understanding dog social behavior than notions of dominance do. How fruitful this will be in the study of dog-human interactions remains to be seen, but many skilled dog trainers have used a variation on this theme. They know that their first task with a dog is to find an object that the dog will "work" for—that is, that the dog values. This is much more productive than a contest for "dominance."

Aggregated Nesting

Sometimes animals clump their nests together in a single location. Why they do this is one of the more intriguing mysteries in animal behavior. Clumped or aggregated nesting is common in mammals such as prairie dogs, birds such as swallows, and ground-nesting wasps and bees. Is it possible that the underside of a single, specific bridge is perfect nesting habitat for swallows (see Chapter 14) while none of the bridges up- or downstream are suitable? What about large clusters of other types of nests (see Figure 13.9)? Living in close proximity increases competition for food and enhances the chances for transmitting diseases and parasites. Clusters of nests might also attract predators. What possible benefits could outweigh these major disadvantages? Two that might immediately come to mind are collective defense and easy availability to mates, along with the previously discussed benefit of finding food by observing other animals foraging.

Beyond these three hypotheses to explain aggregated nests—defense, mates, and information sharing—there is no clear general answer to the question of benefits that outweigh the disadvantages of group living. It may be that microenvironmental variation is more important than humans understand, so a particular nesting site seems superior to the animal, but for reasons that are not obvious to us. (*Umwelt* remains important!) It may also

377

FIGURE 13.9

Aggregations of animals and nests: oropendolas, cormorants, seals, and cliff swallows. *Photos: Ben Pless (seals), Michael Breed (oropendolas, cormorants, and cliff swallows).*

> **KEY TERM** Philopatry is the return, in a later breeding season, to the site where an animal was born.

be that young prefer to nest where their parents nested; the success of the parents may predict success of the next generation in the same spot. This is called *philopatry*, and as generations pass and descendants accumulate, nesting populations may grow in a location, even though nesting together carries no particular advantage.

OF SPECIAL INTEREST: LADYBIRD BEETLE AGGREGATIONS

One of life's mysteries is why some insects, such as ladybird beetles, overwinter in aggregations of hundreds or thousands of individuals (see Figure 13.10). Every year, entomologists are flooded with phone calls and e-mails from curious members of the public who have found such a pile of beetles and want to know why the beetles aggregate. There's no clear answer to this question, but let's consider a couple of hypotheses for why they and other aggregating animals might do this:

1. The beetles may simply all be attracted to the same environmental feature, so there's no particular advantage or selective value in aggregating; they all just like the same place.
2. They may be attracted to each other; this would be an example of how public information provides a key element behind an interesting behavior.

FIGURE 13.10
An aggregation of ladybird beetles. *Photo: Jeff Mitton.*

From studies of the beetles, we know that they like to land on lighter-colored places, but that alone wouldn't explain how they aggregate.[24,25,26] In terms of why they aggregate, the two main ideas are that wintering as a group assists in thermoregulation as they insulate each other, and that the groups help them choose mates, which occurs in the spring before they leave the aggregation.

13.4 EXPLAINING COOPERATION

There are six possible evolutionary explanations for aid-giving behavior and cooperation that are reviewed in this section:

- Kin selection
- Reproductive skew
- Group selection
- Social contract models
- Delayed competition
- Stolen aid/coercion

These theories are not mutually exclusive; many behaviors can be explained equally well by more than one of the theories.

Kin Selection

Kin selection occurs when an animal behaves in ways that add to the reproduction of its close relatives. When William D. Hamilton published his theory of kin selection in 1964, it reshaped how biologists think of animal behavior. His theory forced biologists to see family relationships as a critical element of animal behavior. This, in turn, led to the application of molecular technology to animal behavior studies in order to accurately determine family relationships. In kin selection, the benefits of cooperation and altruism are directed to close relatives. Natural selection then favors the spread of genes for these behaviors. Kin selection can function in extended families or in populations in which related individuals may come into contact after dispersal from their families. Whether or not animals in a group are related is central to investigations of animals living in groups. Groups of unrelated animals are almost certainly held together by individual cost-benefit decisions. While watching a flock of birds, for example, stop and ask what benefits each animal in the group might be gaining, in comparison to those associated with being alone. What costs might each bird incur by being in the group? Does it seem, based on casual observation, that the benefits outweigh the costs? What variables might be measured to gain a more scientific assessment of the cost-benefit equation for group living? The answer often lies in kinship.

Kinship is a key element of most contemporary field studies of social behavior in animals. Kin selection requires and is based on genetic relatedness between the donor and the recipient of the altruistic act. Kin selection is the dominant explanation for the evolution of aid-giving behavior. The other explanations are interesting and important, but are restricted to more limited circumstances. If the group is an extended family, or if the larger group can be subdivided into family groups, then kin selection can be an important dynamic in the costs and benefits of group membership. Kinship can be a powerful force in holding such groups together because all group members have a shared genetic interest in the survival of the group's offspring. Kinship is a part of the cost-benefit equation for some kinds of group living, but the evolutionary dynamics affecting groups of related and unrelated animals are quite similar.

Upon further examination, it becomes clear that an animal has two ways of gaining fitness. The first, and more obvious, is to reproduce. The second is less obvious; an animal can gain fitness by aiding close relatives. This is kin selection. Close relatives share genes, that is, have genes that are identical by descent. In fact, if a donor can act to increase a relative's reproduction enough, it may even benefit the donor to behave in ways that reduce its own reproduction. In the extreme case of the eusocial insects (termites, ants, bees, wasps, and a few other types of insect), workers give up their reproduction to assist their mother in raising her offspring.

Up to this point, aid has been expressed in any currency, such as food or predation risk. What if the currency is restricted to that of reproduction, or fitness? If a person rescues a sibling from drowning, then she or he has protected that sibling's fitness. The rescuer has also gained fitness, because the genes shared by the rescuer and sibling will now be passed on to another generation, something that would not have happened if the sibling had drowned. This added reproduction of a close relative should count in an individual's fitness, but by how much?

The donor pays an evolutionary cost (C) for giving assistance. The cost is the risk involved in giving the aid times the donor's potential for future reproduction. If the donor has a 25% chance of dying when giving aid, and has the potential for producing offspring of its own in the future, then the cost is the 25% chance of dying times the potential future reproduction. The benefit (B) resulting from this aid is the added reproduction the recipient of the aid enjoys as a result of the donor's act. If the recipient was sure to die without aid, then all of its future reproduction is a benefit.

The basic principle of kin selection is captured in Hamilton's equation:

$$B/C > 1/r$$

this can be rearranged as

$$rB > C$$

The elements of cost (C) and benefit (B) and relatedness (r) in this equation have already been introduced. The cost (C) is the lost potential fitness of the donor. The benefit (B) is

379

the *added* fitness for the recipient due to the acts of the donor. The fundamental message of this equation is that aid-giving behavior by the donor should be favored in the course of evolution if the donor-recipient relatedness (r) times the added benefit to the recipient is greater than the cost to the donor.

Now apply the cost-benefit equation to the example of the drowning sibling:

C = the probability of the rescuer drowning in the rescue attempt
B = the subsequent reproduction of the sib
r = 0.50, the relatedness of rescuer and sib

Consider two scenarios. In both scenarios, each sibling will have two children in the future, if it survives. In the first, the risk of the rescuer drowning is small—say, 10%.

Substituting for rB yields $(0.5)(2) = 1$.
Substituting for C yields $(0.1)(2) = 0.2$.

Because $1 > 0.2$, Hamilton's inequality is met, and the plunge to rescue is worthwhile, evolutionarily speaking.

What if the risk of drowning is 0.9?

Doing the arithmetic reveals an alarming increase in C. Hamilton's inequality is not met, and if the world proceeds in accordance with Hamilton's Rule, then the equation predicts that the rescue would not be attempted.

> **KEY TERM** Classical fitness is an animal's own reproductive output.

> **KEY TERM** Inclusive fitness is an animal's reproductive output, plus any added reproduction gained by relatives due to aid-giving by the first animal.

380

Do animals make this cold-hearted calculation? Certainly not consciously, but evolutionary forces drive animals to behave in ways that are consistent with the equation. The rearranged form, $B/C > 1/r$, emphasizes the contrast between the benefit/cost ratio and the relationship between the two animals. To keep this straight, we refer to an animal's own reproduction as *classical fitness*; measures of fitness that include added reproduction due to aid-giving behavior are called *inclusive fitness*. Just to emphasize this point, inclusive fitness includes only the fitness benefits of the aid-giving behavior, not the entire reproduction of the relatives.

Ecological conditions can change B/C ratio so that aid-giving is more or less likely; we discuss this principle in more detail later, but it also merits talking about in the context of kin selection. Suppose that an animal has virtually no chance of reproducing on its own. In this unhappy situation, there is practically no cost to aid-giving behavior; ecological conditions have already extracted the full price from the animal. If the cost is small, then small benefits, even from aiding more distantly related animals, will outweigh the costs. In studies of birds and mammals, often the low probability of successful dispersal of young animals weighs heavily in the likelihood that those animals will stay and help their parents, rather than starting their own nest.

DISCUSSION POINT: KIN SELECTION

Kin selection theory predicts that you should be more likely to undertake a risky rescue, such as saving a drowning person from an icy pond, if that person is a close genetic relative. More creepy, given the choice between saving your child or your grandparent, theory predicts you should save your child because your child's potential future reproduction is higher. If forced to choose between your own child and someone else's, kin selection theory predicts that you would always choose your own. Do people actually behave as theory would predict? Why or why not? What is the role of culture in these decisions?

Reproductive Skew

Behavioral ecologists have developed a set of models, under the general rubric of "reproductive skew," that extend Hamilton's kin-selection models to events in hierarchical social groups characterized by differential reproduction. In such groups, the more highly ranked animals typically account for more than their share of the reproductive output of the social group.[27] In other words, reproductive skew occurs when animals within a social group differ in their reproductive potentials. In cases of skew within kin groups, skew can be viewed as a more modern formulation of kin selection. Skew also occurs when group members are unrelated. Reproductive skew theory attempts to be broad enough to explain aid-giving behavior under a very broad range of possible conditions.

In skew models, reproduction of a subordinate member of the group may actually be "allowed" as a kind of payment that induces the animal to remain in the group. This is called a "concession model of reproductive skew." In contrast, in "restraint" models, the subordinates control reproduction but can remain in the group only if the dominant allows it. The subordinates must "restrain" their reproduction, giving some reproductive rights to the dominant to be able to stay in the group. Finally, in a "tug-of-war" model of reproductive skew, group cohesion is driven by each individual's prospective maximum reproduction if it stays in the group, even though ultimately it would be impossible for all the members to maximize their reproduction. In this theory, the continuing conflict, or tug of war, creates a situation in which the possible benefits of remaining in the group outweigh the potential reproductive benefits of leaving the group.[28]

All of this assumes that the aid-giver is making some kind of choice about its behavior. Instead, the choice has probably been made by natural selection over evolutionary time, and there is little room for an animal to assess its situation and respond. Older theory in this field argued that if kin selection were driving the evolutionary process, then workers should cooperate in accepting the reproductives' role. If this were the case, then mechanisms that allow the reproductives to suppress the workers would not be expected. On the other hand, if the parental generation were coercing or stealing the labor of their offspring, then one would expect to find suppressive mechanisms, such as behavior dominance or pheromonal dominance. More recently, theorists have overlaid the concept of reproductive skew on this dichotomy, while at the same time recasting it slightly.

Group Selection

Group selection is natural selection that acts at the level of the group, so individual survival is dependent on group survival. Group selection has an interesting history as a discredited theory that has come back to life in very different form. In older literature or nonscientific representations of animal behavior, sometimes animals are said to behave "for the good of the species." For example, this is often given as a reason why nonhuman animals supposedly do not fight to the death. It is now clear that this is patently untrue: not only do animals never behave "for the good of the species," but there are many examples of animals of the same species fighting to the death. V. C. Wynne-Edwards attempted to argue for behavior that involved individual sacrifice for species-level benefit on scientific grounds;[29] his work in this area did not gain widespread acceptance at the time it was published and is still not generally accepted. Evolutionary thinking clearly posits that animals behave in ways that result in the passage of their genes to the next generation.[30]

There is another way of looking at group selection that has gained some credibility in recent years. In this type of group, or interdemic, selection, small subgroups within the larger population frequently go extinct. If a gene for altruism—behavior that helps the group to survive even if the behavior is disadvantageous for the carrier of the gene—arises in one of the subgroups, then that gene may be favored in evolution as subgroups without the gene will more rapidly go extinct. David Sloan Wilson, and more recently Edward O. Wilson, have

FIGURE 13.11

In this common model for group selection, a gene may arise for altruism in one of the groups. Individuals in the group do not reproduce as much because of the costs of the altruism, but the group has a higher probability of survival. In the illustration many of the other populations have gone extinct, even though reproduction was higher. This model is designed to explain the evolution of altruism via group selection. Adapted from *Wilson, D.S., Wilson, E.O., 2008. Evolution "for the good of the group". Am. Sci. 96, 380–389.*

championed this theory, but there are few convincing examples of altruism evolving by group selection in the field (see Figure 13.11).[31] This is not surprising, given the strong negative reaction to group selection as good-of-the-species thinking; new theoretical approaches may shift interest once again.[32] In sum, group selection might work if interdemic selection is high, but at this time this is viewed more as an interesting theory than a likely mechanism to explain the evolution of group living and helping within groups.

Social Contract Models for Cooperation

A social contract is an arrangement of trust, in which a donor gives something in return for a promise that the recipient will give back something of equal or greater value at a future date. A simple example of this is the Prisoner's Dilemma, discussed in detail in Chapter 7. Recall that in the Prisoner's Dilemma, both prisoners can cooperate to receive a small reward, or one can rat (defect) on the other and receive a large reward. If both defect, the result is catastrophic for both. Effectively, each promises to cooperate, forgoing the larger reward for defection in return for not being a victim of defection.

Put more broadly, aid given based on promises of future aid is *reciprocal altruism*. The prisoner's dilemma describes a simple, stable form of reciprocal altruism, which can persist in the absence of kinship. A large set of papers has been published on the theory of reciprocal altruism, and some arguments have been made that alarm calls could be an example of such behavior. The big question is enforcement of contract: How would animals be constrained by social conventions alone to reciprocate? Overall, the evidence is weak for reciprocal altruism outside of humans. Nonhuman animals seem to live much more "in the moment" than humans, and most, perhaps all, aid-giving behavior can be explained either by immediate returns or by kin selection. Recent overviews have argued that reciprocity is rare, perhaps absent, in nonhuman animals.[33,34]

Delayed Competition and Selfish Teamwork Models

In quite a few ant species, groups of unrelated queens establish a nest together. They lay eggs and their worker daughters build the nest, forage to feed more larvae, and tolerate each other's presence as the colony grows. The technical word for this behavior is *pleometrosis*. Sometimes these groups of queens are unrelated; in these cases, kin selection does not play a role in cooperation. The skew occurs in the production of the next generation of reproductives.[35]

Why participate in such an arrangement? The explanations are simple: if the queens are sisters, then kin selection can operate. If the queens are not sisters, then the persistence of such groups requires that the groups are much more likely to survive than individuals would be, and all of the members of the group start with a more or less equal chance of "winning." In this case, winning is the absence of reproductive skew (all queens reproduce equally) or the queen having reproductive skew in its favor. Working together in this case is "selfish teamwork" because each individual is in the group to promote its own reproduction. Part of what allows this to function is that the competition phase is delayed; the queens cooperate in colony growth by laying worker eggs and may later compete when the "real" reproduction involving the next generation of queens starts.

DISCUSSION POINT: SELFISH TEAMS AND THE *SURVIVOR* TV SHOW

Here's a quick summary of how the *Survivor* works, in case you've never watched. A group of 16–20 people are recruited to participate in an adventure television series. Participants cooperate, knowing cooperation is necessary for survival, but that only one will get a payoff at the end of the game. When the series begins, the participants are divided into two teams. In some seasons, the division is random; in other seasons, the division follows a theme, such as young versus old, male versus female, street-wise versus naïve. Participants are eliminated, by vote, one by one following team versus team competitions (each week the losing team votes out a member). When each team is down to a few members, the two teams are merged to form a final group; members of this final group then compete for individual advantage in the votes.

The striking result is that members of these randomly composed teams always cooperate and form alliances, and that the cooperation extends beyond the merger of the two teams. These team members have only superficial commonalities, but is there something about human psychology that causes us to see the value of working in teams, even if in the end there can be only one winner? Is this type of group behavior unique to humans, or do animals behave in similar ways?

Stolen Aid and Coerced Cooperation

The final model for aid-giving behavior is more insidious. What if the aid is coerced or stolen? This can clearly be the case in societies in which dominant animals have advantages over subordinates. Another major arena for this type of behavior is *parental manipulation*, in which parents can manipulate the reproductive choices of their offspring using nutritional dominance and social dominance to keep young animals in a pre-reproductive state. When combined with a lack of ecological choice, such as the lack of available territories, parental manipulation can be a powerful dynamic in promoting aid-giving behavior. In mammal societies in which dominant animals take advantage of their position, they manipulate social interactions so they gain food or shelter from subordinate animals.[36] This type of behavior has clear benefits to the dominant and clear costs to the subordinate.

The idea of parental manipulation was first developed in studies of primitively eusocial wasps and bees, in which the sterile workers are usually noticeably smaller than their mother, the queen.[37,38] They are smaller because the queen feeds them less than a potential reproductive will receive; this seems like a clear case of manipulation. Smaller workers then may be more easily physically dominated by their mother, and may also have lower reproductive potential if they were to attempt to start their own nests. This type of parental manipulation probably depends on workers deriving kin selection benefits from their aid-giving behavior;[39] it is an interesting model that needs further testing. Manipulation in parent–offspring relationships often occurs when food is at stake; refer to the discussions of begging and weaning conflict in Chapter 12 for more on this topic.

DISCUSSION POINT: CAN ALTRUISM EXIST?

At first glance, evolutionary theory predicts the complete absence of true altruism, and this might give you pause. You would not be alone. This leads to questions that we all should ask, and that inspire discussion extending far beyond the limits of this book: How should science affect personal behavior? Should science be constrained by its implications for that behavior? What are the roles of culture and learning in creating environments that favor altruistic behavior, and how subtle might those influences be? How well do we understand our own science as we create it? Note that this book is written from and consistent with a scientific worldview, a worldview that has served us remarkably well in many ways. Spend some time thinking about how worldviews create—and limit—perception.

13.5 EXTREME COOPERATION: EUSOCIALITY

Why in the world would an animal yield its reproductive potential to work as a nonreproductive, or sterile worker, in a colony? Put in evolutionary terms, it seems completely counterintuitive that selection would favor the loss of reproductive capacity. At the most extreme, nonreproductives, such as honeybee *workers*, can yield their lives in defense of their colony.

> **KEY TERM** A worker is a sterile or semi-sterile individual within a eusocial colony.

For Charles Darwin, the sterile workers found in termites, bees, ants, and wasps presented one of the largest barriers to his theories of evolution. Remember that species with sterile workers are called *eusocial*. Sterile workers of termites, ants, bees, and wasps represent extreme cooperation, in which an animal will give up some or all of its own reproductive potential to aid others. Recent discoveries add shrimp, aphids, thrips, and naked mole rats to the list of animals with sterile workers. Darwin, though, hinted at the solution: "This difficulty, though appearing insuperable, is lessened, or, as I believe, disappears, when it is remembered that selection may be applied to the family, as well as to the individual, and may thus gain the desired end."[40] The evolutionary underpinnings of extreme cooperation are examined next, followed by the diversity of animals that display aid-giving behavior at a surprising level (see Chapter 14).

Eusocial animals live in parent–offspring social groups in which the young are partially or fully sterile and aid their parents. In eusocial animals, the effort that the offspring might normally exert in parental care for their own young is redirected to care for siblings. These reproductively impaired animals are workers, and in many species the workers have very different physical characteristics than their reproductive parents. The definition of eusociality has been re-engineered in recent years to include animals such as thrips (small plant-eating insects), in which a semi-sterile generation cares for its own young. In fact, in primitively eusocial species, the founding generation of a colony is commonly relatively short-lived; the workers then take over, reproductively, for the founders. Primitively eusocial bees and wasps have workers that retain some reproductive potential. Eusociality differs from the helper-at-the-nest strategy discussed in Chapters 12 and 14 in the potential permanence of the reproductive state of the parents and their offspring.

In addition to having short-lived reproductives, primitively eusocial species, such as paper wasps, sweat bees, and bumblebees, have little physical differentiation between reproductives and workers and relatively small colonies. Highly eusocial species, such as honeybees, hornets, termites, and ants have workers that are well differentiated from the reproductives and tend to have large colonies (see Figure 13.12). Highly eusocial species are ecologically dominant animals in nearly all terrestrial ecosystems.[41]

The big question here is why an animal appears to behave in ways that are extremely detrimental to itself for the apparent benefit of others. All of the explanations for the evolution of cooperation—kin selection, coercion and manipulation, reciprocal altruism, and reproductive skew theory—can be extended as possible explanations for the evolution of eusociality.

As indicated, these hypotheses are not mutually exclusive, and it would be futile to look for a way to test among them. That said, this section focuses on kin selection explanations for the evolution of eusociality.

Diplodiploid Sex Determination

In most animals the receipt of sex-determining chromosomes from the parents determines the sex of offspring from a mating. In mammals, including humans, of course, the sex chromosomes have strikingly different appearance, so that one is called the X chromosome and the other the Y. XX animals are female, XY animals are male. In birds the situation is the opposite: Animals with the same sex chromosomes are male (ZZ), and those with WZ sex chromosomes are female. Other variations, such as X0, are found in some insects.

These modes of sex determination are called *diplodiploidy* because both males and females are diploid—they carry two sets of each chromosome. When sex is determined this way, the relationships among animals in a family are fairly straightforward. When unrelated parents mate, the resulting offspring, whether male or female, hold one-half ($r = 0.50$) of each of their parents' genes (see Figure 13.13).

> **KEY TERM** In diplodiploid animals, both the male and the female are diploid, meaning that each has two sets of each chromosome, one received from its mother and one from its father.

These proportions that represent average genetic identity are the r's used in Hamilton's equation. The second-generation offspring are related to their grandparents by one-fourth ($r = 0.25$), and so on. The relationships are symmetrical: gender does not change the relationship among individuals, either within a generation or between generations. Thus, in diplodiploid organisms, it is easy to calculate r among animals.

Haplodiploid Sex Determination

A less common mode of sex determination, *haplodiploidy*, has cropped up in some animals. In haplodiploid species the females have two sets of each of the chromosomes, but the males have only a single set. In other words, males are haploid and females are diploid. Males come from unfertilized eggs, so a female that has not mated still produces male offspring. The evolutionary history of

> **KEY TERM** In haplodiploid animals, one gender is haploid (one copy of each chromosome) and the other is diploid (two copies of each chromosome). In the Hymenoptera (bees, ants, and wasps), males are haploid, meaning they have no father; their single set of chromosomes comes from their mother.

FIGURE 13.12
This stingless bee species, *Tetragonisca angustula*, is common in the neotropics. It builds large nests within hollow trees; the resinous structure pictured here contains the two entrances to this colony.
Photo: Michael Breed.

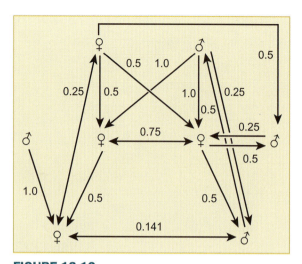

FIGURE 13.13
Relatednesses among relatives in a haplodiploid system.
The mean relationship between full sisters is .75.

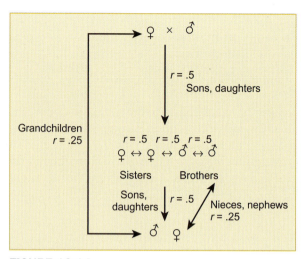

FIGURE 13.14
Relatedness within a diplodiploid family. Mean relatedness
between sibs is .5.

this peculiar adaptation is murky, but haplodiploidy seems to have evolved most commonly in insect species that nest in bark and are exposed to *Wolbachia*, a bacteria that infects the ovaries of insects. In haplodiploid insects, the brothers are related to one another by one-half (0.5) because they develop from an egg that has only one set of each chromosome.

Most importantly, a male haplodiploid insect produces sperm that are all identical, whereas a diplodiploid male produces sperm that vary genetically. In a haplodiploid male, there is only one copy of each chromosome, so meiosis and the accompanying recombination of genetic material cannot occur in males. Because a male's sperm are identical, the offspring of his matings all bear the identical genetic contribution from their father. In haplodiploids, each sperm has 100% of the father's genes. The father has only 50% of the genes of a female, that is, half of the "complete" diploid set of a female. The sperm in a haplodiploid contributes 50% of the genes to each offspring, just as in diplodiploids, but the key difference is that the sperm are all identical. Meanwhile, the contribution of the mother is the same in both haplodiploid and diplodiploid systems; each egg contains 50% of the mother's genes, and the eggs are on average 50% like each other. This yields the same proportions as diplodiploid sperm; the eggs contribute 50% of the genes to each offspring, and if the eggs come from the same mother, they are on average 50% alike (due to meiosis). So again, 0.5×0.5 (50% × 50%) means that eggs that are on average 25% alike ($r = 0.25$ when the mother's contribution alone is considered).

But what happens when the father's contribution is considered? All the sperm are genetically identical, so the average genetic similarity due to the father's contribution is $r = .50$. If we add the egg ($r = 0.25$, on average) to the sperm ($r = 0.50$), the average relatedness among sisters in haplodiploid species is $r = 0.75$! In haplodiploid species, sisters are more related to each other than to their mother or to their daughters.

In contrast, in diplodiploid organisms, the sons or daughters in monogamous families are, on average, 25% like their brothers and sisters due to genes from their mother's egg, and on average 25% like their brothers and sisters due to genes from their father's sperm. Summing the 25% average genetic similarity from the two parents shows that diplodiploid brothers and sisters are on average 50% alike genetically (Figure 13.14).

Haplodiploidy is not always necessary for the evolution of eusociality, but it seems to often prime the evolutionary pump. Some species among sponge shrimp and naked mole rats as well as all termites have all evolved eusociality but have diplodiploid sex determination. Some scientists argue that inbreeding, which raises the level of relatedness of animals within the inbred population, can create a set of circumstances that mirrors the genetic effects of haplodiploidy on relatedness. The most apparent common factor among sponge shrimp, termites, and naked mole rats is that they live in family groups in tunnel systems; this lifestyle constitutes a preadaptation for eusociality, a point that re-emerges in the next section.

This genetic reasoning about haplodiploidy and kin selection yields some tempting predictions:

1. Haplodiploid eusocial colonies should consist of a mother (the queen) and her female offspring (the workers) working together to rear more daughters of the queen.
2. The queen should mate with only one male; if she mates with more than one male, the average relatedness among her daughters will be dramatically reduced.[42,43]
3. The genetic interests of the males do not favor working in the colony to help rear either sisters or nieces; males should not work.
4. There should be conflict between the queen and the workers over whose male offspring are reared. Females are more related to their sons than to either their brothers or their nephews.
5. If workers lay male eggs, then there should be conflict among the workers concerning which male eggs are actually reared. Because genetic relationships between workers and males favor first their own sons, then their brothers, and finally their nephews, workers may prevent each other from rearing worker-laid males; if this happens, then the queen wins the conflict over laying male eggs.

DISCUSSION POINT: WHICH SIDE WOULD YOU TAKE IN A CONTROVERSY?

If you explore kin selection and social evolution on your own, you may come across some recent papers by the Harvard biologist E. O. Wilson, who has leveled attacks on kin selection as an explanation for the dynamics in social groups and the evolution of sterile workers.[44] We invite you to read these papers and then to read counterarguments such as those put forward by Foster and colleagues and Crozier.[45,46] What do you conclude from this controversy?

Meanwhile, you may wish to read more of Wilson. He is one of the most famous biologists of the twentieth century, having written numerous scientific and popular books on social insects, biological diversity, and the relationship between humans and nature. His book *Naturalist* should be required reading for those who want to be one (a naturalist). A famous, often repeated, quotation from Wilson captures his continued sense of wonder with nature: "Most children have a bug period. I never grew out of mine." Did you have a bug period? Have you grown out of it?

Preadaptions for Extreme Cooperation

Given these arguments about kin selection—and particularly with reference to haplodiploidy—is it reasonable to expect that all haplodiploid species will be eusocial and that many diplodiploid species might be eusocial as well? For eusociality to evolve, a number of pre-existing conditions need to be present. Most importantly, the species must have nesting and parental care with extended parent–offspring interactions. But nesting and parental care alone are apparently not sufficient for the evolution of eusociality; most bird species fit this requirement, yet eusociality is unknown among birds. Haplodiploidy, while not essential, is a major facilitator of eusocial evolution. The presence of parental abilities to physically, nutritionally, or chemically dominate their offspring can be necessary preadaptations for the coercion of cooperation. Ecological pressure, in the form of predators, parasites, or seasonal unavailability can be powerful evolutionary factors that tip the balance of young to stay and help their parents. This last point sets the stage for further consideration of the effects of lack of choice on the evolution of helping.

13.6 LACK OF ECOLOGICAL CHOICE IN AID-GIVING DECISIONS

Ecological conditions that make dispersal risky are perhaps undervalued by scientists, but they are nonetheless important in aid-giving decisions. These conditions could range from soil that has become dry and difficult to excavate between the time the nest was founded and the young are ready to disperse to a densely occupied habitat in which all the suitable territories are already occupied by nests. Overall, behavioral ecologists have grossly underestimated the effects of manipulation and ecological limitation of reproductive options on the evolution of cooperation. Ecological limitations enter the kin selection equation by reducing the cost of cooperation. (An animal with little reproductive options of its own stands to gain substantial inclusive fitness by working for a relative, but loses very little classical fitness by not testing its limited options.) The relative contributions of kin selection, ecological limitation, and manipulation to the evolution of extreme cooperation remain a very open question.

Anything that reduces the potential fitness of the offspring if they try to nest on their own increases the benefit they gain from helping. Similarly, parasites or predators that threaten either the parental nest or the nest that could be established by their young may create an evolutionary payoff for offspring that stay and protect the nest; group defense appears to have been an extremely important driving force in young staying in their parental nest. Factors that increase the benefit of helping in Hamilton's equation reduce the threshold required, in terms of genetic relationship between the aid-giver and the recipient, for natural selection to favor aid-giving.

FIGURE 13.15
Honey bee guards defend their nest against intruders from other honey bee colonies. In this case a marked bee from another colony is being presented at the colony entrance, and a guard is poised to attack. *Photo: James Hanken.*

Another often underestimated ecological factor in the evolution of extreme cooperation is the necessity for resource defense. In many primitively eusocial animals, the key element gained from the extreme altruists in the colony is improved colony defense. This is certainly the case in aphids, thrips, shrimp, and sweat bees, all of which face threats from parasites and predators, and all of which feature guards or soldiers in their social organization (see Figure 13.15). Nesting tends to concentrate resources including food, young animals, and the adult colony members, and the nest itself can be an important resource. If the nest becomes an ecological magnet for animals that would take over the nest, steal the food, or eat the colony members, then selection for strong self-defense can easily lead to colony members—generally daughters in haplodiploid species—that specialize in defense.

13.7 SOCIAL RECOGNITION, KIN RECOGNITION, AND COOPERATION WITH CLOSE RELATIVES

How does an animal know where to direct its cooperative behavior? Cooperation often requires that an animal can identify the targets of its behavior. Take a step back and consider how animals might recognize each other. *Social recognition*, the ability to classify or individually recognize members of a social group, works on several levels:

- *Species recognition* is the ability to discriminate between conspecifics and heterospecifics. This is particularly important in mating interactions but may play a role in other types of social behavior.
- *Gender recognition* is the ability to discriminate male from female.
- *Social group recognition* is the ability to discriminate members of one's own group (or social network) from members of other groups.

- *Individual recognition* is the ability to identify other animals as individuals and to discriminate each individual in a social group from each other individual.
- *Kin recognition* is the ability to discriminate close kin from other animals in a population or to make distinctions among kin depending on their degree of relatedness.

Social recognition may evolve as a progression from the most general categorizations (same versus different species, male versus female) to successively narrow categories until finally individual recognition is achieved.[47] In many mammals, social behavior is largely structured around individual recognition, and kin recognition plays little, if any, role in shaping behavior. In all likelihood, the primacy of individual recognition over kin recognition is true for most birds and mammals. Dogs, for example, can clearly identify other dogs as individuals and use information from social encounters to shape future encounters; it is remarkable that dogs can generalize this ability to their interactions with humans, but, of course, they do.

There are two hugely important reasons for animals to recognize their kin. First, kin recognition is the glue that binds together the kin selection and reproductive skew explanations for cooperative behavior. For aid to be directed toward kin, animals must have ways of identifying kin or of discriminating kin from non-kin. Second, most animals avoid mating with close relatives because the genetic cost of inbreeding (inbreeding depression) is potentially very high. Kin are drawn together by helping dynamics but pushed apart by mating dynamics; in either case, a system for recognizing kin is essential for making correct decisions.[48,49,50,51]

What good, then, is kin recognition if it is just another relatively crude classification of animals in a social group? Kin recognition has the special property of being generalizable to previously unmet individuals; an animal that is unfamiliar can still be discriminated as kin or non-kin. This may be particularly important in social systems of animals with discrete litters; nonlittermate siblings can avoid mating only if there is a generalizable mechanism for discriminating kin. For animals that disperse from families in which the parents will have subsequent litters, the ability to classify others as kin or non-kin can be critical in preventing catastrophically inappropriate matings. Similarly aid-giving behavior, such as antipredator alarm calls, can be directed to closely related animals if kin recognition is used.

Most kin recognition relies on a simple mechanism called *phenotype matching*. In phenotype matching, an animal learns its own phenotypic characteristics or those of surrounding animals. This usually takes place when the animal is young and nearby animals are close kin— parents and siblings. The learned phenotypes can include odors, visual cues, vocalizations, or some combination of these. The learned characteristics are called a *template*, and whenever a new animal is met, its characteristics are compared with the template.

> **KEY TERM** Phenotype matching occurs when an animal learns its own phenotype, or the phenotype of animals around it, and then uses that information to classify previously unmet individuals.

Social Recognition in Clonal Invertebrates

Some sea anemone species engage in "wars." The anemones live in clones, with new members produced by budding of old members. When one clone comes into contact with another, the anemones use their stinging capability to "fight," and ultimately one clone comes to dominate an area. This is an example of how animals that are phylogenetically far removed from, say, mammals or birds can nonetheless show well-developed mechanisms of kin discrimination and use these mechanisms to exclude unrelated individuals from habitat patches. Highly polymorphic (hypervariable) genetic systems give members of clones distinctive signatures that differ from those of other clones.[52]

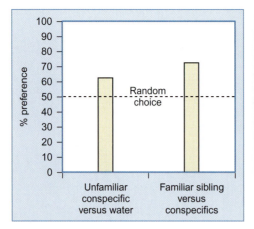

FIGURE 13.16

Preference of tadpoles for the odor in water of a conspecific versus untreated water (left column) and for the odor of a sibling versus a conspecific (right column). *Adapted from Waldman, B. 1985. Olfactory basis of kin recognition in toad tadpoles. J. Comp. Physiol. A. 156, 565–577.*

FIGURE 13.17

Schooling tadpoles. *Photo: Jeff Mitton.*

FIGURE 13.18

Preference for matching MHC genotypes in arctic charr. Fish were allowed to choose between two water sources. The left column is a control, with untreated water for both choices. The middle column shows that the odor of sibling is preferred over untreated water, and the right column shows that siblings with similar MHC alleles to the test fish are preferred over a dissimilar sibling. *Adapted from Olsen, K.H., Grahn, M., Lohm, J., Langefors, A., 1998. MHC and kin discrimination in juvenile Arctic charr, Salvelinus alpinus. Anim. Behav. 56, 319–327.*

390

Social Recognition in Fish

Many fish species use phenotype matching in their social relationships. Most studies have focused on inbreeding avoidance.[53,54,55] Salmonids (trout and salmon) may prefer the water occupied by sibs, even if they have had no prior contact with those sibs (see Figure 13.18). Perhaps schooling is, as a generality, a nonrandom phenomenon involving sibs; much more research is needed to sort this out. Olsen and colleagues genotyped fish for Major Histocompatibility (MHC) loci and tested the fishes' water preferences, showing that water that had been exposed to sibling/MHC matches was more attractive to fish.[56] Mouthbreeding cichlids use phenotype matching in kin recognition.[57]

Social Recognition in Amphibians

Amphibian kin recognition is very similar to that in fish. Many schooling frog and toad tadpoles are known to prefer to school with sibs and choose water that has been exposed to sibs, even if the sibs are not familiar to the individual doing the choosing (see Figure 3.17). Much current interest in amphibian kin recognition is related to cannibalism (see p. 391).[58,59]

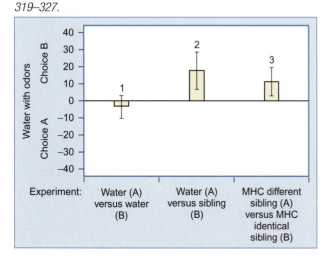

Social Recognition in Mammals

Mammalian social recognition systems seem to rely primarily on odors that reflect MHC variability although some mammals (humans are a good example) use vocalizations and visual characteristics as well (see Figure 13.18).[60,61] Most of the work on kin recognition in mammals has been done on rodents in which phenotype matching on MHC-related odors seems to be the rule, but there are also some interesting studies of primates.[62] Current studies on mammals focus largely on the role of the MHC[63] and on individual recognition as the primary method of social discrimination in carnivores, ungulates, and primates.[64]

DISCUSSION POINT: SOCIAL RECOGNITION

In a broad range of mammals, individual recognition has supplanted kin recognition as the key ingredient in social relationships. This is true of primates, hoofed animals such as horses, and carnivores such as dogs and cats. What are the evolutionary forces that favor individual recognition? How does this affect inbreeding avoidance, which remains an important factor in mate choice?

Social Recognition in Birds

Birds are unique in primarily using vocalizations rather than odor in kin recognition. It was thought for many years that birds were unlikely to have the song variability necessary for discrimination at the level needed to recognize kin, but Michael Beecher and his group led the way in showing that at least some species of birds can do exactly this.[65,66]

Price and colleagues found that some of the wren's calls are learned from older male relatives and are specific to patrilineal families.[67] Palestis and Burger presented an interesting study that suggests a visual as well as an auditory component of recognition in common terns.[68] Individually distinctive calls are significant in social recognition in long-tailed tits, which breed in Europe and Asia.[69]

Social Recognition in Insects

Most studies of insect kin recognition have centered on one of two issues: avoidance of mating with close relatives and nestmate recognition in social insects.[70] Virtually all terrestrial organisms are covered with a mixture of hydrocarbons. The waxy or oily coating of hydrocarbons keeps water in, so the organism does not dessicate. Because the blend of hydrocarbons can vary without interfering with its antidessication function, it provides a perfect tool for identifying individuals or groups of animals. Fruit flies, *Drosophila*, use cuticular hydrocarbons as signals in species recognition and in assortative mating.[71]

Nestmate discrimination in eusocial insects, including the honeybees, depends on odors that are blended among colony members, so that all colony members smell alike (see Figure 13.19). In the honeybee, the wax comb in the nest plays an intermediary role; because all bees in the nest walk around on the same comb, they all smell the same.[72] In paper wasps the nest plays a similar intermediary role. Additionally, in at least some ants, the workers' postpharyngeal glands are critical in the development of a unified chemical recognition signal in the colony.[73]

Social Recognition and Cannibalism

Dining on relatives has a very real inclusive fitness cost, perhaps even higher than if they are also potential mates. Some salamanders (the tiger salamander) and toads (the spadefoot toad) have more than one tadpole morph; most tadpoles have normal, unappealing detritivorous amphibian feeding habits, but some are carnivores.[74] In the tiger salamanders, the frequency of the carnivore morph is dependent on the density of animals relative to the food supply, with the carnivore morph developing more frequently when food per animal is low.

13.8 SOCIAL SYMBIOSES

Symbioses are intimate relationships between pairs of species (see Figure 13.20). One species lives in, on, or around the other so that the evolution of at least one of the species is affected. In *commensal relationships*, one species gains a benefit and the relationship is neutral for the other species. In *parasitism*, one species benefits to the other's detriment. In *mutualism*, both

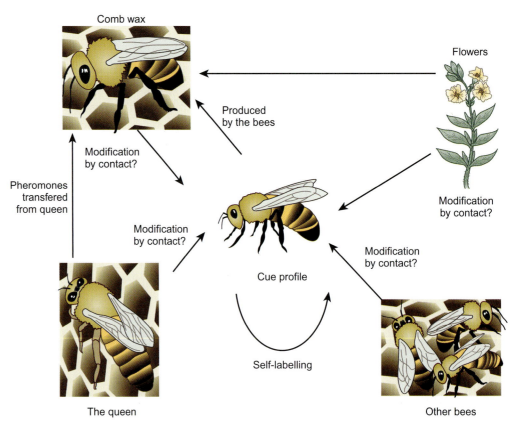

FIGURE 13.19

Nestmate recognition in honeybees. A bee's cue profile could come from flowers (upper right), other bees in the colony (lower right), the queen (lower left), the comb in the colony (upper left) or the bee itself. Systematic testing of these possibilities shows that all the bees in a colony are labeled by walking on the comb. Fatty acids, which vary in concentration from colony to colony, are transferred from the comb to the bees. *Adapted from Breed, M.D., 1998. Recognition pheromones of the honey bee. Bioscience 48, 463–470.*

392

FIGURE 13.20

Magpies on a warthog. The magpies are efficient at gleaning parasites and dead skin from the warthog, which tolerates their presence. *Photo: Michael Breed.*

species benefit. The boundaries among all of these states and parasitism and predation can be a bit fuzzy; parasites generally do not cause the immediate death of their host, whereas predators kill their prey before consuming it. In addition, typical parasites are usually smaller than their hosts and may consume host resources—an indirect kind of robbery—rather than the host itself. Indeed, all of these categories are arbitrary; for instance, sometime the "detriment" caused by a low level of parasite infection is immeasurably small and approaches commensalism. In other instances, one species of parasite can enhance resistance to another species; in that case, parasitism can rightly be called mutualism!

The focus here is on social parasitism. Brood parasitism in birds is covered in Chapter 12, and parasitism as a feeding strategy in Chapter 10. Social parasitism is yet a different type of exploitation. Social parasites take advantage of the benefits of the social group without paying any of the costs that come with membership. They use counterfeit signals or subvert the recognition system of the society to gain entrance and then take advantage of the food, shelter, or protection from predators that members of the group enjoy. Social parasites can come from the same species as the host—intraspecific social parasitism—or from another species—intraspecific social parasitism. These are insect examples, mostly ants, because they are the best-known cases of social parasitism.

Intraspecific Social Parasitism

Given the keys to the kingdom, in the form of appropriate cues for social recognition and behavior, why not use them? At least metaphorically, this is interesting question for any member of a social species. Parasitizing a social group of the same species should be relatively easy because the host and parasite share the same signaling system, and likely the only barrier to be overcome is the host's social recognition system. Social wasp queens sometimes usurp an existing queen, taking over her nest. The larval workers, which are daughters of the first queen, then end up working for the second queen.[75] The barriers to this kind of behavior are relatively low, only requiring that the workers' labor can be harnessed by the second queen.

Interspecific Social Parasitism

SOCIAL PARASITISM BY CLOSELY RELATED SPECIES

Sometimes a social parasite is from a species that is closely related to its host. This relationship confers many, perhaps all, of the tools by which conspecifics can sometimes be parasites. Particularly interesting are two kinds of bees, *Psithyrus* and *Sphecodes*, which are parasites on other closely related bees. Both parasite genera exist only as reproductives (no workers) and lack structures for carrying pollen; to reproduce, these females must find a host nest. The *Psithyrus* species parasitizes bumblebees, which are always eusocial, by entering the nest before the first workers have emerged from their pupae. The *Psithyrus* queen kills the bumblebee queen and takes over the nest. The workers rear her eggs into larvae and ultimately to adults. The key to making this work is that the bumblebee workers treat the *Psithyrus* queen and her larvae as if they belonged in the colony. In *Sphecodes* the female sneaks into a nest and lays eggs that are reared by the resident bees (see Figure 13.21). In some cases, the host species is social; in other cases, the host nest consists only of a reproductive female. (A host nest that consists of only a reproductive female is called a *solitary* species, as opposed to a eusocial one).

> **KEY TERM** A symbiosis is a nonpredatory relationship between two species.

> **KEY TERM** A commensal relationship benefits one species and is neutral for the other.

> **KEY TERM** Parasitism is beneficial to one species and detrimental to the other.

> **KEY TERM** Mutualisms are beneficial to both species in an interspecific relationship.

393

Among ants there are numerous examples of species that, in the course of evolution, have lost their worker caste and depend on gaining entry into colonies of other species, where they lay eggs that are reared by the host species. In most cases, the parasite presents a "blank slate" in terms of social recognition cues and adopts the cues of the host colony as it gains acceptance. (See Chapter 12 for a discussion of nest parasitism in birds.)

SLAVE MAKING IN ANTS

Another parasitic lifestyle that requires evasion of social recognition is *dulosis*, or *slave making*. Dulotic ants steal larvae and pupae from other colonies and bring them into their own colony, where they become part of the workforce. This requires integration of social recognition signals so the workers of the

> **KEY TERM** Dulosis, or slave making, involves capture of a brood from one ant species and the use of that brood as the workforce in the other species' colony.

FIGURE 13.21
A *Sphecodes* bee. *Photo: Jessica Lawrence, Eurofins Agroscience Services, Bugwood.org.*

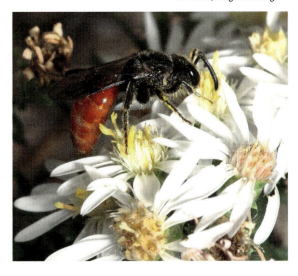

victimized species can work alongside the host species. Ants in the genus *Polyergus* are dulotic parasites on *Formica* ants; both genera are common in Europe and North America. A stream of medium-sized black or reddish ants carrying ant larvae and pupae to their nest is most likely an act of dulosis.

> **KEY TERM** Myrmecophiles are guests in ant nests that take advantage of the food, shelter, and defense provided by the ants.

MYRMECOPHILES

Myrmecophiles are completely unrelated nonsocial species, such as beetles, butterflies, or mites that gain acceptance into an ant colony and benefit from being fed by their hosts. To enter and live unscathed in their host colony, they must be either blank slates or mimics in terms of the social recognition cues they present.

SUMMARY

Animals live in groups primarily for shared protection from predators, to cooperate in discovering and hunting food, and to take advantage of public information. The simplest social groups are often families, consisting of a mother with her offspring and sometimes the father. One way of building larger social groups is by extending family groupings to include more offspring or multiple generations. Another way to obtain a larger group is to have unrelated individuals come together for the benefit of at least some group members.

Herds, flocks, and schools assemble for a variety of reasons, including protection, cooperation in foraging, and public information. Hamilton's theory of the geometry of a selfish herd explains why animals can come together in such groups even if they have no evolutionary interest in the survival of other members of the group. This reasoning explains groupings that include only animals of one species as well as groups that include two or more species.

For many groups the motivations for living together are more complicated and involve aid-giving among animals in the group. This chapter puts forward a series of ideas—group selection, kin selection transactional and social contract models, reproductive skew, delayed competition/selfish teamwork, and stolen aid—as explanations for living in groups. A key concept is that that kinship structure—the web of genetic relationships among animals in such groups—is often important in determining cooperative and competitive interactions within groups. Sometimes groups consist of two or more species. Mutual benefits from the association, exploitation of public information, or parasitism underlie these types of social groups.

Extreme cooperation, or eusociality, in which some group members give up most or all of their own reproduction in order to maximize aid given to other group members is known in kin groups. Kin selection, parental manipulation, and ecological limitations on dispersal all play key roles in the evolution of eusociality. Asymmetric genetic relationships resulting from haplodiploid sex determination set the stage for eusociality; most but not all eusocial species are haplodiploid. Eusocial groups gain efficiency from dividing the labor among colony members so that the group has an advantage over solitary animals that might compete for the same resource.

Social recognition is key to formation and maintenance of many social groups. Social recognition can include recognition by classification, individual recognition, and kin recognition. Because kin groups often form the basis for social behavior, kin recognition, the ability to discriminate close kin from other animals in the population, is an important mechanism known in a variety of animals, from anemones to mammals.

STUDY QUESTIONS

1. How do kin selection and parental manipulation differ as explanations for the evolution of helping behavior?
2. What is phenotype matching? Design a simple experiment that would determine if a species is capable of phenotype matching.

3. Why are the workers in ant, bee, and wasp colonies almost always females?

4. Why might amphibians and fish school in sibling groups? What are some tests of this hypothesis?

5. What are the potential fitness costs and benefits to an individual who "helps at the nest"? Discuss how the various evolutionary models (kin selection, parental manipulation, reciprocal altruism, selfish teams) might aid in explaining helping behavior in young but potentially reproductive animals. How do you think these evolutionary models interact with proximate ecological forces, such as nest site availability and food availability?

6. "Selfish herds" could form in a variety of contexts, such as flocks or herds of migrating animals, feeding aggregations, and mating aggregations. But there may be other evolutionary dynamics that cause animals to group together. If you observe a herd, flock, or aggregation of animals, what alternative hypotheses for their grouping behavior should you consider? How could you test between the selfish herd hypothesis and the alternatives?

Further Reading

Axelrod, R., 1985. The Evolution of Cooperation. Basic Books, New York, 241 pp.

Clutton-Brock, T., 2009. Cooperation between non-kin in animal societies. Nature 462, 51–57.

Costa, J.T., 2010. Social Evolution in 'Other' Insects and Arachnids, In: Michael D. Breed and Janice Moore (Eds.), Encyclopedia of Animal Behavior, Academic Press, Oxford, Vol. 3, 231–241.

Dawkins, R., 1976. The Selfish Gene. Oxford University Press, Oxford. 352 pp.

Linksvayer, T.A., 2010. Subsociality and the Evolution of Eusociality, In: Michael D. Breed and Janice Moore (Eds.), Encyclopedia of Animal Behavior, Academic Press, Oxford, Vol. 3, 358–362.

Queller, D.C., 2010. Kin Selection and Relatedness, In: Michael D. Breed and Janice Moore (Eds.), Encyclopedia of Animal Behavior, Academic Press, Oxford, Vol. 2, 247–252.

Wilson, E.O., 1971. The Insect Societies. Belknap Press of Harvard University Press.

Wilson, E.O., 1975. Sociobiology: The New Synthesis. Harvard University Press.

395

Notes

1. Wilson, E.O., 1975. Sociobiology: The New Synthesis, Harvard University Press.

2. Hamilton, W.D., 1964a,b. The genetical evolution of social behaviour. I, II. J. theor. Biol. 7, 1–52.
   Hamilton, W.D., 1967. Extraordinary sex ratios. Science 156, 477–488.
   Hamilton, W.D., 1971. Selection of selfish and altruistic behavior in some extreme models. In J.F. Eisenberg & W.S. Sillon (Eds.), Man and beast: Comparative social behavior. Washington, DC: Smithsonian Institution Press. pp 59–91.

3. Trivers, R.L., 1971. The evolution of reciprocal altruism. Quart. Rev. Biol. 46, 35–57.
   Wilkerson, G.S., 1984. Reciprocal food sharing in the vampire bat. Nature 308, 181–184.

4. Wey, T., Blumstein, D.T., Shen, W., Jordán, F., 2008. Social network analysis of animal behaviour: a promising tool for the study of sociality. Anim. Behav. 75, 333–344.

5. Croft, D.P., James, R., Krause, J., 2008. Exploring Animal Social Networks. Princeton University Press, 208 pp.

6. Whitehead, H., 2008. Analyzing Animal Societies: Quantitative Methods for Vertebrate Social Analysis. University of Chicago Press, 320 pp.

7. Faria, J.J., Dyer, J.R.G., Tosh, C.R., Krause, J., 2010. Leadership and social information use in human crowds. Anim. Behav. 79, 895–901.

8. Strassmann, J.E., Zhu, Y., Quellar, D.C., 2000. Altruism and social cheating in the social amoeba *Dictyostelium discoideum*. Nature 408, 965–967.

9. Clobert, J., McAdam, A.G., Corrigan, G., Alonzo, S., Lancaster, L., Chaine, A., et al. 2006. Self-recognition, color signals and cycles of greenbeard mutualism and altruism. PNAS 103, 7372–7377.

10. Axelrod, R., Hamilton, W.D., 1981. The evolution of cooperation. Science 211 (4489), 1390–1396.
    Hamilton, W.D., 1971. Selection of selfish and altruistic behavior in some extreme models. In J.F. Eisenberg & W.S. Sillon (Eds.), Man and beast: Comparative social behavior. Washington, DC: Smithsonian Institution Press. pp 59–91.
    Hamilton, W.D., Zuk, M., 1982. Heritable true fitness and bright birds: a role for parasites? Science 218, 384–387.
    Hamilton, W.D., May, R.M., 1977. Dispersal in stable habitats. Nature 269, 578–581.
    Hamilton, W.D., 1964. The evolution of social behavior I. J. Theor. Biol. 7, 1–16.
    Hamilton, W.D., 1964. The evolution of social behavior II. J. Theor. Biol. 7, 17–52.

Hamilton, W.D., 1967. Extraordinary sex ratios. Science 156, 477–488.

Hamilton, W.D., 1971. Geometry for the selfish herd. J. Theor. Biol., 31, 295–311.

Hamilton, W.D., Zuk, M., 1982. Heritable true fitness and bright birds: a role for parasites? Science 218, 384–387.

Hamilton, W.D., Axelrod, R., Tanese, R., 1990. Sexual reproduction as an adaptation to resist parasites (a review). Proc. Nat. Acad. Sci. USA 87 (9), 3566–3573.

11. Creel, S., Winnie Jr., J.A., 2005. Responses of elk herd size to fine-scale spatial and temporal variation in the risk of predation by wolves. 69, 1181–1189.

12. South, J.M., Pruett-Jones, S., 2000. Patterns of flock size, diet, and vigilance of naturalized Monk Parakeets in Hyde Park, Chicago. Condor 102 (4), 848–854.

13. Tringa-Tetanus Cresswell, W., 1994. Flocking is an effective anti-predation strategy in redshanks. Anim. Behav. 47 (2), 433–442.

14. Green, E., 1987. Individuals in an osprey colony discriminate between high and low quality information. Nature 329, 239–241.

15. Charnov, E.L., Krebs, J.R., 1975. Evolution of alarm calls—altruism or manipulation. Am. Nat. 109, 107–112.

16. Van Horn, R.C., Engh, A.L., Scribner, K.R., Funk, S.M., Holekamp, K.E., 2004. Behavioural structuring of relatedness in the spotted hyena (*Crocuta crocuta*) suggests direct fitness benefits of clan-level cooperation. Mol. Ecol. 13, 449–458.

17. Lima, S.L., Bednekoff, P. A., 1999. Back to the basics of antipredatory vigilance: can nonvigilant animals detect attack? Anim. Behav. 58, 537–543.

18. Janson, C.H., Goldsmith, M.L., 1995. Predicting group size in primates: foraging costs and predation risks. Behav. Ecol. 6, 326–336.

19. Blumstein, D.T., Evans, C.S., Daniel, J.C., 1999. An experimental study of behavioural group size effects in Tammar Wallabies, *Macropus eugenii*. Anim. Behav. 58, 351–360.

20. Grand, T.C., Dill, L.M., 1999. The effect of group size on the foraging behaviour of juvenile coho salmon: reduction of predation risk or increased competition? Anim. Behav. 58, 443–451.

21. Pollard, K.A., Blumstein, D.T., 2008. Time allocation and the evolution of group size. Anim. Behav. 76, 1683–1699.

22. Koenig, W.D., 1981. Reproductive success, group size, and the evolution of cooperative breeding in the acorn woodpecker. Am. Nat. 117, 421–443.

23. Bergman, T.J., Beehner, J.C., Cheney, C.L., Seyfarth, R.M., 2003. Hierarchical classification by rank and kinship in baboons. Science 302, 1234–1236.

24. Honek, A., Martinkova, Z., Pekar, S., 2007. Aggregation characteristics of three species of Coccinellidae (Coleoptera) at hibernation sites. Eur. J. Entomol. 104, 51–56.

25. Nalepa, C.A., Kidd, K.A., Ahlstrom, K.R., 1996. Biology of *Harmonia axyridis* (Coleoptera: Coccinellidae) in winter aggregations. Ann. Entomol. Soc. Am. 89, 681–685.

26. Koch, R.L., 2003. The multicolored Asian lady beetle, *Harmonia axyridis*: A review of its biology, uses in biological control, and non-target impacts. J. Insect Sci. 3, 32.

27. Johnstone, R.A., 2000. Models of reproductive skew: a review and synthesis. Ethology 106, 5–26.

28. Scott, M.P., Lee, W.J., van der Reijden, E.D., 2007. The frequency and fitness consequences of communal breeding in a natural population of burying beetles: a test of reproductive skew. Ecol. Entomol. 32 (6), 651–661.

29. Wynne-Edwards, V.C., 1962. Animal Dispersion in Relation to Social Behavior. Oliver & Boyd, London.
Wynne-Edwards, V.C., 1986. Evolution through Group Selection. Blackwell Scientific, Oxford, 340 pp.

30. Dawkins, R., 1976. The Selfish Gene. Oxford University Press, Oxford.

31. Wilson, D.S., Wilson, E.O., 2008. Evolution "for the good of the group." Am. Sci. 96, 380–389.

32. Wilson, D.S., Wilson, E.O., 2007. Rethinking the theoretical foundation of sociobiology. Q. Rev. Biol. 82, 327–348.

33. Hauser, M., McAuliffe, K., Blake, P.R., 2009. Evolving the ingredients for reciprocity and spite. Philos. Trans. R. Soc. Lond., B, Biol. Sci. 364, 3255–3266.

34. Clutton-Brock, T., 2009. Cooperation between non-kin in animal societies. Nature 462, 51–57.

35. Bourke, A.F.G., Heinze, J., 1994. The ecology of communal breeding: the case of multiple-queen leptothoracine ants. Philos. Trans. R. Soc. (Lond), Series B 345, 359–372.

36. Clutton-Brock, T., 2009. Cooperation between non-kin in animal societies. Nature 462, 51–57.

37. Liebig, J, Monnin, T., Turillazzii, S., 2005. Direct assessment of queen quality and lack of worker suppression in a paper wasp. Proc. R. Soc. B 272, 1339–1344.

38. Nonacs, P., 2006. Nepotism and brood reliability in the suppression of worker reproduction in the eusocial Hymenoptera. Biol. Lett. 2, 577–579.

39. Crozier, R.H., 2008. Advanced eusociality, kin selection and male haploidy. Aust. J. Entomol. 47, 2–8.

40. Darwin, C., 1872. The origin of species by the means of natural selection. Chapter 8. Available online at http://www.classicreader.com/book/107/59/

41. Wilson, E.O., 1971. The Insect Societies. Belknap Press of Harvard University Press.

42. Cole, B.J., 1983. Multiple mating and the evolution of social behavior in the Hymenoptera. Behav. Ecol. Sociobiol. 12, 191–201.

43. Sherman, P.W., Seeley, T.D., Reeve, H.K., 1988. Parasites, pathogens and polyandry in social Hymenoptera. Am. Nat. 131, 602–610.

44. Wilson, E.O., Hölldobler, B., 2005. Eusociality: origin and consequences. Proc. Natl. Acad. Sci. U.S.A. 102, 13367–13371.
Wilson, E.O., 2005. Kin selection as the key to altruism: its rise and fall. Soc. Res. 72, 159–168.

45. Foster, K.R., Wenseleers, T., Ratnieks, F.L.W., 2006. Kin selection is the key to altruism. Trends Ecol. Evol. 21, 57–60.

46. Crozier, R.H., 2008. Advanced eusociality, kin selection and male haploidy. Aust. J. Entomol. 47, 2–8.

47. Tibbets, E.A., Dale, J., 2007. Individual recognition: it is good to be different. Tree 22, 529–537.

48. Lacy, R.C., Sherman, P.W., 1983. Kin recognition by phenotype matching. Am. Nat. 121, 489–512.

49. Solomon, N., 1991. Current indirect fitness benefits associated with philopatry in juvenile prairie voles. Behav. Ecol. Sociobiol. 29, 277–282.

50. Sun, L., Muller-Schwarze, D., 1997. Sibling recognition in the beaver: a field test for phenotype matching. Anim. Behav. 54, 493–502.

51. Mateo, J.M., 2003. Recognition in ground squirrels and other rodents. J. Mammal. 84, 1163–1181.

52. Hart, M.W., Grosberg, R.K., 1999. Kin interactions in a colonial hydrozoan (*Hydractinia symbiolongicarpus*): population structure on a mobile landscape. Evolution 53, 793–805.

53. Gerlach, G., Lysiak, N., 2006. Kin recognition and inbreeding avoidance in zebrafish, *Danio rerio*, is based on phenotype matching. Anim. Behav. 71, 1371–1377.

54. Ward, A.J.W., Webster, M.M., Hart, P.J.B., 2007. Social recognition in wild fish populations. Proc. R. Soc. B–Biol. Sci. 274, 1071–1077.

55. HainT., Neff, B., 2009. Promiscuity drives self-referent kin recognition. Curr. Biol. 16, 1807–1811.

56. Olsen, K.H., Grahn, M., Lohm, J., 2002. Influence of MHC on sibling discrimination in Arctic char, *Salvelinus alpinus* (L.). J. Chem. Ecol. 28, 783–795.
Olsen, K.H., Grahn, M., Lohm, J., Langefors, A., 1998. MHC and kin discrimination in juvenile Arctic charr, *Salvelinus alpinus*. Anim. Behav. 56, 319–327.

57. Le Vin, A.L., Mable, B.K., Arnold, K.E., 2010. Kin recognition via phenotype matching in a cooperatively breeding cichlid, *Neolamprologus pulcher*. Anim. Behav. 79, 1109–1114.

58. Blaustein, A.R., Waldman, B., 1992. Kin recognition in anuran amphibians. Anim. Behav. 44, 207–221.

59. Waldman, B., 1987. Mechanisms of kin recognition. J. Theor. Biol. 128, 159–185.

60. Heth, G., Todrank, J., Johnston, R. E., 1998. Kin recognition in golden hamsters: evidence for phenotype matching. Anim. Behav. 56, 409–417.

61. Ehman, K.D., Scott, M.E., 2001. Urinary odour preferences of MHC congenic female mice, *Mus domesticus*: implications for kin recognition and detection of parasitized males. Anim. Behav. 62, 781–789.

62. Yamazaki, K., Beauchamp, G.K., Curran, M., et al., 2000. Parent-progeny recognition as a function of MHC odortype identity. Proc. Natl. Acad. Sci. U.S.A. 97, 10500–10502.

63. Penn, D.J., Potts, W.K., 1999. The evolution of mating preferences and Major Histocompatibility Complex genes. Am. Nat. 153, 145–164.

64. Holekamp, K.E., Boydston, E.E., Szykman, M., Graham, I., Nutt, K.J., Birch, S., et al. 1999. Vocal recognition in the spotted hyaena and its possible implications regarding the evolution of intelligence. Anim. Behav. 58, 383–395.

65. Beecher, M.D., 1990. The evolution of parent-offspring recognition in swallows. In: Dewsbury, D. A. (Ed.), Contemporary Issues in Comparative Psychology. Sinauer, pp. 360–380.

66. Medvin, M.B., Beecher, M.D., 1986. Parent-offspring recognition in the barn swallow (*Hirundo rustica*). Anim. Behav. 34, 1627–1639.

67. Price, J.L., 1999. Recognition of family-specific calls in stripe-backed wrens. Anim. Behav. 57, 483–492.

68. Palestis, B.G., Burger, J., 1999. Individual sibling recognition in experimental broods of common tern chicks. Anim. Behav. 58, 375–381.

69. Sharp, S.P., Hatchwell, B.J., 2005. Individuality in the contact calls of cooperatively breeding long-tailed tits (*Aegithalos caudatus*). Behaviour 142, 1559–1575.

70. Starks, P.T.B., Fischer, D.J., Watson, R.E., Melikian, G.L., Nath, S.D., 1998. Context-dependent nestmate discrimination in the paper wasp, *Polistes dominulus*: a critical test of the optimal acceptance threshold model. Anim. Behav. 56, 449–458.

71. Speiss, E.B., 1987. Discrimination among prospective mates in *Drosophila*. In: Fletcher, D.J.C., Michener, C.D. (Eds.), Kin Recognition in Animals. John Wiley and Sons, Chichester, pp. 75–119.

72. Breed, M.D., 1998. Recognition pheromones of the honey bee. Bioscience 48, 463–470.

73. Soroker, V., Vienne, C., Hefetz, A., Nowbahari, E., 1994. The postpharyngeal gland as a gestalt organ for nestmate recognition in the ant *Cataglyphis niger*. Naturwissenschaften 81, 510–513.

74. Pfennig, D.W., Reeve, H.K., Sherman, P.W., 1993. Kin recognition and cannibalism in spadefoot toad tadpoles. Anim. Behav. 46, 87–94.

75. Klahn, J., 1988. Intraspecific comb usurpation in the social wasp *Polistes fuscatus*. Behav. Ecol. Sociobiol. 23, 1–8.

CHAPTER 14

Comparative Social Behavior

LEARNING OBJECTIVES

Studying this chapter should provide you with the knowledge to:

- See how the evolutionary theories developed in Chapter 13 play out in the social systems of birds, mammals, and invertebrates.

- Understand division of labor in colonies of eusocial animals and the contribution of division of labor to the dominant ecological roles of eusocial animals.

- Have a full knowledge of the methods by which reproductive differences between queens and workers are maintained.

- Learn that colonies of eusocials have novel and highly effective defensive strategies.

- Master the breadth of ecological roles and behavioral adaptations observed in eusocial animals.

14.1 INTRODUCTION

In this chapter prepare to delight in the broad sweep of social behavior, beginning with vertebrate societies and continuing to social invertebrates. The discussion builds from the simple groups such as herds, flocks, and schools discussed in Chapter 13; here, the focus is on more complex types of cooperation. It is easy to cloak social behavior in such a thick theoretical blanket that appreciation of its extraordinary nature all but disappears. This chapter highlights some outstanding examples, such as swallows, lions, naked mole-rats, honeybees, and leafcutter ants. Feel free to skip ahead to these fun pieces and then return to the theoretical layer—perhaps with renewed enthusiasm!

As established in Chapter 13, membership in a group has both benefits and costs. Oftentimes the major expense of group membership is a reduction in reproduction; this can be temporary, as can happen when a subadult animal remains in its family group, or it can be permanent, as in workers in eusocial colonies. The benefits come from access to food, shelter, group defense, and perhaps from the possibility that the animal will "inherit" a resource owned by the group. Much of the challenge of studying social behavior lies in assessing these costs and benefits; the behavioral differences among group members are one of the major focal points in the study of sociality. These differences are often subtle and require many hours of observation to document; they then can be very difficult to interpret.

This cost-benefit approach provides a common theme through social systems that otherwise might seem very different from one another. For instance, what could a worker army ant have in common with a lioness? The answer is that natural selection works with costs and benefits in both species. The theories presented in Chapter 13 apply to how social behavior evolves in any species, albeit with outcomes that can seem vastly different. An understanding of the costs and potential benefits of sociality in a species allows the development of hypotheses about the evolution of that species' social behavior.

A secondary theme that ties social groups together is ecological efficiency. Animals in social groups, working together effectively by coordinating their roles, have distinct advantages over nonsocial animals. Sometimes this efficiency is fairly obvious, as when a pack of wolves dines on an herbivore that would have been an impossible victim for a lone wolf. Other instances, such as very small leafcutter ants riding on leaves carried by larger ants, may initially be puzzling. The efficiency gained from this seemingly lazy behavior becomes apparent only when a parasitic fly arrives and the small ant drives it away.

400

CASE STUDY

Leafcutter Ants and Their Fungus Gardens

Leafcutter ants are herbivorous; their diet consists of pieces of leaf cut from living plants and carried back to their nest. As you probably know, a completely leaf-based diet presents nutritional challenges because it is low in nitrogen and high in undigestible cellulose. To be able to survive on this diet, leafcutter ants depend on symbiotic fungi; they feed their "fungus garden" the pieces of leaf and later eat parts of the fungus. Leafcutter colonies can be huge—more than a million workers—and, as we have seen, have a highly elaborate system for division of labor.

Wilson's detailed analysis of the behavior of leafcutter ant workers shows how size and shape can be well matched with task.[1] Leafcutter ants build immense underground nests (see Figure 14.1) occupied by a single queen and hundreds of thousands or millions of workers. Leafcutters are agriculturalists;[2] they grow fungus gardens and live by eating the fungus. To support the fungi, they harvest leaves from plants, usually trees, in the area surrounding the colony. Processing the leaves involves a series of workers, starting with large-jawed, large-headed ants called *majors*, whose task is to cut semicircular chunks of leaf from the tree. From there, the leaf piece is passed to a medium-sized worker, who carries it, parasol-like, back to the nest. Within the nest, even smaller workers process the leaf into the fungus garden. Each task is performed by workers that are well matched to the demands of the task. In some species, the workers with the largest heads are

(Continued)

CASE STUDY (CONTINUED)

FIGURE 14.1
Leatcutter ants cut pieces of leaves from living plants and carry them to their nest, where they are used to grow fungus gardens. Different-sized ants perform different tasks in this effort, forming the ant equivalent of a "bucket brigade" in their efforts to collect leaves. *Inset Photo: Scott Bauer, USDA-ARS.*

soldiers, active in colony defense and territorial defense, while in leafcutting ants the large-headed workers use their jaws to cut leaves for the colony.[3]

How do leafcutter ants start their fungus garden? As you might guess, a randomly chosen fungus from the environment will not do. In fact, leafcutters are co-adapted with particular kinds of basidomycete (mushroom producing) fungi. When a new queen is produced, she takes a bit of fungus and stores it in a pocket in her oral cavity. When she digs a new nest, she starts the fungus growing by feeding it glandular secretions; her first worker offspring then start collecting leaves to feed to the fungus. One of the challenges of fungus gardening is keeping unwanted fungi and micro-organisms from invading the garden. Leafcutters have an interesting array of behavioral and antimicrobial adaptations to prevent disease and "weeds" from invading.[4]

On the other hand, *social conflict* over allocation of resources or access to reproduction within a group can easily outweigh efficiency gained from division of labor, communication, and protection from predators. Among the animals discussed in this chapter, wild horses are a particularly good example of groups that persist despite very costly conflict. The importance of this balance between conflict and cooperation always needs to be evaluated in analyses of social behavior.

> **KEY TERM** Social conflict occurs when members of a social group don't benefit equally from an outcome.

The behavior captured in this chapter represents, in terms of complexity and sophistication, the culmination of all of the mechanisms and adaptations discussed to this point. Although humans are not discussed until the end of this chapter, the behavioral principles that precede that discussion set the stage for thinking about how human social behavior fits into the larger picture of behavioral evolution and adaptation. Proceeding through this chapter, reflect on how behavioral genetics, learning, communication, cognition, mating systems, and parental care all coalesce when animals become social.

14.2 VERTEBRATE SOCIAL SYSTEMS

This section highlights some examples of social behavior in birds and mammals. The evolutionary pathways to these two major vertebrate groups diverged early, and social evolution proceeded independently in the two groups. Recall from Chapter 11 on mating that most bird species are socially monogamous, whereas mammals are often polygynous. This difference alone appears to make large social groups based on extended families less common among birds than mammals.

Also keep in mind ecological effects on dispersal choices (Chapter 13) while thinking about bird and mammal sociality. If nesting sites or territories are in short supply, then young animals may face a choice between near certain failure in independent life and continued life with their family group. Added to this are the benefits gained by social groups in foraging and defense. Three bird species—cliff swallows, acorn woodpeckers, and sociable weavers—and four mammal species—wild horses, hyenas, lions, and baboons—are highlighted here; they are a small, but representative, sample of the range of social behavior in birds and mammals.

Swallows

Swallows are small insect-eating birds; of the several common species (barn, cave, tree, bank, and cliff), the focus here is on the barn swallow. The barn swallow is widely distributed in Europe and North America, and the males are dark iridescent blue with a light-colored underside and a rusty colored band under their chin. Barn swallows construct their nests using mud; the cup shaped structure protects a brood of two to seven eggs. Barn swallows are monogamous, with both the male and the female contributing to care of the brood. Male ornaments, particularly their elongated tails, play a role in mate choice. Some extrapair copulations take place and, depending on the study population, up to 30–40% of the offspring may be sired by a male from outside the pair bond.

Among the bird species highlighted here, swallows are the least social, but they exemplify the level of sociality observed in many bird species. Barn swallows are opportunistically social, so occasionally solitary nests occur but often they live in colonies ranging from a few nests to several dozen nests (see Figure 14.2). This aggregated nesting (see Chapter 13) stems largely from mutual attraction to suitable nesting habitat and perhaps from some shared advantage in vigilance for predators. Predators on barn swallows include hawks and owls, as well as mammals such as raccoons and weasels. Within a colony, nesting pairs are territorial, and there is no evidence of cooperative construction or brood care. Colony members must be able to identify their own young using calls so that their parental care is directed to the correct nest. Colonial nesting in barn swallows may be aggregation in good habit, but may also allow shared vigilance and defense against predators.

Sociable Weavers

Sociable weavers are small, relatively nondescript, tan-colored birds found in the deserts of southern Africa. Sociable weavers primarily eat insects, diet items that fluctuate considerably in abundance along with rainfall in the region. They are unique among birds in cooperatively constructing a single large nest that houses up to 100 nesting chambers. Cooperation extends to brood care. *Helpers* in colonies are young adults that lack a nest and have no young of their own; they forage and feed nestlings belonging to their parents. Helping may even extend to cooperative brood care for unrelated chicks. Up to nine such helpers can be associated with a single nest, so a colony may include several hundred adult birds, many more than two per nesting chamber.

> **KEY TERM** A helper is an animal that has the capability of reproducing but instead remains with its family group and aids the group in foraging and defense, or plays other roles that benefit the group.

Desert habitats can be extremely hot during the day and equally cold at night, and the Kalahari is no exception. The large nest provides shelter from searing and freezing desert temperatures (see Figure 14.3); the birds thermoregulate the structure so that the maximum temperature inside is 30 °C and the minimum is 15 °C. The main benefit of this cooperation probably comes from protection from cold temperatures; heat production is shared across all the birds in the colony, and the large nest provides insulation from the cold.

FIGURE 14.2
Left: A barn swallow nest, showing an adult bird provisioning her young. Many of these nests may occur within the same barn. Right: A cliff swallow, which also nests in aggregations. *Photos: Matt Wilkins (left) and Ben Pless (right).*

The value of social behavior for this species, then, comes from shared thermoregulation of the nest and shared defense, particularly against venomous snakes, such as cobras and boomslangs that are common in this habitat.[5,6]

Acorn Woodpeckers

Acorn woodpeckers depend on large dead trees for their survival; Walter Koenig and his students[7] noted that the woodpeckers store acorns in their tree and use a larger cavity in the tree for their nest. Such trees can be a relatively rare resource in this bird's native habitat, the oak scrubland and forests of the arid southwest, including parts of Arizona and New Mexico, California, and much of Mexico. In addition to these large dead trees, acorn woodpeckers obviously require the presence of oak trees as a source of their primary food, acorns. Acorn abundance varies greatly from year to year. In a good year, oak trees "mast" with all the trees in a population, producing copious acorns; in a bad year, acorn production is low for all trees. Acorn woodpeckers store thousands of acorns, each in its own little niche carved into the surface of the trunk of their "granary" tree.

The value of the granary tree and the likelihood of failure as a dispersing bird seeks to find and establish a new granary are probably the primary pressures leading to social behavior in the acorn woodpecker. A granary tree may have several breeding males and females, along with subadult helpers; as many as 10–15 birds, plus any nestlings present, can occupy the tree. Both males and females serve as helpers. All eggs are laid in the single nest cavity (see Figure 14.4), and females may destroy each other's eggs. Reproduction is fairly evenly divided among female egg-layers but is typically dominated by one male from among the reproductive males.

403

FIGURE 14.3
Sociable weaver colony and birds. Dozens or hundreds of individual nests are within the colony. *Photo: Kenneth M. Gale.*

FIGURE 14.4
Acorn woodpecker (left) and tree with cached acorns (right). *Left Photo: Gary Kramer, US Fish and Wildlife Service. Right Photo: Tupper Ansel Blake, US Fish and Wildlife Service.*

Helpers could be viewed as waiting to fill a vacancy among the reproductives. However, unrelated birds from outside the group usually fill vacancies, even if one occurs when a helper is present that is the same sex as the vanished bird. This prevents inbreeding and also highlights the complexities of the benefits of helping. In this species, cooperation results in the rearing of nestlings that are related to the helper, supporting the idea that kin selection, combined with lack of ecological choice, are important dynamics in this system.

BRINGING ANIMAL BEHAVIOR HOME: WILD AND DOMESTIC HORSES

Domestic horses derive from smaller ancestor species found in the Americas. They are part of a larger evolutionary radiation of species in the genus *Equus* that includes zebras, donkeys, and asses. The species that gave rise to our domestic horses, *Equus caballus*, went extinct in the Americas between 12,000 and 10,000 years ago and was reintroduced as a domestic animal by Europeans in the 1500s. (The conservation of Przewalski's horse, an endangered subspecies of *Equus caballus* found in Mongolia, is discussed in detail in Chapter 15.) Horses soon escaped and formed the foundation for wild horses, which were once numerous in North America.

Wild horses are now reduced to a smaller number of bands (see Figure 14.5). They have a harem social system, in which a number of mares affiliate with a dominant stallion. In some larger bands, up to five stallions are present, and the band is then subdivided according to the stallion with which each mare is affiliated. While bands move within a familiar home range, they are not territorial, and more than one band may occupy the same general area. In general, grazing herbivores are not territorial, because their resource is widespread and difficult—perhaps even pointless—to defend.

Much of the motivation for membership in bands is reproductive; stallions and mares form long-term associations. For mares, one of the benefits of consorting with a stallion is protection from aggressive harassment by other stallions. Multistallion bands may be best explained by subordinate stallions grouping around a dominant stallion for potential future reproductive benefit, even though the cost of dominance interactions is high. Bands also benefit from public information about location of food and water resources, and from shared vigilance and defense against predators.[8]

Domestic horse behavior reflects the fact that their ancestors were wild, and pastured domestic horses form bands and display combat among stallions. A stallion kept in a stall becomes formidable when the scent of a mare in season is in the air. Horses are highly social animals, and maintaining domestic horses in herds of mares and geldings is a key component of the welfare of these beautiful animals.

FIGURE 14.5
Wild horses. Herds like this are common in the western United States. *Photo: US Department of Interior, Bureau of Land Management.*

Hyenas

There are actually three species of hyena: brown, striped, and spotted (see Figure 14.6). The spotted hyena has the most unusual social behavior of the three. (By the way, hyenas are not closely related to dogs; they are a distinct group within the carnivores.)

Spotted hyenas are highly effective and dangerous predators, and their social behavior has evolved in parallel with a shift from the scavenging habits of their relatives to predation. Perhaps because competition for meat from prey is so high, there is a premium on social aggressiveness in this species. Among newborns, competition for dominance status is quite important, and adult dominance status is determined by a combination of competition among sibs and their mother's status. Spotted hyena females have masculinized genitalia, with a large clitoris that looks like a penis. The canal of the clitoris serves for both copulation and, one might imagine, very painful birth.

Spotted hyenas live in clans consisting of a half dozen to several dozen adult animals, plus their offspring, so that several generations can occur together in a clan. Social structure within the clan is driven by the females, which are organized into a linear dominance hierarchy. Clans fight for territory, and females remain in their clan for life; males can change clans, a behavior that reduces inbreeding. Females communally nurse the young of the clan.

The social structure of hyenas has been characterized as *fission–fusion*, in that small groups may at times break off from the larger clan (fission) and then later rejoin (fusion). This social structure is similar to that of African elephants. Fission and fusion represent a balance between the benefits of the social group (cooperative hunting, defense, and broodcare) and the possibility that food may be too scarce to support a large group.[9,10]

FIGURE 14.6
A pack of spotted hyenas at a kill. *Photo: Kay E. Holekamp.*

KEY TERM Fission–fusion societies are large groups of animals that at times are grouped together (fusion) and at other times split into smaller groups (fission). Fusion may occur when resources are abundant and fission when resources are scarce.

405

Lions

The foundation of a lion pride is a core group of up to 15 females, usually mothers, daughters, sisters, and nieces, which hunt together and cooperatively raise their young (see Figure 14.7). One or a few males associate with the pride. The males participate in pride defense, but not in hunting. Because the females are a kin group, kin selection likely plays a role in the cohesiveness of the group and in cooperative behavior. Prides are territorial, and the ability to hold a feeding territory is key to the pride's survival. Some lionesses are more effective at hunting, group defense, and territoriality; these lionesses are termed *leaders*. *Laggards* are lionesses that share less equally in the tasks necessary for the pride's survival.

A lion pride is a closed social group, with outside males and females being excluded by pride members. As male cubs approach adulthood, they are expelled from the pride. Prides undergo a fission–fusion process like that of hyenas, in which pride size adjusts according to environmental conditions.

Solo males have low survivorship due largely to their poor ability to obtain food, and males may form

FIGURE 14.7
A pair of lions. A pride typically consists of a male and female and their female offspring. *Photo: Michael Breed.*

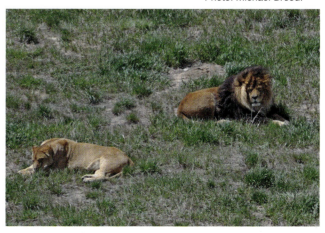

bachelor coalitions. These coalitions improve male survivorship, and ultimately a male coalition may take over a pride, driving the resident males away. Lionesses cooperate to defend their young against infanticide; cubs are particularly at risk when a new coalition of males takes over a pride.[11,12,13,14]

Chimpanzees

Chimpanzee social behavior and ecology have been extensively studied, fascinating the public because of the close evolutionary relationship between chimpanzees and humans.[15,16]

Chimpanzee social behavior varies among populations, and intriguing behavior, such as tool use in food collection, may be seen in some locations but not in others.

The pattern of fission–fusion social groups that occurs in hyenas and lions remains a theme for chimpanzee social organization. Chimpanzees live in loosely organized social groups of 20 to well over 100 animals (see Figure 14.8). Smaller subgroups derived from the community are called *parties*. Males form strong coalitions with other males and tend to stick with their subgroup through fission–fusion processes. Females are more often loners, often moving on their own or with their infants.

Male chimpanzees usually remain in their community, whereas females are likely to disperse to other communities. In this way, chimpanzees are unlike some

FIGURE 14.8
Chimpanzees showing social grooming. Cooperative behaviors like this provide opportunities for reciprocity. *Photo: John Mitani.*

of the other social systems in this text. Males within a community are likely to be related, whereas females are less likely related. Chimpanzee social behavior in large part reflects social relationships among animals in the group and is less structured by kinship than by histories of interactions between animals.[17,18] Males have a strong dominance hierarchy; mating is nonetheless promiscuous, and estrous females mate with many males. Dominance hierarchies are not so apparent among females, but female dominance status influences reproductive success.[19] Given the focus on social relationships in chimpanzee communities, it is not surprising that investigators have observed reciprocity in trading for "political support" within the community.[20] Chimpanzees that have low political status or have violated reciprocity may suffer violent consequences.[21]

DISCUSSION POINT: BONOBOS

The bonobo is closely related to the chimpanzee but much less studied. You might want to do a little research on bonobo social systems and compare them with the better-studied chimpanzee. A good place to start is the 1998 article by Craig Stanford in *Current Anthropology*, "The Social Behavior of Chimpanzees and Bonobos: Empirical Evidence and Shifting Assumptions."

Eusocial Mammals

Permanently sterile worker castes occur in only two species of mammal: the naked mole-rat (see Figure 14.9) and the Damaraland mole-rat.[22,23,24,25,26] These burrowing rodents live in colonies of 50–100 individuals in southern Africa. The colony has one female reproductive, the queen, and a small number (up to three) of male reproductives. The queen, who can live 15 years or more, has a longer body than the workers do, allowing her to support more or less continuous pregnancies. Internally, the queen's vertebrae are lengthened; the lumbar vertebrae grow during pregnancies so that the queen's repeated pregnancies result in elongation of her spine.[27] The rest of the colony members, daughters or sisters of the queen, serve as nonreproductive workers.

The queen uses physical aggression to establish her dominance in the colony, and the colonial dominance hierarchy is linear, with a clear ranking of the colony members. Queen removal experiments show that the loss of a queen is followed by intense competition, with the emerging dominant female assuming the reproductive role in the colony.[28]

FIGURE 14.9
Naked mole-rats are often maintained in zoo exhibits because of their unusual social structure. In addition to being eusocial, they have evolved to live in underground burrow systems, losing most of their hair and the functionality of their eyes in the process.

Male workers have similarly suppressed reproductive capacities and, interestingly, can rapidly mature reproductively if removed from the social environment of the colony. Naked mole-rats are decidedly atypical rodents, with the longest lifespan of any rodent and an inability to regulate body temperature, making them effectively cold-blooded. Worker mole-rats are divided into two subsets: workers who work intensively through their lifespan and workers who work less frequently. These infrequent workers have the physiological capacity to grow into queens, which they can do either by superseding the old queen in their colony or by dispersing to establish a new colony.

Two things set the stage for the emergence of eusociality in mole-rats: colonial living and the high value of the tunnel system (and corresponding low probability of dispersal success for young). Longevity clearly plays a key role in facilitating the existence of a complex society because rapid turnover of reproductives in the colony would disrupt both the social and genetic structures of the colony. Whether the unusual longevity of naked mole-rats is a cause or a consequence of the evolution of eusociality is an interesting question.

As indicated in Chapter 12 on nesting and parenting, young adults in many bird and mammal species work as "helpers at the nest" prior to dispersing and establishing their own nests (or taking over the parental nest). Eusociality in mammals, such as the mole-rats, can be seen as an evolutionary extension of the ecological and evolutionary factors that lead to temporary helping in extended families. The diplodiploidy of mammals shows that the asymmetry of relationships found in haplodiploid species is not essential for the evolution of eusociality. In sum, social behavior in birds and mammals forms a continuum, beginning with minimally cooperative species through a large number of species with complex patterns of cooperation and at least temporary reproductive suppression, and reaching an extreme in some species of mole-rat and the evolution of permanent reproductive suppression.

DISCUSSION POINT: NAKED MOLE-RATS

Naked mole-rats are commonly displayed in zoos and are featured in a number of Internet videos. Visit a zoo display or watch some of the videos to get a better feel for the lifestyle of these animals. Now find some information on the lives of meerkats, another social mammal. Based on your comparisons of these species, what are the similarities and differences in their social lives? Can you hypothesize about key features (preadaptations; see Chapter 13) that may have favored eusociality in the naked mole-rat but not in the meerkat?

14.3 INVERTEBRATE EUSOCIALITY: WORKERS AND THE DIVISION OF LABOR

Caste in eusocial insects can be defined in a couple of ways. The first deals with differences among a colony's *workers* in how they do their work. The second has to do with reproduction; caste in reproduction boils down to the female reproductive(s), or *queens*, the males (sometimes called *drones* in honeybees or *kings* in termites), and the sterile workers. This section focuses on *division of labor* among workers. Reproductive castes are covered in a separate section later in the chapter.

> **KEY TERM** Castes are differentiated morphological or behavioral groups within a society.

> **KEY TERM** Division of labor occurs among members of the worker caste when they specialize in tasks.

Social Efficiency: Why Divide Labor?

Building a house is no easy matter. It requires mastery of carpentry, plumbing, electrical work, painting, laying tile, and so on. Each trade comes with a different set of skills, and some trades may require more physical strength than others to master. For one person to learn each skill and perhaps fail to become highly proficient at all of them can make the construction project slow and inefficient. If, on the other hand, skilled labor can be hired for each trade, the house can be built efficiently and rapidly.

OF SPECIAL INTEREST: SOME CONVENIENT DEFINITIONS

In eusocial insects, the *colony* refers to the animals, and the *nest* is the physical structure in which the colony lives. A *hive* is a honeybee nest (see Figure 14.10). Because colonies can sometimes change locations, and then construct a new nest, it is helpful to maintain this distinction in terms.

FIGURE 14.10
Paper wasp nest and colony (left) showing the small size of this society and its ability to construct a nest by transporting materials to the nesting site. The harvester ant nest (right) is occupied by a much larger colony. Ants bring gravel to the surface for the mound, and also collect gravel from the surrounding area. *Photos: Left, Michael Breed; Right, Ave Bucy.*

The same logic can be applied to division of labor within social insect colonies. Colonies of specialists are more efficient than a collection of unspecialized individuals. How might social insect workers specialize? Some of the principal ways that workers can differ are listed in Table 14.1, and the balance of this section is organized around these types of differences among workers.

Division of Labor Based on Dominance Hierarchies

In a smallish social group, perhaps up to 50 animals, a dominance hierarchy can be effective in establishing social roles. In the primitively eusocial sweat bee (*Lasioglossum zephyrum*), the queen establishes her dominance by nudging the other bees and by refusing to let them pass

TABLE 14.1 Effects of Worker Differences on Social Efficiency in Insect Colonies

Type of Difference among Workers	Potential Efficiency Gain	Examples
Status in dominance hierarchy	Each worker specializes in certain tasks, based on their status in the hierarchy.	Found only in species with small colonies in which dominance hierarchies are effective—paper wasps and sweat bees
Size differences	Size/strength can be matched with task.	Found in termites and ants, but not wasps and bees
Shape differences	Morphology matches task.	Large-headed defenders in termites and ants Nozzle-headed termites Aphid and thrip soldiers
Age differences (also called *temporal division of labor* or *temporal polyethism*)	Can be flexible to match the number of workers performing a task with need. Younger workers can perform lower-risk tasks.	Honeybees, some ants, paper wasps
Learning/experience	Can be a particular advantage for foragers, which may need to learn the landscape and how to return to their nest.	Wasps, some ants, bees
Physiological development (glands ready for task)	Specific glands come into play for specialized tasks (e.g., honeybee wax glands for comb construction).	Almost all eusocial species

her in their tunnels. The second-ranking bee in the colony serves as a guard at the entrance of the nest (see Figure 14.11). Lower-ranking bees specialize in either foraging or nest construction. In *Polistes* paper wasps, there is a similar correlation between dominance status and task performance.[30,31]

409

OF SPECIAL INTEREST: THE UNIT OF SELECTION

What does selection actually act upon? Is it a single nucleotide in the DNA sequence? Or maybe it's the gene? Is it the chromosome? Cells within a multicellular organism? The individual animal? Richard Dawkins asks this question elegantly in his book *The Selfish Gene*. Here, we need to pause and consider the problem of the **unit of selection** in the specific context of eusocial invertebrates.

Up to this point, our discussion of animal behavior has focused on the individual as the unit of selection: evolutionary forces act on individuals and culminate in the level of reproductive success that individuals achieve. We emphasized the major problems with most group selection models in

> **KEY TERM** The unit of selection is the level at which natural selection operates. This can be the individual worker or queen, the colony, or a local population of colonies.

Chapter 13, and we argued that most seemingly altruistic behavior can be explained parsimoniously by individual natural selection acting on the donor animal.

What happens, though, when selection yields animals that are sterile, whose only option is to gain reproductive value through inclusive fitness by working with close relatives? It is easy to see that if an animal's reproductive value is invested in the success of the queen of the colony, and if that queen's success, in turn, depends on the survival of the colony, then selection is acting on the colony as a whole. Given that scenario, colonies which are able to organize their labor most effectively to exploit the environment will transmit the most genes to the next generation. Selection should favor organized, efficient colonies and punish chaotic, inefficient colonies. Colony-level selection is not group selection; it is selection on the queen's phenotype, which extends to include the activities of her workers.[29]

FIGURE 14.11

Sweat bees, and a diagram of a sweat bee nest in soil. The queen constructs a tunnel and rears some workers in the cells that extend from the tunnel. The workers, once they emerge, work to rear additional bees, including the next season's reproductives. *Photos: Joseph Berger, Bugwood.org (right), Russ Ottens, University of Georgia, Bugwood.org (left).*

Size and Shape Differences in Division Of Labor

Some ant and termite species show amazing differences in size and shape among their workers. One intriguing question is why extreme caste differences appear in the two taxonomic groups with flightless workers—ants and termites—but do not occur in bee and wasp workers. Perhaps flight places evolutionary constraints on structural evolution in bee and wasp workers. This section explores how worker size and morphology interact in improving colony efficiency in eusocial insects.

Sometimes fairly subtle size differences are important in division of labor. In some species of eusocial animals, all the workers have the same morphology, but differ in size, with a range of sizes occurring in the colony. One of the most feared animals in the New World tropics is the giant tropical ant. It is sometimes also called the *bala* (or bullet ant because being stung by one feels like being shot) or the *vienticuatro* (24, because its sting hurts for 24 hours). Large workers—20 mm or more in length—are visible outside the nests, foraging, excavating, and when necessary, defending the colony. The very largest workers come to the nest entrance when the colony is disturbed. One might think that all the workers of this ant species are equally gigantic, but excavation of this ant's nests reveals that much smaller workers, down to 1.5 mm in length, toil inside the nest. When the nest is disturbed, it is these small workers that rescue the larvae.[32] This is a typical example of size-based division of labor, a commonly observed mechanism in ants.

FIGURE 14.12
Worker size differences.
Photo: Sanford Porter.

E. O. Wilson envisioned an evolutionary scheme in which selection first favored colonies with workers of highly variable size.[33] Wilson argued that the evolution of a broad range of body sizes allowed body size to be matched to task; if the most appropriate animals engage each task, then the colony becomes more efficient and can outcompete colonies that lack this feature. This size variance would allow specialization of workers in size-appropriate tasks. Typically, small workers spend most of their lives inside the nest, caring for brood, cleaning the nest, and feeding other workers. Larger workers focus more on tasks that take them outside the colony, such as nest and territory defense and foraging. If selection favors colonies with greater variance in worker size, then over evolutionary time, 10-fold or even 100- or 1000-fold size differences between the smallest and largest workers in a colony become possible (see Figure 14.12).

How does shape come into play? The answer lies in *allometric growth*—disproportionate growth of certain body parts as an animal gets larger. To understand this principle, think first of isometric growth. In *isometric growth*, body parts all grow at the same rate, so body proportions are the same in small and large individuals. ("Iso" means the same, "allo" means different.) What if one body part, such as the head, grows at a much faster rate than the rest of the body? Then large animals will have disproportionately large heads (see Figure 14.13). What good is a disproportionately large head? In insects, the head houses the muscles for their jaws (mandibles) and numerous glands, some of which can have important social functions. A large head can support bigger mandibles, larger mandibular muscles, and more glandular development. Allometric growth of the head, and sometimes the thorax, is the nearly universal route to extreme caste differentiation in eusocial insects.

What about the issue of the role of the tiniest of leafcutter workers, mentioned at the beginning of this chapter? Observers of leafcutter workers noticed that the smallest of workers (minims) "hitchhiked" on leaves; recall that a medium-sized worker carrying a leaf piece is often also burdened with one or two very small workers riding on the leaf. This observation was puzzling because the hitchhikers had no apparent function. Detailed tests showed that the hitchhikers play a key role in warding off small parasitic flies (Phoridae) that attempt to lay eggs on the ant carrying the leaf. If the fly succeeds, its larva invades the head of the worker ant and ultimately kills the worker. The minims actively ward off the flies. The fact that much of the activity of the minims is at night, when the flies are absent, suggests that the minims have other functions, such as removing pathogens that might infect the colony's fungus garden from the surface of the leaves. Function and form are closely associated, and the minims play critical roles outside the nest, despite their small size.[34]

How do workers find the appropriate task? Natural selection on task preferences certainly plays a role, but the spatial positioning of workers is also an important part of the dynamic; a worker that is "programmed" to be outside the nest is likely to encounter a set of potential tasks that differs from the tasks encountered by workers who remain inside the nest.

Age Differences

In larger colonies, dominance hierarchies are less likely. The most commonly observed pattern of division of labor is for tasks to be divided by age.[35,36,37] This type of division of labor was first documented in honeybees but also occurs in many ant species, and has been termed *temporal polyethism* (*temporal* = time-based, *polyethism* = many behaviors). In nearly all documented cases, the progression is from young workers performing tasks inside the nest (construction, cleaning, caring for brood), to middle-aged workers busy on the periphery of the nest (guarding, ventilating, undertaking), and then to the oldest workers foraging outside the nest. Seeley built on earlier honeybee studies in Germany to reinforce this point;[38] simple programming of workers to prefer these locations (center of the nest, where brood is likely to occur; periphery of the nest, where nest defense is important; outside the nest, where foraging is paramount), combined with age-related division of tasks, can yield complex patterns of division of labor. Figure 14.14 shows how this plays out

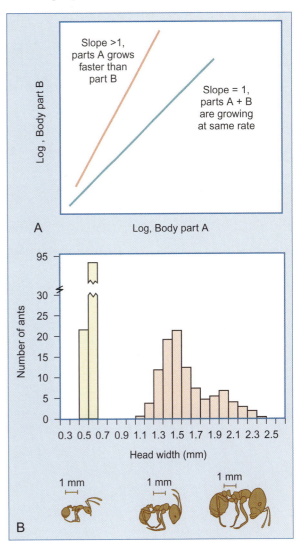

FIGURE 14.13

(A) Allometric growth relationships. The *x* axis shows the size of one body part, such as the head; the *y* axis shows another body part, such as the thorax. If the slope is equal to 1, growth is isometric. If the slope differs from 1, then growth is allometric. (B) A hypothetical size distribution of ants in a colony, with a group of small workers (tan bars on the left) and larger workers (pink bars). The drawings of these ants show allometry, with the head growing larger, proportionately, as body size increases.

411

(A)

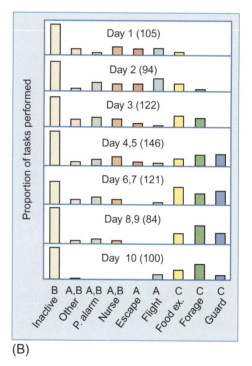

(B)

FIGURE 14.14

Age polyethism. (A) Sequence of events in the life of a worker honey bee. Workers progress from cleaning, to feeding larvae, to storing food, to guarding, and then foraging. (B) A similar analysis of ant behavior, with tasks listed on the x axis, and each part of the plot representing a different age of ants. Young ants are more likely to nurse, and older ants guard and forage. *Adapted from: Muscedere M.L., Willey T.A., & Traniello J.F.A., 2009, Age and task efficiency in the ant Pheidole dentata: young minor workers are not specialist nurses. Animal Behaviour, Vol. 77, pp. 911–918.*

in honeybees and ants. The most elaborate systems of division of labor combine extreme morphological differences with temporal castes (see Figure 14.15).

Based on the observation that the oldest workers engage in the riskiest tasks, biologists have hypothesized that age equates with expendability for the society. In more formal terms, an animal's **marginal value** to the social group, or remaining work that it might do in a lifetime, falls as the animal approaches the end of its normal lifespan. Theory predicts that animals

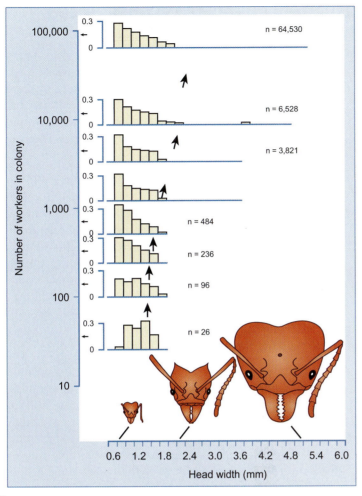

FIGURE 14.15
In this case, age and size combine to correlate with behavior. As a colony grows in size, from 10 to 100, and ultimately to 100,000 (*y* axis in the drawing), the variability of worker size increases, and larger workers become more common. This allows more specialization in task performance by workers. *Adapted from Wilson, E.O., 1976. Behavioral discretization and the number of castes in an ant species. Behav. Ecol. Sociobiol., 1, 63–81.*

413

with little marginal value for the group should engage in the most dangerous occupations. As an aside, note that a different pattern seems to hold true for human societies, in which the oldest people, perhaps as a result of experience, are typically the most risk-aversive members of the society and are often highly valued as sources of wisdom and guidance. Yet another influence might be the fact that in many human societies, taking risks is perceived to confer status, which, in turn, may be age related; status may be important for people who are more interested in attracting mates than in helping to care for an extended family.

> **KEY TERM** The marginal value of life is the amount of work or reproduction expected for an individual of a given age.

DISCUSSION POINT: INSECT AND HUMAN SOCIETIES

Can we learn anything from insect societies that might improve the efficiency of human work? Generally, worker eusocial insects do not require instructions to complete their tasks; the information they need is genetic and is carried with them. Could humans function in a "boss-less" society? Why or why not?

14.4 INVERTEBRATE EUSOCIALITY: QUEENS AND REPRODUCTION

How do workers in a colony partition the labor necessary to complete the tasks facing the colony? By definition, reproductive labor is divided (or skewed) among members of eusocial colonies. One or a few females lay the bulk of the eggs in the colony. By convention, these egg-layers are called *queens*, a term that can be a bit misleading because in most species of ant and social wasp the "queen" does little in the way of coordinating or directing activity within the colony.

Queens are usually well adapted to their job, with large ovaries contained in a very enlarged abdomen. In primitively eusocial species, such as sweat bees, bumblebees, and paper wasps, the queen does not differ morphologically from the workers, although she may be somewhat (5–10%) larger than her worker coterie. In primitively eusocial species, the queen is capable of establishing a nest on her own, without workers, and of rearing her first brood of workers. In highly eusocial insects, queens are much differentiated, morphologically, from their workers; for example, the honeybee queen lacks the hairs necessary for collecting pollen, she has relatively small eyes, and her sting is not barbed (see Figure 14.16).

FIGURE 14.16
A fire ant queen with workers. The queen is well-adapted for her primary function of laying eggs. *Photo: Sanford Porter.*

In many species, an evolutionary interplay continues between the queen and workers, even if substantial morphological differences have evolved so that the queens are specialized for reproduction and the workers are specialized for brood care, foraging, and colony defense. Workers usually retain small, but potentially functional ovaries, and although they lack the apparatus for mating, ant, wasp, and bee workers can lay unfertilized (haploid) eggs that develop into males. Trivers and Hare pointed out that in a haplodiploid society kin selection theory predicts different outcomes for workers that help to rear young, depending on the sex of those young.[39] For instance, workers should cooperate with queens in the production of female offspring; assuming the queen is the mother of the workers, then the workers should prefer to assist the queen in producing their sisters, rather than producing their own daughters. But production of males is a different matter because theory predicts that workers will benefit more, evolutionarily, from producing their own sons than from helping the queen to produce their brothers.

Trivers and Hare used this idea to develop an argument that if kin selection was responsible for the evolution of extreme altruism, then there should be worker–queen conflict over the laying of male eggs. In fact, in many species of ant, bee, and wasp, the ovaries of at least some workers develop to the point that they can lay male eggs. Queens and workers may respond to this competition by eating each other's male eggs.

Ratnieks and Visscher realized that this is not the complete story;[40] they put the focus on how a worker responds to another worker's reproductive efforts. A worker should prefer its own sons, but given the choice between rearing a male egg laid by the queen (this egg is the worker's brother), and a male egg laid by a sister (this egg is the worker's nephew), the worker should prefer the queen's egg. This reasoning leads to the prediction of worker policing, in which workers monitor and control the reproductive efforts of their fellow workers by eating the worker-laid eggs. Worker policing has been documented in a number of species, including honeybees.

How Many Matings and How Many Queens?

Kin selection explanations of extreme altruism rest on the assumption that a queen mates no more than once, so her daughters are, indeed, supersisters. If a queen mates more than

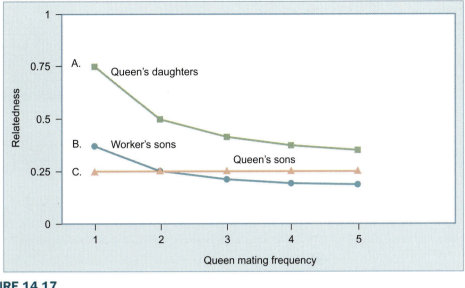

FIGURE 14.17

This illustration shows how the number of times a queen mates affects genetic relationships within an ant, bee, or wasp colony. The upper line shows the average relatedness among daughters of the queen; if she has mated only once, this is .75; added matings rapidly reduce this average. The same pattern applies to workers' sons, but not to the queen's sons, which are always related to each other by .25 (they have no father!).

once, the genetic relatedness among workers in the colony is diluted (see Figure 14.17), and the inclusive fitness benefits from working are reduced. Selection could favor multiple mating by queens for several reasons, including (1) genetic diversity among workers in the colony facilitates division of labor,[41] (2) genetic diversity among workers increases colony-level resistance to diseases and parasites, (3) multiple mating insures the queen against the costs of inbred mating (more matings at least some of the matings won't be inbred), and (4) increased sperm supply allows the queen to lay more eggs over her lifetime.

If workers in primitively eusocial species have the evolutionary option of leaving the social group and reproducing solitarily, but workers in highly eusocial species lack this option (largely because they have lost the morphology necessary for mating), then the prediction is that multiple mating by queens would be more common in highly eusocial species than in primitively eusocial species. Once workers are "trapped" into that role, without the possibility of evolutionary reversal out of the worker role, selection can then act on the queens to gain the benefits of worker genetic diversity and insurance against inbreeding.

Similarly, kin selection theory predicts that a colony should have only one queen; multiple queens dilute the genetic relatedness in the colony, even if the queens are sisters. In most primitively eusocial species, such as sweat bees, bumblebees, and paper wasps, colonies do, in fact, have only one primary reproductive female, and even if workers are capable of occasionally laying eggs, the workers have not mated, so their offspring are all males. (See the earlier section on worker policing for more on the topic of eggs laid by workers.)

In highly eusocial species, multiple queens are more common. There are two theoretical models for the evolution of multiple queens in colonies. One is the delayed competition model discussed previously, in which groups of queens must cooperate to found colonies; they later become competitors, and only one of the queens actually contributes reproductives to the next generation. In this case, the queens need not be genetically related because inclusive fitness has no role in their cooperation. In the other model, related queens (generally sisters) remain together, and each contributes to colony reproduction. This is particularly advantageous in pioneering or "weedy" species that invade new habitats, such as Argentine ants.

415

Reproductive Suppression

Kin selection theory predicts that workers voluntarily give up their own reproduction in order to aid their mother. If this were strictly the case, conflict over reproduction would not occur in colonies, and there would be no need for the queen to actively suppress worker reproduction. As already discussed, in ants, bees, and wasps, the asymmetries in genetic relationships caused by the haplodiploid method of sex determination make conflict over production of males likely. Also, there is a distinct possibility that eusociality could be explained by coercion, rather than by kin selection.

The existence of dominance hierarchies in a broad range of primitively eusocial species (or, in the case of the naked mole-rat, a highly eusocial species) suggests that reproductive suppression may not be an entirely voluntary act on the part of the workers. In some highly eusocial insects, there is a distinct possibility that queens engage in chemical suppression of the workers in the colony. Scientists have not yet sorted out the underlying evolutionary forces that lead to the presence of dominance hierarchies in so many eusocial species; division of labor, competition for production of males, and suppression of production of both males and females are the leading hypotheses.

A mystery in ant behavioral biology is the question of how the queen's reproductive dominance is maintained in the colony. Unlike primitively eusocial wasps and bees (see the following section), in which behavioral dominance is clearly expressed, or honeybees and termites, in which chemical signals from the queen affect worker reproduction and activities, in ants the queen seems to have little organizing influence on the colony. Behavioral dominance should be obvious and, if it were present, it probably already would have been discovered. It is possible that as yet unidentified chemical signals from the queen play an organizing role in ants. Or, intriguingly, perhaps worker function has evolved so that workers perform their tasks quite independently of any knowledge of the presence of the queen. This model challenges stereotypic assumptions that complex societies require a central directing figure for efficient functioning.

Dominance Hierarchies

The dominance hierarchies of *Lasioglossum* sweat bees and *Polistes* paper wasps described in the discussion of division of labor also have reproductive implications, as studies of ovary development in workers consistently show that higher-ranking workers in the hierarchy have larger ovaries than lower-ranking workers.[42,43] Workers usually do not mate, so they lack the potential for laying female eggs, but ovarian development could be indicative of an ability to lay male eggs. Dominance hierarchies are also known in bumblebees and in some ant species that have small colonies.[44] Body size differences can facilitate the ability of a queen to dominate her workers; size confers an advantage in physical conflict. Thus, it is important to know that the workers of sweat bees, bumblebees, and paper wasps are all smaller than the queens and that, at least in sweat bees and bumblebees, smaller workers tend to have lower status in the hierarchy.

Chemical Suppression

Honeybee queens produce powerful pheromones that are distributed among the colony members. These pheromones could serve either to pharmacologically suppress worker reproduction or could simply act to inform workers of the queen's presence.[45] Pheromones from the queen as well as from the larvae in the colony may act together to suppress worker reproduction and to maintain worker behavior (see Figure 14.18). In the older literature, the queen pheromone is called ***mandibular pheromone*** because some of its components were first identified from the mandibular glands of queens. ***Retinue pheromone***

FIGURE 14.18
A honeybee queen with workers. Note that the queen is relatively hairless and has small eyes, relative to the workers that surround her. *Photo: Scott Bauer, USDA Agricultural Research Service, Bugwood.org.*

may be more accurate because the chemical mix in the pheromone comes from several glands, and one of the behavioral effects of the pheromone is to attract a circle of bees (a "retinue") around the queen. Chemical communication of termite queen and king presence has also been suggested.[46,47]

> **KEY TERM** Mandibular pheromone is a set of chemicals released from the mandibular glands of the queen honeybee. This pheromone affects worker behavior.

"Willing" Workers

In the vast majority of ant species and in the highly eusocial wasps, there is no evidence of either dominance or chemical reproductive suppression of the workers. Negative results like this are as tricky here as they are throughout science. Is the absence

> **KEY TERM** Retinue pheromone includes mandibular pheromone and other glandular products; the entire retinue pheromone is necessary for normal functioning in a honeybee colony.

of evidence for behavioral or chemical dominance due to faulty experimental approaches? Is the investigator asking the right question? In this case, kin selection theory predicts that workers should willingly give up their reproductive "rights" to help their mothers rear their sisters, so the lack of evidence for reproductive suppression could support a kin selection hypothesis for eusociality.

14.5 INVERTEBRATE EUSOCIALITY: COLONY DEFENSE

Eusocial colonies face a variety of threats, but the common theme in studies of colony defense is that the lives of individual workers may be expended in efforts to protect the colony. This is not to say that workers are always expendable; instead, the claim is that the cost-benefit equation favors combat if the cost of the threat (and failing to oppose it) is greater than the cost of losing the workers needed to block the threat. Physical defenses usually include biting and in some cases stinging, along with many different types of chemical defenses.

While most ants and eusocial bees and wasps can sting, one large and ecologically important group of ants, the Formicinae, or *formic acid* ants, has lost, in the course of evolution, the ability to sting. Instead, these ants rely on an acid secretion for their defense. Another ecologically prominent group of eusocial insects, the stingless bees, has also lost the ability to sting, but these bees can readily defend their nests by biting and, in some cases, with chemical secretions. Because the sting is a modification of the egg-laying apparatus in the insect order Hymenoptera, it is found only in ants, bees, and wasps; other eusocial invertebrates, including termites, aphids, and shrimp, lack the ability to sting.

Often the major threats to a social insect colony are workers from other colonies of the same species, or workers from closely related species. Theft of brood or food is a continuing threat, and members of the same or closely related species are ideally adapted to use a colony's food or to employ workers as slaves. The key to preventing entry by larcenous workers is the ability to discriminate who belongs, and who does not belong, in the colony. In most species, odor differences among colonies are the key to this recognition, but in some wasps, visual cues apparently serve this role.

Eusocial colonies have three types of resources that may need defending. The first is their brood, which can be a valuable source of food for other animals or a source of workers that are used as labor by raiding eusocial species (these are sometimes called *slave-makers*). Second, colonies may have stored food, such as honey, which also has high value for other species. Finally, colonies may have feeding territories that are subject to seizure by other colonies. Brood and stored food are attractive to many other species, while foraging territories are usually defended against conspecifics or other species with very similar ecological requirements. Because the value of brood and stored food is so high, even relatively small eusocial insect colonies can attract vertebrate predators, so many eusocial insect species have defenses that are geared toward warding off vertebrates, a lesson that many humans learn in painful ways.

insect called a *thrip*, an insect with wings that look like bird feathers (see Figure 14.20). Thrips have another, less obvious, distinguishing feature; they are among the few insects other than Hymenoptera to have haplodiploid sex determination.[53] Haplodiploidy and living in a nest-like structure—in the case of thrips, a *gall* on a plant—give some thrips the necessary preadaptations for eusociality, which has been discovered in a half dozen species of Australian thrips.

> **KEY TERM** A gall is a growth of plant tissue stimulated by an insect. The insect lives within the gall and eats the plant tissue.

On the surface, thrips (insect order Thysanoptera) seem like an extraordinarily unlikely candidate for eusociality. These minute plant-feeders seem to have a limited behavioral repertoire and, if one expects complex communication as a part of eusocial life, the thrips are even more unlikely candidates. However, they have the two critical preadaptations for extreme altruism. They are haplodiploid and some species of thrips live in galls formed by their host plants, probably in response to thrip secretions. The presence of a mother and her daughters living together in a protected structure (the gall) sets the stage for the evolution of extreme altruism. Add to this haplodiploidy, which tips the evolutionary balance in favor of aid-giving behavior, and ecological pressure from species of thrips that are specialized to invade and take over the galls, and it becomes the perfect recipe for the evolution of extreme altruism.[54]

FIGURE 14.20
A eusocial thrip, *Kladothrips habrus*.
Photo: Bernie Crespi.

Thrip galls are initiated by a single, large-winged (macropterous) female thrip. She has previously mated and lays a batch of eggs inside her gall. The hatchlings, which in thrips look like miniature wingless versions of the adults, include both males and females. Some of these young develop into large-winged males and females, which may mate with females moving on to found new galls of their own. (Read more on the importance of brother-sister matings below.) Other young males and females develop very differently, with small (micropterous), useless wings. These are the extreme altruists; they work for the benefit of their mother and their sibs by defending the family's gall from invasion by socially parasitic species of thrip. The parasitic thrips attempt to exploit the work of the founding family by stealing the gall and the food and protection it provides.

Why might both males and females serve as soldiers when a central prediction of kin selection theory is that, in haplodiploid species, the worker castes should be female? High levels of inbreeding reduce the effects of the asymmetry in relatedness due to haplodiploidy, making the potential benefits to male soldiers higher than might have been expected.[55]

Females of some thrip species invade the stems of plants and then cause the plant to form a chamber, or gall, around them. The gall protects the female, and she feeds on plant cells on the inner surface of the gall. The female, or foundress, then lays a batch of eggs that all develop into a soldier caste with small wings and enlarged forelegs. The signals that cause the foundress's eggs to develop as soldiers, rather than the winged form, remain a mystery.

These soldiers are effective defenders of the gall against parasites and predators, and improve the foundress's chances for survival, but because they are unable to fly, the soldiers cannot disperse to start new galls. Soldiers are particularly adept at defending the gall against another genus of thrip, *Koptothrips*, which attempts to take over the gall for its own use.[56] Unlike many eusocial species, the soldiers retain the ability to reproduce. Brother and sister soldiers mate, and the female soldiers lay eggs inside the gall.

Thrip families are small: in *Kladothrips hamiltoni*, the number of soldiers is about 25 per gall. The next generation, which comes from eggs produced by the soldiers, is fully winged and capable of dispersing to establish new galls. Because the soldiers' matings are inbred, the winged dispersers coming from the soldiers are more related to the foundress than the foundress is to her own sons and daughters.[57] This supports a kin-selection argument for

FIGURE 14.21
Eusocial aphids, showing
aphids feeding on a
stem (left) and a colony
with larger soldiers
and smaller, winged,
reproductives (right).
Photos: Patrick Abbott.

the evolution of soldiers in these thrips. The winged dispersers, most of which are female, overwinter, mate, and initiate new galls the next year.

The discovery of eusociality in the thrips shows that given the right set of preadaptations, extreme altruism can arise in surprising places. The eusocial thrips lack many of the intriguing adaptations for colony member communication and coordination found in termites, ants, bees, and wasps, yet the addition of the soldier generation greatly improves the chances that the foundress's genes will be passed to the next generation. With the thrips, evolutionary pressure from parasites and predators, combined with nesting in a confined space and haplodiploid sex determination, produces the perfect mix for the evolution of division of labor in colonies.

About the same time that Crespi uncovered eusociality in thrips, Aoki uncovered a sterile soldier caste in some colonial aphids. (Aphids are also small, plant-feeding insects.) Like thrips, a few species of aphids that inhabit galls are eusocial (see Figure 14.21). The eusocial aphids are diplodiploid; the similarities between thrip and aphid colonies represent convergent evolution. Aphid soldiers defend their colony against possible predators, and can stab potential predators with their mouthparts (stylets).[58,59,60,61,62]

Termites

Termites probably evolved from colonial, wood-eating cockroaches, and in many ways their biology is still strongly linked to this evolutionary root. Nearly all termites use plant material as their primary dietary component. Depending on the species, termites may specialize on dead wood, grasses, or leaves. This food, while abundant in many ecosystems, is notably lacking in nutrients and is rich in near-indigestible cellulose. Termites make up for their nutrient-poor diet by employing intestinal symbionts to assist in digesting cellulose.

Behavioral adaptations for the passage of the *intestinal symbionts* among workers are critical to termite social biology. These symbionts are a mixture of flagellated protozoa and bacteria in the ancestral termites and bacteria in the derived termites. The symbionts reside in the termite hindgut; at hatching, a termite infects itself with symbionts by feeding from the anus of another termite, repeating this process each time it molts and thus loses the hindgut contents. The need for anal-oral infection with symbionts may have been a major factor in the evolution of social cohesion of families in the wood-eating cockroach ancestors of termites.

> **KEY TERM** Intestinal symbionts are microorganisms that aid in digestion and are found in the gut of many animals. In termites, flagellates or bacteria digest cellulose, making the carbon available to the termites.

Termites are among the master architects of the animal world, and their large mounds, essentially castles constructed of dirt, plant material, and feces, rise prominently from the

FIGURE 14.22

(A) Termite nest (Nasutitermes) in a tropical rainforest. (B) Termites from the nest pictured in A build covered walkways. When the covering is broken away, guards, which have spray nozzles and can shoot a gluey substance at ants, converge on the break. (C) A Formosan termite queen (Reticulitermes). *Photos: (A), (B) Michael Breed; (C), (D) USDA/ARS Formosan termite image gallery.*

tropical and subtropical savannas in Africa, Australia, and South America (see Figure 14.22). Although not all termites are such expert builders (North American termites simply excavate cavities in wood or soil), this is such a noticeable feature of termite behavioral biology that it deserves special discussion. This elaborate construction raises two sets of questions: First, what is the functional significance of the architecture, and second, how can termites "know" how to construct these complex fortresses?

Termites, like all animals, have to contend with the physiological impacts of water, oxygen, carbon dioxide, temperature, and direct sunlight. They must also deal with potential predators, particularly ants, but also vertebrates. The nest serves to mitigate all of these problems. First, it protects termites from environmental extremes. Termite workers are typically soft-bodied and relatively unprotected from heat and direct sun by their exoskeletons; the nest compensates for this vulnerability. Even North American termites construct coverings over their walkways when there is a risk of exposure to sun or predators.

Second, the nest helps solve problems that are created by life in an enclosed space. Animals produce heat, water, and carbon dioxide as byproducts of their metabolic activities. Metabolic heat and water can be put to use to help to maintain favorable conditions in the nest, but excesses of either need to be vented from the nest. In particular, carbon dioxide must not be allowed to build up in the nest because it can act as a metabolic poison, and instead, oxygen must be circulated into the nest. Because of this, large numbers of termites living in a closed space need a ventilation system that retains adequate heat and moisture, brings in needed oxygen, and gives the option to vent carbon dioxide, heat, and moisture when needed.

The nest also serves a third function—that of protection against predators. The hardened exterior—termite feces are a great glue—protects the termites from all but the most determined vertebrates. By limiting the number and size of entrances and exits, concentrated soldier forces can repulse encroachments by ants.

Architecturally, it seems highly unlikely that termite workers individually have the cognitive ability to step back, look at the overall structure of the nest, and make decisions concerning what and how to build next. There is also no evidence for "foreman" or "supervisor" workers directing the building efforts of others. This means that the ultimate construction, no matter how elaborate, arises from individual workers' responses to conditions at a given location and moment of time. This is seen most simply in the repair of covered walkways; if a covering is damaged, the damage or sunlight serves as a stimulus for workers to add construction material to the remaining walkway; if they follow the pattern established by the undamaged portions, the gap in the walkway is soon closed. Construction effort in response to a specific element of the existing structure is called *stigmergy*, and this explains much about how elaborate structures can arise from workers that use simple rules. If other simple rules for workers are added—plug up air flow if it is too cool or too dry, make an opening if carbon dioxide is too high, and so on—it becomes apparent how collective behavior yields complex buildings in the absence of either cognition or centralized direction. Computer models that are built on such simple rules can successfully mimic the construction behavior of termites.[63]

> **KEY TERM** Stigmergy is construction that follows a pattern already established by previous building. For instance, if a small deposit of dirt stimulates further deposits at the same spot, a pillar results.

Termite castes differ from those of ants, bees, and wasps in several dramatic ways. Termites are diplodiploid and, consistent with Hamilton's theory, both males and females serve as workers in the colony, following parallel developmental pathways (see Figure 14.23).

423

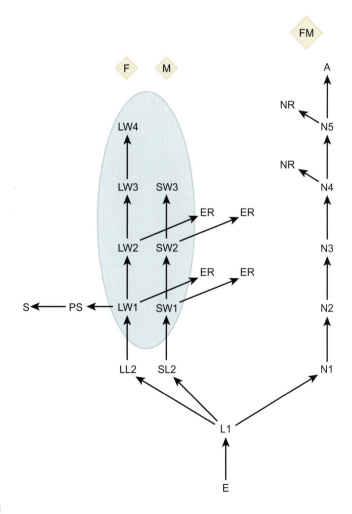

FIGURE 14.23
The typical fate of males and female termites. Eggs and first molt termites of this species are termed larvae (L1) and may be either male or female. Both males and females may develop through nymphal stages to become the next generation of reproductives (N1 to N5, then A—adult). Females may become large workers, molting through stages LW1 through LW4, while males may become small workers (SW1 to SW3). Either worker sex can also molt to become a supplemental (ergatoid) reproductive (ER). *Adapted from: Stewart, A.D., and Zalucki, M.P., 2006. Developmental pathways in Microcerotermes turneri (Termitidae :Termitinae): Biometric descriptors of worker caste and instar. Sociobiology. 48: 727–740.*

Termite colonies are headed by a queen and a king; these individuals are winged as young adults and mate following a nuptial flight. The mated pair sheds their wings and lives the balance of their lives within the nest; the king and queen mate again each time the queen runs low on sperm. Termites have incomplete metamorphosis; a termite hatchling, or nymph, looks like a miniature wingless adult. While the smallest hatchings cannot contribute much to colonial welfare, as an immature termite grows through several molt stages, its ability to work grows as well. Tasks within a termite colony include feeding and caring for the queen, excavation and construction of the nest, and colony defense.[64]

In most species, the largest females are major workers, whereas the largest males are soldiers, although in some cases these roles are reversed. Soldiers typically have enlarged heads and mandibles, and can use the mandibles as effective weapons in territorial combat with other termites or in battles with predators, which are often ants. The role of immatures as workers and the presence of both male and female workers make termite caste systems more complex than those of ants, bees, and wasps.

In addition to the nymphs that develop as workers, some male and female nymphs grow into the next generation's reproductives. Colonies, in response to environmental triggers such as seasonal changes, rainfall, and temperature, may release hundreds or thousands of winged reproductives on a single night. To ensure outbreeding, these flights of reproductives are usually synchronized among colonies, so there may be hundreds of thousands of winged termites flying at once. The winged females (potential queens) attract males with sex pheromones from a gland on their backs; once a male finds a female, he follows her as she runs on the ground, searching for a nest site. Very few of the termite pairs actually succeed in establishing nests; most become prey for ants, birds, rodents, and even humans. The mass release of winged termites is probably an example of predator saturation (see Chapter 9 on foraging); the winged termites are poorly defended, but so many are produced that the predators cannot possibly eat all of them. Once the nest is established, the queen's abdomen expands to contain her growing ovaries; in some species the abdomen is so large that the queen is immobilized. In very large termite societies, supplemental reproductives may develop; these females are intermediate in size between the queen and the workers and contribute by laying eggs that will develop into workers. Termite colonies of some species can become immense, containing hundreds of thousands or millions of workers and occupying underground cavities that span several hectares. (A hectare is 100 meters on a side, about twice the length of an American football field.)

Ants

Ants are all highly eusocial and, in many ways, are the most diverse group of eusocial animals. This diversity is especially apparent in caste structure and in diet. Ant diets range from strictly carnivorous (army ants) to strictly herbivorous (leafcutter ants). The evolutionary ability of ants to exploit this wide range of diets has led to their ubiquity in terrestrial habitats.[65]

In castes, ants range from species such as the common wood ants, *Formica sp.*, that show very little size variation in workers and no noticeable differences among workers in form, to species such as leafcutter ants, *Atta sp.*, and members of the genus *Pheidole*, which have extreme physical worker castes (see Figure 14.24). As discussed in the section on castes, caste differentiation allows task specialization, a key element in gaining efficiency for the social unit. The high level of eusociality in ants is established by the major morphological differences between ant queens and workers and the lack of reproductive potential (beyond laying male eggs) in workers of most ant species.

Ant queens exhibit two different patterns of nest founding, depending on the species. In some less-advanced species, the queen mates and then finds a nest site, to be used as a base

FIGURE 14.24
Ants: (A) Worker ants with brood. (B) Worker ants tending eggs. (C) An ant colony filled with plaster and then excavated (next to scientist Walter Tschinkel). (D) An ant tending aphids. *Photos: (A) USDA APHIS PPQ Archive, USDA APHIS PPQ, Bugwood.org; (B) Susan Ellis, Bugwood.org; (C) Charles F. Badland, with permission from Walter Tschinkel; (D) David Cappaert, Michigan State University, Bugwood.org.*

for her foraging activities. Her first batch of eggs, which are all female, develop into workers, but to feed them she needs to actively forage and bring food back to the nest. This style of nest establishment is exemplified by the giant tropical ant, *Paraponera clavata*, in which the queens, during the initial stages of nest foundation, forage on insect prey. The obvious downsides to this strategy are that the queen must retain morphological adaptations for foraging and that she is exposed to predators while foraging.

Alternatively, ant queens may begin nests claustrally or semiclaustrally.[66] In claustral species, the queen relies solely on her metabolic resources, including digestion of her now-unnecessary wing muscles, to feed her first batch of larvae. Semiclaustral species rely largely on the queen's metabolic resources, but the queen occasionally forages. Claustral founding avoids the risk of exposure to predation, but the queen is limited by how much fat and muscle she can use as nutritional resources for her offspring. Semiclaustral founding, which is seen in leafcutter and harvester ants, bridges the gap between the two extremes, allowing the queen to collect some critical nutrients while she avoids the risks involved in full-time foraging.

Army Ants

Army ants are not the terror for humans portrayed in movies, but their biology is extraordinary and well worth discussing. Army ants lack a permanent nest; instead, colonies alternate between a stationary stage, during which the queen is enlarged and laying eggs, and a nomadic phase, during which the colony often moves in search of food for their voracious larvae (see Figure 14.25). Armed with large mandibles and a painful sting, army ants are highly effective predators. Most army ants specialize on preying on other ants, but some conspicuous species eat a broad range of insects and even frogs, lizards, and nestling birds.

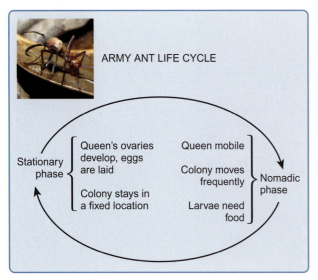

FIGURE 14.25

The life cycle of an army ant colony. *Photo: California Academy of Sciences.*

ARMY ANT LIFE CYCLE

Stationary phase → Queen's ovaries develop, eggs are laid / Colony stays in a fixed location → Queen mobile / Colony moves frequently / Larvae need food → Nomadic phase

426

Wasps

The painful stings of eusocial wasp workers get the attention of humans. Many, perhaps most, "bee stings" in late summer were actually delivered by a worker from a eusocial wasp colony. In North America, the eusocial wasps are easily divided into the ***primitively eusocial*** paper wasps (species in the genera *Polistes* and *Mischocyttarus*), and the ***highly eusocial*** yellowjackets and hornets. The colony cycle is annual, so new queens found nests in the spring; by mid- to late summer, from 20 to 100 or so workers populate the nest.

Paper wasps build a single comb using plant fibers (see Figure 14.26). The paper wasp comb is open to the environment and, when full of larvae and pupae, is a tempting food item for rodents and birds, hence the strong defensive reaction of the workers. Primitively eusocial wasp colonies have a dominance hierarchy, in which the queen is quite physically aggressive to subordinates. Worker larvae may also receive inadequate nutrition for full reproductive development.

Yellowjackets and hornets construct paper nests consisting of a stack of combs that are covered by a spherical paper sheath (see Figure 14.27). In highly eusocial wasps, as with ants, the mechanism of queen dominance in the colony is not understood. Queens are much larger than workers, but as yet neither dominance nor pheromones have been implicated in regulation of these colonies, which can have hundreds or even thousands of workers. Yellowjackets and hornets have extremely aggressive and intimidating defensive responses to disturbances in the area around their nests.

The beginning stages of wasp colonies present an interesting evolutionary puzzle. Many wasp species found new nests with multiple queens. From a kin selection perspective, this seems counterintuitive: even if the queens are sisters, if one attains reproductive dominance, the reduction of the rest to the production of nieces may be a losing proposition, evolutionarily. How might multiple foundresses in wasp

FIGURE 14.26

Paper wasps, Polistes spp. *Photos: (A) Michael Breed, (B) Joan Strassman.*

(A)

(B)

colonies be explained? Possibly, they are not really working but are waiting on the periphery for the chance to take over the nest, should the dominant foundress die. Or perhaps the foundresses work together to ensure the early success of the nest but then later compete intensely over reproductive rights. Finally, the option of founding a nest alone may be so unfavorable that ecological constraints force grouping of sisters. Work by Joan Strassmann of Rice University and other researchers has shown that foundresses are typically sisters and that kin selection is the best model to explain this cooperation.[67,68] However, ecological constraints are quite important, and under the right ecological conditions, nest founding by solitary wasps, rather than by groups, is favored.

Bees

Most of the world's 25,000 or so bee species are solitary, but the eusocial species get the most notice. All but a few bee species specialize on nectar and pollen diets; the evolutionary history of bees is inextricably interwoven with their association with flowers. Bees, both solitary and eusocial, are the most important pollinators in nearly every ecosystem. In North America, the primitively eusocial bumblebees are among the most commonly observed native bees. The highly eusocial honeybee, which was introduced to North America from Europe in the 1600s, is now the dominant pollinator in all but the highest elevation or latitude ecosystems.

FIGURE 14.27
A yellowjacket nest. The colony is started by a single queen in the spring and grows to hundreds or thousands of highly defensive wasps by late summer. *Photo: Howard Ensign Evans, Colorado State University, Bugwood.org.*

427

Bumblebee queens leave their natal nest in the late summer or early fall and mate, usually with only one male. These queens then find a safe place, often underground, to spend the winter. In the spring, the queen establishes a nest, usually in an abandoned rodent burrow but sometimes in an old bird's nest or just in an underground cavity. The adopted nesting material from the rodent provides insulation and support for the bumblebee's nest, which is constructed of wax and cocoons spun by the larvae. In the early stages of colony life, the queen does all of the tasks in the nest; she constructs the nest, lays the eggs, forages for food, and defends the nest against predators and parasites. Her first batch of larvae, usually 4 to 8 or 10 in number, are all female and grow to be much smaller adults than the queen. These small adults, first seen in May or June, are workers and take over the work of the colony, except for egg laying. The queen maintains her reproductive dominance of the society by physical means, forming a dominance hierarchy in the nest once the workers are present. In warmer climates, additional female worker brood may be reared so that a bumblebee colony can grow to include several dozen workers. In late summer or early fall, the queen lays a mixed batch of male and female eggs; the female larvae hatching from these eggs are better fed than the males are and develop into large adults, which will complete the cycle by becoming the next summer's queens. The males are short-lived, serving only to mate with the queens. There are more than two dozen species of bumblebee in North America. The annual colony cycle, physical dominance hierarchy, overlap of queen and workers in their activities, and small colony size are all characteristics of primitive eusociality.

Honeybees are highly eusocial and have a perennial colony life cycle (see Figure 14.28); the key to honeybee survival is the storage of large amounts of nectar (in the form of honey) and pollen so that the colony can survive the winter. Honeybee nests are constructed in cavities, such as hollow trees. Bees excrete the wax, which is shaped into extensive combs. The queen honeybee lacks the hairs needed to carry pollen and never forages; her roles in the society are restricted to egg-laying and secreting pheromones, which are spread through the colony. Honeybee queen pheromones inform the workers of the presence of the queen and may inhibit workers from laying eggs; if the queen is removed, the absence of her pheromones

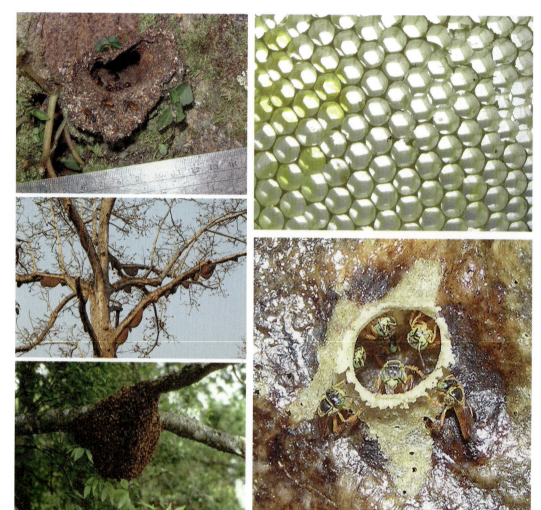

FIGURE 14.28
(Upper left and lower right) Nest entrances of tropical stingless bees, showing guards. (Middle and lower left) Nests of the giant honey bee, *Apis dorsata*, in Southeast Asia. (Upper right) Honeybee comb, showing the repeated hexagonal structure of cells in which larvae are reared and food is stored. *Photos: Michael Breed.*

stimulates efforts by the workers to rear replacement queens and to lay haploid male eggs, something they can do in the absence of being mated.

All female eggs have the potential to become either workers or queens; the vast majority of eggs are reared as workers. Larvae receiving a carbohydrate-rich diet develop into workers; the workers are uniform in size, have hairs over their bodies for collecting pollen, have very small ovaries, and have barbed stings. Queen larvae receive a diet rich in lipid and protein (royal jelly, which is produced by glands in the workers' heads); this rich diet stimulates development of a larger, relatively hairless female with small eyes, an elongated abdomen, large ovaries, and no barbs on her sting. Queens are reared in the spring and leave their natal colony on mating flights; a queen mates between 10 and 20 times when she is young and stores the semen for use throughout her life, which can last 5 or 6 years. When a swarm begins, the old queen in a colony leaves with about half the workers; this swarm searches for a new nesting location. Meanwhile, the new queen then takes over reproduction in the nest in which she hatched. The clear physical differentiation of queens and workers and queen dominance via pheromones are characteristics of a highly eusocial society.

SUMMARY

Chapter 13 shows how evolutionary forces of selfishness, reciprocity, and kin selection can bring groups of animals together. This chapter includes cooperation in birds, starting with aggregated nesting in swallows. In swallows, a distinct monogamous pair maintains each nest, and the aggregation results from widespread attraction of birds to (limited) good nesting sites, as well as possible benefits from shared vigilance against predators. Sociable weavers and acorn woodpeckers both nest cooperatively, with shared nest construction and maintenance and, particularly in acorn woodpeckers, cooperative rearing of nestlings. In both of these species, helpers at the nest also contribute to the survivorship of the group.

A similar progression occurs in mammals, ranging from the relatively independent lives of the members of a horse herd, which nevertheless gain antipredator benefits from group membership, to highly organized hyena and baboon societies. Among vertebrates, the permanently sterile workers in eusocial colonies of naked mole-rat colonies represent the extreme of helping behavior and of differential roles within a society.

In invertebrates, there are numerous examples of eusociality. They include gall-occupying plant feeders, shrimp, termites, ants, wasps, and bees. Invertebrate eusocial societies typically feature division of labor in colonies that makes the social group ecologically efficient. Division of labor contributes much to the dominant ecological roles of these animals. Reproductive differences between queens and workers are maintained, highlighting evolutionary implications of worker reproductive suppression. Eusocial invertebrates hold substantial resources in their nest and become targets for predation. This has stimulated novel and highly effective defensive strategies. Stinging, biting, and chemical defenses are all present in eusocial invertebrates.

The breadth of ecological roles and behavioral adaptations in eusocial animals correlates with their ecological dominance. Termites are important consumers of cellulosic material. Bees pollinate plants and play key roles in the maintenance of many communities. Ants have a tremendous range of feeding niches and are among the most numerous animals in most terrestrial ecosystems. In sum, social behavior provides a set of remarkable adaptations for the ecological success of many animal species.

429

STUDY QUESTIONS

1. Our examples of bird social behavior represented a range from aggregated nests (swallows) to groups with cooperative brood care (sociable weavers and acorn woodpeckers). What common themes exist among these species? Are there key differences in their social behavior?
2. Group living has many advantages but can have the downside of focusing competition for food within the group. In the examples of bird and mammal social behavior, what balances have been struck between the value of living in a group and competition for food?
3. How has social competition affected female morphology and behavior in the spotted hyena?
4. Can you explain the differences between division of labor based on a dominance hierarchy and temporal division of labor?
5. Why are eusocial insects so successful ecologically?
6. Why do the eusocial insects have such strong defensive mechanisms? Give some examples of particularly effective defenses.

Further Reading

Breed, M.D., 2010. Honeybees, In: Michael D. Breed and Janice Moore (Eds.), Encyclopedia of Animal Behavior, Academic Press, Oxford, Vol. 1, pp. 89–96.

Brown, J.H., 2010. Swallow Summer. Encyclopedia of Animal Behavior, Academic Press, Oxford, Vol. 1, pp. 548–552.

Dawkins, R., 1976. The Selfish Gene. Oxford University Press, New York, 368 pp.

Fewell, J.H., 2010. Division of Labor, In: Michael D. Breed and Janice Moore (Eds.), Encyclopedia of Animal Behavior, Academic Press, Oxford, Vol. 1, pp. 548–552

Gadau, J., Fewell, J.H. (Eds.), 2009. Organization of Insect Societies: From Genome to Sociocomplexity. Harvard University Press, Cambridge, Mass, 640 pp.

Holldobler, B., Wilson, E.O. 1990. The Ants. Belknap Press of Harvard University Press, Cambridge, Mass., 752 pp.

Lubin, Y., 2010. Spiders: Social Evolution, In: Michael D. Breed and Janice Moore (Eds.), Encyclopedia of Animal Behavior, Academic Press, Oxford, Vol. 2, pp. 329–334.

Stanford, C.B., 1998. The social behavior of chimpanzees and bonobos: empirical evidence and shifting assumptions. Current Anthropology 39(4), 399–420.

Wilson, E.O., 1974. The Insect Societies. Belknap Press of Harvard University Press, Cambridge, Mass., 562 pp.

Wilson, E.O., 1975. Sociobiology: The New Synthesis. Harvard University Press, Cambridge, Mass., 720 pp.

Notes

1. Wilson, E.O., 1975. Sociobiology: The new synthesis. Cambridge, Mass.: Belknap Press of Harvard University Press. 697 pp.

2. Schultz, T.R., Brady, S.G., 2008. Major evolutionary transitions in ant agriculture. PNAS 105, 5435–5440.

3. Holldobler, B., Wilson, E.O., 2011. The leafcutter ants: Civilization by instinct. W. W. Norton and Company, Inc., New York. 160 pp.

4. Rodrigues, A., Bacci, M., Mueller, U.G., Ortiz, A., Pagnocca, F.C., 2008. Microfungal "weeds" in the leafcutter ant symbiosis. Microb. Ecol. 56, 604–614.

5. Covas, R., Brown, C.R., Anderson, M.D., Brown, M.B., 2004a. Juvenile and adult survival in a southern temperate colonial cooperative breeder. Auk 121, 1199–1207.
Covas, R., Doutrelant, C., du Plessis, M.A., 2004b. Experimental evidence of a link between breeding conditions and the decision to breed or to help in a colonial cooperative breeder. Proc. R. Soc. Lond. B 271, 827–832.

6. Doutrelant, C., Covas, R., 2007. Helping has signalling characteristics in a cooperatively breeding bird. Anim. Behav. 74, 739–747.

7. Koenig, W.D., Walters, E.L., Haydock, J., 2009. Helpers and egg investment in the cooperatively breeding acorn woodpecker: testing the concealed helper effects hypothesis. Behav. Ecol. Sociobiol. 63, 1659–1665.
Koenig, W.D., Mumme, R.L., 1987. Population Ecology of the Cooperatively Breeding Acorn Woodpecker. Princeton University Press. 462 p.
Koenig, W.D., Haydock, J., Stanback, M.T., 1998. Reproductive roles in the cooperatively breeding acorn woodpecker: Incest avoidance versus reproductive competition. Am. Nat. 151, 243–255.
Haydock, J., Koenig, W.D., 2002. Reproductive skew in the polygynandrous acorn woodpecker. Proc. Natl. Acad. Sci. U.S.A. 99, 7178–7183.

8. Linklater, W.L., Cameron, E.Z., 2000. Tests for cooperative behaviour between stallions. Anim. Behav. 60, 731–743.
Linklater, W.L., Cameron, E.Z., Minot, E.O., Stafford, K.J., 1999. Stallion harassment and the mating system of horses. Anim. Behav. 58, 295–306.

9. Frank, L.G., 1986. Social organization of the spotted hyaena (Crocuta crocuta): I and II. Anim. Behav. 34, 1500–1527.

10. Holekamp, K.E., Sakai, S.T., Lundrigan, B.L., 2007. Social intelligence in the spotted hyena (Crocuta crocuta). Philos. Trans. R. Soc. B, Biol. Sci. 362, 523–538.

11. Grinnell, J.C., Packer, A.E., Pusey, 1995. Cooperation in male lions: kinship, reciprocity or mutualism? Anim. Behav. 49, 95–105.

12. Bygott, J.D., Bertram, B.C.R., Hanby, J.P., 1979. Male lions in large coalitions gain reproductive advantages. Nature 282, 839–841.

13. Packer, C., Pusey, A.E., Eberly, L., 2001. Egalitarianism in female African lions. Science 293 (5530), 690–693.
Packer, C.H., Robert, 1996. Lioness Leadership. Science 271 (5253), 1215–1216.

14. Spong, G., Creel, S., 2004. Effects of kinship on territorial conflicts among groups of lions, Panthera leo. Behav. Ecol. Sociobiol. 55, 325–331.

15. Goodall, J., 1986. The Chimpanzees of Gombe. Belknap Press, Cambridge, MA.

16. Nishida T (ed.), 1990. The Chimpanzees of the Mahale Mountains. University of Tokyo Press, Tokyo.

17. Langergraber, K., Mitani, J., Vigilant, L., 2007. The limited impact of kinship on cooperation in wild chimpanzees. Proc. Natl. Acad. Sci. U.S.A. 104, 7786–7790.

18. Muller, M., Mitani, J., 2005. Conflict and cooperation in wild chimpanzees. Adv. Study Behav. 35, 275–331.

19. Pusey, A., Williams, J., Goodall, J., 1997. The influence of dominance rank on the reproductive success of female chimpanzees. Science 277, 828–831.

20. Duffy, K., Wrangham, R., Silk, J., 2007. Male chimpanzees exchange political support for mating opportunities. Curr. Biol. 17, R586–R587.

21. Silk, J.B., 2007. Chimps don't just get mad, they get even. PNAS 104, 13537–13538.

22. Reeve, H.K., 1992. Queen activation of lazy workers in colonies of the eusocial naked mole-rat. Nature 358 (6382), 147–149.

23. O'Riain, M.J., Jarvis, J.U.M., Alexander, R., et al., 2000. Morphological castes in a vertebrate. Proc. Natl. Acad. Sci. U.S.A. 97 (24), 13194–13197.

24. Bromham, L., Harvey, P.H., 1996. Behavioural ecology: naked mole-rats on the move Curr. Biol. 6 (9), 1082–1083.

25. Cooney, R., Bennett, B.C., 2000. Inbreeding avoidance and reproductive skew in a cooperative mammal. Proc. R. Soc. Lond. B 267, 801–806.

26. Scantlebury, M., Speakman, J.R., Oosthuizen, M.K., Roper, T.J., Bennett, N.C., 2006. Energetics reveals physiologically distinct castes in a eusocial mammal. Nature 440, 795–797.

27. Henry, E.C., Dengler-Crish, C.M., Catania, K.C., 2007. Growing out of a caste—reproduction and the making of the queen mole-rat. J. Exp. Biol. 210 (2), 261–268.

28. Clarke, F.M., Faulkes, C.G., 1997. Dominance and queen succession in captive colonies of the eusocial naked mole-rat, *Heterocephalus glaber*. Proc. R. Soc. Lond., B, Biol. Sci. 264 (1384), 993–1000.

29. Dawkins, R., 1982. The Extended Phenotype. Oxford University Press.

30. O'Donnell, S., 1995. Division of labor in postemergence colonies of the primitively eusocial wasp *Polistes instabilis de Saussure* (Hymenoptera, Vespidae). Insectes Soc. 42, 17–29.

31. Markiewicz, D.A., O'Donnell, S., 2001. Social dominance, task performance and nutrition: implications for reproduction in eusocial wasps. J. Comp. Physiol. A Neuroethol. Sens. Neural. Behav. Physiol. 187, 327–333.

32. Breed, M.D., Harrison, J., 1988. Caste in the giant tropical ant, *Paraponera clavata*. J. Kans. Entomol. Soc. 61, 285–290.

33. Wilson, E.O., 1974. The Insect Societies. Belknap Press of Harvard University Press. 562 pp.

34. Feener, D.H., Moss, K.A.G., 1990. Defense against parasites by hitchhikers in leaf-cutting ants: a quantitative assessment. Behav. Ecol. Sociobiol. 26, 17–29.

35. Seeley, T.D., 1982. Adaptive significance of the age polyethism schedule in honeybee colonies. Behav. Ecol. Sociobiol. 11, 287–293.

36. Muscedere, M.L., Willey, T.A., Traniello, J.F.A., 2009. Age and task efficiency in the ant *Pheidole dentata*: young minor workers are not specialist nurses. Anim. Behav. 77, 911–918.

37. Shorter, J.R., Tibbetts, E.A., 2009. The effect of juvenile hormone on temporal polyethism in the paper wasp *Polistes dominulus*. Insectes Soc. 56, 7–13.

38. Seeley, T.D., 1982. Adaptive significance of the age polyethism schedule in honeybee colonies. Behav. Ecol. Sociobiol. 11, 287–293.

39. Trivers, R.L., Hare, H., 1976. Haploidploidy and the evolution of the social insect. Science 191, 249–263.

40. Ratnieks, F.L.W., Visscher, P.K., 1989. Worker policing in the honey bee. Nature 342, 796–797.

41. Tarpy, D., Page, R.E., 2002. Sex determination and the evolution of polyandry in honey bees (*Apis mellifera*). Behav. Ecol. Sociobiol. 52, 143–150.

42. Kukuk, P.F., May, B., 1991. Colony dynamics in a primitively eusocial halictine bee *Lasioglossum (Dialictus) zephyrum* (Hymenoptera: Halictidae). Insectes Soc. 38, 171–188.

43. Pratte, M., 1997. Recognition and social dominance in *Polistes* wasps. J. Ethol. 15, 55–59.

44. van Doorn, A., Heringa, J., 1986. The ontogeny of a dominance hierarchy in colonies of the bumblebee, *Bombus terrestris* (Hymenoptera, Apidae). Insectes Soc. 33, 3–25.

45. Slessor, K.N., Winston, M.L., Le Conte, Y., 2005. Pheromone communication in the honeybee (*Apis mellifera L.*). J. Chem. Ecol. 31, 2731–2745.

46. Lüscher, M., 1960. Hormonal control of caste differentiation in termites. Ann. N. Y. Acad. Sci. 89, 549–563.

47. Korb, J., Weil, T., Hoffmann, K., Foster, K.R., Rehli, M., 2009. A gene necessary for reproductive suppression in termites. Science 324, 758.

48. Prestwich, G.D., 1983. Chemical systematics of termite exocrine secretions. Annu. Rev. Ecol. Syst. 14, 287–311.

49. Jones, T.H., Clark, D.A., Edwards, A.A., Davidson, D.W., Spande, T.F., Snelling, R.R., 2004. The chemistry of exploding ants, *Camponotus spp.* (cylindricus complex). J. Chem. Ecol. 30, 1479–1492.

50. Duffy, J.E., 1996. Eusociality in a coral-reef shrimp. Nature 381, 512–513.
Duffy, J.E., Morrison, C.L., Macdonald, K.S., 2002. Colony defense and behavioral differentiation in the eusocial shrimp *Synalpheus regalis*. Behav. Ecol. Sociobiol. 51 (5), 488–495.
Duffy, J.E., Morrison, C.L., Rios, R., 2000. Multiple origins of eusociality among sponge-dwelling shrimps (*Synalpheus*). Evolution 54 (2), 503–516.

51. Toth, E., Duffy, J.E., 2005. Coordinated group response to nest intruders in social shrimp. Biol. Lett. 1 (1), 49–52.

52. Crespi, B.J., 1992. Eusociality in Australian gall-thrips. Nature 359, 724–726.
Crespi, B.J., 1994. Three conditions for the evolution of eusociality: are they sufficient? Insectes Soc. 41, 393–400.

53. Normark, B.B., Jordal, B.H., Farrell, B.D., 1999. Origin of a haplodiploid beetle lineage. Proc. R. Soc. Lond. B, Biol. Sci. 266 (1435), 2253–2259.

54. Chapman, T.W., Crespi, B.J., Kranz, B.D., et al., 2000. High relatedness and inbreeding at the origin of eusociality in gall-inducing thrips. Proc. Natl. Acad. Sci. U.S.A. 97 (4), 1648–1650.
Chapman, T.W., Kranz, B.D., Bejah, K.L., et al., 2002. The evolution of soldier reproduction in social thrips. Behav. Ecol. 13 (4), 519–525.

55. Bono, J.M., Crespi, B.J., 2008. Cofoundress relatedness and group productivity in colonies of social *Dunatothrips* (Insecta: Thysanoptera) on Australian Acacia. Behav. Ecol. Sociobiol. 62, 1489–1498.
Bono, J.M., Crespi, B.J., 2006. Costs and benefits of joint colony founding in Australian Acacia thrips. Insectes Soc. 53, 489–495.

56. Perry, S.P., Chapman, T.W., Schwarz, M.P., Crespi, B.J. 2004. Proclivity and effectiveness in gall defence by soldiers in five species of gall-inducing thrips: benefits of morphological caste dimorphism in two species (*Kladothrips intermedius* and *K. habrus*). Behav. Ecol. Sociobiol. 56 (6), 602–610.

57. Kranz, B.D., Chapman, T.W., Crespi, B.J., Schwarz, M.P., 2001. Social biology and sex ratios in the gall-inducing thrips, *Oncothrips waterhousei* and *Oncothrips habrus*. Insectes Soc. 48 (4), 315–323.
   Kranz, B.D., Schwarz, M.P., Morris, D.C., Crespi, B.J., 2002. Life history of *Kladothrips ellobus* and *Oncothrips rodwayi*: insight into the origin and loss of soldiers in gall-inducing thrips. Ecol. Entomol. 27 (1), 49–57.
   Kranz, B.D., Schwarz, M.P., Mound, L.A., Crespi, B.J., 1999. Social biology and sex ratios of the eusocial gall-inducing thrips *Kladothrips hamiltoni*. Ecol. Entomol. 24 (4), 432–442.
   Kranz, B.D., Schwarz, M.P., Wills, T.E., et al., 2001. A fully reproductive fighting morph in a soldier clade of gall-inducing thrips (*Oncothrips morrisi*). Behav. Ecol. Sociobiol. 50 (2), 151–161.

58. Ijichi, N., Shibao, H., Miura, T., et al., 2005. Comparative analysis of caste differentiation during embryogenesis of social aphids whose soldier castes evolved independently. Insectes Soc. 52 (2), 177–185.
   Ijichi, N., Shibao, H., Miura, T., et al. 2005. Soldier caste with morphological and reproductive division in the aphid tribe *Nipponaphidini*. Insectes Soc. 52 (2), 132–138.

59. Aoki, S., 1991. Aphid soldiers discriminate between soldiers and non-soldiers, rather than between kin and nonkin, in *Ceratoglyphina bambusae*. Anim. Behav. 42, 865.
   Aoki, S., Imai, M., 2005. Factors affecting the proportion of sterile soldiers in growing aphid colonies. Popul. Ecol. 47 (2), 127–136.
   Aoki, S., Kurosu, U., 2004. How many soldiers are optimal for an aphid colony? J. Theor. Biol. 230 (3), 313–317.

60. Benton, T.G., Foster, W.A., 1992. Altruistic housekeeping in a social aphid. Proc. R. Soc. Lond. B, Biol. Sci. 247 (1320), 199–202.

61. Pike, N., Foster, W., 2004. Fortress repair in the social aphid species *Pemphigus spyrothecae*. Anim. Behav. 67, 909–914.

62. Fukatsu, T., Sarjiya, A., Shibao, H., 2005. Soldier caste with morphological and reproductive division in the aphid tribe *Nipponaphidini*. Insectes Soc. 52, 132–138.

63. Ladley, D., Bullock, S., 2005. The role of logistic constraints in termite construction of chambers and tunnels. J. Theor. Biol. 234 (4), 551–564.

64. Miura, T., 2004. Proximate mechanisms and evolution of caste polyphenism in social insects: from sociality to genes. Ecol. Res. 19, 141–148.

65. Sendova-Franks, A.B., Hayward, R.K., Wulf, B., Klimek, T., James, R., Planqué, R., et al., 2010. Emergency networking: famine relief in ant colonies. Anim. Behav. 79, 473–485.

66. Reber, A., Meunier, J., Chapuisat, M., 2010. Flexible colony-founding strategies in a socially polymorphic ant. Anim. Behav. 79, 467–472.

67. Queller, D.C., Peters, J.M., Solís, C.R., Strassmann, J.E., 1997. Control of reproduction in social insect colonies: individual and collective relatedness preferences in the paper wasp, Polistes annularis. Behav. Ecol. Sociobiol. 40, 3–16.

68. Goodnight, K.F., Strassmann, J.E., Klingler, C.J., Queller, D.C., 1996. Single mating and its implications for kinship structure in a multiple-queen wasp. Ethol. Ecol. Evolution. 8, 191–198.

Conservation and Behavior

LEARNING OBJECTIVES

Studying this chapter should provide you with the knowledge to:

- Know that conservation behavior offers a broad range of tools for understanding and possibly manipulating animal behavior to further conservation.
- Learn how maintenance of animals in captivity can serve the goals of preserving endangered species, conserving genetic resources, or growing populations of animals

15.2 SPECIES PROTECTION IN NATURAL HABITATS

How is species protection status assessed and described? The answer to this question helps to direct this discussion of conservation behavior. Conservationists usually begin this process with a response to reports or surveys that show a species has a limited population size or has undergone rapid population declines. Some species are rare or limited in range simply as a reflection of their biology. *Endemics*, populations found in restricted geographic areas, often present particularly serious conservation challenges and are very susceptible to habitat destruction and climate change. For example, the Tahiti monarch is a bird that is endemic to Tahiti; fewer than 100 individuals of this endangered species persist in the wild. This illustrates the larger point that island animals are particularly at risk; for instance, many Hawaiian animals are threatened or endangered endemics.[8] An interesting contrast is that while two North American bird species, Carolina parakeets and passenger pigeons, have been pushed to extinction in the past 150 years and two others, ivory-billed woodpeckers and Bachman's warbler, may be extinct, more than 25 Hawaiian bird species have disappeared (http://hbs.bishopmuseum.org/endangered/ext-birds.html). The land area of Hawaii is about 0.2% that of the continental United States; endemism in a small land area makes Hawaii's species particularly vulnerable to extinction.

OF SPECIAL INTEREST: ISLAND BIOGEOGRAPHY, BEHAVIOR, AND CONSERVATION

The theory of *island biogeography*, first proposed by Robert MacArthur and E. O. Wilson, is another example of "academic" science that has profound implications for conservation, ranging from extinctions on islands to the design of the wildlife reserve—another type of "island."[9] MacArthur and Wilson focused on (1) how island size affects diversity (using what are called *species-area curves;* see Figure 15.2); (2) how distance from the mainland and other islands affects dispersal to the island (see Figure 15.3); (3) how persistence of a population on an island and species' *life history strategies*, which MacArthur and Wilson characterized along a continuum of *r and K selection*, interact with colonization and extinction (see Figure 15.4). Many of the elements that make up the theory of island biogeography are behavioral; perhaps the most significant

KEY TERM Island biogeography is a field of scientific inquiry, proposed by Robert MacArthur and E. O. Wilson, that integrates the study of the distribution of animals on islands with knowledge of the behavioral and ecological processes that determine that distribution.

KEY TERM A species-area curve depicts the relationship between island size and the number of species found on a given island.

KEY TERM A life history strategy is the combination of the age at first reproduction, the number of offspring per clutch, the number of clutches per year, and the number of years of reproduction for any one animal in a population. These factors together determine r, or the rate of population growth.

KEY TERM The terms r and K selection refer to a continuum of life history strategies, ranging from very high reproductive rates in colonizing species (r-selected) to low reproductive rates in territorial species (K-selected).

FIGURE 15.2
Example of a species-area curve. As the area sampled within a habitat increases (*x*-axis), the number of species increases as well (*y*-axis), but eventually nearly all of the species have been found, so further effort yields few new species. Some sites, such as A, require more sampling than others, such as B, to reach the point at which the curve flattens.

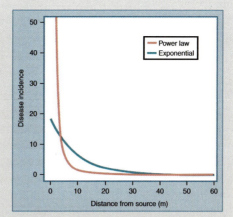

FIGURE 15.3
The effect of distance from a source population to a new habitat on dispersal. As distance increases (*x*-axis), the likelihood of successful dispersal diminishes rapidly.

FIGURE 15.4
The *r–K* continuum of life history strategies; r-selected species are good dispersers, but often not behaviorally competitive in established communities.

is dispersal (see Chapter 8). Because human activities fragment large tracts of habitat into smaller habitat islands, and because reserves are habitat islands, embedded in a "sea" of less welcoming places, this field of study has great relevance to conservation.[10]

Once a species is thought to be in trouble, more precise surveys are usually conducted. They can result in classification of a species or subspecies as being a conservation concern. Political arguments have surrounded the identification and classification process, and as a result, classifications by government agencies may differ from those of conservation organizations.

The listing of a species by the U.S. Fish and Wildlife Service (USFWS) as endangered or threatened is a politicized process that may pit landowners against conservationists. This means that interpretation of USFWS listings needs to be tempered by knowledge of political issues. Presently, the International Union for the Conservation of Nature (IUCN) appears to provide the most credible evaluations; the IUCN has the advantage of taking a global, rather than national, view. The *IUCN Red List*, which is more politically independent and contains conservation assessments for many species, can be accessed online. Another good source of information about North American birds is the Audubon Society's watch list, which highlights birds with declining populations. Awareness of threatened or endangered birds and mammals tends to be much higher than that for other organisms; this leads to criticisms of listing processes as being biased to protect charismatic animals and consequently being not only incomplete, but also ill considered with respect to animals that are less conspicuous but perhaps foundational in ecosystems. The IUCN Red Listing process is more inclusive than the USFWS process, in terms of including invertebrates and species that are less obvious to the public eye; nonetheless, the reality is that the conservation status of the vast majority of the planet's species is simply unknown.

Three categories of species in trouble are particularly important in conservation—*threatened*, *endangered*, and *extinct in the wild*. Species that are extinct in the wild are divided into those that are completely extinct and those that are maintained only in captive populations. It is these captive species, whose last hope lies in the maintenance of captive populations and reintroductions in the wild, that have been the focus of the most intensive programs of behavioral research. Famous extinctions within the past two centuries include the dodo, a large nonflying bird that was endemic to Mauritius, and the passenger pigeon and Carolina parakeet in North America. The abundance of these two North American birds surpassed the descriptive abilities of most observers until early in the twentieth century; hunting was a major factor in their extinction. No North American mammals have suffered complete extinction, although the American bison and the black-footed ferret came perilously close. The USFWS defines "endangered species" as ones that are in danger of extinction and "threatened" species as those that are at risk of becoming endangered. These definitions are necessarily nebulous, a factor that adds to contention in conservation discussions. In addition, the USFWS sometimes recognizes subspecies as threatened or endangered, even if other populations of the species are healthy.

> **KEY TERM** A threatened species is one that is at risk for becoming endangered.

> **KEY TERM** An endangered species is one that is at imminent risk to become extinct.

> **KEY TERM** A species that is extinct in the wild no longer exists in its natural habitat or in other noncaptive environments.

438

DISCUSSION POINT: CONSERVATION PRIORITIES

Is it better to focus conservation efforts and funding on identifying the potential for species loss and campaigning to save individual species, or to target habitat areas to save intact ecosystems? With classmates or friends, take turns defending each strategy.

Conservation efforts tend to focus on large and/or colorful birds and mammals because these species capture the interest of the public, and consequently government funding and private donations are available to support their conservation. In North America, many of large mammals, such as mountain lions, wolves, and grizzly bears, are carnivores, and this sets up politically charged situations because such carnivores are regarded by private landowners as behavioral threats to livestock, pets, and perhaps to themselves. Animals do not recognize park or reserve boundaries as dispersal barriers. Populations in parks, such as wolves in the Greater Yellowstone Ecosystem inevitably serve as sources of dispersing animals into surrounding areas. Also, landowners may fear that legal protection of critical habitat will reduce the value of their land or prevent them from changing the use of the land.[11]

Animals maintained in parks or reserves can be reservoirs for disease that might be passed to livestock in surrounding areas. Bison and elk in the Greater Yellowstone Ecosystem may possibly harbor brucellosis, a bacterial disease that is devastating to livestock. Treating animals in wild populations for diseases such as brucellosis, rabies, or plague is difficult or impossible; when behavioral interactions between these populations and livestock, pets, or humans create risk for transmission of disease, landowner and public sentiment often turns against maintenance of the wild populations of animals. Many bats are the subject of serious conservation concerns; they are also stereotyped as rabies-carrying vermin by much of the public, reducing the pressure for conservation of bats.

OF SPECIAL INTEREST: BAT CONSERVATION

Bats that relied on forest elements, such as hollow trees, for roosting have suffered major impacts from deforestation. It may surprise you to learn that, like bluebirds and purple martins (discussed later in this chapter), bats can use nest boxes, and many people are involved in providing bat roosts by constructing nest boxes for bats (http://www. batcon.org/). Clearly, preserving natural roost sites is the highest priority, but when those have been removed, artificial roosts are an interesting conservation alternative.[12] (Figure 15.5).

In tropical forests, bat species fill a wide variety of feeding niches; in addition to the familiar insectivores, bats serve as pollinators and dispersers of seed. Vampire bats, of course, are common in the tropics as well. Temperate zone bats are more typically insect eaters. Whatever their feeding niche, bats play important ecological roles that are disrupted by loss of nesting sites. Nest boxes in Costa Rican pastures support a range of bat species, an encouraging finding for bat conservation.[13] Artificial roosts also show promise in temperate zone areas.[14]

FIGURE 15.5
Bat boxes, designed to assist conservation of species such as the big brown bat, which suffers from habitat destruction. *Photo: Rick Adams.*

439

Unfortunately for some species that lack captive populations and are extinct in the wild, and that lack captive populations, the only hope is that remnant undiscovered populations remain. This hope is not unfounded, as a few species thought to be extinct, such as po'ouli honeycreeper (Hawaii), have been "rediscovered." A similar phenomenon occurred with the ivory-billed woodpecker. Until 2005, the last sightings of ivory-billed woodpeckers were in Louisiana in 1948 and in Cuba in 1988, and the species was thought to be extinct. Sightings reported in 2005 suggest that remnant populations of this species may exist in Arkansas. The search for this species continues, with behavioral traits (a distinctive call) giving biologists a major tool for recognizing the species. Another rediscovery is that of a golden-crowned manakin, spotted in Brazil in 2002 after a 45-year gap in sightings.[15] In contrast, the Golden Toad, a native to the mountains of Costa Rica, was never seen outside an area of about 4 square kilometers. This toad disappeared in 1989 and is presumed to be extinct, although the search continues for remnant populations. Understanding behavioral signs of species that are potentially extinct in the wild, such as scent marks, nests, food caches, flight patterns for birds spotted at a distance, and distinctive vocalizations, can play important roles in rediscoveries of species.

OF SPECIAL INTEREST: THE PO'OULI HONEYCREEPER

What about species that are rare and elusive to begin with? The po'ouli honeycreeper was endemic to the Hawaiian island of Maui and was not known to scientists until 1973 (see Figure 15.6). The range of this species was a 1,300 hectare tract of forest. When discovered, this species' population was thought to be fewer than 150 birds; habitat destruction contributed to a decline, so that in 2000, only three birds, one male and two females, could be located in the wild. Each bird occupied a separate territory, and they did not seem likely to form a mating pair. This presented an extreme conservation dilemma; could the species be preserved based on such a limited population? Or was the species' fate already determined, making human intervention fruitless?

FIGURE 15.6
The po'ouli honeycreeper. *Photo: Paul E. Baker, USFWS.*

In a desperate effort to save the species, conservation biologists caught a female and translocated her to the male's territory. The female left the male's territory soon after she was moved and flew back to her previous territory. A male captured in an attempt to form a captive breeding population died, and the remaining two birds disappeared. Nevertheless, hope remains that this species will be rediscovered and that information gained from these experiences will facilitate a captive breeding program.

Other species, such as the Rodrigues fody, have recovered from extremely low population sizes. The continuing argument among conservationists is whether identifying and preserving (or reforesting) habitat areas is the best approach, or if the more extreme measures of translocation or captive breeding are more likely to succeed. This question is probably best resolved on a case-by-case basis, with the answer depending on the threats to the species and the likelihood of success of either approach. A glaring fault in the process is that decision making by governmental agencies can be excruciatingly slow, hampering emergency actions.

FIGURE 15.7
A pile of bison skulls.

Many species that are extinct in the wild survive in populations in zoos or small reserves. Prior to the twentieth century, the American bison lived in vast herds in the United States, forming a key element of the plains ecosystem. Bison grazing, combined with fire and migratory locust swarms, prevented woody plants from invading grasslands. Bison were easy to kill with guns, and in the late nineteenth century, large numbers of bison were shot, often for their hides, but sometimes for no purpose other than the "pleasure" of the kill (see Figure 15.7).

Perhaps as many as 50,000,000 bison were killed, leaving a few small remnant populations on isolated ranches and Indian reservations; the National Park Service estimates that at the low point, the bison population numbered 600. A few ranchers and conservationists, including President Theodore (Teddy) Roosevelt, developed an interest in the preservation of this species, and captive and semicaptive breeding have generated a modern-day population in excess of 100,000. The American bison became federally protected in 1894; it is one of the first examples of using laws for conservation protection of a species.

Behaviorally, bison are indeed wild animals—unpredictable, strong, and largely unmanageable; their wild behavior is part of their charisma. Probably much of the genetic variation of bison was lost in the population bottleneck of the late 1800s.

An example of the remaining variation is the Wood Bison (see Figure 15.8), found in northern Alberta, Canada; this subspecies is larger than Plains Bison found in the United States, with males weighing about a ton (900 kg). One behavioral concern in the conservation of bison is their ability to hybridize with domestic cattle; beginning with the initial stages of the bison recovery, bison have been behaviorally managed in an effort to maintain the populations' genetic integrity. Another concern is that ranchers raising bison for slaughter might select for more behaviorally manageable (domesticated) stock. Plains and Wood subspecies of bison now live in several national parks in the United States and Canada, as well as on private ranches and reserves.

FIGURE 15.8
American bison, a species that was re-established after near extinction. *Photo: US Fish and Wildlife Service.*

441

... management of a much-reduced population and ... model that conservationists have applied ... scussed in detail in Section 15.5 on captive ... ith bison proceeded largely through trial ... pulses to save the species, to provide stock ... neat to food markets. A saving grace for the ... of the American West, embodied by larger- ... uffalo Bill Cody. This theme of romanticizing ... f conservation efforts continues to this day; ... phasize the significance—and in some cases, ... species in maintaining ecosystems.

... eeds for species is to identify why a particular ... of critical habitat, reduction of population ... nbreeding and maintains genetic diversity, ... species may have more than one of these ... utions. For example, black-footed ferrets, ... nbination of reduction of populations below ... er, and habitat loss. (The ferrets live in prairie

... cians have not only quarreled over the ... endangered, but they have also engaged in ... forts should be focused on the preservation of ... eservation of critical ecosystems. The study of ... s will become apparent, species conservation ... w well, or poorly, they embrace an understanding ... central to reserve design because both the ... at components are dictated by the behavioral ... nity.

15.3 EXTINCTIONS AND BEHAVIOR

The loss of a species from a community can have far-reaching effects on the behavior of remaining species in that community. The extirpation of wolves and mountain lions from much of the United States, for example, has led to changed dietary habits and social behavior of the coyote. In the absence of both wolves and mountain lions, coyotes typically become the top mammalian predator in their communities. Coyote social groups become larger, and these larger social groups, by hunting cooperatively, can handle larger prey. Coyotes are also bolder and more commonly seen in daylight when wolves are absent. In the Greater

Yellowstone Ecosystem, where wolves have been reintroduced, coyotes have been pushed back to the behavioral margin, subsisting on smaller prey and themselves risking predation from wolves.

More generally speaking, top predators are commonly among the most at-risk species in any community. The loss of top predators releases second-tier predators, allowing their populations to grow and exploit broader ranges of food. However, larger herbivores, which the second-tier predators cannot handle because of their size, may also experience a release of their populations, resulting in extensive loss of vegetation. In North America, when wolves and mountain lions are absent, the largest herbivores, such as deer (mule and white tailed), elk, and bison live virtually predator-free. How does this affect herbivore behavior? Hypothetically, herbivores without predators should be bolder, more willing to forage during daylight, and spend less time being vigilant. These effects have been demonstrated in elk, for example.[16]

Disruption of Migration: High Risks for Extinction

Long-distance migrants that depend on small patches of stopover habitats are very susceptible to habitat disruption. For many migratory birds, wetlands serve as important intermediate habitats; one of the compelling motivations for legal protection of wetlands is their key habitat role for migratory birds. Identification and protection of essential stopover points for birds other than waterfowl, such as warblers, is more challenging because of the warblers' small size, nocturnal migratory flights, and secretive habits. Conservation of the monarch butterfly, which migrates from the United States and Canada to the mountains of Mexico, has required international cooperation in attempts to prevent deforestation of wintering habitat in Mexico.

442

OF SPECIAL INTEREST: CONSERVATION MISTAKES

Sometimes big mistakes are made in well-intentioned conservation programs. An attempt to establish a new migratory population of Whooping Cranes is a great example of how not understanding bird behavior led to a failed conservation program. This species is highly endangered and is placed at greater risk because the cranes exist as a single population with only one overwintering site—Aransas National Wildlife Refuge on the Texas coast. The well-intentioned idea was simple: Whooping Crane eggs would be collected and placed into Sandhill Crane nests, a species that would accept and rear Whooping Cranes. Whooping Cranes would then, in theory, learn feeding and migratory habits from the Sandhills and establish a new population. Unfortunately, the program did not take into account the fact that birds typically imprint on the adults that care for their nest, and most importantly, that males then court females that look like their adopted mother. (See the section "Imprinting" in Chapter 5.) The young Whooping Cranes survived well, but failed to mate because they were imprinted on the Sandhills.

Intentional Extinctions

Humans have often responded to perceived predation risk for livestock or competition for agricultural products with programs that attempt to eradicate entire species. Coyotes present an excellent example of a species that has been shot, trapped, and poisoned in an all-out effort at eradication. Many states offered bounties on coyotes; while coyotes were not exterminated, their populations were marginalized in many areas. Ranchers and agricultural economists failed to understand that the impacts of coyotes on livestock were potentially outweighed by the fact they limited herbivore populations.[17] Even now the relative balance of the impacts of coyotes on livestock and on potential competitors with livestock is not fully understood. The general result of the campaign against coyotes is probably an increase in rodent and rabbit (particularly jackrabbit) populations, which in turn directly compete for forage with cattle and sheep. A similar eradication program aimed at wolves nearly succeeded in driving them to extinction in North America.

Intentional eradication is not a modern idea. Some medieval Europeans believed that cats were associated with Satan, and they were systematically slaughtered, a decimation hypothesized to have contributed to the spread of the Black Plague. This disease, carried by rats and their fleas, killed approximately one-third of the human population of Europe and changed the course of history.

Extinction Due to Hunting and Fishing

Many animals are hunted for meat, as well as for the sale of pelts or feathers to the fashion trade. Historically, hunting has been a major cause of reduction of bird and mammal species; hunters have extirpated many local populations, although few of these species have been completely eliminated by hunting. In North America, beavers were once one of the most abundant species; hunting pressure greatly reduced beaver populations, which have only recently begun to recover in some parts of their previous range. Beavers are an interesting example because their comeback has generated vexing issues. Some behavioral attributes of beavers—their toppling of trees for food and construction, their construction of dams or blocking drains with tree branches, and their building of burrows in riverbanks—can cause land managers to view them as pests. Colonizing beavers are often euthanized or translocated away from urban and suburban habitats to preserve streamside plants or to protect dams or dikes built for flood control.

Animals for which hunting actually contributed to extinction include passenger pigeons and Carolina parakeets. Passenger pigeons were probably the most common North American birds prior to the arrival of Europeans on the continent. They numbered in the billions and were hunted for their meat. Carolina parakeets numbered in the millions and were hunted for their feathers. Both species were extinct by the early 1900s.

The greatest behavioral contributor to the extinction of the passenger pigeon was its apparent dependence on large, gregarious, flocks and the resulting ease with which they were hunted; small captive populations maintained after its extinction in the wild failed to reproduce and ultimately dwindled to a single bird, "Martha," which died in the Cincinnati Zoo in 1914 (http://www.si.edu/Encyclopedia_SI/nmnh/passpig.htm; see Figure 15.9). The last captive Carolina parakeet, "Incas," also died in Cincinnati, in 1918, and the last small wild flock was seen in Florida in 1920. Should this be taken as a cautionary note when the population sizes of other highly social species are reduced?

FIGURE 15.9
The passenger pigeon.
Source: John James Audubon painting (passenger pigeons were common during Audubon's lifetime).

443

OF SPECIAL INTEREST: THE PASSENGER PIGEON

The passenger pigeon is yet another example of how behavior may be a strong influence in species conservation. In the seventeenth century, this species may have accounted for 25–40% of all birds in what was to become the United States (http://web4.audubon.org/bird/Boa/F29_G3a.html); nesting groups could occupy over 800 square miles. Cotton Mather described one flock's flight as stretching over a mile wide and taking several hours to pass overhead. The highly social habits of the birds—effective against most predators—became their undoing, as they were easy prey for professional hunters, who netted them at nesting sites or simply knocked young birds out of the nest and sold them to urban meat markets. There were no provisions for maintaining a breeding population. The species declined noticeably by 1860; nonetheless, in 1878 in one location alone, 50,000 birds were killed every day for almost 5 months. Within 300 years of Mather's observations, the passenger pigeon was extinct. While the deforestation of the eastern United States would have doomed the passenger pigeon—there are no remaining forests that can accommodate flocks of over 130,000,000!—the professional hunter and the birds' inability to thrive in small groups were the immediate cause of its demise.

FIGURE 15.10
A brown tree snake.
Photo: Gordon H. Rodda, USFWS.

Invasive Species and Extinctions

Invasive species can exhibit ecologically dominant behavior, causing extinctions of indigenous species. The brown tree snake, native to the South Pacific, was accidentally introduced into Guam after World War II; prior to that introduction, only one species of snake existed on Guam, and that snake was a specialized resident of termite nests. In contrast, the brown tree snake is a voracious predator on birds, against which the birds of Guam have no evolved defenses. Ten of the 12 forest birds that were native to Guam are now extinct; because forest plants on Guam depended on these birds for pollination and seed dispersal, the effects of the snake reverberate through the ecosystem. The brown tree snake has proven unstoppable, a clear indicator of the extinction dangers that can come with species introductions (see Figure 15.10).[18]

Successful invasive species often outcompete native species for critical niche components, such as food and nesting sites. Other events such as *habitat fragmentation* sometimes open opportunities for colonization by non-native species, especially if native species do not adapt to modified conditions.[19]

> **KEY TERM** Habitat fragmentation is the breakdown, as the result of human activities, of large expanses of habitat into small habitat islands that are separated by human development.

Extinction threats posed by invasive species underscore the importance of understanding the behavioral predictors of invasiveness. Are there behavioral commonalities among brown tree snakes, fire ants, Argentine ants, cane toads, house sparrows, and Norway rats (see Figure 15.11) that make them successful invaders of so many habitats? One trait is the ability to adapt easily to novel environments. Another is dispersal behavior that increases the probability that invasives will hitchhike on ships, airplanes, and automobiles. If invasives competitively overwhelm native species, then invasions can contribute to population collapses of native species and resulting community degradation. These two behavioral attributes—adaptability and a tendency to disperse—are at least in part due to lack of neophobia, that is, a lack of caution that, in turn, may be fatal to an animal introduced into an open, wild, habitat.

FIGURE 15.11
The Norway rat, a common habitant of urban environments.
Photo: Douglas E. Norris.

Neophobia, the fear of new objects or animals in the environment, can be a critical element in animal's survival. Neophobia keeps animals from consuming potentially poisonous foods, from approaching predators, and from entering dangerous human-modified environments. In birds and mammals, feeding preferences are often shaped by juvenile experiences with food items; there is a temporal window during development when the young animals shape their diets. During this period, young need to be more open to novel foods simply because everything is new (see Figure 15.12). In many species, the exposure of juveniles to new food is driven by parental diet or is carefully monitored by adults in a social group.[20,21,22,23] In contrast, *neophilia*, the desire for new or novel experiences, is rare in animals and usually limited to periods of dispersal.[24] Domestic dogs tend to be neophilic in their choice of toys, a behavioral trait that may have been favored in the evolution of dogs as companions to humans.[25]

> **KEY TERM** Neophobia is the fear of the new or unusual.

> **KEY TERM** Neophilia is an attraction to the new or unusual.

444

FIGURE 15.12
Young robin (left) and fox (right) exploring for food. *Photos: Left, Sean Varner, U.S. Fish and Wildlife Service; Right, Lamar Gore, U.S. Fish and Wildlife Service.*

Summary: Extinctions and Conservation Behavior

Knowing about the behavior of imperiled animals can inform conservation decisions, such as protecting migratory stopover habitats or designing replacement habitats for migrating birds. Sometimes knowing the proximate behavioral causes of animal population declines, such as the predatory behavior of raccoons on bluebird nests, assists in the design of appropriate conservation plans; raccoons can be foiled by mechanical baffles that prevent their entry into a nest box (see Figure 15.13). Biological invasions, and consequent extinctions, may be prevented if the dispersal behavior of the invader is appreciated. Although a fairly new subdiscipline of animal behavior, conservation behavior is already providing powerful tools for understanding and preventing animal extinctions.

15.4 RESERVE DESIGN

This section is necessarily far-ranging. The questions of how large reserves should be, how many reserves are needed to protect species against extinction, and how reserves should be managed are complex and far from resolved. Here the topic is divided into four issues that conservation behaviorists address:

- Population genetics and reserve design
- Fragmentation and edge effects in reserves
- Special habitat requirements and reserve design
- Human–wildlife interactions in reserves

In an ideal world, large tracts of land would be placed into absolute reserves, meaning that the species occupying the ecosystem would be free to survive through time in the absence of human interference. In the real world, reserves are limited in space and often influenced by human activities. Conservation behavior provides a major tool in the design of reserves that maximizes the conservation impact of those reserves.

Population Genetics and Reserve Design

A fundamental tenet of biology is that genetic diversity is the basis for adaptation—and as such, is the basis for survival of a population through evolutionary time. Environmental changes may favor new genetic combinations, certain genetic combinations may be better

445

FIGURE 15.13
Mechanical barriers such as this device designed to keep predators away from this bluebird box, can be effective in programs to support threatened or endangered species. *Photo: U.S. Fish and Wildlife Service.*

dispersers, and genetic diversity helps a population to avoid inbreeding effects. When population size decreases, the statistical chances of rare alleles surviving in the population also diminish. The reduction of a population to near extinction and then recovery of that population is called a *genetic bottleneck*. The analogy is simple; the population at its smallest point resembles the narrow neck of a bottle, in that not all of the previous genetic variation passes through.

> **KEY TERM** A genetic bottleneck is a low point in the number of animals in a population that results in decreased genetic diversity.

Genetic bottlenecks have been documented in a variety of species. Perhaps the two most famous examples are elephant seals and cheetahs. Elephant seals live on the Pacific coast of North America. They were hunted to very low numbers, about 50 to 100, and have since recovered and recolonized coastal areas. The cheetah in Africa and Asia is highly endangered at this time due to a combination of hunting pressure and competition with lions and hyenas; the genetic diversity of its wild populations is quite low. The largest known population is in Iran and numbers about 100 animals; approximately 15,000 animals remain in the wild in Africa.[26]

There are at least three potential behavioral effects of genetic bottlenecks: lack of ability to express behavioral adaptations to environmental changes, expression of maladaptive behavior due to inbreeding effects, and possible disruption of the mating system due to lack of suitable mates. Presently, these effects are mostly hypothetical because they are difficult to measure in wild populations, particularly if those populations are small. Most captive populations of extinct-in-the-wild animals are undergoing a population bottleneck, but managers of these populations can plan pairings and use artificial insemination to conserve genetic resources.[27] This type of management is usually not feasible in a wild population.

446

How large does a wild population need to be to provide sustainability? This is a difficult question. Conservationists use a *population viability analysis* (PVA) that takes into account the age structure, rate of population growth, mating structure, and spatial distribution of the animals. Populations that fall below a critical number may, effectively, be extinct even if a few animals remain.[28,29]

> **KEY TERM** Population viability analysis (PVA) takes into account the age structure, rate of population growth, mating structure, and spatial distribution of a species to determine the likelihood that the population will persist over time.

African elephants provide a good example of viability analysis. With an average of 3.1 African elephants per square mile, the minimum reserve size for these elephants is 1,000 square miles, or about half the size of the state of Delaware.[30] Over time, this elephant population would fluctuate between about 1,000 animals and over 10,000, along with changes in ecological conditions, but it would be unlikely to go extinct. Approximately 2,750 Rio Grande cutthroat trout are needed in a population to maintain viability and genetic diversity; this number yields about 500 breeding adults and a population with less than a 10% chance of going extinct over 100 years.[31] Finding space for populations this size of small animals, such as insects, is nowhere near as problematical as finding space for viable populations of large carnivores.[32]

Given the lack of spaces large enough for conservation of many species, conservation biologists talk about maintaining *metapopulations*, which are networks of interlinked subpopulations. Good reserve design can preserve species by allowing dispersal of animals among subpopulations. This allows both retention of the reservoir of underlying genetic variation and recolonization if local populations go extinct, a likely event if they are small. The behavioral biologist can provide knowledge of how dispersal affects the movement of genes among subpopulations of a species and how its mating system might maintain or reduce genetic variation.[33] For instance, given the mating system of the scimitar-horned oryx (discussed later), if the core reserve contains only one herd with one breeding male, interventions or dispersal interactions with other subpopulations will be necessary to retain genetic variation. Some tropical bird species have extremely limited dispersal capabilities, amplifying the isolating effects of living in forest fragments.[34] For conservation efforts to be successful, much more needs to be discovered about dispersal behavior and how animals living in metapopulations move among fragments in fragmented habitats.

If dispersal among subpopulations in metapopulations is cut off by distance, highways, or other factors, human-engineered *translocations* of animals among sites may be necessary. But how successful are translocations? Translocation, or moving an animal from one natural setting to another, is similar in concept to reintroduction, but usually involves moving wild adult animals to habitats that are already occupied by the species. Examples of commonly translocated animals include beavers, prairie dogs, and bears. Translocations of animals are costly and in many cases ineffective; in deer and gray squirrels, for which some data are available, more than 70% of the translocated animals die soon after they are moved. Relocated rattlesnakes suffer higher mortality than undisturbed conspecifics; over half of them return to the site where they were captured.

> **KEY TERM** Translocation is the movement of an animal from one location to another. This may be done to re-establish a population that is extinct, to supplement a small population, or to move a troublesome animal.

Oftentimes translocations are made to palliate public feelings even though the governmental agencies involved in the relocation know that the animals are likely to perish in their new habitat. The failure of relocated animals is typically due to stress from the translocation process, to inability of the animals to find territorial space in the habitat, or to their efforts to return to their original home ranges.[35] Stressful events in the translocation process include capture, transport, and release.[36] In some cases, the lack of documented success of translocations has led managers to favor euthanasia, or culling, despite strong public resistance to killing animals. In sum, considerable research needs to be done to develop methods of monitoring the results of translocation and to discover successful translocation techniques.

Fragmentation and Edge Effects in Reserves

Habitat edges differ in key ways from core habitats.[37,38] Under natural conditions, edges are called *ecotones*; the margin between a forested area and a grassland is a good example of an ecotone. Human modifications create a wide variety of edges, which are often much sharper than natural ecotones. Highways, parking lots, and grassy lawns all create well-defined edges with surrounding habitats. Some species do well in edges and, in fact, may be adapted for ecotonal life in undisturbed habitats, whereas others must have core habitat, away from edges, for nesting and foraging. Researchers determined habitat preferences of 31 bird species at a heath-woodland edge in southeastern Australia; of those species, 11 were conspicuous in edges but not limited to those areas, and 3 were "shy" of entering the ecotone.[39] The conservation effects of edges vary considerably among species; some species do better than they did before fragmentation and some do worse, but fragmentation always changes the community.[40,41]

Conservationists who are interested in managing metapopulations often ask if one large reserve is better or worse than a number of smaller reserves. An important factor in this is the relationship between reserve size and edge effects. Many small reserves have more edge than one large reserve, even if the overall areas are the same. Open-cup nesting birds have much lower nesting success in a fragmented rainforest habitat, where there is more edge relative to the size of the habitat,[42] than in larger tracts of forest. If the goal is to conserve a

447

community that represents the core habitat, rather than ecotones, then enough core habitat needs to be preserved to maintain minimum viable populations of core habitat species. A minimum viable population is the number of animals required to maintain genetic diversity and avoid the effects of inbreeding, as well as to withstand normal fluctuations in population size without going extinct. Detailed research on community members and their responses to edges is necessary for a well-informed consideration of edge effects in reserve design.[43,44]

> ## OF SPECIAL INTEREST: FRAGMENTED CULTURE?
>
> Michael Huffman, a primatologist who studies chimpanzee behavior, told one of us about the hypothesis that young chimpanzees learned the medicinal use of leaf-swallowing from their mothers. All across the chimpanzees' range, the animals employ leaf-swallowing in response to gastrointestinal upset associated with nematode infection; the leaves are not taxonomically related, but they all have similar features, including a rough, "hairy" surface, perfect for "scrubbing" the parasites from the intestine. Dr. Huffman then wondered aloud about what the impact of the chimpanzees' fragmented habitat might be on the possible transmission of this or other information; after all, chimpanzees had once lived in a continuous swath of land stretching almost across Africa. Now, groups of chimpanzees, if they survive at all, do so in isolation from most other groups. Might this result in a depauperate chimpanzee culture, much as similar fragmentation has diminished the songs of some birds?[45]

Special Habitat Requirements and Reserve Design

Part of reserve management is maintenance or supplementation of specific habitat features to ensure survival of key species. Sunken ships can serve as artificial reefs in marine reserves, downed trees left in reservoirs serve the same function in freshwater systems, cut scrub and brush provide cover and habitat variation in grasslands and woodlands, and periodic managed fire maintains key grassland habitat features. Reserves in habitat fragments may not contain all the needed features for all species in the community, so active habitat management, aimed at meeting the behavioral requirements of the animals, has high importance.[46]

FIGURE 15.14
Purple martin at an artificial nest. Populations of this species are supported by a large public effort to provide nesting sites. *Photo: Thomas G. Barnes, US Fish and Wildlife Service.*

Competition for nest cavities is a good example of how reserve design can be informed by knowledge of behavior. For instance, hollow trees are a premium nesting site. Protected from the elements and usually easily defended, a cavity inside a hollow tree is attractive to a wide range of insects, birds, and mammals. Cavity occupants include honeybees, woodpeckers, squirrels, and bats. Human activities such as logging and burning of forests to clear land for agriculture can eliminate older, hollow trees in ecosystems.

In the southwestern United States, these dynamics have affected acorn woodpeckers, which

depend on nesting trees for their survival. In addition, if an introduced species relies on tree cavities, it may exclude native species from those cavities; one alternative hypothesis for the extinction of the Carolina parakeet states that honeybees, introduced from Europe, outcompeted the parakeets for nesting cavities in trees. Understanding how competition affects cavity-nesting species and what cavity-nesters are looking for when they search for nests can be critical in conservation plan designs. Two excellent examples of public involvement in providing nest sites for cavity-nesters in North America are the purple martin (http://purplemartin.org/main/mgt.html) and the Eastern bluebird (http://www.sialis.org/), shown in Figure 15.14.

Human–Wildlife Interactions in Reserves

Many parks, reserves, zoos, and open space systems depend on taxpayer support or paying visitors; the financial stability of conservation often depends on conservation co-existing with recreation or with opportunities for humans to observe and interact with animals. Indeed, in the United States, the national park system owes its very existence in large part to President Theodore Roosevelt's enthusiasm for hunting and, by extension, outdoor activities.

Tourists—either "ecotourists" or the regular kind of tourists—play a critical role in most conservation plans. Tourists bring much-needed cash to local economies; if they come solely to see a reserve, this provides a strong incentive to maintain the reserve. This is a stereotype of conservation in developing nations, but it is also true in North America, where tourism is the economic engine that maintains communities on the fringes of national parks. Unfortunately, in addition to cash, tourists bring air pollution, water pollution, and noise pollution, all potentially major disruptors of wildlife behavior. One basic element of reserve design is to incorporate mechanisms for the economic support of reserve protection. In many cases, this involves the simultaneous challenges of both attracting tourists and controlling the impacts of tourism to minimize stress on wildlife. Because many of the ways that humans affect animals in reserves are behavioral, animal behaviorists should be influential in designing reserves.

The obvious solution to the double-edged problem of visitor support and visitor impact is to provide adequate buffers between humans and animals in the reserve.[47] Unfortunately, for many species, the data that would allow scientists to determine this distance are not available. Behavioral data about sensitivity of animals to motion, noise and odors, their ability to habituate to these factors, and other human impacts on their behavioral environment are generally lacking. For example:

- How closely can a person approach a deer or elk before disrupting its foraging?
- Do these species habituate to human presence? If so, is that desirable?
- How do songbirds in the same environment differ in their behavioral responses to humans, if, in fact, they differ?

Off-road vehicle use—motorcycles, snowmobiles, ATVs, or trucks—brings noise and air pollution into core reserve areas. Vehicles crossing streams can cause serious damage to aquatic systems. Solutions such as restricting use of motorized vehicles, providing adequate wastewater treatment, and enforcing noise regulations often are expensive or meet public resistance.

In the United States, much public land management emphasizes multiple uses, so wildlife shares space with recreational uses. The downsides of mixed use, in terms of behavioral disruption of wildlife, are obvious. In the United States, the public has not generally supported "absolute" reserves, where no human encroachment is allowed. Designated wilderness areas are regarded as recreational sites; and even if the general public is excluded, scientists often want to exempt themselves from exclusion. Scientific access to otherwise closed areas can be valuable, because it allows for monitoring of populations, but it can create an image of scientific elitism and increase public opposition to closures. In the developing world, immediate needs for food, shelter, and ways to generate cash can exert substantial pressure on reserves. Whether in the developed world or in developing nations, reserves are likely to experience recreational, hunting (including poaching), fishing, and perhaps logging and mining pressures.

In many parts of the world, reserves partly function as game parks for hunting and fishing. Often introduced species, such as pheasants and rainbow trout in North America, or native species, whose populations are augmented through translocations or supplemental feeding, are maintained to supply hunters and fishers. The coalition of hunting/fishing enthusiasts and conservationists that began in Teddy Roosevelt's time continues today with the conservation initiatives of groups such as Ducks Unlimited and Trout Unlimited. The bedfellows that inhabit the world of conservation can seem strange, but the economic and

449

political power of hunters and fishers is considerable, and they are often a positive force in the establishment and maintenance of reserves.

Hunting typically targets herbivores, such as deer or elk, but sometimes carnivores are taken for trophies. Behavioral impacts of hunting include shifts by herds or flocks of animals to suboptimal habitat to avoid hunters, loss of dominant males (often hunted for the size of their horns or antlers) from social groups, and lowered food availability for carnivores.

Fishing usually targets carnivores, hence the efficacy of bait such as artificial flies, minnows, or worms; fishing tends to remove the top carnivores from a community, sometimes increasing competitive interactions among herbivores, whose populations can grow, limited only by food availability in the absence of carnivores. Removal of the largest fish also can affect mating systems because these fish are at their reproductive prime. Catch-and-release fishing, popular among fly-fishers and some bass fishers, mitigates some of the behavioral impacts of fishing.

DISCUSSION POINT: FISHING, HUNTING, AND CONSERVATION

Some of the most ardent enthusiasts for conservation and habitat preservation are hunters and fishers, who are presented by groups such as Ducks Unlimited and Trout Unlimited. Find out a bit more about these groups by looking at their websites. Do you support the marriage of hunting and fishing and conservation, or are you uncomfortable with this? Can hunting and fishing be used as effective population management tools? Many students of animal behavior view themselves as advocates for animals but are also very committed to conservation, so this is a question that can spark lively discussion.

Observing the community effects of the loss of top carnivores led Crooks and Soule to develop the *mesopredator release hypothesis*,[48] an important way to explain how the removal of top carnivores affects community structure. Top carnivores, through a combination of direct consumption of middle-sized carnivores (mesopredators) and competition with them for food, control the population sizes of mesopredators. The Crooks and Soule study focused on the effects of coyotes, as a top carnivore, on smaller carnivores such as raccoons and domestic cats. These mesopredators are major threats to nesting songbirds; when coyotes are present, songbirds are protected because cat and raccoon populations are held in check. In the absence of coyotes, predation on songbirds increases, potentially resulting in a conservation threat to the songbirds.

KEY TERM The mesopredator release hypothesis postulates that the removal of middle-sized predators results in the release of population growth of the animals they were eating.

Despite the clearcut need for spatial separation of humans and animal populations, it is surprising how often such measures are controversial or individuals attempt to evade regulations. A large body of literature addresses "the tragedy of the commons," that is, the fact that many people feel less responsibility about taking care of shared, common space than they do about caring for their private space.[49] We clearly have much to learn about how to keep people from loving reserve space so much that they "love it to death."

A Bike Path Runs through It

Modern suburban development often leaves greenways—strips of vegetation that separate groups of houses, or houses from shopping areas. In older developments or urban areas, riparian areas along waterways serve similar functions as greenways. Conservationists often view greenways as conservation areas, even though uses of greenways are typically shared among conservation, flood control, and recreation. In suburban areas, open land may be viewed as ripe locations for soccer fields, and some members of the public have trouble distinguishing between the conservation value of a weedy-appearing open space and a field of planted and watered bluegrass. Greenways in urban areas often attract transients, a situation that can cause public officials to clear vegetation and take other measures to discourage homeless people from camping in greenways, actions that are often detrimental to conservation values. This section addresses the values and drawbacks of greenways as conservation areas, and suggests ways that conservation behaviorists can help to shape greenway management with an eye to maximizing conservation values.[50]

Riparian corridors are often the most "natural" appearing habitat in urban areas. Because these strips of land along the sides of streams and rivers are prone to flooding, they tend not to be occupied by buildings. In recent years, the laudable environmental and recreational goal of creating bike paths has resulted in the encroachment of pathways into riparian corridors. These pathways create more edge and bring humans and their dogs into closer contact with wildlife that shelters or nests in the riparian areas. If, for example, a road runs parallel to but at some distance from a creek or river, conservation goals are better met if the bike path is built as close as possible to the road; this concentrates the disturbance away from the waterway and can help to minimize the behavioral impacts of recreation on animals. Unfortunately, bikers and joggers like the feeling of riding or running "in the woods" and often bike paths are built close to the waterway, even if this separates a wooded or shrubby riparian area from the creek or river. This design adds edges and is likely to lead to greater behavioral disruption of wildlife populations.[51,52,53,54]

In general, the density of trails (meters of trail per hectare of protected area) decreases as park size increases.[55] This means that small reserves are more likely to be designed in ways that put wildlife at risk of habitat disruption by human recreational activities. In aquatic and shoreline reserves, buffer zones from boat activity are important in protecting shorebirds and sea life, such as whales, manatees, and dolphins, from disturbance.[56,57]

BRINGING ANIMAL BEHAVIOR HOME: DOGS, TRAILS, AND CONSERVATION

Most dog owners enjoy taking their quadruped best friends for a walk along a scenic trail, preferably close to a creek. As much fun as that is, you may think differently about it after reading the following study. We have noted how creekside trails, while lovely, do not function well in the world of conservation. Do dogs make a difference?

To answer this question, Benjamin Lenth and colleagues used five indicators of wildlife activity—pellets, tracks, remote-triggered cameras, on-trail scat, and maps of prairie dog burrow location—to ask if the presence of dogs affected wildlife around trails.[58] They compared trails that prohibited dogs completely with trails that allowed off-leash dogs under "voice and sight" control. The trails were chosen to be as similar as possible with regard to ecological characteristics and visitation levels and were not within 300 meters of roads or structures.

Lenth and coworkers found that wildlife activity was greatly altered by trails and even more altered in areas that permitted dogs on trails. For instance, in areas that prohibited dogs, mule deer were nonetheless less active within 50 meters of trails. That boundary increased to 100 meters in areas that were visited by dogs. Small mammals such as squirrels, rabbits, mice, and chipmunks reacted in similar ways, and bobcats tended to avoid areas with dogs. (Red foxes showed no aversion.) In areas where trails are within 100 meters of each other or where their margins overlap considerably, it is possible that much of the area between trails is unsuitable for a wide range of animals. This situation is exacerbated by the disinclination of some people to remain on trails; the creation of "social" (unofficial) trails can generate a network of trails so close together that even in the absence of dogs, wildlife will find the area unattractive.

Of course, this study simply addressed the spatial aspects of wildlife reaction to trails and trail users. Depending on visitation patterns, the timing of wildlife activity may change in response to people and/or their pets. The sad news is that a natural area with a poorly designed trail is likely not all that natural after all.

Ecotourism: A Two-Edged Sword

Ecotourism can bring needed money to reserve areas, but tourists can disrupt the behavior of animals in a reserve. The most severe criticisms of ecotourists have been leveled at whale watchers, snorkelers at coral reefs, and various "adventure" activities, such as river rafting, all of which can bring human disturbances in close proximity to easily disturbed wildlife. Whale watching provides a good set of examples of the pros and cons of ecotourism. Some coastal communities rely on whale watching tours as a significant source of income, and whale watching can economically supplant the hunting of whales. Of course, not all whale watching takes place on a boat, and land-based whale watching is the least intrusive on whales. In contrast, motorized boats making close passes at foraging whales may be the most disruptive because noise pollution interferes with the ability of whales to echolocate and communicate.

> **KEY TERM** Ecotourism is activities in reserves, parks, or other relatively natural habitats that is rationalized by its educational benefits for the tourists and its economic benefits for local communities around the reserve or park.

DISCUSSION POINT: ECOTOURISM AND ADVENTURE TOURISM

Have you been on an outing that was billed as ecotourism but was really less "eco" and more "adventure" tourism? In retrospect, how big an ecological footprint did this activity have? How can you resolve the possible conflicts between economic needs of local communities for tourism dollars and the ecological impacts of ecotourism and adventure tourism?

Flight Initiation Distance: A Tool for Evaluating Human-Wildlife Interactions

Tarlow and Blumstein evaluated techniques for measuring how much stress human activity puts on wildlife.[59] They focused on birds, but the behavioral measures they used have broad applicability. Behavior plays a key role among the assessments they discussed; breeding success and flight initiation distance are particularly important. Managers assess breeding success by observing the number of nests originated, the number producing eggs, and the number of offspring that fledge. *Flight initiation distance* (FID) describes how close a bird can be approached before it flies away; analogous measures can be applied to fleeing mammals.[60] If FIDs are documented, then perimeters around sensitive nesting and feeding areas can be established and humans can be prevented from approaching an animal so closely that it is disturbed. Of course, animals may be disturbed and not flee—they may stop feeding and watch an intruder, for example—so perimeters can be established based on changes in vigilance as well as in actual flight. FIDs vary by nature of the disturbance (walking human, biking human, automobile noise), context, season, and population, so applying this principle is sometimes more difficult than it sounds, but overall this is an excellent tool for minimizing disturbance.

> **KEY TERM** Flight initiation distance (FID) describes how close an animal can be approached before it takes off.

15.5 CAPTIVE BREEDING PROGRAMS AND REINTRODUCTIONS

The basic concept of captive breeding and reintroduction is simple: Gather up enough animals of an endangered species, breed a reasonably large base population in protected conditions, and then release some of this base population into a protected wild or semiwild reserve.[61] This is Noah's idea, writ large, and for species on the brink of extinction, this may be their only hope.

A few such programs have been unqualified successes, but many have faltered because either captive breeding failed or the reintroduced animals did not survive. One of the more difficult cases, as detailed previously, involved a Hawaiian honeycreeper in which a translocation failed and a captive individual died. Mortality is typically high in reintroduced animals, and discovering the correct methods of reintroduction for any species at this time has a substantial trial-and-error component.

Many of the problems of reintroductions surround the ability of captive-reared animals to function in the wild. Some animals carry with them, in genetic form, much of the information they need to survive. Others, including most birds and mammals, rely on information learned from parents, sibs, and social group members, as well as on their own experiences, to shape the behaviors necessary for their survival.[62]

This section shows that managing imprinting, recognizing the importance of neophobia in shaping behavior, and learning about food preferences and predator avoidance can all play key roles in the success of a reintroduction program. It highlights six species, all of which were either endangered or extinct in the wild, that have been bred in captivity; most of them have successfully been reintroduced into the wild. These species are discussed in the order of difficulty they have presented in captive management and reintroduction; herbivores are probably easier to work with than carnivores, but all of these species have presented interesting behavioral questions and problems.

Scimitar-Horned Oryx

The scimitar-horned oryx (see Figure 15.15) is a good model for discussing conservation of ungulates (hoofed mammals). There may have been as many as 1,000,000 oryx in North Africa as recently as 1900; the rapid extinction of this species in the wild, in many ways parallel to the extinction of the wild American buffalo, is shocking. Ungulates are capable of fairly rapid population growth, and in nature, populations are typically limited by predators and food resource availability, with major causes of decline being habitat destruction and hunting. Ungulates either graze on grass and herbaceous vegetation (cattle) or browse on leaves and buds of shrubs and low trees (deer, giraffes). They seem to easily adapt to any plant-based digestible diet, so feeding animals maintained in captivity is not challenging, and reintroduced animals do not seem to face problems with learning food preferences.

A fair amount is known about vigilance in ungulates, and shared vigilance in groups is probably an important part of the time budgets of these animals. Less is known about how ungulates learn about the identities of predators in their environments and whether reintroduced animals are more vulnerable to predators because they lack experience. Deer, for example, adjust their behavior seasonally to the presence of human hunters,[63] suggesting that learning of predators could be important. A key element of ungulate behavior is a harem-type mating system; in the scimitar-horned oryx, a single male leads a herd of 20–40 females and young. This reduces effective population size (N_e, a measure that takes into account the number of animals actually reproducing. If only a few males reproduce, even though they inseminate many females, the N_e is low), which is accounted for in a genomic resource program organized by zoos for captive breeding of this species.

453

FIGURE 15.15
Scimitar-horned oryx.

The Smithsonian National Zoological Park has used artificial insemination to overcome problems with behavioral incompatibility between pairs of scimitar-horned oryx. Progeny have successfully been reintroduced into the wild at Ben-Hedma National Park in Tunisia, where a population had grown to about 130 animals in 2005. The reintroduced population was initially maintained in a 10 hectare (a hectare is 100 meters × 100 meters) enclosure for the first 16 months; this allowed supplemental feeding with hay while the animals adapted to the natural vegetation. Supplemental foods were available to the animals following their release as well. The park, which is 2,400 hectares, is fenced so the oryx are confined within the park; this space may be adequate for only one herd (one breeding male at a time). If this is the case, then clearly the reserve population, on its own, is not sustainable. However, this population and zoo animals may form the basis for future reintroductions at other sites; with management, a viable population could be maintained through translocations.

Przewalski's Horse

Przewalski's horse, a native of Mongolia (see Figure 15.16), at one point extinct in the wild, is the closest living relative of the domestic horse, and as a result, its greatest lure for behavioral biologists comes from the insights it gives into the roots of domestic horse behavior.[64] Sometimes regarded as a subspecies of the domestic horse and sometimes as a distinct species, Przewalski's horse has never been domesticated. Early in the 1900s, fewer than two dozen of these horses were captured; this small group of horses was the foundation stock for captive populations, which are a popular exhibit in many zoos, and for a population of reintroduced horses in Hustai National Park, Mongolia. While Przewalski's horses are not nearly as manageable as their domesticated relatives, knowledge of horse management has eliminated much of the guesswork often involved in captive management and reintroduction programs. The low genetic diversity of this species, due to the population bottleneck suffered in the early 1900s, has not had any visible negative effects on the species, and the major issue in the reintroduction has been development of a program to protect the population. Like the domestic horse, a single stallion moves with a harem of females over a relatively large home range. Home ranges of bands of Przewalski's horses can be overlapping, but a relatively large reserve is needed to maintain the number of bands required to preserve the species' remaining genetic diversity.

FIGURE 15.16
Przewalski's horses are bred in zoos, but a small wild population persists in Mongolia. *Photo: Michael Breed.*

454

Black-Footed Ferret

FIGURE 15.17
Black-footed ferrets are the target of a breeding and reintroduction program. *Photo: Ryan Haggerty, US Fish and Wildlife Service.*

Endemic to the western United States and specialized as a predator that lives only in prairie dog towns, the black-footed ferret (see Figure 15.17) was thought to be extinct in the wild until 1981, when a Wyoming rancher's dog caught one and brought it to the rancher's house. From 1985 to 1987, all individuals in this wild population were caught, and they formed the basis for a captive breeding program, with the goal of re-establishing this species in the wild. The decline of black-footed ferrets paralleled the loss of prairie dog towns. Black-footed ferrets are highly susceptible to canine distemper, a vulnerability that may have contributed to their decline; this was one of the rationales for capturing the only known wild population. As with the scimitar-horned oryx, initial breeding failures in captivity led to use of artificial insemination to ensure reproduction. Reintroductions into the wild began in 1991 and continue to the present; to protect the long-term survival of the species and its genetic diversity, scientists hope to maintain several wild populations.

A number of behavioral issues had to be addressed during the black-footed ferret project; these issues included the ferrets' willingness to mate, given the available animals in the initial population, and the ferrets' ability to learn predatory strategies and food acceptability, as well as predator avoidance. Initial reintroductions failed because of a combination of disease and lack of skills that the ferrets needed to survive in the wild. Simulations of wild conditions for captive animals, such as exposing them to hunting opportunities within their captive confines and exposing them to cues of potential predators, have increased the success of reintroduction programs. Minimization of human interaction is also important for black-footed ferrets, as habituation to humans (such as in bottle feeding) can lead to diminished neophobia. A humorous, but telling, anecdote is that when ferrets were given a moving model of a predator in an attempt to teach them predator avoidance, they instead habituated to the model and started riding it around their enclosure. This suggests that even the best-laid plans for preparing animals for reintroduction to the wild can go awry, and that much trial and error is required to find strategies to prepare animals for reintroduction. Presently, the black-footed ferret reintroduction program is starting to succeed, with animals in several wild populations rearing young.

Golden Lion Tamarin

Because the golden lion tamarin (see Figure 15.18) captive breeding project is largely based in zoos, it is one of the best-known breeding and release programs. Tamarins are small monkeys found in the Americas; the golden lion tamarin is native to the Atlantic Rainforest of Brazil, a unique ecosystem, separate and fairly distinct from the more recognized Amazonian Forest. Because the Atlantic Rainforest is close to Brazil's population centers, such as Rio de Janeiro, the survival of this forest and the animal species it shelters has been a major conservation concern. Between 1,000 and 2,000 golden lion tamarins live in the wild; about half of these are reintroduced captive animals. Another 500 or so live in zoos, and are maintained with a carefully managed breeding program to avoid inbreeding and maintain genetic resources.

455

The success of golden lion tamarin reintroductions has given behavioral biologists the opportunity to explore how to best ensure survival of captive-bred animals in the wild.[65] Prerelease training of adult golden lion tamarins to handle foods they might find in the wild or to avoid certain predators is relatively ineffective, suggesting that prerelease training or reintroductions of young animals is preferable to prerelease training of adults. After reintroduction, golden lion tamarins' locomotory skills improved dramatically; after a few months, they fell less, climbed higher, and began to prefer natural habitat components. Foraging improved less dramatically, perhaps because the reintroduced animals received supplemental food. Animals with improved locomotory skills survived better and, not surprisingly, juvenile animals showed more behavioral improvement than adults. Animals that ranged freely in a seminatural environment prior to reintroduction and animals reintroduced directly from caged environments did not differ in survival.

FIGURE 15.18
Golden lion tamarin; this species has been the subject of an intensive breeding program in zoos and has been reintroduced in its native habitat in Brazil. *Photo: Michael Breed.*

The primary lesson from golden lion tamarin experience is that a combined effort of zoos and conservation agencies, combined with a charismatic animal that captures public attention, can generate a very effective reintroduction program. The need for postrelease feeding and veterinary care of released tamarins is an expensive drawback for the program and brings into question just how "wild" the reintroduced animals are. The behavioral lessons are that investment in pretraining should be carefully evaluated with follow-up field data on the performance and survival of animals. The difference between young and adult animals in development and adaptability may be generalized beyond tamarins and is an important point to consider in any captive breeding–reintroduction program.

FIGURE 15.19
A California condor in
a breeding and release
program. *Photo: Gary
Kramer, USFWS.*

California Condor

One of the most controversial conservation decisions of
the twentieth century was the capture of the few remaining
California Condors in the wild to create a core population
for captive breeding (see the case study for this chapter). The
California Condor (see Figure 15.19) is a large vulture, with
a wingspan of nearly 3 meters. By 1982, California Condor
populations had diminished to fewer than 30 birds. Captive
breeding and release programs have increased the wild
population of this species to about 200 individuals. Condors
are highly susceptible to lead poisoning; because they scavenge
on animal carcasses, they risk coming into contact with lead
shot or bullet fragments. In response to this, California has banned the use of lead bullets in
areas occupied by condors.

The major behavioral issues in the capture-and-release program have concerned imprinting
of species recognition. Great care is used in the captive rearing program, through the use
of feeding puppets, to imprint condor chicks on appropriate cues. Chicks are initially
hand fed, using the puppets, and as they grow are visually imprinted on adult condors.
Condor releases employ a double trap-door technique, in which a caged condor is given the
opportunity to move into a second enclosure; when it does, a door closes behind it and a
release door opens, so it can then fly, if motivated to do so. This process minimizes human
contact and disturbance of the bird, but makes the timing of actual releases unpredictable.
Appropriately imprinted condors, released using this technique, have succeeded in the wild,
finding nesting sites, food, and mates. The wild population of condors is slowly growing
but is still imperiled by the potential for lead poisoning, not to mention the microtrash
described previously.

Peregrine Falcons

Peregrine falcons (see Figure 15.20), like many birds of prey, suffered dramatic population
declines due to the use of DDT as an insecticide in the middle part of the twentieth
century. In a phenomenon known as *biomagnification*, the effects of DDT are compounded
at the upper levels of food webs because the insecticide is not metabolized or excreted; it
concentrates in the fatty tissues of animals. If, over time, a predator eats 1,000 herbivores,
then it ends up with 1,000 times the individual herbivore's concentration of DDT in its

FIGURE 15.20
Peregrine falcons
suffered greatly as
the result of DDT
accumulation in the
food chain. Successful
reintroductions have
been made in both native
habitats and urban
environments, where they
prey on pigeons. *Photos:
U.S. Fish and Wildlife
Service.*

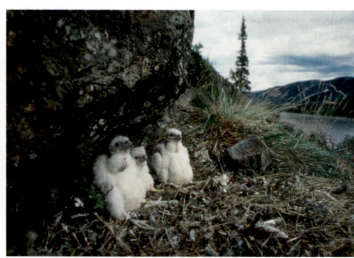

system. A predator that eats predators ends up with hugely magnified concentrations of the poison in its system. In birds, DDT interferes with the calcification of eggshells; females lay thin-shelled eggs that are susceptible to breakage and lethal exposure of the embryo. By the mid-1960s, peregrine falcons had disappeared from much of their range in the United States.

Under natural conditions, peregrine falcons nest on cliffs and ledges. They prey on birds and mammals on the ground, diving from great heights at high speeds (up to 200 mph), surprising the prey. The fairly narrowly specialized nesting habits of the peregrine falcon meant that they were rare even prior to the pesticide problems; thus, DDT diminished already-small populations.

Reintroductions of peregrine falcons, starting in 1980, have been spectacularly successful. A detailed report published by the Colorado Division of Wildlife gives insight into the intricacies of the falcon recovery program.[66] A trait that favors captive breeding is the tendency of birds to replace eggs collected from a wild nest with a replacement clutch. (This is termed *recycling*.) Thus, eggs can be removed from the wild without eliminating reproduction of the wild population. Captive-reared chicks are gradually "hacked" (partially liberated) from hand feeding so that they can learn how to hunt on their own. Hacking is a technique developed by falconers over centuries (perhaps millennia), and the peregrine reintroduction programs benefited greatly from taking advantage of existing methods for handling and releasing birds. Perhaps the most interesting aspect of the reintroductions, from a behavioral standpoint, is that peregrines show unexpected adaptability in nesting and feeding habits, accepting ledges on bridges and tall buildings as nesting sites. In urban environments, they prey on pigeons, which are typically plentiful. However, many of the released peregrine falcons fail to survive hazards such as power lines and transmission towers, and others disperse from their release points. The peregrine falcon experience, like that with black-footed ferrets, points out how trial and error can be used improve release programs and how important it is to release large numbers of animals to achieve success. (In the case of peregrines, 14–16 birds are needed to establish one wild pair.) As a result of the ban on DDT use, peregrine eggshell thickness has increased over the past 20 years, leading to increased reproductive success and expanding populations in both urban and natural cliff settings.

457

OF SPECIAL INTEREST: CAPTIVE BREEDING AND REINTRODUCTIONS

Here is a list of critical behavioral points for captive breeding and reintroduction programs. The role of animal behaviorists in captive breeding and release programs is to ensure that enough is known about the behavior of the species to maximize the chances that the program will succeed.

- Match important habitat characteristics of wild and captive environments.
- Mimic teaching that occurs between parents and offspring in the wild to shape food preferences and foraging strategies
- Understand dispersal patterns and recognize that introduction into the wild may trigger long-distance dispersal.
- Minimize human contact and habituation to potential threats.
- Manipulate neophobia so that it does not impair foraging but provides needed protection from threats.
- Understand the role of imprinting in the species' biology and provide species recognition cues that will be necessary later in life for mating, parenting, and social behavior.
- Monitor and evaluate the fates of reintroduced animals so that the captive rearing program receives useful feedback.

Ethical Considerations

A scientific approach to animal behavior allows discovery of the knowledge that managers need to create appropriate living environments. With so many imperiled species, maintenance of captive populations with an eye to re-establishment of populations in the wild at a later time is becoming a critical strategy in maintaining biodiversity. Understanding and attending to behavioral indicators of *animal welfare*, such as repetitive movements or excessive self-grooming, (see Chapter 4), are key aspects of this.

> **KEY TERM** Animal welfare involves the consideration of the effects of human activities and behavior on animal well-being.

DISCUSSION POINT: ETHICS AND CAPTIVE POPULATIONS

What are the ethics of maintaining animals in captive populations for conservation purposes? Is it better to be extinct than captive? Some extreme advocates of "animal rights" have argued that captivity is so cruel, or that captive conditions are so unnatural, that from the animals' point of view, extinction is a better option than captivity. Others say that captive-born animals may adapt easily to captive conditions. Once again, this points to the interface of innate tendencies, cognition, and umwelt.

Of course, welfare, in the sense of good health, is easier to measure than "happiness," a concept that eludes definition even for many humans. Anyone who has watched a puppy eagerly seize upon some horse droppings left in the middle of a trail knows that there are considerable, and possibly unfathomable, interspecific differences in what is apparently experienced as a source of delight.

To the extent that people can actually "i nterview" some of the great apes, through the use of a variety of communication tools, this gap may be bridged for at least a few species. This is a tantalizing possibility, but one that still suffers from great limitations. The great apes that have been taught to communicate with their human research partners may not be representative of animals, even apes, in general. In general, for most species, it may be better to focus on stress indicators such as cortisol and the correlation of such indicators with specific behaviors to get an idea of how "happy" an animal is. Moreover, given the variation in perceived happiness among humans, even if a "happiness indicator" is discovered for some species, it may not be possible to devise an environment that produces happiness across a wide range of individuals.

A related question concerns the appropriateness of zoos for the task of managing populations for captive breeding. Before the prevalence of television and web-based information resources, zoos played critical roles in allowing the public to see the behavior of animals, albeit under highly modified circumstances. Zoos were the premier venues for public education and wildlife awareness, albeit at a high cost to animals. With video available of many animals and of almost every conceivable behavior, one might conclude that much of the informative value of zoos has been superseded. There is nonetheless something about the presence of a living animal that video does not replace, particularly for children. Scientific knowledge gives zookeepers strategies for behavioral assessments and habitat enrichment that reduce some of the harsher aspects of captivity,[67] but the main advantage of zoos remains the fact that the public funds them, while funding for captive breeding programs that do not include public displays is much more difficult to maintain. Of our primary examples, only the peregrine falcon program has not involved zoos as a major platform for captive breeding.

The human behavior that supports zoos also shapes conservation priorities. The public's greatest affinity is for vertebrates. This means that the conservation status of the vast majority of animal species, which are invertebrates, is unknown and that much of the hope for invertebrate conservation lies in reserve design, rather than captive rearing and release.

15.6 THE HUMAN–WILDLIFE INTERFACE IN THE SUBURBS

There is no sharp dividing line between a city and relatively undisturbed habitat; instead, densely built and populated cities give way to suburbs, with more space between houses, which then give way to suburbanized countryside, where farms or ranches have been subdivided in larger tracts (often 30–50 acres), each with a house. Mountainous areas are typically dotted with cabins and summer homes. This type of development means that, from the point of view of the wildlife, there is rarely an area that is clearly unpopulated by humans, and there is no clear boundary that demarks areas to be avoided. This section explores what happens when human settlement and wildlife habitats overlap, or when wildlife find their way into more dense human settlements.

Garbage Dump and Other Feeding Effects

Some wildlife, particularly skunks, raccoons, and bears, are attracted to human garbage. In the early history of the U.S. national parks, garbage dumps actually became wildlife viewing posts, particularly for bears; wildlife officials eventually realized that for bears, the behavioral effects of feeding at dumps, including dependency on a human-supplied food and hazards to human observers from habituated bears, outweighed any educational advantage gained by the viewing public. This has led to the development of considerable technology to protect garbage from these bears (and also from skunks and raccoons); a surprising level of sophistication in design is needed to keep these clever and strong foragers from accessing garbage (see Figure 15.21). It has been even more challenging to regulate human behavior in suburban settings, and dealing with bears that come into yards is a major preoccupation for wildlife managers in many states. Because bears can be hazardous to humans, the typical approach is to trap or tranquilize the bear and release it in a remote location. This approach is often ineffective, because once a bear has habituated to humans and has learned to take advantage of easy pickings in suburbs, it is highly likely to return, even from great distances. In Colorado, first offenders are marked before release, and second-time offenders are euthanized. This should be a strong incentive for humans to manage their garbage and other attractions.

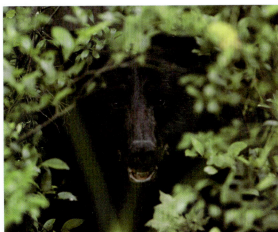

FIGURE 15.21
Black bears are frequent visitors to suburban habitats and may be attracted by garbage. Their presence is often unwelcome because people view them as threats. *Photo: Steve Hillebrand, U.S. Fish and Wildlife Service.*

459

In general, feeding pets outdoors creates behavioral issues with wildlife, attracting animals that may pose hazards, facilitating habituation, and often ending with tragic consequences for the animals. In Florida, pet food attracts alligators (obviously a serious hazard for both pets and humans) and cane toads, a nuisance invasive species. Animal behaviorists can help to educate the public about simple measures in managing pet food and garbage that maintain separation between people and wildlife. This approach supports conservation goals because it makes reserve management more compatible with human neighbors.

BRINGING ANIMAL BEHAVIOR HOME: BIRD FEEDERS

Garbage and dog food are obviously detrimental to wildlife, but what about bird feeders? Members of the general public and some biologists view feeding birds as a providing a benefit to wildlife, as well as something that brings interest and entertainment into backyards (see Figure 15.22). The bulk of bird food is seed. In the United States and the United Kingdom together, 500,000 metric tons of birdseed are put in feeders each year.[68] To place this amount into perspective, a gondola railcar carries a little over 100 metric tons; a coal train of average length, about 100 cars, carries around 10,000 metric tons of coal. The amount of birdseed provided annually in the United States and the United Kingdom is

FIGURE 15.22

Birds at feeders. The upper feeder is intended to attract birds, and the lower feeder is for domestic animals and helps to support invasive species, such as these English sparrows. *Photos: Michael Breed.*

equal in weight to the loads of 45–50 coal trains. This huge supplement to natural food sources has contributed to population gains and range expansions of many seed-eating birds, such as the northern cardinal and tufted titmouse in the northeastern United States and the house finch in the southwest. The house finch was also introduced into the eastern United States as a pet. Birds supported by feeders have also delighted many cats. Seed generalists such as the cardinal, willing to consume a variety of seeds, benefit the most from backyard feeders. In urban environments, feeders support populations of house sparrows, allowing them to succeed in this manmade habitat. The ultimate results of bird feeding may include selection for birds that have behavioral attributes for success at feeders but would do less well in unsupplemented environments. Range expansions of seed generalists may also impact, through competition for food, populations of less-flexible seed specialists; the house finch affects Cassin's and Purple finches in this way. Feeding birds is a huge, and interesting, experiment, and conservation behaviorists do not yet fully understand the consequences.

Public Fear and Overcoming Zoophobia

Public fear of wild animals, or *zoophobia*, often leads to pressures for translocating or killing animals. People are particularly fearful of large mammalian carnivores, snakes, stinging insects, and spiders. Biologists are sometimes arrogantly dismissive of these fears, and it is hard for an educated scientist to have much sympathy for someone who has a great desire to live in a house "in the forest" but then wants to kill the animal residents of the forest—a forest without animals is just an empty set of trees. For conservation behaviorists to remain credible, though, they need to be sympathetic to members of the public who lose their dog or cat to a fox, coyote, or mountain lion. Public education needs to focus on managing the behavior of both pets and predators to keep domestic animals separated from potential predators; in the long run, this is a major element of enlisting public support for the presence of wildlife.

Conservation behaviorists can help to build public sympathy for important species, such as venomous snakes, that are generally reviled. Timber and prairie rattlesnakes, once abundant throughout the eastern and plains states, have suffered tremendously from a kill-on-sight mentality. This has extended to nonvenomous snakes as well; many people cannot distinguish among snakes and therefore feel obliged to kill them all. Snakes are important in controlling rodent populations, and understanding their behavior helps humans to avoid snakebites (Figure 15.23).

FIGURE 15.23
A rattlesnake in a defensive pose.
Photo: Jeff Mitton.

Other species are less hated but create significant problems when interfacing with humans. Like their rural ranching and farming counterparts, suburban gardeners often suffer from herbivory by animals such as rabbits and deer on ornamental and vegetable plants. Many people hold an image of deer as "Bambi," and public resistance to management by culling deer herds comes into strong conflict with gardeners who would like to protect their plants. Landscapers turn to behavioral barriers such as high fences to exclude deer or fine mesh screens around plants to foil rabbits, but these barriers are unsightly and often ineffective. Considerably mythology surrounds the use of odorous repellents to drive herbivores away; among the solutions sprayed on vegetation are predator urine, including mixtures thought to mimic territorial markings and mixtures that attempt to make vegetation taste unpalatable to the herbivores. Some of these treatments, particularly the urine-based mixes, bring unpleasant odors to gardens. Hydrolyzed casein (milk protein) shows promise as a repellent;[69] when combined with latex, which makes the protein stick to the plants, deer substantially reduce their consumption of plant seedlings. Hydrolyzed casein also has shown good results in Australia, where marsupials present many of the same problems that deer cause in North America and Europe. Whether the use of repellents will ultimately have value as deer overcome their neophobia is open to question; the spectacle of crows roosting on a scarecrow spotlights the power of habituation in defeating efforts to repel wildlife.

Perceived direct threat of animals to humans or pets often turns human residents of areas against animal inhabitants. Threats can be predation: foxes, coyotes, and mountain lions all seem to enjoy snacking on household pets. Mountain lions, albeit rarely, can turn to humans as potential prey. Foxes are particularly adept at preying on cats, and coyotes may entice dogs into play or courtship prior to turning the tables and consuming the dogs. In the summer of 2008, a mountain lion entered a home in Colorado and captured a dog while its owners slept in the same room.

The conservation consequences of human/predator interactions often play out in calls for trapping or killing of predators; as developments encroach into wooded areas or farmland, homeowners who are intolerant of predators work to destroy the natural values that initially motivated them to live in semirural areas. Many homeowners in suburban southern California view coyotes as intolerable threats to their children and pets. Very little can be done to modify predator behavior once it is established, but humans can learn to take simple precautions to not attract predators. These include supervising pets and not leaving windows and doors open. Education of homeowners concerning predator behavior and ways to minimize risk to children and pets is an important conservation goal.

Diseases in wildlife populations are even more likely to stimulate harsh public responses. In addition to the potential for wildlife to carry diseases such as brucellosis that affect ranch animals (see earlier description), urban and suburban wildlife populations can

461

bring serious disease risk to their human neighbors and their pets. Prairie dog colonies can harbor plague; pigeons in urban environments may carry a respiratory fungus, *Cryptococcosis;* many mammalian species have the potential to harbor rabies; and some wildlife (most notably foxes and raccoons in suburban environments) can carry canine distemper. Heated debates among homeowners, some of them conservationists who enjoy seeing wildlife in their neighborhood and others who are terrified of disease, are extremely difficult to resolve.

OF SPECIAL INTEREST: MARK TWAIN AND WILE E. COYOTE

One of the first non-native-American descriptions of the coyote was penned by the Prince of American Letters, Mark Twain. He had a grudging admiration for an animal that, over the intervening years, has proven to be an adaptable survivor (Figure 15.24). When Chuck Jones, a highly successful animation artist, encountered the passage below, he was inspired to create the unforgettable cartoon character—Wile E. Coyote.

. . . he was not a pretty creature or respectable either, for I got well acquainted with his race afterward, and can speak with confidence. The cayote is a long, slim, sick and sorry-looking skeleton, with a gray wolf-skin stretched over it, a tolerably bushy tail that forever sags down with a despairing expression of forsakenness and misery...He is always hungry. He is always poor, out of luck and friendless. The meanest creatures despise him, and even the fleas would desert him for a velocipede. He is so spiritless and cowardly that even while his exposed teeth are pretending a threat, the rest of his face is apologizing for it. And he is so homely!—so scrawny, and ribby, and coarse-haired, and pitiful. When he sees you he lifts his lip and lets a flash of his teeth out, and then turns a little out of the course he was pursuing, depresses his head a bit, and strikes a long, soft-footed trot through the sage-brush, glancing over his shoulder at you, from time to time, till he is about out of easy pistol range, and then he stops and takes a deliberate survey of you; he will trot fifty yards and stop again—another fifty and stop again; and finally the gray of his gliding body blends with the gray of the sage-brush, and he disappears...The cayote lives chiefly in the most desolate and forbidding desert, along with the lizard, the jackass-rabbit and the raven, and gets an uncertain and precarious living, and earns it. He seems to subsist almost wholly on the carcasses of oxen, mules and horses that have dropped out of emigrant trains and died, and upon windfalls of carrion, and occasional legacies of offal bequeathed to him by white men who have been opulent enough to have something better to butcher than condemned army bacon… He does not mind going a hundred miles to breakfast, and a hundred and fifty to dinner, because he is sure to have three or four days between meals, and he can just as well be traveling and looking at the scenery as lying around doing nothing and adding to the burdens of his parents.[70]

FIGURE 15.24

A coyote. This species has been reviled by ranchers but also revered for its cleverness and ability to elude eradication. *Photo: Jeff Mitton.*

Maintaining Neophobia in Support of Conservation Goals

Social learning appears to help some species adapt to the presence of humans, cars, houses, and other features of the suburban landscape. Usually, this takes several generations of animals, as each new generation learns to be a little less fearful than previous generations.

The persistence of coyotes in suburban settings is probably in part due to loss of neophobia, but a variety of animals, including deer, mountain lions, raccoons, foxes, and porcupines, have taken up residence in suburban settings. Among birds, the starlings, house (English) sparrows, house finches, robins, magpies, and crows all seem to easily adapt to suburban environments, overcoming any neophobia that might keep other species of birds in more natural settings. A yard in Denver, Colorado, may be frequented by Eastern fox squirrels, starlings, house sparrows, and house finches, none of which are native species to the area, but all of which adapt, behaviorally, to close co-existence with humans. Lack of neophobia may be an important part of such adaptability.

As mentioned previously, for many animals neophobia is a protective adaptation. Benign interactions with humans, particularly if they persist across several generations, seem to result in progressively bold wildlife. Feeding, including the availability of pet food and garbage, only accelerates this process of habituation. This does not matter much if the fed animal is a songbird but can have tragic results, for human and animal alike, if the fed animal is a raccoon, coyote, or bear. Obviously, eliminating human-provided food is an important goal.[71]

More methods of maintaining neophobia include use of loud noises; introduction of predator models, calls, or scents; and flashing lights. All of these have startle value, but if produced regularly and without physical consequence to the animal, habituation is the near-certain result. Deer respond to rifle hunting with shyness, not because they fear the sound of gunfire but because they are observant and see other animals injured or killed. Using a variety of aversive stimuli in random order is more effective than repeating a single stimulus. In species that are socially observant and can learn from the experiences of other animals, observing injuries or deaths is a powerful reinforcement of neophobia, but this approach is often unacceptable to the general public.

Developing techniques for encouraging neophobia is a current challenge for animal behaviorists. An increased understanding of how species of concern react to aversive stimuli will result in more effective manipulation of the human–wildlife interface, allowing people and wild animals to live in close proximity without unfortunate interactions.

SUMMARY: THE FUTURE AND CONSERVATION BEHAVIOR

Conservation behavior provides tools for understanding and possibly manipulating animal behavior to further conservation goals. Keeping animals in captivity can help to preserve endangered species, conserve genetic resources, or obtain populations of animals for reintroduction. To do this requires research on mating systems, territoriality, and foraging behavior. The introduction of captive animals into natural habitats must be informed by knowledge of dispersal behavior and of how animals obtain critical information for survival.

Neophobia, the fear of new situations, protects animals from risky situations and can be manipulated to keep human and animal populations apart. The lack of neophobia is a key element of success in invasive animals and contributes to negative human–wildlife interactions. Activities such as hunting, feeding birds, and leaving pet food outdoors have behavioral effects that often result in public pressure to relocate or kill wildlife. Human activities in conservation areas also affect wildlife behavior, and future design of reserves must take these effects into account.

This chapter, more than any other chapter in this book, highlights large and important gaps in current knowledge. There is some irony here, because conservation behavior is also the culmination of almost everything behavioral biologists have learned in the past few decades. From life history traits to foraging behavior, from observational learning to auditory signals and noise pollution, every aspect of animal behavior has the potential to help meet the challenge of conserving the natural world.

For conservation programs to be effective, far more knowledge is needed about habitat choice, mating systems, learning and imprinting, the functions of neophobia in animals' lives, as well as how to educate the public about interacting with wildlife. As the field of conservation behavior matures, it will play a key role in the protection of endangered species, reintroductions of animals that are extinct in the wild, and the success of reserve designs. Learning about animal behavior and its relationship to conservation is a critical field of biology for the twenty-first century. In fact, as human populations continue to invade and alter animal habitats, such understanding is more than an essential conservation tool—it is our ethical obligation.

STUDY QUESTIONS

1. Habitat fragmentation—the breakdown of a previously large area of habitat into small pieces by roads, buildings, and the like—has serious conservation implications. What effects might fragmentation have on mating systems? Think about all the different aspects of behavior that we've discussed—signaling, territoriality, types of mating systems—as you formulate your answer.

2. Captive rearing of animals for release in the wild has been successful with California Condors and black-footed ferrets, but these efforts have been criticized because they take conservation dollars away from habitat protection and other priorities. Discuss the pros and cons of rearing animals in captivity for later release in natural habitats. If you were trying to predict the success of a captive rearing/release program, what do you think the most important behavioral factors would be? How would a behaviorist contribute to designing such a program?

3. You are given the job of designing a conservation plan for a large mammalian species. What would you need to know about the species' behavior before designing your plan? If money for research was limited, what information would you choose to gain first?

4. You are given a grant to study the conservation of an endangered animal species. Knowing only that the animal is rare and has a limited habitat, what, in your view, are the most important things to know about an animal's behavior? If your time and funds are limited, what would you want to discover first (about its behavior), and why?

5. You are an elected official and you have an opportunity to invest tax dollars in some land to be kept open and in a natural state, so that the wildlife species composition can be maintained. The public you serve has a variety of concerns about this—some want bike trails, some want horse trails, dog-owners want an off-leash space, and some constituents question this use of tax dollars when other needs are pressing. (You can imagine other public input!) Think carefully about these issues and their possible effects on the land in question and on future public support of natural areas. How do you respond, and why?

Further Reading

Angeloni, L., Schlaepfer, M.A., Lawler, J.J., Crooks, K.R., 2008. A reassessment of the interface between conservation and behaviour. Anim. Behav. 75, 731–737.

Angeloni, L., Crooks, K.R., Blumstein, D.T., 2010 Conservation and Behavior: Introduction, In: Michael D. Breed and Janice Moore (Eds.), Encyclopedia of Animal Behavior, Academic Press, Oxford, Vol. 1, 377–381.

Brumm, H., 2010. Anthropogenic Noise: Implications for Conservation, In: Michael D. Breed and Janice Moore (Eds.), Encyclopedia of Animal Behavior, Academic Press, Oxford, Vol. 1, 89–93.

Caro, T., 2007. Behavior and conservation: a bridge too far? Trends Ecol. Evol. 22, 394–400.

Curio, E., 1996. Conservation needs ethology. Trends Ecol. Evol. 11, 260–263.

Swaisgood, R.R., 2007. Current status and future directions of applied behavioral research for animal welfare and conservation. Appl. Anim. Behav. Sci. 102, 139–162.

Notes

1. Curio, E., 1996. Conservation needs ethology. Trends Ecol. Evol. 11, 260–263.

2. Angeloni, L., Schlaepfer, M.A., Lawler, J.J., Crooks, K.R., 2008. A reassessment of the interface between conservation and behaviour. Anim. Behav. 75, 731–737.

3. Caro, T., 2007. Behavior and conservation: a bridge too far? Trends Ecol. Evol. 22, 394–400.

4. Swaisgood, R.R., 2007. Current status and future directions of applied behavioral research for animal welfare and conservation. Appl. Anim. Behav. Sci. 102, 139–162.

5. Stamps, J.A., Swaisgood, R.R., 2007. Someplace like home: experience, habitat selection and conservation biology. Appl. Anim. Behav. Sci. 102, 392–409.

6. Kearney, M., Shine, R., Porter, W.P., 2009. The potential for behavioral thermoregulation to buffer "cold-blooded" animals against climate warming. PNAS 106, 3835–3840. doi:10.1073/pnas.0808913106.

7. Mandel, J.T., Donlan, C.J., Armstrong, J., 2010. A derivative approach to endangered species conservation. Front. Ecol. Environ. 8, 44–49.

8. Vanderwerf, E.A., Rohrer, J.L., Smith, D.G., Burt, M.D., 2001. Current distribution and abundance of the O'ahu 'Elepaio. Wilson Bull. 113, 10–16.

9. MacArthur, R.H., Wilson, E.O., 1967. The Theory of Island Biogeography. Princeton University Press, Princeton, 203 pp.

10. Quammen, D., 1996. The Song of the Dodo: Island Biogeography in an Age of Extinctions. Scribner, New York, 704 pp.

11. Simberloff, D., 1987. The spotted owl fracas: mixing academic, applied and political ecology. Ecology 68, 766–772.

12. Ritzi, C.M., Everson, B.L., Whittaker Jr., J.O., 2005. Use of bat boxes by a maternity colony of Indiana Myotis (*Myotis sodalist*). Northeastern Naturalist 12, 217–220.

13. Kelm, D.H., Wiesner, K.R., von Helversen, O., 2008. Effects of artificial roosts for frugivorous bats on seed dispersal in a neotropical forest pasture mosaic. Conserv. Biol. 22, 733–741.

14. Whitaker, J.O., Sparks, D.W., Brack, V., 2006. Use of artificial roost structures by bats at the Indianapolis International Airport. Environ. Manage. 38, 28–36.

15. Groombridge, J., Massey, J.G., Bruch, J.C., Malcolm, T., Brosius, C.N., Okada, M.M., et al., 2004. An attempt to recover the Po'ouli by translocation and an appraisal of recovery strategy for bird species of extreme rarity. Biol. Conserv. 118, 365–375.

16. Liley, S., Creel, S., 2008. What best explains vigilance in elk: characteristics of prey, predators, or the environment? Behav. Ecol. 19, 245–254.

17. Henke, S.E., Bryant, F.C., 1999. Effects of coyote removal on the faunal community in western Texas. J. Wildlife Manage. 63, 1066–1081.

18. Henderson, R.W., 2004. Lesser Antillean snake faunas: distribution, ecology and conservation concerns. Oryx 38, 311–320.

19. Coulon, A., Fitzpatrick, J.W., Bowman, R., Lovette, I.J., 2010. Effects of habitat fragmentation on effective dispersal of Florida scrub-jays. Conserv. Biol. 24, 1080–1088

20. Apfelbeck, B., Raess, M., 2008. Behavioural and hormonal effects of social isolation and neophobia in a gregarious bird species, the European starling (*Sturnus vulgaris*). Horm. Behav. 54, 435–441.

21. Echeverria, A.I., Vassallo, A.I., 2008. Novelty responses in a bird assemblage inhabiting an urban area. Ethology 114, 616–624.

22. Martin, L.B., Fitzgerald, L., 2005. A taste for novelty in invading house sparrows, *Passer domesticus.* Behav. Ecol. 16, 702–707.

23. Mettke-Hofmann, C., Winkler, H., Leisler, B., 2002. The significance of ecological factors for exploration and neophobia in parrots. Ethology 108, 249–272.

24. Aronsen, G., 2007. Animals behaving badly and the people who love/hate them. J. Mammal. Evol. 14, 71–73.

25. Kaulfuss, P., Mills, D.S., 2008. Neophilia in domestic dogs (*Canis familiaris*) and its implication for studies of dog cognition. Anim. Cogn. 11, 553–556.

26. Marker, L.L., Wilkerson, A.J.P., Sarno, R.J., Martenson, J., Breitenmoser-Wuersten, C., O'Brien, S.J., et al., 2008. Molecular genetic insights on cheetah (*Acinonyx jubatus*) ecology and conservation in Namibia. J. Hered. 99, 2–13.

27. Saura, M., Perez-Figueroa, A., Fernandez, J., Toro, M.A., Caballero, A., 2008. Preserving population allele frequencies in ex situ conservation programs. Conserv. Biol. 22, 1277–1287.

28. Agetsuma, N., 2007. Minimum area required for local populations of Japanese macaques estimated from the relationship between habitat area and population extinction. Int. J. Primatol. 28, 97–106.

29. Fujiwara, M., 2007. Extinction-effective population index: incorporating life-history variations in population viability analysis. Ecology 88, 2345–2353.

30. Peter, A., Lande, R., 1993. A population viability analysis for African elephant (*Loxodonta africana*): how big should reserves be? Conserv. Biol. 7, 602–610.

31. Cowley, D.E., 2008. Estimating required habitat size for fish conservation in streams. Aquatic Conservation–Marine and Freshwater Ecosystems 18, 418–431.

32. Wiersma, Y.F., Nudds, T.D., 2009. Efficiency and effectiveness in representative reserve design in Canada: the contribution of existing protected areas. Biol. Conserv. 142, 1639–1646.

33. O'Brien, C.M., Crowther, M.S., Dickman, C.R., Keating, J., 2008. Metapopulation dynamics and threatened species management: why does the broad-toothed rat (*Mastacomys fuscus*) persist? Biol. Conserv. 141, 1962–1971.

465

34. Moore, R.P., Robinson, W.D., Lovette, I.J., Robinson, T.R., 2008. Experimental evidence for extreme dispersal limitation in tropical forest birds. Ecol. Lett. 11, 960–968.

35. Fischer, J., Lindenmayer, D.B., 2000. An assessment of the published results of animal relocations. Biol. Conserv. 96, 1–11.

36. Teixeira, C.P., de Azevedo, C.S., Mendl, M., Cipreste, C.F., Young, R.J., 2007. Revisiting translocation and reintroduction programmes: the importance of considering stress. Anim. Behav. 73, 1–13.

37. Hinam, H.L., Clair, C.C.S., 2008. High levels of habitat loss and fragmentation limit reproductive success by reducing home range size and provisioning rates of Northern saw-whet owls. Biol. Conserv. 141, 524–535.

38. Fuentes-Montemayor, E., Cuaron, A.D., Vazquez-Dominguez, E., Benitez-Malvido, J., Valenzuela-Galvan, D., Andresen, E., 2009. Living on the edge: roads and edge effects on small mammal populations. J. Anim. Ecol. 78, 857–865.

39. Baker, J., French, K., Whelan, R.J., 2002. The edge effect and ecotonal species: bird communities across a natural edge in southeastern Australia. Ecology, 83, 3048–3059.

40. Burger, L.D., Burger, L.W., Faaborg, J., 1994. Effects of prairie fragmentation on predation on artificial nests. J. Wildlife Manage. 58, 249–254.

41. Baguette, M., Van Dyck, H., 2007. Landscape connectivity and animal behavior: functional grain as a key determinant for dispersal. Landscape Ecol. 22, 1117–1129.

42. Young, B.E., Sherry, T.W., Sigel, B.J., Woltmann, S., 2008. Nesting success of Costa Rican lowland rain forest birds in response to edge and isolation effects. Biotropica 40, 615–622.

43. Mbora, D.N.M., Meikle, D.B., 2004. Forest fragmentation and the distribution, abundance and conservation of the Tana River red colobus (*Procolobus rufomitratus*). Biol. Conserv. 118, 67–77.

44. Robinson, S.K., Thompson III, F.R., Donovan, T.M., Whitehead, D.R., Faaborg, J., 1995. Regional forest fragmentation and the nesting success of migratory birds. Science 267, 1987–1990.

45. Laiola, P., Tella, J.L., 2007. Erosion of animal cultures in fragmented landscapes. Front. Ecol. Environ. 5, 68–72.

46. Kazmaier, R.T., Hellgren, E.C., Ruthven III, D.C., 2001. Habitat selection by the Texas tortoise in a managed thornscrub ecosystem. J. Wildlife Manage. 65, 653–660.

47. Smith, H., Samuels, A., Bradley, S., 2008. Reducing risky interactions between tourists and free-ranging dolphins (*Tursiops sp.*) in an artificial feeding program at Monkey Mia, Western Australia. Tourism Manage. 29, 994–1001.

48. Hardin, G., 1968. The tragedy of the commons. Science 162, 1243–1248.

49. Crooks, K.R., Soule, M.E., 1999. Mesopredator release and avifaunal extinctions in a fragmented system. Nature 400, 563.

50. Taylor, A.R., Knight, R.L., 2003. Wildlife responses to recreation and associated visitor perceptions. Ecol. Appl. 13, 951–963.

51. Fletcher, R.J., McKinney, S.T., Bock, C.E., 1999. Effects of recreational trails on wintering diurnal raptors along riparian corridors in a Colorado grassland. J. Raptor Res. 33, 233–239.

52. Keyser, A.J., 2002. Nest predation in fragmented forests: landscape matrix by distance from edge interactions. Wilson Bull. 114, 186.

53. Marzluff, J.M., Ewing, K., 2001. Restoration of fragmented landscapes for the conservation of birds: a general framework and specific recommendations for urbanizing landscapes. Restoration Ecol. 9, 280–292.

54. Mason, J., Moorman, C., Hess, G., Sinclair, K., 2007. Designing suburban greenways to provide habitat for forest-breeding birds. Landsc. Urban Plan. 80, 153–164.

55. McKinney, M.L., 2005. Scaling of park trail length and visitation with park area: conservation implications. Anim. Conse. 8, 135–141.

56. Packard, J.M., Frohlich, R.K., Reynolds III, J.E., Wilcox, J.R., 1989. Manatee response to interruption of a thermal effluent. J. Wildl. Manage. 53, 692–700.

57. Rodgers, J.A., Schwikert, S.T., 2002. Buffer-zone distances to protect foraging and loafing waterbirds from disturbance by personal watercraft and outboard-powered boats. Conserv. Biol. 16, 216–224.

58. Lenth, B.E., Knight, R.L., Brennan, M.E., 2008. The effects of dogs on wildlife communities. Nat. Areas J. 28, 218–227.

59. Tarlow, E.M., Blumstein, D.T., 2007. Evaluating methods to quantify anthropogenic stressors on wild animals. Appl.Anim. Behav. Sci. 102, 429–451.

60. Stankowich, T., 2008. Ungulate flight responses to human disturbance: a review and meta-analysis. Biol. Conserv. 14, 2159–2173.

61. McDougall, P.T., Reale, D., Sol, D., Reader, S.M., 2006. Wildlife conservation and animal temperament: causes and consequences of evolutionary change for captive, reintroduced, and wild populations. Anim. Conserv. 9, 39–48.

62. Keith-Lucas, T., White, F.J., Keith-Lucas, L., Vick, L.G., 1999. Changes in behavior in free-ranging *Lemur catta* following release in a natural habitat. Am. J. Primatol. 47, 15–28.

63. Benhaiem, S., Delon, M., Lourtet, B., Cargnelutti, B., Aulagnier, S., Hewison, A.J.M., et al., 2008. Hunting increases vigilance levels in roe deer and modifies feeding site selection. Anim. Behav. 76, 611–618.

64. Souris, A.C., Kaczensky, P., Julliard, R., Walzer, C., 2007. Time budget-, behavioral synchrony- and body score development of a newly released Przewalski's horse group *Equus ferus przewalskii*, in the great Gobi B strictly protected area in SW Mongolia. Appl. Anim. Behav. Sci. 107, 307–321.

65. Stoinski, T.S., Beck, B.B., 2004. Changes in locomotor and foraging skills in captive-born, reintroduced golden lion tamarins (*Leontopithecus rosalia rosalia*). Am. J. Primatol. 62, 1–13.

66. Craig, G.R., Enderson, J.H., 2004. Peregrine Falcon Biology and Management in Colorado 1973–2001. Technical Publication No. 43, Colorado Division of Wildlife.

67. Maple, T.L., Finlay, T.W., 1989. Applied primatology in the modern zoo. Zoo Biol. 8, 101–116.

68. Robb, G.N., McDonald, R.A., Chamberlain, D.E., et al., 2008. Food for thought: supplementary feeding as a driver of ecological change in avian populations. Front. Ecol. Environ. 6, 476–484.

69. Kimball, B.A., Russell, J.H., DeGraan, J.P., Perry, K.R., 2008. Screening hydrolyzed casein as a deer repellent for reforestation applications. West. J. Appl. Forestry 23, 172–176.

70. Twain, M., 1872. Roughing It. (Chapter 5, available online at: http://www.mtwain.com/Roughing_It/6.html).

71. Restani, M., Marzluff, J.M., Yates, R.E., 2001. Effects of anthropogenic food sources on movements, survivorship, and sociality of common ravens in the Arctic. Condor 103, 399–404.

469

475